Ferroelectric Materials for Energy Applications

Ferroelectric Materials for Energy Applications

Edited by
Haitao Huang and James F. Scott

WILEY-VCH

Editors

Prof. Haitao Huang
Hong Kong Polytechnic University
Department of Applied Physics
BC620
Hung Hom, Kowloon
Hong Kong

Prof. James F. Scott
University of St Andrews
School of Chemistry
Purdie Building
North Haugh
KY16 9AJ St Andrews
United Kingdom

Cover image: © nadla/iStockphoto

Library of Congress Card No.: applied for

British Library Cataloguing-in-Publication Data
A catalogue record for this book is available from the British Library.

Bibliographic information published by the Deutsche Nationalbibliothek
The Deutsche Nationalbibliothek lists this publication in the Deutsche Nationalbibliografie; detailed bibliographic data are available on the Internet at <http://dnb.d-nb.de>.

Print ISBN: 978-3-527-34271-6
ePDF ISBN: 978-3-527-80747-5
ePub ISBN: 978-3-527-80749-9
oBook ISBN: 978-3-527-80750-5

Typesetting SPi Global, Chennai, India
Printing and Binding C.O.S. Printers Pte Ltd, Singapore

Printed on acid-free paper

10 9 8 7 6 5 4 3 2 1

Contents

Preface

Ferroelectric materials are well-known for their switchable spontaneous polarizations that are responsive to external stimuli, such as stress, temperature, electric field, and magnetic field (for multiferroic materials). Hence, ferroelectrics are often called "smart materials" and have found wide applications in sensors and actuators, as well as memory devices and field effect transistors, etc. Recent years have witnessed the emerging of ferroelectric materials in various kinds of energy harvesting, conversion, and storage devices. This is a relatively new and rapidly developing area, which we would like to call it "*ferro-energy*." In order to attract more researchers and engineers to work in this fascinating area, we invited a group of leading scientists to give comprehensive and timely reviews on various topics of ferroelectric materials for energy-related applications. For the benefit of young graduate students and novices in this area, the book starts from the fundamentals of ferroelectric materials. The next few chapters are mainly focused on applications including, but not limited to, piezoelectric energy generation, ferroelectric photovoltaics, pyroelectric energy harvesting, electrocaloric cooling, electric energy storage, photocatalysis, etc. First-principles calculations are reviewed in a separate chapter, which can be used as a powerful tool to explain or even predict the material properties for energy-related applications. Finally, the future perspectives of ferroelectrics for energy are summarized.

Due to their unique responsive spontaneous polarizations, ferroelectric materials will play an important role in energy-related applications that cannot be replaced by other conventional energy materials. This is also one of our purposes for the compilation of this book. This book is aimed at researchers working on ferroelectric materials and energy materials, where experts in this area can find it a useful resource book, engineers can have a broadened view after reading and beginners can be led quickly to the forefront of this area.

Haitao Huang
Hong Kong Polytechnic University

James F. Scott
University of St Andrews

1

Fundamentals of Ferroelectric Materials

Ling B. Kong[1,2], Haitao Huang[3], and Sean Li[4]

[1] Shenzhen Technology University, College of New Materials and New Energies, 3002 Lantian Road, Shenzhen, Guangdong 518118, PR China
[2] Nanyang Technological University, School of Materials Science and Engineering, 50 Nanyang Avenue, Singapore 639798, Singapore
[3] Hong Kong Polytechnic University, Department of Applied Physics, 11 Yuk Choi Road, Hung Hom, Kowloon, Hong Kong
[4] The University of New South Wales, School of Materials Science and Engineering, High Street, Kensington, Sydney, NSW 2052, Australia

1.1 Introduction

Ferroelectricity is defined as the property of a material, with two characteristics, i.e. (i) spontaneous polarization is present and (ii) it is reversible when subjected to external electric fields [1]. The property was first observed in Rochelle salt and is named so because of its analogy to ferromagnetism, which is a magnetic property of a material that has a permanent magnetic moment [2, 3]. Other similarities include hysteresis loop, Curie temperature (T_C), domains, and so on. The prefix, *ferro*, meaning iron (Fe), was used at that time because of the presence of the element in the magnetic materials. However, ferroelectricity has nothing to do with Fe. Even though some ferroelectric materials contain Fe, it is not the originating factor.

Generally, as a material is polarized by an external electric field, the induced polarization (P) is linearly proportional to the magnitude of the applied external electric field (E), which is known as dielectric polarization. Above the Curie temperature T_C, ferroelectric materials are at a paraelectric state. In this case, a nonlinear polarization is present versus an external electric field. As a result, electric permittivity, according to the slope of the polarization curve, is not a constant. At the ferroelectric state, besides the nonlinearity, a spontaneous nonzero polarization was present, as the applied field (E) is zero. Because the spontaneous polarization can be reversed by a sufficiently strong electric field, it is dependent on the currently applied electric field and the history as well, thus leading to the presence of the hysteresis loop.

The electric dipoles in a ferroelectric material are coupled to the crystal lattice of the material, so that the variation in lattice could change the strength of the dipoles, i.e. the strength of the spontaneous polarization. The change in the spontaneous polarization, in turn, leads to a change in the surface charge, which causes

Ferroelectric Materials for Energy Applications, First Edition.
Edited by Haitao Huang and James F. Scott.

current flow when a ferroelectric material is made into a capacitor without the application of an external field across it. There are two stimuli that can be used to change the lattice structure of a ferroelectric material, i.e. (i) mechanical force and (ii) temperature. The generation of surface charge as a result of the application of an external stress is known as piezoelectricity, while the change in spontaneous polarization in response to the change in temperature is named as pyroelectricity.

To understand ferroelectricity, it is necessary to link with piezoelectricity and pyroelectricity, because they have interesting interrelationships in terms of crystal structures. All crystals can be categorized into 32 different classes. In the theory of point groups, these classes are determined by using several symmetry elements: (i) center of symmetry, (ii) axis of rotation, (iii) mirror planes, and (iv) several combinations of them. The 32 point groups are subdivisions of seven basic crystal systems that are, in order of ascending symmetry, triclinic, monoclinic, orthorhombic, tetragonal, rhombohedral (trigonal), hexagonal, and cubic. Out of the 32 point groups 21 classes are non-centrosymmetric, which is a necessary condition for piezoelectricity to exist. Twenty of them are piezoelectric.

Among the 20 piezoelectric crystal classes, 10 crystals have pyroelectric properties. Within a given temperature range, this group of materials is permanently polarized. Compared to the general piezoelectric polarization produced under stress, pyroelectric polarization develops spontaneously and remains as permanent dipoles in the structure. Because this polarization varies with temperature, the response is named as pyroelectricity. Within the pyroelectric group, there is a subgroup that has spontaneous polarization, which is the ferroelectric materials. On one hand, the polarization in a ferroelectric material is similar to that in a pyroelectric one. On the other hand, these two polarizations are different, because ferroelectric polarization is reversible by an external applied electric field. Therefore, ferroelectricity is defined as the presence of spontaneous polarization that is reversible by an external electric field [4, 5]. Figure 1.1 summarizes the interrelationship among piezoelectric, pyroelectric, and ferroelectric materials, together with general dielectrics. This relationship

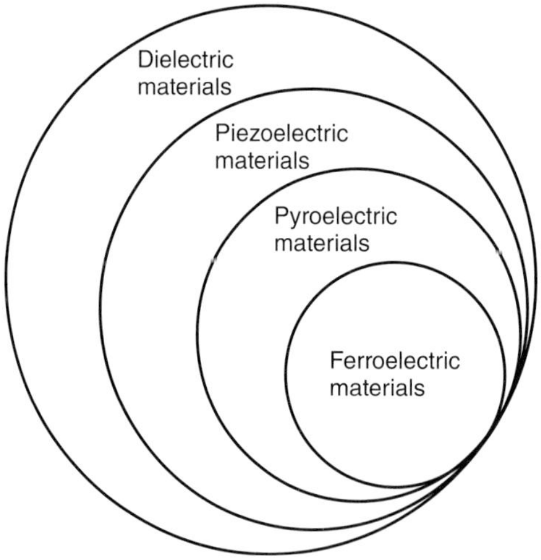

Figure 1.1 Interrelationship among piezoelectric, ferroelectric, pyroelectric, and dielectric materials. Ferroelectric materials have superior pyroelectric, piezoelectric, and dielectric properties than non-ferroelectric materials.

implies that ferroelectric materials have the highest piezoelectric performance compared to non-ferroelectric materials.

According to Ginzburg–Landau theory, the free energy of a ferroelectric material, without the application of an external electric field and stress, can be expressed as a Taylor expansion in terms of the order parameter, polarization (P) [6]. When a sixth order expansion is used, i.e. the eighth order and higher terms are neglected, the free energy is given by

$$\begin{aligned}\Delta G = &\frac{1}{2}\alpha_0\left(T - T_0\right)\left(P_x^2 + P_y^2 + P_z^2\right) + \frac{1}{4}\alpha_{11}\left(P_x^4 + P_y^4 + P_z^4\right)\\ &+ \frac{1}{2}\alpha_{12}\left(P_x^2P_y^2 + P_y^2P_z^2 + P_z^2P_x^2\right) + \frac{1}{6}\alpha_{111}\left(P_x^6 + P_y^6 + P_z^6\right)\\ &+ \frac{1}{2}\alpha_{112}\left[P_x^4\left(P_y^2 + P_z^2\right) + P_y^4\left(P_x^2 + P_z^2\right) + P_z^4\left(P_x^2 + P_y^2\right)\right] + \frac{1}{2}\alpha_{123}P_x^2P_y^2P_z^2\end{aligned} \tag{1.1}$$

where P_x, P_y, and P_z are the components of the polarization vector in the x, y, and z directions, respectively, while α_i, α_{ij}, α_{ijk} are the coefficients that are related to the symmetry of the crystals. These coefficients could be either obtained through experimental measurement or derived from *ab initio* calculation. Generally, in ferroelectric materials, both α_0 and α_{111} are positive. For ferroelectrics with a first-order phase transition, $\alpha_{11} < 0$, while for those having a second-order phase transition, $\alpha_{11} > 0$.

During the phase transition from cubic to tetragonal phase, the spontaneous polarization, P_s, of a ferroelectric can be derived in a relatively simple way, if the free energy is described as a 1D expression, which is given by

$$\Delta G = \frac{1}{2}\alpha_0\left(T - T_0\right)P_x^2 + \frac{1}{4}\alpha_{11}P_x^4 + \frac{1}{6}\alpha_{111}P_x^6 \tag{1.2}$$

The solution of the free energy is characterized with a shape that has a double well potential, with two free energy minima at $P = \pm P_s$, where P_s is the spontaneous polarization. At these two minima, the derivative of the free energy is zero, i.e.

$$\frac{\partial(\Delta G)}{\partial P_x} = \alpha_0\left(T - T_0\right)P_x + \alpha_{11}P_x^3 + \alpha_{111}P_x^5 = 0 \tag{1.3}$$

Then,

$$P_x\left[\alpha_0\left(T - T_0\right) + \alpha_{11}P_x^2 + \alpha_{111}P_x^4\right] = 0 \tag{1.4}$$

Because $P_x = 0$ corresponds to a free energy maxima in the ferroelectric phase, the spontaneous polarization, P_s, is obtained from the solution of the equation

$$\alpha_0\left(T - T_0\right) + \alpha_{11}P_s^2 + \alpha_{111}P_s^4 = 0 \tag{1.5}$$

which is

$$P_s^2 = \frac{-\alpha_{11} \pm \sqrt{\alpha_{11}^2 - 4\alpha_0\alpha_{111}(T - T_0)}}{2\alpha_{111}} \tag{1.6}$$

Obviously, for either the first- or the second-order phase transitions, the negative square root should be eliminated, thus leading to

$$P_s = \sqrt{\frac{-\alpha_{11} + \sqrt{\alpha_{11}^2 - 4\alpha_0\alpha_{111}(T - T_0)}}{2\alpha_{111}}} \tag{1.7}$$

If $\alpha_{11} = 0$, the spontaneous polarization is similarly given by

$$P_s = \sqrt{-\frac{\alpha_0(T - T_0)}{\alpha_{111}}} \tag{1.8}$$

The hysteresis loop, i.e. P_x versus E_x, can be readily derived from the free energy expansion by adding an electrostatic term, E_xP_x, as follows:

$$\Delta G = \frac{1}{2}\alpha_0\left(T - T_0\right)P_x^2 + \frac{1}{4}\alpha_{11}P_x^4 + \frac{1}{6}\alpha_{111}P_x^6 - E_xP_x \tag{1.9}$$

Therefore,

$$\frac{\partial(\Delta G)}{\partial P_x} = \alpha_0\left(T - T_0\right)P_x + \alpha_{11}P_x^3 + \alpha_{111}P_x^5 - E_x = 0 \tag{1.10}$$

Accordingly,

$$E_x = \alpha_0\left(T - T_0\right)P_x + \alpha_{11}P_x^3 + \alpha_{111}P_x^5 = 0 \tag{1.11}$$

Plotting E_x as a function of P_x and reflecting the graph about the 45° line will result in a curve with "S" shape. The central part of the "S" corresponds to a free energy local maximum, because there is $\partial^2(\Delta E)/\partial P_x^2 < 0$. By eliminating this region and connecting the top and bottom portions of the "S" curve by vertical lines at the discontinuities, the hysteresis loop is obtained.

Since ferroelectric materials have the highest performances in terms of piezoelectric and pyroelectric properties, they are promising candidates for related applications of mechanical and thermal energy harvesting. At the same time, ferroelectric materials also have very interesting electrocaloric effect (ECE), which can be used for refrigeration applications, while their photovoltaic effect has been paid new attention in recent years. Therefore, in the following sections, these four types of applications of ferroelectric materials, closely related to energy conversion and harvesting, will be discussed, with a focus on more theoretical considerations.

1.2 Piezoelectric Mechanical Energy Harvesting

1.2.1 Piezoelectricity

The word "piezoelectricity" is derived from the Greek "*piezein,*" which means to "squeeze" or "press" [4]. There are two piezoelectric effects: (i) direct effect and (ii) converse effect. Direct effect (also known as generator) is identified as the phenomenon where electrical charge (variation in polarization) is generated due to the application of an external mechanical stress, while the converse effect (also known as motor) is defined as the mechanical deformation because of the application of an electrical field. In this respect, piezoelectric energy harvesting is the use of the direct effect (generator).

Properties of piezoelectric materials are generally characterized by k_p, k_{33}, d_{33}, d_{31}, and g_{33}. The k factors (e.g. k_{33}, k_{31}, and k_p), which are also called piezoelectric coupling factors, are convenient and direct measurements of the overall strength of the electromechanical effects, i.e. the ability of the ceramic transducer to convert one form of energy to another. They are defined as the square root of the ratio of energy output in electrical form to the total mechanical energy input (direct effect), or the square root of the ratio of the energy available in mechanical form to the total electrical energy input (converse effect). Because the conversion of electrical to mechanical energy (or vice versa) is always incomplete, k is always less than unity. Commonly used as a figure-of-merit for piezoelectrics, the higher k values are most desirable and constantly sought after in new materials. For ceramics, k_p is a typical measure used to compare piezoelectric properties of ferroelectric materials – values ranging from 0.35 for $BaTiO_3$ to as high as 0.72 for PLZT [4].

The d coefficients are called piezoelectric coefficients, having magnitudes of $\times 10^{12}$ C N^{-1} (or pC N^{-1}) for the direct effect and $\times 10^{-12}$ m V^{-1} (or pm V^{-1}) for the converse effect, respectively. The subscript is used to describe the relative direction of inputs and outputs. For example, d_{31} means that this piezoelectric coefficient relates to the generation of polarization (direct effect) in the electrodes perpendicular to the vertical direction (3) and to the stress mechanically applied in the lateral direction (1), while d_{33} indicates the polarization generated in the vertical direction (3) when the stress is applied in the same direction. There are also other similar symbols [4].

The g factors are called open-circuit coefficients, another parameter used to evaluate piezoelectric ceramics for their ability to generate a large voltage per unit of input stress. The g constant is related to the d constant: $g = d/K\varepsilon_0$ (K is the relative dielectric constant and ε_0 is the dielectric constant of free space that equals unity). High-g-constant piezoelectric ceramics are usually ferroelectrically hard materials whose polarizations are not readily switched and thus they possess lower K values.

Ferroelectric materials can be grouped into four subcategories according to their crystal structures: perovskite group, pyrochlore group, tungsten-bronze group, and bismuth layer structure group, among which the perovskite group is the most important and thus the most widely studied. Perovskite is usually expressed as ABO_3. A typical ABO_3 unit cell structure is shown in Figure 1.2, taking $PbTiO_3$ as an example [4]. It consists of a corner-linked network of oxygen octahedron, creating an octahedral cage (B-site) and the interstices (A-sites). Ti^{4+} ions occupy the B-site while Pb^{2+} ions occupy the A-site. Figure 1.2 also shows the paraelectric and ferroelectric states of $PbTiO_3$. Most ferroelectric materials undergo a structural phase transition from a high temperature paraelectric phase into a low-temperature ferroelectric phase. The paraelectric phase always has a higher symmetry than the ferroelectric phase. The temperature of the phase transition is called the Curie temperature (T_C). Different ferroelectric materials have different values of T_C, which can be either lower than liquid nitrogen (LN)

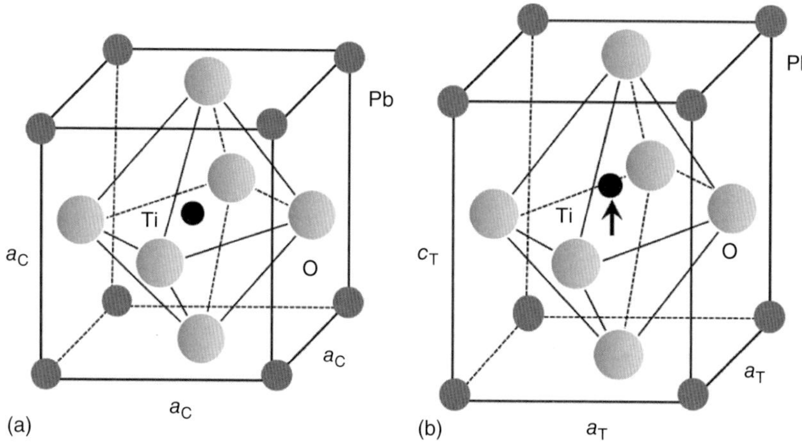

Figure 1.2 Schematic perovskite structure of $PbTiO_3$, with cubic (C) structure in the paraelectric state ($P_s = 0$) and tetragonal (T) structure in the ferroelectric state ($P_s \neq 0$).

temperature or higher than 1000 °C. For a given material (composition), the T_C is closely related to the microstructure (grain size and distribution, density, porosity and pore size and distribution, impurity, and so on). Generally, the T_C of a given material decreases with decreasing grain size [7].

In the ferroelectric state, the ability of displacement of the central Ti^{4+} ion causes reversibility of polarization. The switch of many adjacent unit cells is referred to as domain reorientation or switching. The homogeneous areas of the material with the same polarization orientation are referred to as domains, with domain walls existing between areas of unlike polarization orientation [5]. For as-prepared piezoelectric ceramics, the domains are randomly oriented and thus the net polarization of materials is zero because of their cancellation effect. Therefore, the as-prepared piezoelectric ceramics have no piezoelectric effect. To show piezoelectric properties, the newly obtained polycrystalline piezoelectric ceramics must be poled at strong external DC electric fields (10–100 kV cm^{-1}). Poling is to electrically align the orientation of the randomly distributed domains within the grains and to make ferroelectric ceramics act like a single crystal possessing both ferroelectric and piezoelectric properties. Poling is usually conducted at elevated temperatures, because polarization would be more compliant at high temperatures [5]. After poling during cooling, the applied voltages should be retained until the temperature is sufficiently low.

1.2.2 Brief History of Modern Piezoelectric Ceramics

The history of modern piezoelectricity and piezoelectric ceramics is closely related to the history of ferroelectricity and ferroelectric ceramics. The history of ferroelectrics can be tracked back to Rochelle salt (sodium potassium tartrate tetrahydrate, $KNa(C_4H_4O_6)\cdot 4H_2O$), which was synthesized more than 400 years ago, initially for medicinal purpose. It is from this same crystalline material that pyroelectric (thermal-polar), piezoelectric (stress-polar), and ferroelectric properties were discovered subsequently. Before this discovery, ferroelectricity was only a hypothetical property of solid materials at the turn of the twentieth century. However, the practical application of this material is largely limited due to its water solubility. It was after the discovery of ferroelectric ceramics (barium titanate, $BaTiO_3$) that this class of materials became extremely useful for a variety of applications.

The first ferroelectric ceramic material is $BaTiO_3$ (or BT), which was discovered in the mid-1940s. Before the discovery of $BaTiO_3$, the most widely used materials for capacitors were steatite, mica, TiO_2, $MgTiO_3$, and $CaTiO_3$, with the dielectric constant not higher than 100. During the World War II, there was a pressing need for high-dielectric-constant materials to fabricate high-capacitance capacitors. Before any publications were available in the literature, $BaTiO_3$ had already been studied as a high-dielectric-constant material concurrently. In the later open publications, it was concluded that the high dielectric constant in $BaTiO_3$ is due to its ferroelectric properties [4].

The history of ferroelectric ceramics also includes the discovery of lead zirconate titanate ($PbZr_{1-x}Ti_xO_3$, or PZT) piezoelectric ceramics, the development of transparent electro-optical lead lanthanum zirconate titanate ($Pb_{1-x}La_xZr_{1-y}Ti_yO_3$, or PLZT), the research on lead magnesium niobate ($PbMg_{1/3}Nb_{2/3}O_3$, or PMN) relaxor ferroelectric ceramics, and the discovery of many other non-perovskite ferroelectric ceramics [4]. Among these, PZT has been demonstrated to possess the best performance as piezoelectric ceramics.

Recently, there has been a concern with PZT due to the toxicity of Pb. Regulations and legislations have been established globally to restrict the use of lead-containing materials. For example, according to the directive for the Restriction of the use of certain Hazardous Substances (RoHS) in electrical and electronic equipment adopted by the European Parliament in the year 2006, the maximum allowed concentration of lead is established to be 0.1 wt% in homogeneous materials for electrical and electronic equipment used in households as well as industry. Therefore, lead-containing piezoelectric materials will be prohibited eventually. Similar regulations have been established worldwide. As a consequence, there is significantly increasing interest in developing lead-free piezoelectric ceramics all around the world. Although major progress has been made in materials research [8], there are still obstacles blocking the successful industrial implementation of lead-free piezoelectric ceramics. Key problems are their poor piezoelectric properties and fatigue degradation. Therefore, PZT will still be a dominant material for piezoelectric applications in most areas.

1.2.3 Principle of Piezoelectric Effect for Mechanical Energy Harvesting

The harvesting of mechanical energy is to convert it into electrical energy, which requires a mechanical system that couples motion or vibration to a transduction mechanism. The mechanical system should be designed to be able to maximize the coupling between the mechanical energy sources and the transduction mechanism, depending on the characteristics of the environmental motions. For example, energy due to vibration can be converted by using inertial generators, with the mechanical component attached to an inertial frame that acts as a fixed reference. The inertial frame transmits the vibrations to a suspended inertial mass to produce a relative displacement between them. Systems such as this usually have a resonant frequency, which can be designed to match the characteristic frequency of the environmental motions. Detailed analysis on mechanisms of mechanical energy harvesting can be found in the open literature [9, 10]. A brief description is presented here.

These inertia-based generators can be well described as second-order spring-mass systems. For a system with a seismic mass of m on a spring with a stiffness of k, its energy loss, consisting of parasitic loss, c_p, and electric energy generated by the transduction mechanism, c_e, can be represented by the damping coefficient, c_T. The system is excited by an external sinusoidal vibration, $y(t) = Y\sin(\omega t)$. At resonant frequency, there is a net displacement, $z(t)$, between the mass and the frame. If the mass of the vibration source is greatly larger than that of the seismic mass, the latter can be ignored. If the external excitation is harmonic, the differential equation of the motion is given by

$$m\ddot{z}(t) + c\dot{z}(t) + kz(t) = -m\ddot{y}(t) \tag{1.12}$$

Standard solution for the mass displacement will be

$$z(t) = \frac{\omega^2}{\sqrt{\left(\frac{k}{m} - \omega^2\right)^2 + \left(\frac{c_T\omega}{m}\right)^2}} y\sin(\omega t - \varphi) \tag{1.13}$$

where ϕ is the phase angle, given by

$$\varphi = \tan^{-1}\left(\frac{c_T\omega}{k - \omega^2 m}\right) \tag{1.14}$$

Energy conversion can be maximized when the excitation frequency matches the natural frequency of the system, ω_n, given by

$$\omega_n = \sqrt{\frac{k}{m}} \tag{1.15}$$

The power dissipated with the system is

$$P_d = \frac{m\zeta_T Y^2\left(\frac{\omega}{\omega_n}\right)^3 \omega^3}{\left[1 - \left(\frac{\omega}{\omega_n}\right)^2\right]^2 + \left[2\zeta_T\left(\frac{\omega}{\omega_n}\right)\right]^2} \tag{1.16}$$

where ζ_T is the total damping ratio, which is $\zeta_T = c_T/2m\omega_n$. Maximum power is achieved when the system is operated at ω_n, while P_d can be expressed as

$$P_d = \frac{mY^2\omega_n^3}{4\zeta_T} \tag{1.17}$$

$$P_d = \frac{mA^2}{4\omega_n\zeta_T} \tag{1.18}$$

where A is the excitation acceleration level, with $A = \omega_n^2 Y$. Noting that these are steady-state solutions, power will not tend to infinity when the damping ratio approaches zero. The maximum power generation can be evaluated by considering the parasitic and system damping ratio, which is given by

$$P_e = \frac{m\zeta_e A^2}{4\omega_n(\zeta_p + \zeta_e)^2} \tag{1.19}$$

P_e is maximized at $\zeta_p = \zeta_e$. When there is sufficient acceleration, increased damping effects will lead to a response with broadened bandwidth, so that the generator will be less sensitive to frequency. An excessive device amplitude can lead to nonlinear behavior of the generator, which will make it difficult to keep the generator working at resonance frequency. For specific applications, both the frequency of the generator and the level of damping should be specifically designed to maximize the power output. The power generation can also be maximized by maximizing the mass of the mechanical structure.

The piezoelectric damping coefficient can be estimated by using the following equation:

$$c_e = \frac{2m\omega_n^2 k^2}{2\sqrt{\omega_n^2 + \left(\frac{1}{R_{load}C_{load}}\right)^2}} \tag{1.20}$$

where k is the piezoelectric coupling factor of the materials, while R_{load} and C_{load} are load resistance and capacitance. At maximum power generation, there is optimal load, given by

$$R_{opt} = \frac{1}{\omega_n C}\frac{2m\omega_n^2 k^2}{\sqrt{4\zeta_p^2 + k^4}} \tag{1.21}$$

Specifically, for piezoelectric materials, the mechanical and electrical behaviors can be described by using the following linear constitutive equations:

$$S_{ij} = s_{ijkl}^E T_{kl} + d_{kij}E_k \tag{1.22}$$

$$D_i = d_{ikl}^E T_{kl} + \varepsilon_{ik}^T E_k \tag{1.23}$$

Each of the subscripts i, j, k, and l can take values of 1, 2, and 3. S and T are strain and stress tensors, respectively. The stresses, represented by T with a unit of $\mathrm{N\,m^{-2}}$, are induced by the mechanical and electrical effects. D and E are the electric displacement and electric field vectors, with units of $\mathrm{C\,m^{-2}}$ and $\mathrm{V\,m^{-1}}$, respectively. Also, s^E is the elastic compliance matrix evaluated at a constant electric field with a unit of $\mathrm{m^2\,N^{-1}}$, d is a matrix of piezoelectric strain coefficients with a unit of $\mathrm{m\,V^{-1}}$, and ε^T is a matrix of permittivity values that are evaluated at a constant stress with a unit of $\mathrm{N\,V^{-2}}$. In Eqs. (1.22) and (1.23), d represents the charge created by an external force in the absence of an electric field (short circuit electrical condition) or the displacement caused by an applied voltage in the absence of an applied force (stress-free mechanical condition).

For piezoelectric composite materials, effective electromechanical coupling factor is used, which is given by

$$k_{eff} = \sqrt{1 - \left(\frac{F_r}{F_a}\right)^2} \tag{1.24}$$

where F_r is the resonance frequency (Hz) and F_a is the antiresonance frequency (Hz) of a piezoelectric cantilever beam. The voltage coefficient, g (V m N^{-1}), is given by

$$g = \frac{d}{\varepsilon_T} \tag{1.25}$$

The mechanical quality factor, Q_M, is defined as [11]

$$Q_M = 2\pi \frac{\text{energy stored/cycle}}{\text{energy dissipated/cycle}} \tag{1.26}$$

The amount of energy stored in a piezoelectric device, E_C, is given by

$$E_C = \frac{1}{2}CV^2 \tag{1.27}$$

where C is capacitance of the piezoelectric element and V is the voltage produced.

The maximum efficiency of the piezoelectric devices can also be estimated by the following equation:

$$\eta = \frac{\frac{1}{2}\left(\frac{k^2}{1-k^2}\right)}{\frac{1}{Q_M} + \frac{1}{2}\left(\frac{k^2}{1-k^2}\right)} \tag{1.28}$$

This equation indicates that the efficiency can be increased by increasing k and Q_M, which are properties of the piezoelectric materials. Therefore, selection of materials is a very important step toward energy harvesters with high efficiencies. This is also the reason why PZT has been the most promising piezoelectric material for mechanical energy harvesting applications.

1.3 Pyroelectric Thermal Energy Harvesting

1.3.1 Principle of Pyroelectric Effect

As stated earlier, pyroelectric effect is defined as the phenomenon of change in polarization of a material induced by change in temperature [12, 13]. For a single domain ferroelectric material, there are localized charges that are present on the two end surfaces, due to the alignment of the polarization. At thermal equilibrium state, these localized charges are shielded by the free charges with same quantity but opposite signs, so that there is no net electricity produced in the ferroelectric material. However, when there is a change in temperature, the polarization of the material is changed. As a result, the free charges cannot completely shield the localized charges and thus there will be free charges at the surfaces, which leads to the formation of an electric field nearby. The presence of the electric field is indicated by the fact that the material can attract or repel charged particles. If the surfaces are connected with external circuit, there will be electric current flowing through it. The directions of the electrical currents corresponding to heating and cooling are opposite.

Similar to the piezoelectric of ferroelectric materials, the charge of current of pyroelectric effects is closely related to the change in polarization, which thus breaks the equilibrium with free charges, so that the amount of surface charges is increased and decreased corresponding to heating and cooling, respectively. The difference is that the change in polarization of pyroelectric effect is caused by the change in temperature, while that of piezoelectric effect is caused by mechanical stress. The physics of pyroelectric effect has been well described in the open literature [12–15]. A brief description of various physical quantities related to pyroelectric effect and pyroelectric materials is presented in this subsection.

The pyroelectric effect of a pyroelectric material is characterized by the pyroelectric coefficient, a parameter that measures the pyroelectric efficiency of the material. With a small change in temperature, ΔT, the variation induced can be given by

$$\Delta P = p\Delta T \tag{1.29}$$

where p is the pyroelectric coefficient, which is a vector and has three nonzero components, with unit of $\mathrm{C\,m^{-2}\,K^{-1}}$:

$$p_m = \frac{\partial P_m}{\partial T}, \quad m = 1, 2, 3 \tag{1.30}$$

The sign of the pyroelectric coefficient is determined with respect to the piezoelectric axis of the piezoelectric crystal. According to IRE standards, the end of a crystal axis with positive charge under tension is defined as the positive end. When a pyroelectric crystal is heated, if there is positive charge produced at the positive side, the pyroelectric coefficient is defined to be positive; otherwise, it is negative. Generally, because the spontaneous polarization of ferroelectric materials decreases with increasing temperature, the pyroelectric coefficient is usually negative. However, there are also exceptions; for example, the spontaneous polarization of Rochelle salt increases with increasing the temperature slightly below its Curie point.

Because the best pyroelectric materials are piezoelectric materials, the deformation of a pyroelectric material due to change in temperature can induce variation in polarization, which also contributes to pyroelectric effect. Therefore, it is important to pay attention to mechanical boundary conditions and the ways of change variation.

Under uniform heating or cooling, depending on mechanical boundary conditions, there are two types of pyroelectric effects. Under constant strain (sample is clapped), only the variation in polarization induced by temperature change contributes to pyroelectric effect. It is called primary pyroelectric effect or constant strain pyroelectric effect. Usually, it is difficult to maintain a constant strain. In other words, samples are in free-strain state or constant stress state. In this case, the variation in polarization induced by the mechanical deformation of the sample due to thermal expansion will be added on top of the primary pyroelectric effect. This additional pyroelectric effect at uniform temperature is called secondary pyroelectric effect. Therefore, the pyroelectric coefficient of a material under constant stress is the summation of the primary and secondary coefficients, and is known as the total pyroelectric coefficient. However, in practice, the pyroelectric element in a pyroelectric device is neither under constant strain nor under constant stress [16]. For example, it is only strained in one specific direction and free in other directions. In this case, the pyroelectric coefficient is called a partially strained coefficient.

If a pyroelectric sample is heated or cooled in a nonuniform manner, it will experience a stress gradient, which also contributes to the pyroelectric effect, through piezoelectric effect. This nonuniform heating- or cooling-induced pyroelectric effect is called a tertiary effect or false effect. It is named false effect because piezoelectric materials, including those outside the 10 polar point group, have such an effect under nonuniform heating or cooling. Those piezoelectric materials actually have no pyroelectric effect. As a result, the heating or cooling should be uniform to exclude the false pyroelectric effect.

1.3.2 Pyroelectric Coefficient and Electrocaloric Coefficient

The thermodynamic states of elastic dielectrics can be described with temperature T, entropy S, electric field $\boldsymbol{E}$, electric displacement D, strain X, and stress x. With T, E, and X as independent variables, the differential form of electric displacement can be expressed as

$$\begin{aligned}\mathrm{d}D_m &= \left(\frac{\partial D_m}{\partial X_i}\right)_{E,T} \mathrm{d}X_i + \left(\frac{\partial D_m}{\partial E_n}\right)_{X,T} \mathrm{d}E_n + \left(\frac{\partial D_m}{\partial T}\right)_{E,X} \mathrm{d}T \\ &= \mathrm{d}_{mi}^{E,T}\,\mathrm{d}X_i + \varepsilon_{mn}^{X,T}\,\mathrm{d}E_n + p_m^{E,X}\,\mathrm{d}T \end{aligned} \tag{1.31}$$

where the subscript $m = 1 - 3$ and $i = 1 - 6$, while the superscript means that the physical quantities are kept constant. On the right-hand side, the first and the second terms are piezoelectric and dielectric characteristics, whereas the third term is the pyroelectric property of a material.

When $\mathrm{d}X_i = 0$ and $\mathrm{d}E_n = 0$,

$$\mathrm{d}D_m = p_m^{E,X}\,\mathrm{d}T \tag{1.32}$$

With independent variables, T, E, and X, the characteristic functions of Gibbs free energy become

$$G = U - TS - X_i x_i - E_m D_m \tag{1.33}$$

According to thermodynamic first and second laws,

$$\mathrm{d}G = -x_i\,\mathrm{d}X_i - D_m\,\mathrm{d}E_m - S\,\mathrm{d}T \tag{1.34}$$

The differential form of the Gibbs free energy, with respect to T, E, and X, can be expressed as

$$\mathrm{d}G = \left(\frac{\partial G}{\partial X_i}\right)_{E,T} \mathrm{d}X_i + \left(\frac{\partial G}{\partial E_m}\right)_{X,T} \mathrm{d}E_m + \left(\frac{\partial G}{\partial T}\right)_{E,X} \mathrm{d}T \tag{1.35}$$

By comparing Eqs. (1.34) and (1.35), we have

$$\left(\frac{\partial G}{\partial E_m}\right)_{X,T} = -D_m \tag{1.36}$$

$$\left(\frac{\partial G}{\partial T}\right)_{E,X} = -S \tag{1.37}$$

$$\left(\frac{\partial^2 G}{\partial E_m\,\partial T}\right)_X = -\left(\frac{\partial D_m}{\partial T}\right)_{E,X} = -p_m^{E,X} \quad (1.38)$$

$$\left(\frac{\partial^2 G}{\partial E_m\,\partial T}\right)_X = -\left(\frac{\partial S}{\partial E_m}\right)_{X,T} \quad (1.39)$$

The pyroelectric coefficient is thus defined in Eq. (1.38), whereas Eq. (1.39) describes the change in entropy induced by an external electric field, which is known as EC coefficient. Therefore, ECE is inverse pyroelectric effect, and thus

$$p_m^{E,X} = -\left(\frac{\partial S}{\partial E_m}\right)_{X,T} \quad (1.40)$$

It means that the pyroelectric coefficient under constant electric field and stress is equal to the ECE under constant stress and temperature.

With independent variables, T, E, and x, the differential form of electric displacement can be expressed as

$$\begin{aligned} \mathrm{d}D_m &= \left(\frac{\partial D_m}{\partial x_i}\right)_{E,T} \mathrm{d}x_i + \left(\frac{\partial D_m}{\partial E_n}\right)_{x,T} \mathrm{d}E_n + \left(\frac{\partial D_m}{\partial T}\right)_{E,x} \mathrm{d}T \\ &= e_{mi}^{E,T}\,\mathrm{d}x_i + \varepsilon_{mn}^{x,T}\,\mathrm{d}E_n + p_m^{E,x}\,\mathrm{d}T \end{aligned} \quad (1.41)$$

Under constant strain and electric field,

$$\mathrm{d}D_m = p_m^{E,x}\,\mathrm{d}T \quad (1.42)$$

Similarly, with independent variables, T, E, and x, the characteristic functions of Gibbs free energy are written as

$$\mathrm{d}G_2 = X_i\,\mathrm{d}x_i - D_m\,\mathrm{d}E_m - S\,\mathrm{d}T \quad (1.43)$$

Similar to Eq. (5.7),

$$\mathrm{d}G_2 = \left(\frac{\partial G_2}{\partial x_i}\right)_{E,T} \mathrm{d}x_i + \left(\frac{\partial G_2}{\partial E_m}\right)_{x,T} \mathrm{d}E_m + \left(\frac{\partial G_2}{\partial T}\right)_{E,x} \mathrm{d}T \quad (1.44)$$

As a result,

$$\left(\frac{\partial G_2}{\partial E_m}\right)_{x,T} = -D_m \quad (1.45)$$

$$\left(\frac{\partial G_2}{\partial T}\right)_{E,x} = -S \quad (1.46)$$

According to partial derivatives,

$$p_m^{E,x} = -\left(\frac{\partial S}{\partial E_m}\right)_{x,T} \quad (1.47)$$

Therefore, the pyroelectric coefficient under constant electric field and strain is equal to the ECE under constant strain and temperature.

The change in entropy of a ferroelectric material is due to the change in ordering of polarization. The higher the degree of ordering, the lower the value of entropy of the system will be. As a consequence, depolarization of a ferroelectric material leads to an increase in the value of entropy because the degree of ordering of polarization is decreased. Under adiabatic conditions, depolarization results in a decrease in temperature, which is the reason why ECE can be used for refrigeration through depolarization.

1.3.3 Primary and Secondary Pyroelectric Coefficient

At zero electric field (constant electric field), electric displacement is only a function of strain and temperature, while strain is only a function of stress and temperature, i.e.

$$\mathrm{d}D_m = \left(\frac{\partial D_m}{\partial x_i}\right)_T \mathrm{d}x_i + \left(\frac{\partial D_m}{\partial T}\right)_x \mathrm{d}T \tag{1.48}$$

$$\mathrm{d}x_i = \left(\frac{\partial x_i}{\partial X_j}\right)_T \mathrm{d}X_j + \left(\frac{\partial x_i}{\partial T}\right)_X \mathrm{d}T \tag{1.49}$$

With $\mathrm{d}X_j = 0$, combining Eqs. (5.20) and (5.21) leads to

$$\mathrm{d}D_m = \left[\left(\frac{\partial D_m}{\partial x_i}\right)_T \left(\frac{\partial x_i}{\partial T}\right)_X + \left(\frac{\partial D_m}{\partial T}\right)_x\right] \mathrm{d}T \tag{1.50}$$

Rearranging Eq. (5.22),

$$\left(\frac{\partial D_m}{\partial T}\right)_X = \left(\frac{\partial D_m}{\partial T}\right)_x + \left(\frac{\partial D_m}{\partial x_i}\right)_T \left(\frac{\partial x_i}{\partial T}\right)_X \tag{1.51}$$

where the left-hand side term is the total pyroelectric coefficient, while the first term on the right-hand side is the primary pyroelectric coefficient and the second term is the secondary coefficient.

In Eq. (1.51),

$$\left(\frac{\partial D_m}{\partial x_i}\right)_T = e_{mi} \tag{1.52}$$

$$\left(\frac{\partial x_i}{\partial T}\right)_X = \alpha_i \tag{1.53}$$

where e_{mi} and α_i are the piezoelectric constant and thermal expansion coefficient, respectively. Therefore, Eq. (1.51) becomes

$$p_m^X = p_m^x + e_{mi}^T \alpha_i^X \tag{1.54}$$

where the second term on the right-hand side means that the secondary pyro electric coefficient is the product of the piezoelectric constant and the thermal expansion coefficient.

If Eq. (1.51) is rewritten as

$$\left(\frac{\partial D_m}{\partial T}\right)_X = \left(\frac{\partial D_m}{\partial T}\right)_x + \left(\frac{\partial D_m}{\partial X_i}\right)_T \left(\frac{\partial X_i}{\partial x_j}\right)_T \left(\frac{\partial x_j}{\partial T}\right)_X \tag{1.55}$$

the following expression for total pyroelectric coefficient results:

$$p_m^X = p_m^x + d_{mi}^T c_{ij}^T \alpha_j^X \tag{1.56}$$

where the second term on the right-hand side means that the secondary pyroelectric coefficient is the product of piezoelectric constant, elastic stiffness, and thermal expansion coefficient. In most cases, the primary coefficient is the main contributor of pyroelectric materials.

1.3.4 Tertiary Pyroelectric Coefficient and Other Aspects

As stated before, under the condition of nonuniform heating or cooling without mechanical stress, the total pyroelectric coefficient consists of primary, secondary, and tertiary coefficients. The tertiary pyroelectric coefficient is attributed to thermal stress, whose contribution to the change in polarization is $d_{mnp}X_{np}(r,t)$, where d_{mnp} and X_{np} are the piezoelectric constant and thermal stress component, while r and t are displacement and time, respectively [17]. Owing to the difficulty in characterization of thermal stress, which is a function of location and time, it is a challenge to accurately measure the tertiary pyroelectric coefficient.

When elastic Gibbs free energy G_1 is used as a characteristic function, with variables such as stress and electric field being one dimensional,

$$\mathrm{d}G_1 = -S\,\mathrm{d}T + x\,\mathrm{d}X + E\,\mathrm{d}D \tag{1.57}$$

Taking the derivative of Eq. (1.57),

$$\frac{\partial G_1}{\partial D} = E \tag{1.58}$$

$$\frac{\partial^2 G_1}{\partial D^2} = \frac{1}{\varepsilon} \tag{1.59}$$

At temperatures near Curie point,

$$\frac{\partial G_1}{\partial D} = \alpha D + \beta D^3 + \gamma D^5 \tag{1.60}$$

which can be rewritten as

$$E = \alpha_0(T - T_0)D + \phi(D) \tag{1.61}$$

where $\varphi(D)$ represents all higher order terms and $\alpha_0 = 1/(\varepsilon_0 C)$, with C being the Curie constant, which leads to

$$\frac{\partial E}{\partial T} = \alpha_0 D \tag{1.62}$$

Therefore, if $E = 0$ and $D = P_s$, at temperatures near the Curie point,

$$\alpha_0(T - T_0)P_s + \phi(P_s) = 0 \tag{1.63}$$

i.e.

$$\left.\frac{\partial G_1}{\partial D}\right|_{D=P_s} = \alpha_0(T - T_0)P_s + \phi(P_s) = 0 \tag{1.64}$$

$$\frac{\mathrm{d}}{\mathrm{d}T}\left(\frac{\partial G_1}{\partial D}\right)_{D=P_s} = \left.\frac{\partial^2 G_1}{\partial D^2}\right|_{D=P_s}\frac{\mathrm{d}D}{\mathrm{d}T} + \left.\frac{\partial^2 G_1}{\partial T\,\partial D}\right|_{D=P_s} = 0 \tag{1.65}$$

Comparing Eqs. (1.59) and (1.62),

$$\left.\frac{\partial^2 G_1}{\partial T\,\partial D}\right|_{D=P_s} = \left.\frac{\partial E}{\partial T}\right|_{D=P_s} = \alpha_0 P_s \tag{1.66}$$

Substituting Eq. (1.66) in Eq. (1.65),

$$\frac{p}{\varepsilon_r} + \alpha_0 P_s = 0 \tag{1.67}$$

where $p = \mathrm{d}D/\mathrm{d}T$, which is the pyroelectric coefficient, thus leading to

$$\frac{p}{\varepsilon_r} = \frac{P_s}{C} \tag{1.68}$$

which correlates the pyroelectric coefficient with Curie constant, spontaneous polarization, and permittivity of a pyroelectric material. With this relationship, the pyroelectric coefficient can be estimated by using the Curie constant, and vice versa. It is necessary to mention that the temperature range of validity of the formula should not be too wide.

It has been found that although there is a significant difference in the pyroelectric coefficient and permittivity among various pyroelectric/ferroelectric materials, the value of the ratio $p\varepsilon_r^{-1/2}$ is nearly constant for most ferroelectric materials, with a temperature range across room temperature. This phenomenon has been well explained by using ferroelectric phenomenological theories. With Devonshire's assumption, the elastic Gibbs free energy G_1 can be expressed as

$$\frac{\partial G_1}{\partial D} = \alpha_0 (T - T_0) D + \beta D^3 + \gamma D^5 = E \tag{1.69}$$

Near the Curie point, with $E = 0$ and $D = P_s$,

$$\alpha_0 (T - T_0) P_s + \beta P_s^3 + \gamma P_s^5 = 0 \tag{1.70}$$

The solutions of this equation are

$$P_s^2 = -\frac{\beta}{2\gamma}\left\{1 + \left[1 - 4\alpha_0\gamma\beta^{-2}(T - T_0)\right]^{1/2}\right\} \tag{1.71}$$

$$P_s^2 = -\frac{\beta}{2\gamma}\left\{1 - \left[1 - 4\alpha_0\gamma\beta^{-2}(T - T_0)\right]^{1/2}\right\} \tag{1.72}$$

where Eq. (1.71) is for $\beta < 0$ and Eq. (1.72) is for $\beta > 0$. For second-order phase transition with $\beta > 0$, Eq. (1.71) becomes

$$P_s^2 = -\frac{\beta}{2\gamma} - \frac{\beta}{2\gamma}\left[1 + \frac{4\gamma(T_0 - T)}{\varepsilon_0 C \beta^2}\right]^{1/2} \tag{1.73}$$

For first-order phase transition, it is

$$P_s^2 = -\frac{|\beta|}{2\gamma} - \frac{|\beta|}{2\gamma}\left[1 - \frac{4\gamma(T_0 - T)}{\varepsilon_0 C \beta^2}\right]^{1/2} \tag{1.74}$$

The pyroelectric coefficient can be derived from their derivatives with respect to temperature, given by

$$p = (2P_s\varepsilon_0 C|\delta|)^{-1}\left[1 + \frac{4\gamma(T_0 - T)}{C\beta^2}\right]^{1/2} \tag{1.75}$$

With Eq. (1.68), on multiplying the two sides of Eq. (1.75) by p/ε_r and P_s/C, respectively,

$$\frac{p^2}{\varepsilon_r} = P_0^2(2\varepsilon_0 CT_0)^{-1}\left[1 + \frac{4P_0^2\gamma}{|\beta|}\left(1 - \frac{T}{T_0}\right)\right]^{1/2} \tag{1.76}$$

where

$$P_s^2 = T_0\frac{\partial P_s^2}{\partial T}\bigg|_{T\leq T_0} = \frac{2T_0}{\varepsilon_0 C|\beta|} \tag{1.77}$$

When $T \approx T_0$, Eq. (1.76) can be simplified as

$$p\varepsilon_r^{-1/2} \approx P_0(2CT_0)^{-1/2} \tag{1.78}$$

The equation can be used to understand the abovementioned phenomenon, which has been taken as a reference to estimate the performance of pyroelectric materials.

1.3.5 Pyroelectric Effect versus Phase Transition

For ferroelectric pyroelectric materials, there are two types of phase transition, where a peak is observed in the pyroelectric coefficient curve with temperature [18]. Ferroelectric–paraelectric transition is the most important phase occurring in ferroelectric pyroelectric materials. During this phase transition, there is a sharp change in spontaneous polarization.

For first-order ferroelectrics, at temperatures slightly higher than Curie point T_C, an external electric field may induce the appearance of a ferroelectric phase, where the temperature of the maximized variation in polarization, corresponding to the peak of pyroelectric coefficient, increases with increasing magnitude of external applied electric field [19]. This phenomenon can be readily explained with Eq. (1.69). According to reduced electric displacement d, electric field e, and temperature t,

$$d = -(2\gamma/|\beta|)^{1/2}D \tag{1.79}$$

$$e = -8(2\gamma^3/|\beta|^5)^{1/2}E \tag{1.80}$$

$$t = 4\alpha_0\beta^{-2}(T - T_0) \tag{1.81}$$

Then, Eq. (1.69) can be simplified as

$$e = 2d^5 - 4d^3 + 2td \tag{1.82}$$

According to the definition of pyroelectric coefficient, we have reduced the pyroelectric coefficient, given by

$$\frac{\partial d}{\partial t} = \left(4d - 4d^3 - \frac{e}{2d^2}\right)^{-1} = \frac{d}{6d^2 - 5d^4 - t} \tag{1.83}$$

For second-order phase transition ferroelectrics, there is no electric-field-induced phase transition [18]. The application of an external electric field only suppresses the curve of dielectric constant versus temperature and thus the temperature corresponding to the pyroelectric coefficient peak is kept unchanged. However, the magnitude of pyroelectric coefficient decreases with increasing magnitude of electric field. By omitting the higher order term, Eq. (1.69) is simplified as

$$E = \alpha_0(T - T_0)D + \beta D^3 \tag{1.84}$$

where $T_0 = T_C$. Because the electric field is constant, the derivative of Eq. (1.84) is

$$0 = \left[\alpha_0(T - T_0) + 3\beta D^2\right] \mathrm{d}D + \alpha_0 D\,\mathrm{d}T \tag{1.85}$$

Therefore, the pyroelectric coefficient is given by

$$p = \left(\frac{\partial D}{\partial T}\right)_{E,X} = \frac{-\alpha_0 D_0}{\alpha_0(T - T_0) + 3\beta D^2} \tag{1.86}$$

where D_0 is the electric displacement induced by the electric field E at temperature T, which can be calculated by using Eq. (1.84). There are three special cases, where we have

$$D_0 \approx E\left[\alpha_0(T - T_0)\right]^{-1}, \quad T \gg T_0 \tag{1.87}$$

$$D_0 \approx \left[\alpha_0\beta^{-1}(T_0 - T)\right]^{-1/2}, \quad T \ll T_0 \tag{1.88}$$

$$D_0 \approx (\alpha_0/\beta)^{1/3}, \quad T = T_0 \tag{1.89}$$

Substituting Eqs. (1.87)–(1.89) in Eq. (1.86), with p_0 being the pyroelectric coefficient at $T = T_0$, we have

$$p = E\alpha_0^{-1}(T - T_0)^{-2}, \quad T \gg T_0 \tag{1.90}$$

$$p = \frac{1}{2}\alpha_0^{1/2}\beta^{-1/2}(T_0 - T)^{-1/2}, \quad T \ll T_0 \tag{1.91}$$

$$p_0 = \frac{1}{3}\alpha_0/\beta^{-2/3}E^{-1/3}, \quad T = T_0 \tag{1.92}$$

Therefore, below Curie point, the pyroelectric coefficient increases with temperature with $(T_0 - T)^{-1/2}$, while above Curie point, the pyroelectric coefficient decreases with $(T - T_0)^{-2}$. At Curie point, the pyroelectric coefficient is maximized, which is affected by the electric field in terms of magnitude, but without shift in temperature.

Another phase transition is ferroelectric–ferroelectric phase transition, where both magnitude and direction of the spontaneous polarization do not change in most cases. There are seldom cases where the direction of the polarization does change. In this case, other phase transition related parameters will experience only negligible variation, but pyroelectric effect still has a peak value. Therefore, pyroelectric coefficient has been a sensitive indicator of ferroelectric–ferroelectric phase transition of ferroelectric materials.

When a pyroelectric material is heated or cooled from one temperature to another, it takes time for the polarization to reach a new equilibrium state. Therefore, the response of pyroelectric charges to the variation in temperature is not instantaneous, but with a certain degree of delay, depending on a number of factors, including thermal conductivity and heat capacity of the material, as well as sample dimension and shape. Generally, thermal relaxation time can be expressed as

$$\tau_T = \frac{L^2 c'}{\sigma_T} \tag{1.93}$$

where c' is the heat capacitance per unit volume in $\mathrm{J\,K^{-1}\,m^{-3}}$, L is the length in the direction of heat conduction in meters (m), and σ_T is thermal conductivity in $\mathrm{J\,K^{-1}\,m^{-1}\,s^{-1}}$. Therefore, it is important to have an estimate on τ_T before discussing the pyroelectric response. It has been acknowledged that, during both the ferroelectric–ferroelectric and ferroelectric–paraelectric phase transitions, ferroelectric materials usually experience a peak in the value of their specific heat, which will also lead to a significant increase in τ_T [20].

1.4 Electrocaloric (EC) Effect of Ferroelectric Materials

When an electric field is applied to a dielectric material it will induce a change in the polarization and consequently a change in the entropy and temperature in the material. Such an electric-field-induced temperature and entropy change in a dielectric material is known as the ECE [1–3].

The ECE is attractive because it may provide a very efficient means to realize solid-state cooling devices for a broad range of applications such as on-chip cooling and temperature regulation for sensors, electronic devices, and medical specimens. Furthermore, refrigeration based on the ECE approach is more environmentally friendly and hence may also provide an alternative to the existing vapor-compression approach.

More recently, ECE of ferroelectric materials has received intensive attention, due to its potential application in the area of energy saving and conversion, with progress that has been well overviewed in the literature [21–24]. It is well known that as a dielectric material is subjected to an external applied field, polarization is induced in the material, thus leading to a variation in entropy and temperature, which is called the ECE [25]. Owing to its potential of achieving solid-state cooling, the ECE has become a worldwide hot research topic. Solid-state cooling could find a wide range of applications, including on-chip cooling and temperature management of electronic devices. Moreover, it can be used to realize refrigeration, with more environment friendliness, so as to replace the currently used vapor-compression technique in the future.

The study of ECE can be traced back to the 1930s when the ECE was first measured experimentally [26]. However, the ECE was too weak to be used for practical applications. It was after the discovery of ferroelectricity in the 1920s that ECE-derived applications started to draw more attention, due to the large polarization required by real device design [27–29]. Besides large polarization, ferroelectric materials for ECE applications should be workable near room temperature, which requires them to have a suitable ferroelectric (F)–paraelectric (P) transition point (Curie temperature or T_C), because ECE is maximized just above T_C. For example, a $PbZr_{0.95}Ti_{0.05}O_3$ thin film, at temperatures above its F–P phase transition of about 222 °C, exhibited an adiabatic temperature change of as high as 12 °C [30]. Such a large ECE, with an adiabatic temperature change of 12 °C, was also observed in the ferroelectric polymer, poly(vinylidene fluoride–trifluoroethylene) [P(VDF–TrFE)] copolymer [31].

Theoretically, ECE can be related to the pyroelectric effect of ferroelectric materials, through the Maxwell relation, in terms of electric field (E) induced adiabatic temperature change, given by [32]

$$\left(\frac{\partial T}{\partial E}\right)_S = \frac{T p_E}{c_E} \tag{1.94}$$

where T is the temperature, c_E is the volumetric specific heat at a constant electrical field, S is entropy, and p_E is the pyroelectric coefficient at a constant field, $(\partial D/\partial T)_E$, with D being the electric displacement. According to Eq. (1.94), the isothermal entropy change ΔS and adiabatic temperature change ΔT, as the external applied field is changed from E_1 to E_2, can be expressed as

$$\Delta S = -\int_{E_1}^{E_2} \left(\frac{\partial D}{\partial T}\right)_{S,E} \mathrm{d}E \tag{1.95}$$

$$\Delta T = -\int_{E_1}^{E_2} \frac{T}{c_E}\left(\frac{\partial D}{\partial T}\right)_{S,E} \mathrm{d}E \tag{1.96}$$

where $D = \varepsilon_0 E + P$, with P being the polarization and ε_0 the vacuum dielectric constant. To achieve large values of ΔS and ΔT, a sufficiently high pyroelectric coefficient should be available over a wide range of both electric field and temperature. Usually, ferroelectric materials exhibit high pyroelectric effect at temperatures that are near the F–P transition point. Also, it is possible to obtain a large ΔT, even though ΔS is not very high, as long as the material has a sufficiently low value of c_E. However, practical refrigeration applications require a relatively high ΔS. Currently, ECE has been mostly studied according to the Maxwell relationships, as described in Eqs. (1.95) and (1.96), i.e. electric displacement D as a function of temperature T at different electric fields.

During the cycling of refrigeration, thermal entropy is absorbed by the working material or refrigerant from the cooling load, which is in thermal contact with the load, i.e. isothermal entropy change ΔS. Then, the material is isolated from the load, whereas the temperature is increased due to the application of an external field, i.e. an adiabatic temperature change ΔT is present. Meanwhile, the material is in thermal contact with a heat sink, to which the entropy absorbed from the cooling load is ejected. Because the working material is isolated from the heat sink, the temperature is decreased to that of the cooling load, as the field is reduced and the process is repeated. As a result, the isothermal electrical entropy change ΔS and the adiabatic temperature change ΔT are the two key parameters that are used to characterize the ECE.

The performances of different refrigerants can be characterized by using a parameter known as refrigerant capacity (RC), which is expressed as $\Delta S_c \Delta T_{hc}$ [33]. Here, ΔS_c is the isothermal entropy change at T_c and $\Delta T_{hc} = T_h - T_c$ is the temperature span of the refrigeration cycle, where T_h and T_c are the hot end and cold end temperatures, respectively. Figure 1.3 shows a schematic diagram demonstrating the refrigeration cycle [21]. For the ECE of a dielectric material to be used as the refrigerant, ΔT_{hc} and ΔS_c can vary as a function of electric field. Also, the largest value of $\Delta S_c \Delta T_{hc}$ at a selected operation temperature should be used as a measure to characterize the performance of a refrigerant. Phenomenological theory has been used widely in the study of ferroelectric materials, which is also useful in guiding the development of ferroelectrics to achieve a large ECE.

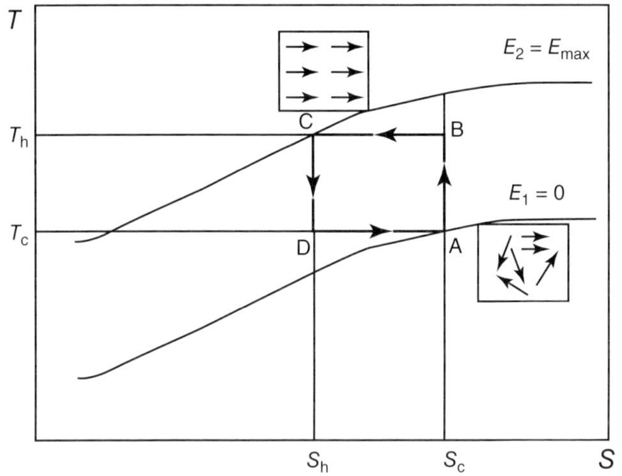

Figure 1.3 Schematic diagram of a potential thermodynamic refrigeration cycle based on EC effect that is similar to a Carnot cycle. From the path of (A)–(B) and then (B)–(C), an applied electric field is raised from E_1 to E_2, which induces a polar-ordered phase. From (A) to (B), the EC effect material experiences an adiabatic temperature change ΔT_{hc} from T_c to T_h, and then from (B) to (C) it ejects heat (entropy) to the heat sink at T_h while the material entropy is reduced from S_c to S_h (isothermal entropy change ΔS). From the path of (C) to (D) and then (D) to (A), the applied electric field is reduced from E_2 to E_1 and the EC effect material loses polar ordering. Accompanying this, the EC effect material experiences a decrease in temperature from T_h to T_c from (C) to (D) (adiabatic temperature change) and then absorbs heat (entropy) from the cold load (isothermal entropy change). Because the EC effect materials of interest are insulators, the above electric field (refrigeration) cycle has the potential of very low electric loss, and consequently, the cooling devices based on the EC effect have the potential to reach very high efficiency. *Source:* Lu and Zhang 2009 [21]. Reproduced with permission of John Wiley & Sons.

According to the general form of the Gibbs free energy related to polarization as discussed earlier, the entropy and adiabatic temperature changes are given by [32]

$$\Delta S = -\frac{1}{2}\beta D^2 \tag{1.97}$$

$$\Delta T = -\frac{1}{2c_E}\beta T D^2 \tag{1.98}$$

As expected, as the material develops a polar phase its entropy will decrease, and Eq. (1.97) indicates that this entropy change is proportional to the square of the electric displacement change. Furthermore, Eq. (1.97) also indicates that a dielectric material with a large b will exhibit a large ΔS.

It is expected that the change in entropy at an external electric field can be treated as the alignment of dipoles. Therefore, below a critical point, the degree of alignment is proportional to the electrical field. However, once approaching the critical point, all the dipoles are aligned, so that saturation is reached. In this case, there will be no change in entropy as the electric field is further increased [22]. That critical point corresponds to a physical upper limit for a given material, which is accompanied by an EC temperature change, known as the saturated EC temperature change, ΔT_{sat}. According to thermodynamic and statistical mechanics, ΔT_{sat} can be expressed as [34]

$$\Delta T_{\text{sat}} = \frac{kT \ln \Omega}{3\varepsilon_0 \Theta C_E} \tag{1.99}$$

where Ω is the number of discrete equilibrium orientations of dipolar entities, C_E is the volume specific heat, Θ is the Curie constant, k is the Plank constant, and T is the absolute temperature. However, most of the experimental values of EC are lower than the calculated ΔT_{sat}, while much higher values have been observed in some oxide ferroelectric thin films. This observation has not been well clarified, which could be attributed to either the inaccuracy of the experimental measurement or prematurity of the theoretical estimation.

When comparing a 1-μm thin film and a 1-mm-thick ceramic sheet, with ΔT of 10 and 1 K, respectively, the cooling capacity of the ceramic sample is 100 times that of the thin film. Therefore, from a practical application point of view, thin films cannot be used for mid- and large-scale cooling applications. In this regard, it is still necessary to explore ferroelectric ceramics or single crystals for large-scale EC applications. Additionally, high dielectric strength is another important requirement to achieve high-performance ECEs. This is the main reason why relatively high EC temperature changes have been observed in ferroelectric thin films, because the dielectric strength is inversely proportional to thickness of the samples. The dependence of dielectric strength (E_{b}) on the thickness of a material has the following relationship [35]:

$$E_{\text{b}} \propto t^{-n} \tag{1.100}$$

where t is the thickness and n is a fitting parameter that is closely related to microstructure and charge transfer of the material. Generally, in an $E_{\text{b}} - t$ curve, the dielectric strength abruptly decreases with increasing thickness initially and the decreasing rate slows down, which is attributed to the change in the breakdown mechanism [36]. For instance, electron ionization avalanche mechanism is responsible for the breakdown for thin films, while electromechanical breakdown becomes dominant as the thickness of the films is increased [37]. It is expected that if the microstructural features inside the bulk ceramics could be eliminated, high dielectric strength and thus high EC temperature changes could be achieved. Therefore, increasing the breakdown strength of bulk ceramics should be a challenge in developing EC materials for practical applications.

1.5 Ferroelectric Photovoltaic Solar Energy Harvesting

Recently, the coupling of polarization with optical properties in ferroelectric materials with small bandgaps has drawn much attention, due to the potential photovoltaic effects [38–40]. Because of their non-centrosymmetric crystal structure, ferroelectric materials were found to exhibit an unconventional photovoltaic behavior, which is known as bulk photovoltaic effect (BPVE) [41]. A strikingly high power conversion efficiency (PCE) of 8.1% has been achieved in multilayered Bi_2FeCrO_6 thin film solar cells [42]. Because Si-based solar cells work on the principle of p–n junction, they are governed by the Shockley–Queisser limit [43]. In contrast, the BPVE does not encounter such a problem. Besides the BPVE, there are other factors that can be considered when using ferroelectric-based photovoltaics. For instance, according to a semiconductor-based model, with a charge distribution in the ferroelectric, which could be polarization surface charges, Schottky-induced space charges, and screening charges in the electrodes, a high photovoltaic efficiency of 19.5% has been predicted in thin films with a thickness of 1.2 nm [44].

Another important parameter is called fill factor (FF), which is defined as the maximum power in the current–voltage curve that is divided by the product of the short-circuit photovoltaic current and the open-circuit photovoltage. As compared with the conventional semiconductor solar cells that have FF of up to 80%, the values of FF of photo-ferroelectrics are only in the range of 10–30% [11, 45]. Such low FF values have been ascribed to the resistive losses of the devices, due to the poor contact of electrodes and the presence of defects in the materials. Other intrinsic resistances could be from grain boundaries and even the domain walls. Therefore, there is still plenty of room to improve the performances of ferroelectric photovoltaic materials.

The photovoltaic effect of ferroelectric materials was first observed in $BaTiO_3$, with a steady photovoltaic current, together with a weak pyroelectric current above the Curie temperature [46]. It was ascribed to the presence of a surface charged layer, which was trapped as the material was heated above the Curie temperature. Following that, the effect was quantitatively analyzed, eventually leading to the establishment of models for the photovoltaic effect of ferroelectrics – for example, a steady photovoltaic current, i.e. electrical current flowing in a short-circuited sample under illumination in $BaTiO_3$ [47] and Fe^{2+}-doped $LiNbO_3$ [48]. The photovoltaic current in Fe^{2+}-doped $LiNbO_3$ was attributed to the asymmetric scattering of the excited electrons related to the Fe impurities [49]. Because the Fe impurities in $LiNbO_3$ crystal occupied the non-centrosymmetric sites, the potential barriers for electrons were different in different directions. For instance, if one direction is more favorable than others, there will be more electrons statistically in this direction with a definite momentum vector, thus leading to the presence a nonvanishing photovoltaic current. As a consequence, there are more photoexcited electrons, which in turn increase the imbalance and thus the photovoltaic current.

Besides the asymmetric scattering of free electrons, Franck–Condon relaxation of excited ions is another fact that could be responsible for the generation of the photovoltaic current. According to the asymmetric scattering model, the short-circuit photovoltaic current (J_{sc}) is proportional to the intensity (I) of the monochromatic incident light, which is given by [49, 50]

$$J_{SC} = \kappa_1 \alpha I \tag{1.101}$$

where α is the absorption coefficient and κ_1 is a constant that is dependent on factors including the characteristics and concentration of the impurity center, mean free path of the excited electron, and the energy level of the incident photons.

Although this model was further elaborated by using a fully quantum calculation to account for the scattering of excited electrons through the introduction of an asymmetric quantum well [51], it could not be used to explain the sinusoidal variation in the photovoltaic current observed in the Fe:$LiNbO_3$ crystal, as the linear polarization of the incident light was rotated [41]. To explain this observation, a second-order optical response has to be employed to describe the BPVE by using a third-rank tensor [52–54]. As a non-centrosymmetric material is irradiated with an incident electromagnetic plane wave E_{inc}, with frequency ω, wavevector $\boldsymbol{k}$, unit polarization vector $\boldsymbol{e}$, and amplitude E_0, i.e.

$$E_{inc} = eE_0 e^{i(k,r-\omega t)} + e^* E_0 e^{-i(k,r-\omega t)} \tag{1.102}$$

the current density j is usually expressed as a Taylor expansion in terms of both the applied electric field E_{ext} and the incident light electric field having intensity $I = E^*_{inc} \cdot E_{inc}$, as follows [55]:

$$j_i = \sigma_{ij} E_j^{ext} + \sigma_{ijl} E_j^{ext} E_l^{ext} + \sigma_{ijlm} E_j^{ext} e_l e_m^* I + \chi_{ijlm} q_j e_l e_m^* I + \beta_{ijl} e_j e_l^* I + \cdots \tag{1.103}$$

The first term is known as the basic Ohm's response to an external applied electric field, which is given by the conductivity tensor σ_{ij}. The second one is the nonlinear correction to Ohm's law, corresponding to the tensor σ_{ijl}. The third term stands for the photoconductivity effect, corresponding to the photoconductivity tensor σ_{ijlm}. The fourth term is called the light entrainment effect, which generally can be neglected. The last term describes the BPVE. Since the current density is a real quantity, while the photovoltaic tensor β_{ijl} is a complex value, the latter should be separated into real and imaginary parts. The real part β^L describes the linear photovoltaic effect (LPVE) caused by the linearly polarized incident light, whereas the imaginary part β^C represents the circular photovoltaic effect

(CPVE) due to the circularly polarized incident light. The short-circuit current, j_{sc}, due to the BPVE, is given by

$$j_i^{SC} = \beta_{ijl}^{L} e_j e_l^* I + i\beta_{il}^{C}[e \times e^*]_l I \tag{1.104}$$

Because the BPVE is expressed as a third-rank tensor, it is only observed in materials without inversion center. Therefore, materials belonging to the 21 piezoelectric crystal classes, as stated earlier, could potentially have BPVE effect, which has been confirmed by experimental measurement results of various materials.

The BPVE photovoltaic current can be expressed in terms of microscopic quantities, given by [52]

$$j_{SC} = e\alpha(\omega)\left(\frac{I}{\hbar\omega}\right)\xi(\omega) \tag{1.105}$$

where e is the elementary charge of the electron, ω is the incident light pulsation, and $\xi(\omega)$ is the typical mean free path that the coherent carriers traveled in the favored direction, which is a measure of the distance that the non-thermalized or ballistic carriers would travel in a given anisotropy direction. In addition, a shift current theory was developed to describe the BPVE of ferroelectric materials [56]. Accordingly, the BPVE tensor is expressed as the product of two items: (i) transition intensity that is proportional to the absorption between the ith and jth directions of incident light, I_{ij}, and (ii) shift vector in the lth direction, R_l, i.e.

$$\beta_{ijl}^{L}(\omega) = e\sum_{n',n''}\int dkI_{ij}(n',n'',k,\omega)R_l(n',n'',k) \tag{1.106}$$

In this case, the summation over n' and n'' is over the different bands, whereas that over k is over the whole Brillouin zone. Equation (1.105) indicates that the necessary asymmetry in k-space could be developed in different ways, in order to achieve high photovoltaic currents. Meanwhile, the summation over the Brillouin zone can be asymmetric, especially when the degrees of freedom of spin are involved, because of the breaking of symmetry of time inversion. Furthermore, the shift vector could be anisotropic in k-space, so that the scattering of the states associated with an asymmetric potential well would be different with different wavevectors.

The significant progress of photovoltaic effect in ferroelectrics was the observation of the switchable diode behavior in $BiFeO_3$ (BFO) single crystals [38]. At the same time, another important discovery involves the periodic array of 71° domain walls in BFO thin films, as shown in Figure 1.4 [57]. As the thin films samples were illuminated with a white light of 28.5 mW cm^{-2}, an above-bandgap open-circuit photovoltage of 16 V and an in-plane photovoltaic current of 120 μA cm^{-2} were obtained. Furthermore, the photovoltage had a linear relation to the spacing of the electrode, which implied that it increased with increasing number of domain walls. At the same time, it was found that single domain BFO samples exhibited no photovoltaic effect. Moreover, photovoltaic effect was only observed as the electrodes were parallel to the domain walls (PLDW), whereas there was no photovoltaic effect if the electrodes were perpendicular to the domain walls (PPDW).

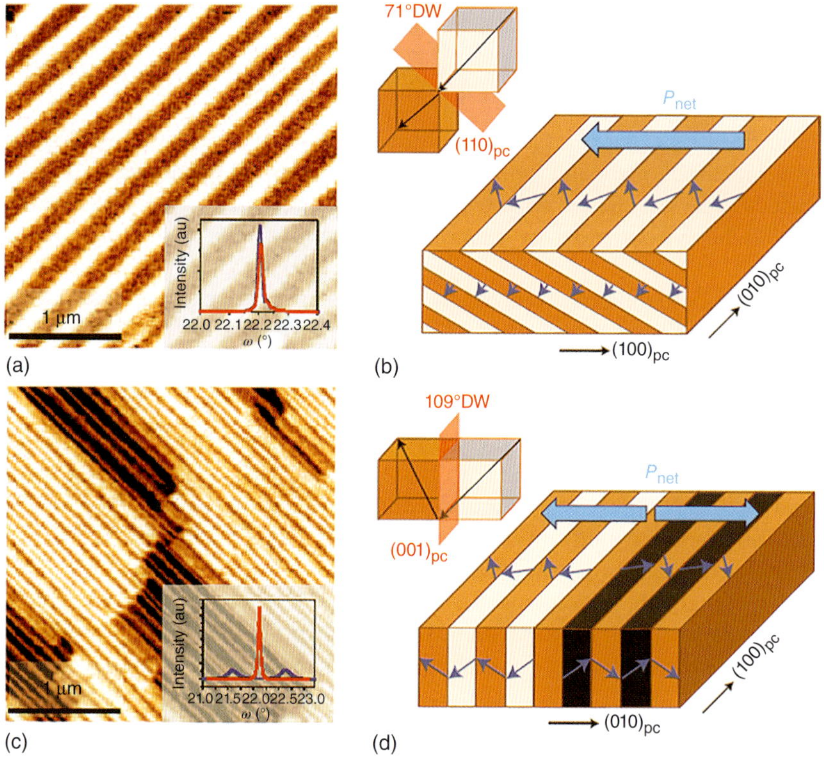

Figure 1.4 (a) Piezoresponse force microscopy (PFM) image of the ordered arrays of 71° domain walls. *Inset:* Corresponding X-ray rocking curves, along two orthogonal crystal axes. (b) Schematic diagram of the 71° domain wall arrays. The arrows indicate the different components of polarization (in-plane and out-of-plane), as well as the net polarization direction (large arrow) in the samples. Samples are found to have net polarization in the plane of the film. (c) PFM image of the ordered arrays of 109° domain walls. Inset shows the corresponding X-ray rocking curves, along two orthogonal crystal axes. (d) Schematic diagram of the 109° domain wall arrays. *Source:* Yang et al. 2010 [57]. Reproduced with permission of Springer Nature.

The domain wall photovoltaic effect (DWPVE) has been explained by considering the domain wall potential. Generally, there is a drop in potential across a domain wall in $BiFeO_3$, which is related to the rigid rotation of the polarization [58]. The level of such a drop is in the order of 180° DWs > 109° DWs > 71° DWs [59]. Accordingly, it was assumed that the potential drop could build up a high electric field in the DWs, so that electrons would be effectively separated from holes [60]. Meanwhile, recombination could also occur in the domains, since the internal electric field might not be sufficiently strong. As a result, carriers would be accumulated on different sides of the DWs, thus making them behave in a manner similar to a p–n junction.

More recently, the DWPVE model encountered serious challenges. For instance, first-principle calculations indicated that the bandgap variations predicted at the walls of BFO could not be generated [61]. Also, the assumption in the model that there is a recombination in the domains, due to the insufficiently strong electric field to dissociate the photogenerated electron–hole pairs, has not been confirmed experimentally. In an attempt to probe the local photoelectric effect of a single crystal of $BiFeO_3$ by using atomic force microscopy (AFM), the $I-V$ curves revealed no significant difference in responses between domains and domain walls [62]. Instead, the origination of PV effect of ferroelectric materials from BPVE was further confirmed [63].

However, a recent study of photovoltaic, photoconductive, and electrical conductive properties of 109° and 71° DW arrays in BFO, as a function of temperature, has clarified the roles of domain walls [64]. To avoid the effects of asymmetric contacts, symmetric Pt electrodes were employed in the study, where a photovoltage of as high as 7 V was observed for the 71° DWs with PPDW and PLDW configurations. Meanwhile, for 109° DWs, a photovoltage of 5.4 V was achieved with the PLDW configuration, whereas the PPDW configuration had no PV effect. Additionally, in all cases, the open-circuit voltage was decreased exponentially with increasing temperature, while the electrical conductivity and the photoconductivity were exponentially increased. If the photovoltaic current is assumed to be constant, the photovoltage is related to both the electrical conductivity σ and the photoconductivity σ_{ph}, which is given by

$$V_{\propto} = \frac{L}{\sigma + \sigma_{ph}} J_{SC} \tag{1.107}$$

where L is the spacing between the electrodes. It can be used to describe the exponential decrease in the photovoltage with temperature. The domain walls seemed only to modulate the global conductivity of the materials. In the PLDW configuration, the domain walls increased the resistance by adding an additional resistance to the bulk one, whereas in the PPDW configuration, the domain walls behaved as a shunt, because the domain walls of $BiFeO_3$ have higher electrical conductivity than the bulk [65–67]. It was experimentally demonstrated that the photovoltaic current was varied as the direction of polarization of the light was rotated [64], which is in good agreement with the tensor model of the BPVE, as mentioned above [54]. In summary, the photovoltaic effect observed in ferroelectric materials is intrinsic and has a tensorial nature, which is a bulk property. Besides the contribution of domain walls, the photovoltaic effect in ferroelectric materials can also be tailored by altering external parameters, such as electric field, configuration of electrodes, and creation of defects.

1.6 Concluding Remarks

Ferroelectric materials form a special group of materials with various unique properties that can be used for a wide range of applications in different areas. With the recent global research focus on energy generation and harvesting, ferroelectric materials have drawn renewed attention and interest worldwide, because of their potential energy related applications. Piezoelectric properties of ferroelectric materials enable the harvesting of waste mechanical energy in various ways, with the key issues laying on design of devices based on the principles to have higher efficiencies. Various devices have been developed to harvest different types of mechanical energy. More work is needed to further improve their performances. Pyroelectric effect has been employed to harvest waste thermal energy, but with similar problems as the piezoelectric ones. Developing new materials with higher pyroelectric coefficients seems to be a difficult task, while there is still plenty of room to improve the thermal energy harvesting efficiency through device design. The study of ECE of ferroelectric materials should be focused on the development of bulk ceramics with higher dielectric breakdown strengths, whereas their photovoltaic effects still have unsolved issues.

References

1 Lines, M.E. (1979). *Glass AM. Principles and Applications of Ferroelectrics and Related Materials*. Oxford: Clarendon Press.

2 Valasek, J. (1921). Piezo-electric and allied phenomena in Rochelle salt. *Physical Review* 17: 475–481.

3 Valasek, J. (1927). Note on the piezo-electric effect in Rochelle salt crystals. *Science* 65: 235–236.

4 Haertling, G.H. (1999). Ferroelectric ceramics: history and technology. *Journal of the American Ceramic Society* 82: 797–818.

5 Damjanovic, D. (1998). Ferroelectric, dielectric and piezoelectric properties of ferroelectric thin films and ceramics. *Reports on Progress in Physics* 61: 1267–1324.

6 Chandra, P. and Littlewood, P.B. (2007). A Landau primer for ferroelectrics. In: *Physics of Ferroelectrics: A Modern Perspective* (ed. K.M. Rabe, C.H. Ahn and J.M. Triscone), 69–115. Springer.

7 Randall, C.A., Kim, N., Kucera, J.P. et al. (1998). Intrinsic and extrinsic size effects in fine-grained morphotropic-phase-boundary lead zirconate titanate ceramics. *Journal of the American Ceramic Society* 81: 677–688.

8 Rodel, J., Jo, W., Seifert, K.T.P. et al. (2009). Perspective on the development of lead-free piezoceramics. *Journal of the American Ceramic Society* 92: 1153–1177.

9 Cook-Chennault, K.A., Thambi, N., and Sastry, A.M. (2008). Powering MEMS portable devices – a review of non-regenerative and regenerative power supply systems with special emphasis on piezoelectric energy harvesting systems. *Smart Materials and Structures* 17: 043001.

10 Beeby, S.P., Tudor, M.J., and White, N.M. (2006). Energy harvesting vibration sources for microsystems applications. *Measurement Science and Technology* 17: R175–R195.

11 Chakrabartty, J., Nechache, R., Li, S. et al. (2014). Photovoltaic properties of multiferroic $BiFeO_3/BiCrO_3$ heterostructures. *Journal of the American Ceramic Society* 97: 1837–1840.

12 Whatmore, R.W. (1986). Pyroelectric devices and materials. *Reports on Progress in Physics* 49: 1335–1386.

13 Bowen, C.R., Taylor, J., LeBoulbar, E. et al. (2014). Pyroelectric materials and devices for energy harvesting applications. *Energy & Environmental Science* 7: 3836–3856.
14 Lang, S.B. (2004). A 2400 year history of pyroelectricity: from Ancient Greece to exploration of the solar system. *British Ceramic Transactions* 103: 65–70.
15 Lang, S.B. (2005). Pyroelectricity: from ancient curiosity to modern imaging tool. *Physics Today* 58: 31–36.
16 Liu, S.T. and Long, D. (1978). Pyroelectric detectors and materials. *Proceedings of the IEEE* 66: 14–26.
17 Kosorotov, V.F., Kremenchugskij, L.S., Levash, L.V., and Shchedrina, L.V. (1986). Tertiary pyroelectric effect in lithium niobate and lithium tantalate crystals. *Ferroelectrics* 70: 27–37.
18 Poprawski, R. (1981). Investigation of phase-transitions in NH_4HSeO_4 crystals by pyroelectric method. *Ferroelectrics* 33: 23–24.
19 Devonshire, A.F. (1954). Theory of ferroelectrics. *Advances in Physics* 3: 85–130.
20 Zeyfang, R.R., Sehr, W.H., and Kiehl, K.V. (1976). Enhanced pyroelectric properties at a FE-FE phase transition. *Ferroelectrics* 11: 355–358.
21 Lu, S.G. and Zhang, Q.M. (2009). Electrocaloric materials for solid-state refrigeration. *Advanced Materials* 21: 1983–1987.
22 Valant, M. (2012). Electrocaloric materials for future solid-state refrigeration technologies. *Progress in Materials Science* 57: 980–1009.
23 Li, X.Y., Lu, S.G., Chen, X.Z. et al. (2013). Pyroelectric and electrocaloric materials. *Journal of Materials Chemistry C* 1: 23–37.
24 Scott, J.F. (2011). Electrocaloric materials. *Annual Review of Materials Research* 41: 229–240.
25 Lang, S.B. (1976). Cryogenic refrigeration utilizing the electrocaloric effect in pyroelectric lithium sulfonate monohydrate. *Ferroelectrics* 519–523.
26 Kobeko, P.P. and Kurchatov, I.V. (1930). Dielectric characteristics of Seignette's salts. *Zeitschrift fuer Physik* 66: 192 205.
27 Thacher, P.D. (1968). Electrocaloric effects in some ferroelectric and antiferroelectric $Pb(Zr,Ti)O_3$ compounds. *Journal of Applied Physics* 39: 1996–2002.
28 Radebaugh, R., Lawless, W.N., Siegwarth, J.D., and Morrow, A.J. (1979). Feasibility of electrocaloric refrigeration for the 4-15 K temperature range. *Cryogenics* 19: 187–208.
29 Shebanovs, L., Borman, K., Lawless, W.N., and Kalvane, A. (2002). Electrocaloric effect in some perovskite ferroelectric ceramics and multilayer capacitors. *Ferroelectrics* 273: 2515–2520.
30 Mischenko, A.S., Zhang, Q., Scott, J.F. et al. (2006). Giant electrocaloric effect in thin-film $PbZr_{0.95}Ti_{0.05}O_3$. *Science* 311: 1270–1271.
31 Neese, B., Chu, B.J., Lu, S.G. et al. (2008). Large electrocaloric effect in ferroelectric polymers near room temperature. *Science* 321: 821–823.
32 Lines, M.E. and Glass, A.M. (1977). *Principles and Applications of Ferroelectrics and Related Materials*. Oxford: Clarendon Press.

33 Wood, M.E. and Potter, W.H. (1985). General analysis of magnetic refrigeration and its optimization using a new concept – maximization of refrigerant capacity. *Cryogenics* 25: 667–683.

34 Pirc, R., Kutnjak, Z., Blinc, R., and Zhang, Q.M. (2011). Upper bounds on the electrocaloric effect in polar solids. *Applied Physics Letters* 98: 021909.

35 Forlani, F. and Minnaja, N. (1964). Thickness influence in breakdown phenomena of thin dielectric films. *Physica Status Solidi* 4: 311–324.

36 Kim, H.K. and Shi, F.G. (2001). Thickness dependent dielectric strength of a low-permittivity dielectric film. *IEEE Transactions on Dielectrics and Electrical Insulation* 8: 248–252.

37 Agarwal, V.K. and Srivastava, V.K. (1971). Thickness dependence of breakdown field in thin films. *Thin Solid Films* 8: 377–381.

38 Choi, T., Lee, S., Choi, Y.J. et al. (2009). Switchable ferroelectric diode and photovoltaic effect in $BiFeO_3$. *Science* 324: 63–66.

39 Guo, Y.P., Guo, B., Dong, W. et al. (2013). Evidence for oxygen vacancy or ferroelectric polarization induced switchable diode and photovoltaic effects in $BiFeO_3$ based thin films. *Nanotechnology* 24.

40 Paillard, C., Bai, X.F., Infante, I.C. et al. (2016). Photovoltaics with ferroelectrics: current status and beyond. *Advanced Materials* 28: 5153–5168.

41 Barsukova, M.L., Karimov, B.K., Kuznetsov, V.A. et al. (1980). Photo-voltaic effect in $Bi_{12}TiO_{20}$ piezoelectrics. *Fizika Tverdogo Tela* 22: 2870–2871.

42 Nechache, R., Harnagea, C., Li, S. et al. (2015). Bandgap tuning of multiferroic oxide solar cells. *Nature Photonics* 9: 61–67.

43 Shockley, W. and Queisser, H.J. (1961). Detailed balance limit of efficiency of p–n junction solar cells. *Journal of Applied Physics* 32: 510–519.

44 Qin, M., Yao, K., and Liang, Y.C. (2009). Photovoltaic mechanisms in ferroelectric thin films with the effects of the electrodes and interfaces. *Applied Physics Letters* 95: 022912.

45 Nechache, R., Harnagea, C., Licoccia, S. et al. (2011). Photovoltaic properties of Bi_2FeCrO_6 epitaxial thin films. *Applied Physics Letters* 98: 202902.

46 Chynoweth, A.G. (1956). Surface space-charge layers in barium titanate. *Physical Review* 102: 705–714.

47 Fridkin, V.M., Volk, T.R., Grekov, A.A., and Kosonogov, N.A. (1972). Photodomain effect in $BaTiO_3$. *Ferroelectrics* 4: 169–175.

48 Glass, A.M., Linde, D.V.D., and Negran, T.J. (1974). High-voltage bulk photovaltaic effect and the photorefractive process in $LiNbO_3$. *Applied Physics Letters* 25: 233–235.

49 Glass, A.M., Von der Linde, D., Auston, D.H., and Negran, T.J. (1975). Excited state polarization, bulk photovoltaic effect, and the photorefractive effect in electrically polarized media. *Journal of Electronic Materials* 4: 915–943.

50 Chanussot, G. and Glass, A.M. (1976). A bulk photovoltaic effect due to electron-phonon coupling in polar crystals. *Physics Letters A* 59: 405–407.

51 Von Baltz, R. and Kraut, W. (1978). A model calculation to explain the existence of bulk photocurrents in ferroelectrics. *Solid State Communications* 26: 961–963.

52 Kraut, W. and Von Baltz, R. (1979). Anomalous bulk photovoltaic effect in ferroelectrics: a quadratic response theory. *Physical Review B: Condensed Matter and Materials Physics* 19: 1548–1554.
53 Von Baltz, R. and Kraut, W. (1981). Theory of the bulk photovoltaic effect in pure crystals. *Physical Review B: Condensed Matter and Materials Physics* 23: 5590–5596.
54 Fridkin, V.M. (2001). Bulk photovoltaic effect in noncentrosymmetric crystals. *Crystallography Reports* 46: 654–658.
55 Sturman, B.I. and Fridkin, V.M. (1992). *The Photovoltaic and Photorefractive Effects in Noncentrosymmetric Materials.* Gordon and Breach Science Publishers.
56 Young, S.M. and Rappe, A.M. (2012). First principles calculation of the shift current photovoltaic effect in ferroelectrics. *Physical Review Letters* 109: 116601.
57 Yang, S.Y., Seidel, J., Byrnes, S.J. et al. (2010). Above-bandgap voltages from ferroelectric photovoltaic devices. *Nature Nanotechnology* 5: 143–147.
58 Lubk, A., Gemming, S., and Spaldin, N.A. (2009). First-principles study of ferroelectric domain walls in multiferroic bismuth ferrite. *Physical Review B* 80: 104110.
59 Seidel, J., Yang, S.Y., Alarcon-Llado, E. et al. (2012). Nanoscale probing of high photovoltages at 109° domain walls. *Ferroelectrics* 433: 123–126.
60 Seidel, J., Fu, D.Y., Yang, S.Y. et al. (2011). Efficient photovoltaic current generation at ferroelectric domain walls. *Physical Review Letters* 107: 126805.
61 Dieguez, O., Aguado-Puente, P., Junquera, J., and Iniguez, J. (2013). Domain walls in a perovskite oxide with two primary structural order parameters: first-principles study of $BiFeO_3$. *Physical Review B* 87: 024102.
62 Alexe, M. and Hesse, D. (2011). Tip-enhanced photovoltaic effects in bismuth ferrite. *Nature Communications* 2: 256.
63 Ji, W., Yao, K., and Liang, Y.C. (2011). Evidence of bulk photovoltaic effect and large tensor coefficient in ferroelectric $BiFeO_3$ thin films. *Physical Review B* 84: 094115.
64 Bhatnagar, A., Chaudhuri, A.R., Kim, Y.H. et al. (2013). Role of domain walls in the abnormal photovoltaic effect in $BiFeO_3$. *Nature Communications* 4: 2835.
65 Seidel, J., Martin, L.W., He, Q. et al. (2009). Conduction at domain walls in oxide multiferroics. *Nature Materials* 8: 229–234.
66 Farokhipoor, S. and Noheda, B. (2011). Conduction through 71° domain walls in $BiFeO_3$ thin films. *Physical Review Letters* 107: 127601.
67 Seidel, J., Singh-Bhalla, G., He, Q. et al. (2013). Domain wall functionality in $BiFeO_3$. *Phase Transitions* 86: 53–66.

2

Piezoelectric Energy Generation

Hong G. Yeo and Susan Trolier-McKinstry

Penn State University, Materials Research Institute, Materials Science and Engineering Department, N-227 Millennium Science Complex, University Park, PA 16802, USA

There are several technologies for energy harvesting that are being extensively studied, including harvesting from ambient light, temperature, temperature gradients, and mechanical motion. As the capabilities of harvesting systems grow, it is becoming increasingly plausible to envision self-powered electronics systems. The need for self-powering can be seen readily for the field of structural health monitoring. If thousands of sensor systems are mounted on bridges and buildings to monitor their conditions, then each of those systems will require power. Changing batteries in remote or difficult-to-access locations requires labor and expense, and can be physically hazardous. For such applications, energy harvesting could supply electrical energy to extend the lifetime of the monitoring system.

Each of the technologies for energy harvesting has advantages and disadvantages. Solar cells typically provide the largest power per area in cases where direct sunlight is available as an energy source. Where sunlight is not available, mechanical energy harvesting from sources such as car engines, microwave ovens, and human motion offers a viable alternative [1, 2]. This chapter focuses on mechanical harvesting using thin films, rather than nanowires.

2.1 Kinetic Energy Harvesting

2.1.1 Theory of Kinetic Energy Harvesting

Mechanical modeling is used to predict the performance of energy harvesting systems. This is often done using a mass–spring–damper system to describe the harvester system, as shown in Figure 2.1.

The model of a mass–spring system with a linear damper can be expressed as

$$m\ddot{z} + c_{\mathrm{T}}\dot{z}(t) + k_{\mathrm{s}}z(t) = -m\ddot{y}(t) \tag{2.1}$$

where m is a seismic mass and c_{T} is the combined damping constant ($c_{\mathrm{T}} = c_{\mathrm{e}} + c_{\mathrm{m}}$; c_{m} is the mechanical damping and c_{e} is the electrical damping coefficient), k_{s} is

Ferroelectric Materials for Energy Applications, First Edition.
Edited by Haitao Huang and James F. Scott.

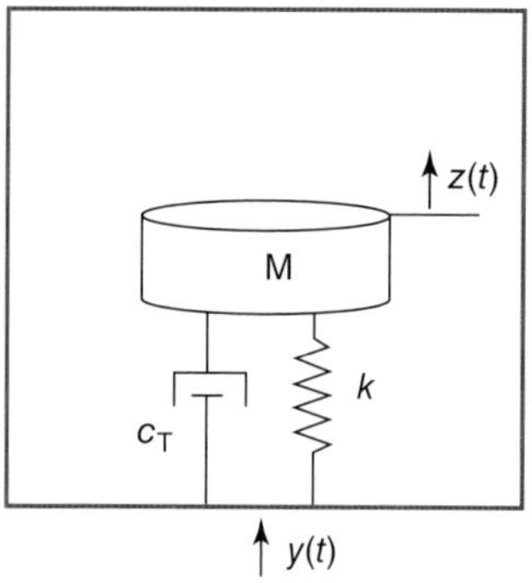

Figure 2.1 Model of a linear mass–spring damper system with a mechanical component and inertial frame for a vibration-based generator.

the spring stiffness, $z(t)$ is the net displacement of the mass, and $y(t) = Y\sin(\omega t)$ is an external sinusoidal vibration of the form within the inertial frame. Y is the amplitude of vibration. The instantaneous power generated from the resonant system is simply the product of force ($F = -m\ddot{y}(t)$) and velocity ($v = \dot{y}(t) + \dot{z}(t)$) [3]:

$$p(t) = -m\ddot{y}(t)[\dot{y}(t) + \dot{z}(t)] \tag{2.2}$$

A complete analytical expression for power can be derived by solving Eq. (2.1) for $\dot{z}(t)$ and substituting the result into Eq. (2.2). The standard steady-state solution for the displacement of the mass $z(t)$ yields the following expression [3, 4]:

$$z(t) = \frac{\omega^2}{\sqrt{\left(\frac{k}{m} - \omega^2\right)^2 + \left(\frac{c_T\omega}{m}\right)^2}} Y\sin(\omega t - \phi) \tag{2.3}$$

where ω is the frequency of the driving vibrations, and the phase angle ϕ is expressed as

$$\phi = \tan^{-1}\left(\frac{c_T\omega}{k - \omega^2 m}\right) \tag{2.4}$$

Mechanical power is converted to electrical power when damping is present. The electrical power generated, $P(t)$ at a single sinusoidal frequency, is given by [4–6]

$$P(t) = \frac{m\zeta_t Y^2\left(\frac{\omega}{\omega_n}\right)^3\omega^3}{\left[1 - \left(\frac{\omega}{\omega_n}\right)^2\right]^2 + \left[2\zeta_t\frac{\omega}{\omega_n}\right]^2} \tag{2.5}$$

where ζ_t is the total damping ratio ($\zeta_t = c_T/2m\omega_n$), and ω_n is the natural frequency of the mass–spring system. This system can extract maximum energy when the excitation frequency matches the natural frequency of the system (ω_n) given by

$$\omega_n = \sqrt{k_s/m} \tag{2.6}$$

Maximum power occurs when the resonant frequency of the system matches the input frequency; Eq. (2.5) can be re-expressed by the following equations [4]:

$$P(t) = \frac{mY^2\omega_n^3}{4\zeta_t^2} \tag{2.7}$$

$$P(t) = \frac{mA^2}{4\omega_n \zeta_t} \tag{2.8}$$

where A ($A = \omega_n^2 Y$) is the magnitude of the acceleration corresponding to the input vibrations. Based on Eq. (2.8), the output power is inversely proportional to the natural frequency of the generator when the acceleration is held constant. Practically, the acceleration levels of ambient vibration tend to reduce with increasing frequency. Thus, the design of the generator (related to the generator size and maximum tip displacement) should be appropriate for the input vibration source. Thus, the spectrum of available vibrational energy should be analyzed before designing the generator [4].

2.1.2 Kinetic Vibration Source in the Ambient

There are numerous environmental sources of kinetic vibration, in a wide range of frequencies, available for energy harvesting. However, not all ambient mechanical sources convert effectively into electrical energy. Table 2.1 lists the maximum acceleration amplitudes and frequencies of several environmental vibration sources [7, 8].

Practically, most kinetic mechanical sources produce a low frequency vibration, typically below 200 Hz. The energy generated from these sources ranges from microwatts to milliwatts [3]. It is noted that most vibration sources have small amplitudes and low peak frequencies. Figure 2.2 shows an example of the low vibrational frequency emitted from a microwave oven as studied by Roundy [7]. In this example, the highest acceleration peak is around 120 Hz, a fairly low frequency, with a few higher frequency harmonics. These conditions prevent the effective operation of many piezoelectric energy harvesters.

Table 2.1 Acceleration amplitude and frequency of ambient vibration sources from commercial devices.

Vibration source	Acceleration (ms^{-2})	$Frequency_{peak}$ (Hz)
Car engine compartment	12	200
Base of three-axis machine tool	10	70
Blender casing	6.4	121
Clothes dryer	3.5	121
Person tapping their heel	3	1
Car instrument	3	13
Door frame just after door closes	3	125
Small microwave oven	2.5	121
HVAC vents in office building	0.2–1.5	60
Windows next to a busy road	0.7	100
CD on notebook computer	0.6	75
Second story floor of busy office	0.2	100

Source: Roundy et al. 2003 [7]. Reproduced with the permission of Elsevier.

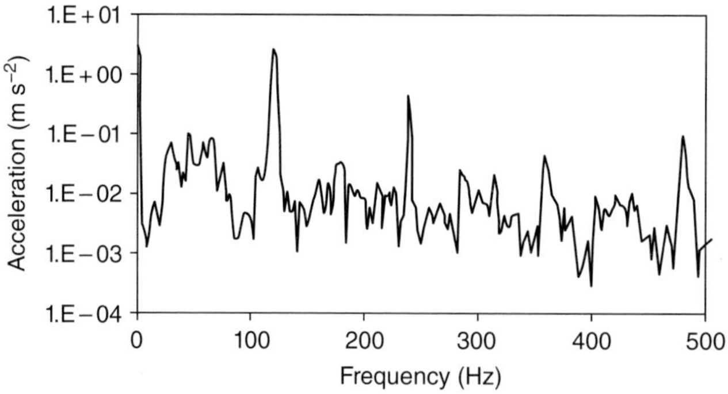

Figure 2.2 Acceleration as a function of frequency for the top of a microwave oven. *Source:* Roundy et al. 2003 [7]. Reproduced with the permission of Elsevier.

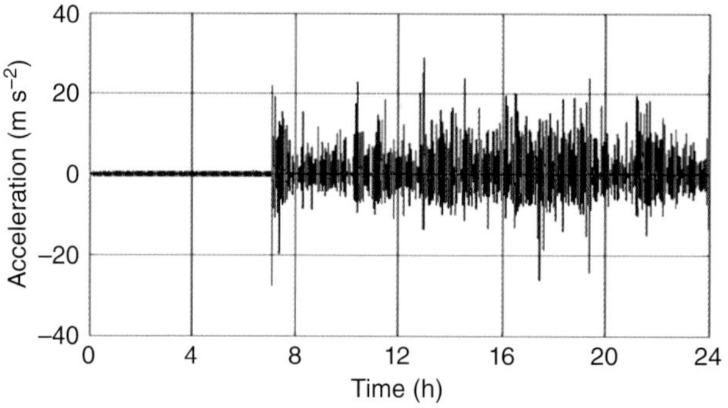

Figure 2.3 Acceleration as a function of time plot evaluated from data using a tri-axis accelerometer with a sensing unit located on the arm for an entire day. *Source:* Yun et al. 2011 [9]. Reproduced with the permission of IEEE.

Recently, wearable devices to monitor human health are a growing market. Harvesting energy from the human body for these devices is of particular interest. Most human motion are characterized by large amplitudes of motion at low frequencies, less than 10 Hz. These conditions lead to difficulties in microscale devices. For example, the acceleration $a(t)$ versus time over a 24-hour period, captured by a tri-axis accelerometer mounted on the arm is plotted in Figure 2.3, where the overall magnitude of the acceleration is given by $a(t) = \sqrt{a_x(t)^2 + a_y(t)^2 + a_z(t)^2}$ [9]. This result, coupled with the fact that most human motion is not strongly tonal, may discourage the application of resonant energy harvesters for this purpose.

2.1.3 Transducers for Mechanical Energy Harvesting

Three different types of electromechanical transducers are utilized to convert mechanical energy to electrical energy. Schematics of electrostatic (capacitive), electromagnetic (inductive), and piezoelectric harvesters are shown in Figure 2.4.

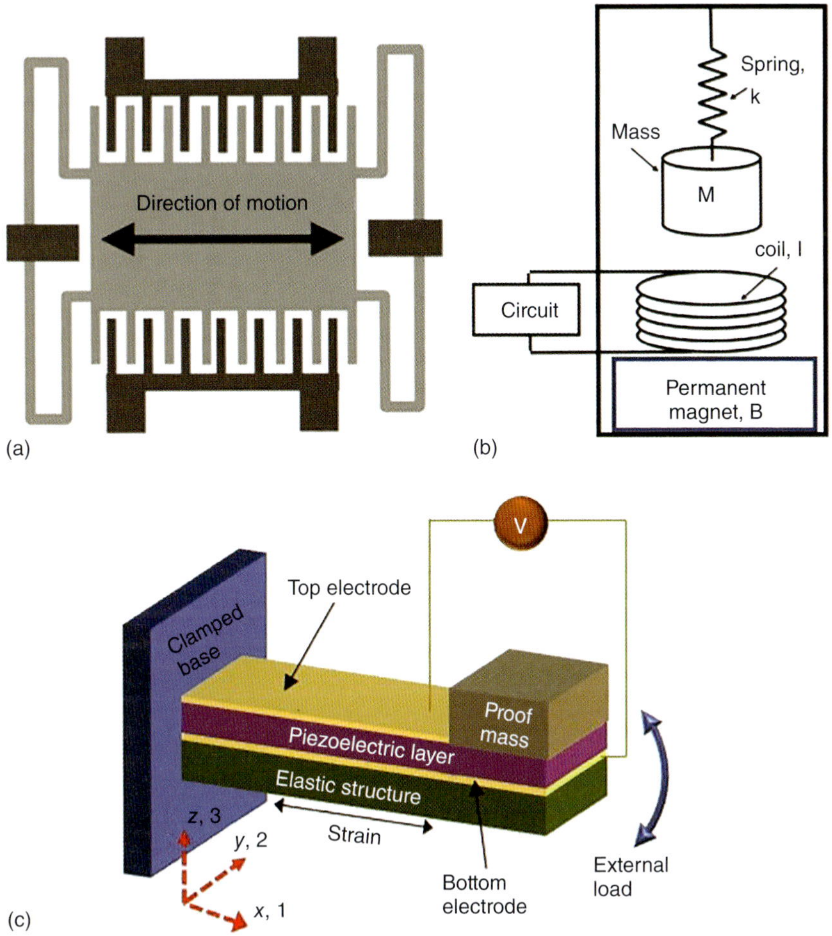

Figure 2.4 Schematic of the three types of electromechanical generators: (a) electrostatic, (b) electromagnetic, and (c) cantilever piezoelectric. *Source:* Roundy et al. 2003 [7]. Reproduced with the permission of Elsevier.

The electrostatic generator in Figure 2.4a consists of two conductive plates separated by a capacitor, typically using air as the dielectric. If the distance between plates with constant charge is changed, the voltage is also modulated; this can be used to scavenge energy. Similarly, if the voltage between the plates is held constant, the charge on the electrodes depends on the distance between them [4]. Typically, as shown in Figure 2.4a, electrostatic generators utilize a comb electrode structure to improve the power density. Also, electrostatic systems require an external voltage source to create the initial voltage or charge. In an electrostatic generator (shown in Figure 2.4a), mechanical movement of two oppositely charged electrodes separated by a dielectric provides electrical energy conversion. Fundamentally, the voltage on the plates is expressed by Eq. (2.9).

$$V = Q/C \tag{2.9}$$

where Q is the charge and V is the voltage on the plate. C is the capacitance given by $C = \varepsilon_0 A/d$, where ε_0 is the dielectric constant of free space (8.85×10^{-12} F m^{-1}), A is the area of the plates in m^2, and d is the distance between the plates in m. It has two operation modes (e.g. voltage and charge constrained conversion). For parallel plate structures, when voltage is held constant, a decrease in d leads to increases in the charge. When the charge is held constant, decreasing the capacitance by increasing d leads to increases in the voltage. According to the principle of an electrostatic generator, it is necessary to establish an initial voltage for the conversion process to start. One primary advantage is that the electrostatic converters integrate with electronics and microsystems using MEMS technology more readily than electromagnetic converters do.

Electromagnetic generators typically generate a current in the conductive coil by the relative movement of permanent magnets or a coil, as shown in Figure 2.4b. The open circuit voltage generated depends on the number of turns in the coil, the length of one coil, the strength of the magnetic field, and the velocity of the relative motion. Table 2.2 presents a comparison of the power density of a variety of electromagnetic inertial energy harvesters [10]. However, the efficiency of these microscale devices is limited by their low induced voltage (180 mV); this is not easily rectified by a diode for electronics, as reported by Amirtharajah et al. [2, 11]. Another disadvantage of the electromagnetic transducer is that it is hard to integrate with a microsystem with typical output voltage levels. This is due to the limited strength of the magnetic field and the number of turns in the coil for device sizes below 0.5–1 cm^3 [2, 5]. Unlike electrostatic devices, electromagnetic harvesters do not require an external voltage source to extract electricity.

The third type of mechanical generator is based on a piezoelectric transducer. Piezoelectric materials deform in response to an applied electrical field and produce charge or current in response to a mechanical deformation. The piezoelectric response depends upon the orientation of the strain, or electrical field,

Table 2.2 Comparison of reported electromagnetic energy harvesters.

Authors	Volume (cm^3)	Proof mass (g)	Frequency (Hz)	Acceleration ($G = 9.8$ m s^{-2})	Power density (μW G^{-2} cm^{-3})
Li et al. [72]	1	0.22	60	2.9	11.9
Ching et al. [73]	1	—	190	8.3	0.07
Williams et al. [74]	0.02	0.0024	4400	39	0.014
El-hami et al. [75]	0.24	0.5	322	10.4	20.3
Ching et al. [76]	1		60	2.9	99
Mizuno et al. [77]	2.1	0.54	700	1.3	0.0001
Lee et al. [78]	7.3	0.14	85	4.4	6
Glynne-Jones et al. [79]	0.84		322	5.4	1.5
Beeby et al. [80]	0.06	0.44	350	0.3	507
Hung et al. [81]	0.04	0.03	100	2	9.86
Perpetuum Limited [82]	131		99	0.1	2685

Source: Adapted from Mitcheson et al. [10].

Table 2.3 Comparison of three types of transducers for vibration energy harvesting.

Mechanism	Advantages	Disadvantages
Piezoelectric	No voltage source required Large voltage generated with simple mechanical structures; depends on the film/sample thickness	Limitation of strain level related to mechanical properties in the piezoelectric materials
Electrostatic	Precise control over the resonance	Separate external voltage source needed
Electromagnetic	No external voltage source needed	Complex design needed for microsystems

Source: Roundy et al. 2003 [7]. Reproduced with the permission of Elsevier.

with respect to the crystallographic axes. Unlike an electromagnetic generator, a piezoelectric generator provides the needed voltage and current levels for the electrical load circuit [5, 11]. Another major advantage is that the piezoelectric films can be used in the microfabrication processes for wafer scale systems with relatively high output voltages [2]. The three families of transducers, along with their advantages and disadvantages, are summarized in Table 2.3. Although mechanical energy harvesting using electromagnetics exhibits high performance with bulk magnets and macroscale coils in large systems, it is not suitable for integration with MEMS systems [2, 5]. Piezoelectric generators show most of the advantages of electromagnetic converters. Additionally, they can be employed in MEMS systems using piezoelectric films. Thus, piezoelectric energy harvesters could be useful for portable electronics.

2.2 Piezoelectric Vibration Harvesting

2.2.1 Piezoelectricity

Piezoelectricity is the linear coupling between an applied stress and an induced polarization (direct effect), or an applied electric field and an induced strain (converse effect). Among the 32 crystallographic point groups, 20 of the 21 non-centrosymmetric groups are potentially piezoelectric [12]. Piezoelectricity is symmetry-forbidden in centrosymmetric structures. Figure 2.5a illustrates the direct piezoelectric effect, while Figure 2.5b shows the converse effect.

For many piezoelectric applications, Pb-containing perovskite piezoelectrics (e.g. $Pb(Zr,Ti)O_3$ (PZT)) are used because of their high piezoelectric effect, and the ability to reorient the remanent polarization with an applied electric field, unlike wurtzite and quartz [13]. Figure 2.6 shows the ABO_3 crystal perovskite structure in its cubic prototype form (no spontaneous polarization) above the Curie temperature (T_C). Below T_C, the material distorts into tetragonal or rhombohedral ferroelectric phases (having [001]/[111] polarization directions, respectively) over most of the phase diagram. In the tetragonal perovskite ferroelectrics (e.g. $PbTiO_3$), the Ti atom moves close to one of its six oxygen neighbors; this produces a spontaneous polarization directed in one of six possible directions. In

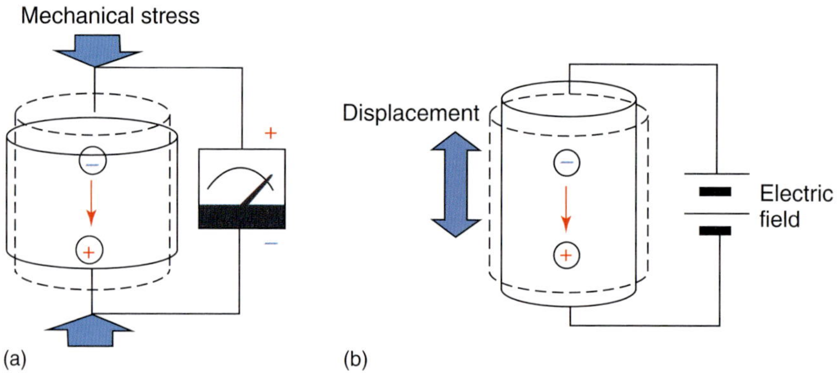

Figure 2.5 (a) Direct piezoelectric effect: a polarization is generated by an applied stress; (b) the converse piezoelectric effect: physical displacement is caused by the applied electric field.

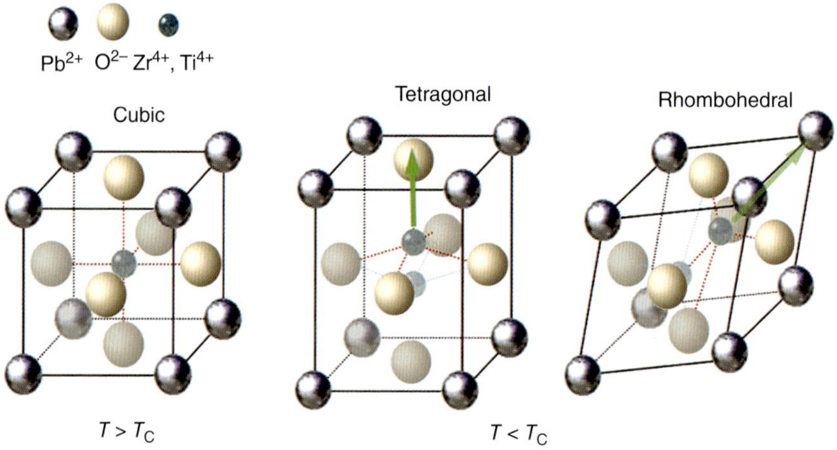

Figure 2.6 Perovskite structure.

rhombohedral perovskite ferroelectrics, the polarization direction favors ⟨111⟩ directions; this shifts the Ti atom close to three adjacent oxygens. There are eight possible polarization directions.

2.2.2 Theory of Piezoelectric Vibration Energy Harvesting

Based on the linear inertial model described in Section 2.1.1, the piezoelectric constitutive equations (relating strain (S) and electrical displacement (D), or stress (T) and the electric field (E)) can be used to derive the electromechanical coupling coefficient (k) and maximum output power (P_{max}). In the constitutive equations s, d, and ε are used for the mechanical compliance, the piezoelectric coefficient, and the dielectric permittivity. These equations are

$$S_{ij} = d_{ijk}E_k + s^E_{ijkl}T_{kl} \tag{2.10}$$

$$D_i = \varepsilon_{ij}^T E_j + d_{ijk} T_{jk} \tag{2.11}$$

(using Einstein notation where *i, j, k, l* can each take the values 1, 2, or 3). The efficiency (coupling coefficient) for a piezoelectric transducer is defined as $k^2 = \frac{d_{ijk}^2}{s\varepsilon}$.

For a linear transducer, when the AC power signal is sinusoidal, the maximum extracted power (P_{max}) can be expressed as

$$P_{max} = \lambda_{max} \omega U_{in} \tag{2.12}$$

where ω is the angluar frequency of the driving vibration, U_{in} is the input energy, and λ_{max} is the maximum transmission coefficient $\left(\lambda_{max} = \frac{U_{max}}{U_{in}}\right)$. Based on the linear relationship of the output variables D and S in Figure 2.7, D' is the value of the through variable when S is constrained to be zero, and vice versa. When $D = D'/2$ and $S = S'/2$, the maximum output energy can be expressed by

$$U_{max} = 1/4D'E' \tag{2.13}$$

If D is zero, E' is given by Eq. (2.11).

$$E' = -\frac{d}{\varepsilon} T \tag{2.14}$$

From Eqs. (2.10) and (2.11),

$$\frac{1}{2} D' = \mathrm{d}T + \frac{1}{2} \varepsilon E' \tag{2.15}$$

Substituting Eqs. (2.14) and (2.15) into Eq. (2.13) yields

$$U_{max} = -\frac{1}{4} \frac{d^2}{\varepsilon} T^2 \tag{2.16}$$

From a standard two-port model of a linear piezoelectric transducer, the maximum transmission coefficient can be expressed by [14]

$$\lambda_{max} = \frac{d^2}{4s\varepsilon - 2d^2} \tag{2.17}$$

From Eq. (2.17), the maximum transmission coefficient can be simplified by use of the coupling coefficient $\left(k^2 = \frac{d^2}{s\varepsilon} = \frac{d^2 E}{\varepsilon}\right)$, where E is the elastic (Young's) modulus.

$$\lambda_{max} = \frac{k^2}{4 - 2k^2} \tag{2.18}$$

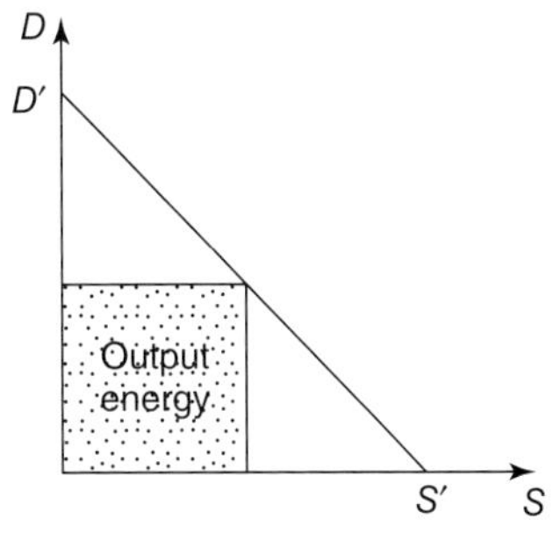

Figure 2.7 Relationship of the output variables *D* and *S* for piezoelectric transducer.

In the piezoelectric system, the input energy density (u_{in}) is given in terms of the stress (T):

$$u_{in} = \left(1 - \frac{k^2}{2}\right)\frac{T^2}{E} \tag{2.19}$$

When an AC force of magnitude $|F| = mQA$ is applied to the piezoelectric transducer, the stress in the piezoelectric can be written as $T = |F|/a_p$, where Q is the total quality factor of the resonator and a_p is the cross-sectional area of the piezoelectric element.

$$u_{in} = \left(1 - \frac{k^2}{2}\right)\frac{(mQA)^2}{Ea_p^2} \tag{2.20}$$

Spring stiffness can be expressed as

$$k_s = \frac{Ea_p}{h} \tag{2.21}$$

where h is the height of the piezoelectric element. Based on Eq. (2.21), the elastic modulus can be written as $E = \omega_n^2 hm/a_p$. Thus, the input energy ($U_{in} = ha_p u_{in}$) is given by the following expression:

$$U_{in} = \left(1 - \frac{k^2}{2}\right)\frac{(QA)^2 m}{\omega_n^2} \tag{2.22}$$

Substituting Eqs. (2.22) and (2.18) into Eq. (2.12) yields the maximum extracted power from a vibration source using a cantilever structure with a piezoelectric material at resonance (ω_n) [14, 15].

$$P_{max} = \frac{k^2 m Q^2 A^2}{4\omega_n} \tag{2.23}$$

Piezoelectric thin films for MEMS scale energy harvesters are governed by the $e_{31,f}$ coefficient for films polarized out of plane with top and bottom planar electrodes. In the case of the 31 mode, the coupling coefficient of piezoelectric films is given by $k_{31}^2 = d_{31}^2/(s_{11}\varepsilon_{33})$. The $e_{31,f}$ is $d_{31}/(s_{11} + s_{12})$ under the condition of zero strain (but finite stresses) in the plane of the film, while being stress free out of the plane [16]. Thus, output power can be expressed as Eq. (2.24) including material terms, where ν is Poisson's ratio.

$$\text{Power}_{max} = \frac{m}{4\omega}\left(\frac{e_{31,f}^2}{\varepsilon_r\varepsilon_0} \cdot \frac{(1-\nu)^2}{E}\right) Q^2 A^2 \tag{2.24}$$

If most of the device volume is occupied by the substrate, the effective mechanical properties of the device are governed by the substrate stiffness [17]. In this case, the output power relies on piezoelectric material properties such as the piezoelectric coefficient and dielectric permittivity. Thus, the figure of merit (FoM) for energy harvesting in thin-film piezoelectric cantilevers is approximated as [15, 18, 19]

$$\text{FoM} = \frac{e_{31,f}^2}{\varepsilon_r} \tag{2.25}$$

The term $(e_{31,f}^2)/\varepsilon_r$ is used to compare the FoM between materials.

Additionally, since the performance of the piezoelectric harvester is governed by the dielectric loss tangent (tan δ), tan δ of the piezoelectric material should be low for use in energy harvesters. In the off-resonance condition for a 31 mode, an alternative figure of merit (FoM_{off}) for a piezoelectric energy harvester can be expressed by the following equation: $FoM_{off} = d_{31} \times g_{31} / \tan\delta$ (effective piezoelectric voltage constant, g) [20].

Two main approaches have been widely studied to improve the efficiency of power conversion from mechanical stress/strain to electric field/current: selection of the piezoelectric material and optimization of the mechanical design of the harvester.

2.3 Choice of Materials for Energy Harvesting

2.3.1 Materials for Piezoelectric MEMS Harvesting

As was shown in the previous section, the piezoelectric coefficient and dielectric constant are key parameters that influence the performance of piezoelectric energy harvesters. Piezoelectric single crystals, ceramics, thick and thin films, and polymers such as poly-vinylidene fluoride (PVDF) have all been explored for energy harvesting for piezoelectric energy harvesters [3, 4, 21].

Figure 2.8 shows a comparison of reported piezoelectric constants and FoM ($(e_{31,f})^2/\varepsilon_r$) for a variety of thin films for 31 mode MEMS energy harvesters [22, 23]. Two major groups are apparent: non-ferroelectric: e.g. ZnO, AlN; ferroelectric: e.g. (K,Na)NbO_3 (KNN), $BiFeO_3$, Pb(Zr,Ti)O_3. Among the non-ferroelectric piezoelectric materials, AlN has a large FoM due to the low dielectric constant in spite of its modest piezoelectric coefficient [24, 25].

In recent work, Al substitution by Sc was shown to enhance the piezoelectric coefficient (e_{33}) by decreasing the stiffness (c_{33}), which leads to increase in the longitudinal piezoelectric coefficient ($d_{33,f} = e_{33}/c_{33}$) [26]. Akiyama et al. reported that the FoM of $Sc_{0.41}Al_{0.59}N$ is five times higher than that of undoped AlN, in part due to the fact that the transverse piezoelectric coefficient ($e_{31,f}$) also increased with decreasing elastic modulus [25].

Many researchers have reported improvements in the energy harvesting FoM of ferroelectric materials such as PZT, KNN, and $BiFeO_3$ through control of composition (MPB), orientation control (texturing), reducing compositional gradients, doping, control of residual stresses, and introduction of strong levels of imprint [15, 18, 19, 27, 29–33]. It has been shown through the research of Jaffe et al. that the Zr/Zr + Ti ratio is a major factor in determining the properties of lead zirconate titantate [34]. The composition of $Pb(Zr_{0.52}Ti_{0.48})O_3$, which corresponds to the MPB – the nearly temperature independent boundary between rhombohedral and tetragonal phases of PZT – is known to exhibit enhanced dielectric and piezoelectric coefficients. For a given composition, the crystallographic orientation and domain state also influence the magnitude of the remanent polarization and the piezoelectric response. These two factors may be governed by various factors such as the thermal expansion coefficient (TEC) of the substrates, templating from a seed layer (e.g. $PbTiO_3$), or an orienting

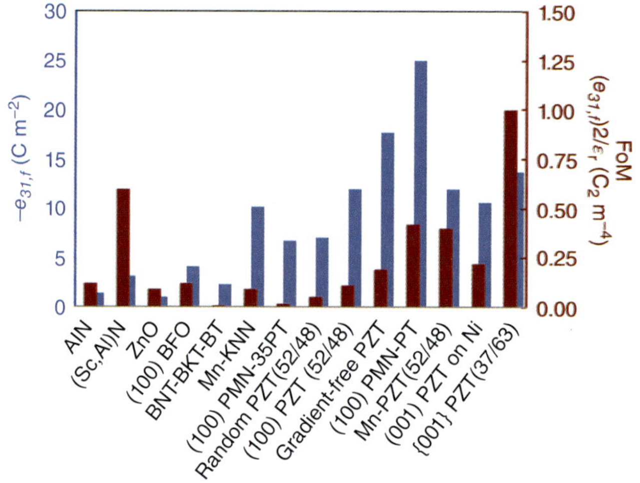

Figure 2.8 Comparison of $e_{31,f}$ and FoM of piezoelectric films for MEMS energy harvesting with different orientations on various substrates [22–30]. (Data from: left AlN, $(Sc_{0.41}Al_{0.59})N$ ((Sc,Al)N), ZnO, (100) $BiFeO_3$ on $SrTiO_3$ ((100)BFO), epitaxial $Bi_{0.5}Na_{0.5}TiO_3$-$Bi_{0.5}K_{0.5}TiO_3$-$BaTiO_3$ (BNT-BKT-BT), 0.5 mol% Mn-doped $(K_{0.5}Na_{0.5})NbO_3$ (Mn-KNN), (100) $0.65Pb(Mg_{0.33}Nb_{0.67})O_3$ -$0.35PbTiO_3$ ((100)PMN-35PT), random $Pb(Zr_{0.52},Ti_{0.48})O_3$, (Random PZT(52/48)), (100) $Pb(Zr_{0.52},Ti_{0.48})O_3$ on Si ((100)PZT(52/48)), gradient-free (100) $Pb(Zr,Ti)O_3$ on Si (gradient-free PZT), epitaxial (100) $0.67Pb(Mg_{0.33}Nb_{0.67})O_3$-$0.33PbTiO_3$ on $SrRuO_3$/$SrTiO_3$/Si ((100) (100)PMN-PT, (001) $Pb(Zr_{0.52},Ti_{0.48})O_3$ on Ni foil ((001) PZT on Ni). Epitaxial (001) $Pb(Zr_{0.52},Ti_{0.48})O_3$ on (100) MgO with 1 mol% Mn (Mn-PZT(PZ/48)), {001} $Pb(Zr_{0.37},Ti_{0.63})O_3$ on CaF_2 substrate ({001} PZT(37/63))). Redrawn after Fig. 6 from Ref. [35].

bottom oxide (e.g. $LaNiO_3$, $SrRuO_3$) electrode, and optimized nucleation conditions (Pt(111)/Ti/TiO_2) [15, 19, 29, 30, 34, 36–38].

Recently, Yeager et al. demonstrated the dependence of FoM on *c*-domain volume fraction (f_{001}) and $x = $ Zr/(Ti + Zr) with tetragonal PZT films grown on CaF_2, MgO, $SrTiO_3$, or Si substrates [15, 30]. It was noted that a superior FoM (0.8–1.0 C^2 m^{-4}) was exhibited for *c*-axis PZT films with $f_{001} > 95\%$ [30].

Also, reducing the compositional gradient and increasing the process-induced imprint in the film are critical means to ameliorate the FoM for energy harvesting [18, 29, 39]. As described by Ledermann, the transverse piezoelectric coefficient is degraded by the existence of any gradients in the Zr/Zr + Ti ratio produced in the PZT film during crystallization [19]. Gradient-free (100) textured films on Si exhibited nearly uniform Zr/Zr + Ti ratios (±2.5%) at the MPB with superior $e_{31,f}$ (−17.7 C m^{-2}) and dielectric constant (1620) as compared to the Zr/Zr + Ti gradient (±12%) PZT film [29]. Another means to enhance the FoM is a strong level of imprint. Imprint is a manifestation of the built-in internal field, and can be quantified by a shift of the *P*–*E* hysteresis loop along the field axis. The internal field tends to increase the remanent polarization while suppressing the dielectric permittivity. These useful features certainly contribute to an improved FoM. Recently, PMN-PT films were reported with an extremely high piezoelectric coefficient, combined with suppressed permittivity caused by the high levels of imprint [18].

Because mechanical harvesters produce ac current outputs, it is essential to rectify the output. This, in turn, imposes additional constraints. Efficient rectification is possible only when the device generates a voltage that is high enough, but not so high that a buck converter is required. For a MEMS harvester, the first of these conditions is often the most problematic. Decreasing the permittivity is helpful, as the output voltage is inversely proportional to the permittivity.

2.3.2 Effect of Stress Induced by Substrate

The piezoelectric and dielectric responses of mono-domain PZT are anisotropic, and so depend on film orientation, the electrode type (e.g. interdigitated, parallel plate), and the domain states configuration [40]. In particular, the type and level of stress caused by the difference in the TECs of films and substrate influence the domain orientation.

When a thin-film material is mechanically constrained to a substrate, the total stress present consists of three terms (intrinsic stress, extrinsic stress, and thermal stress) and can be written in the following form:

$$\sigma_{\mathrm{tot}}(T) = \sigma_{\mathrm{th}} + \sigma_{\mathrm{i}} + \sigma_{\mathrm{e}} \tag{2.26}$$

Intrinsic stress (σ_{i}) is determined by deposition parameters.

Extrinsic stress (σ_{e}) arises from the dimensional changes as the material undergoes either crystallization of the amorphous phase or a phase transition. The most important contribution of film stress in thick films is the thermal stress (σ_{th}) due to the thermal expansion mismatch between substrate and film. Thermal stresses due to cooling from the crystallization temperature (T_{growth}) to room temperature (T_{RT}) can be expressed as

$$\sigma_{\mathrm{th}} = (\alpha_{\mathrm{f}} - \alpha_{\mathrm{s}})\frac{E_{\mathrm{f}}}{1 - \nu_{\mathrm{f}}}\left(T_{\mathrm{growth}} - T_{\mathrm{RT}}\right) \tag{2.27}$$

α_{f} and α_{s} are the TECs of the film and substrate. E_{f} and ν_{f} are the Young's modulus and Poisson's ratio of the thin film, respectively.

Especially, the thermal stress/strain generated during cooling from the growth (or crystallization) temperature to the Curie temperature is directly correlated with the volume fractions of *c*- and *a*-domains in a tetragonal film; for a *c*-domain, the spontaneous polarization is aligned *out of plane* and for an *a*-domain, the spontaneous polarization is aligned *in plane*. Below the cubic to tetragonal transition, the volume fraction of *c*- and *a*-domains influence the thermal stress, because partial stress relief occurs at the Curie temperature as favorable domain orientations develop. Figure 2.9 shows the nonlinear TEC of PZT, Si, and Ni respectively as a function of temperature; these values were used to calculate the thermal stress and strain in Refs. [41–43]. The thermal strain (ξ_{th}) that develops during cooling from T_{growth} to T_{C} is given by

$$\xi_{\mathrm{th}} = (\alpha_{\mathrm{f}} - \alpha_{\mathrm{s}})(T_{\mathrm{growth}} - T_{\mathrm{C}}) \tag{2.28}$$

Figure 2.10 illustrates the domain response to different stress types in the film. When the ferroelectric film is under tensile stress, *a*-domain development is

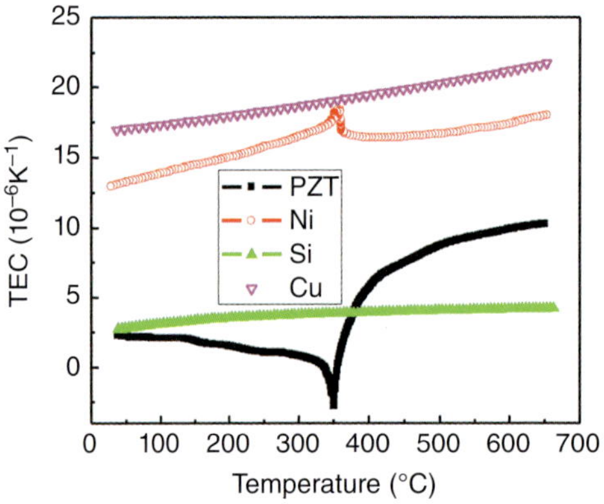

Figure 2.9 Thermal expansion of PZT, Si, Cu, and Ni in the temperature range between crystallization temperature of PZT and room temperature for the residual stress calculation [41–43].

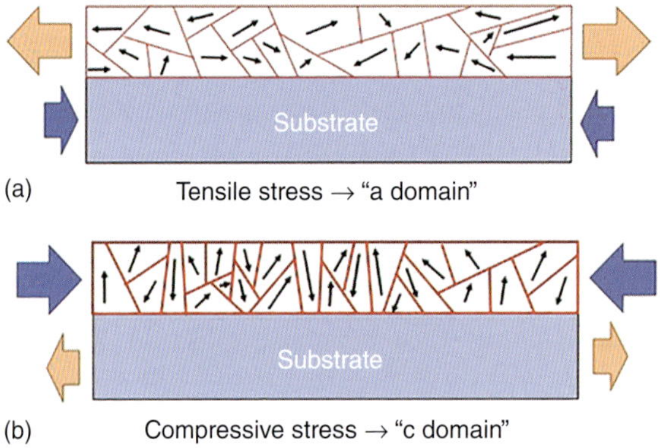

Figure 2.10 Illustration of domain structure in the film under (a) tensile stress and (b) compressive stress, in the cases where the stresses are controlled by thermal expansion mismatch.

preferred to *c*-domain. Conversely, *c*-domain structures are preferred for compressive stress. This phenomenon allows the film–substrate thermal expansion mismatch to govern the domain state at T_C.

For example, when the substrate TEC is larger than that of the PZT film, it causes compressive stress in the film and preferred *c*-axis orientation, as shown in Figure 2.10b. In contrast, substrates with low TEC, such as Si, induce tensile stresses in the PZT film and preferred *a*-domain states. The resulting domain structure affects film properties such as dielectric constant and remanent polarization [15, 44–46].

Figure 2.11 shows the dielectric constant of PZT films as a function of calculated thermal strain. Yeager et al. demonstrated experimentally that the dielectric constant of (001) $Pb(Zr_{0.48},Ti_{0.52})O_3$ film (PZT 48/52, blue diamonds) is a function of thermal strain by growing PZT films on three different substrates (e.g. silicon, $SrTiO_3$, MgO) [15].

PZT films under compression (grown using substrates with large TECs) have lower dielectric constants due to a large volume fraction of *c*-domains. Yeo et al. also confirmed that a lower dielectric constant is obtained in the PZT (52/48)

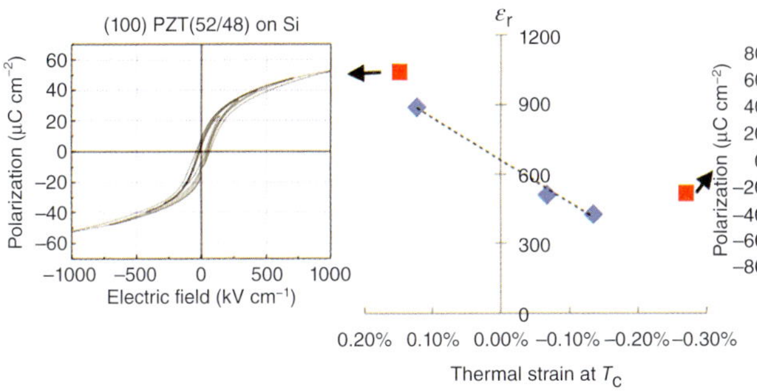

Figure 2.11 Effect of thermal strain on the dielectric constant (blue diamond [15]: PZT 48/52, red cubic [46]: PZT 52/48) and P–E hysteresis loops in the PZT 52/48 films. *Source:* Yeo and Trolier-McKinstry 2014 [46]. Reproduced with the permission of AIP Publishing.

film under compressive stress on Ni substrates [46]. As a result, (001) PZT films on Ni have larger remanent polarizations than those on (100) PZT on Si due to the differences in the volume fraction of *a* and *c*-domains caused by the thermal stress [46]. Such films have been demonstrated to be superior for energy harvesting relative to PZT films grown under conditions producing tensile stress, as is discussed below.

2.4 Design and Configuration of Piezoelectric Harvester

2.4.1 Option of Piezoelectric Configuration

The relative direction of the applied strain/stress from the environment and the electrical polarization play another important role in the performance of microscale vibration energy harvesters using piezoelectric thin films. There are two configuration options, 31 mode and 33 mode, for piezoelectric microdevices.

Figure 2.12 exhibits the relationship between the direction of the applied stress and the poling direction in a PZT layer. In the 31 mode, the film has top and bottom electrodes (TBEs). In this case, the poling direction, which is typically

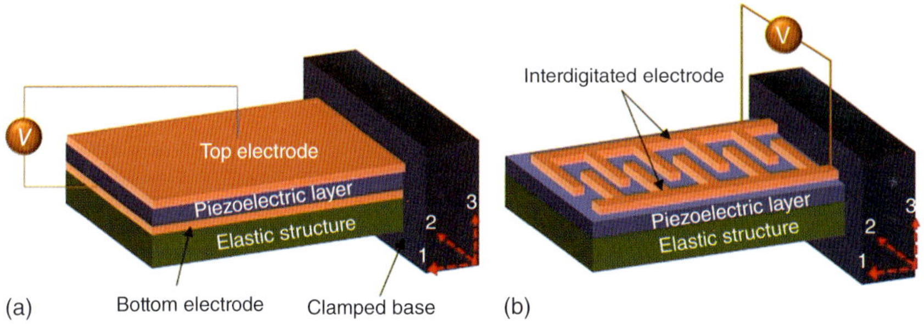

Figure 2.12 Cantilever piezoelectric energy harvesters with proof masses: (a) 31 mode device using top and bottom electrodes (TBEs) and (b) 33 mode device using interdigitated electrodes (IDE).

referred to as the "3" direction, is perpendicular to the in-plane direction ("1" direction) of the applied strain: On the other hand, for the 33 mode, interdigitated electrodes (IDE) are utilized. In this case, the poling direction and the strain are both in plane, as shown in Figure 2.12b [2, 47, 48].

For a 31 mode energy harvester with TBEs the generated open voltage (V_{OC}) is determined by the thickness of the piezoelectric layer (t_{xx}), the effective piezoelectric constant, $g_{31,f}$ (V m N^{-1}), and the applied stress [2].

$$V_{\text{OC}} = \sigma_x g_{31,f} t_{xx} \tag{2.29}$$

In the case of the 33 mode, t_{xx} represents the spacing between the IDE.

In general, the longitudinal piezoelectric coefficient $e_{33,f}$ is about two times higher than $|e_{31,f}|$ [24]. Moreover, the open-circuit voltage can be easily increased by increasing the electrode spacing, and is thus relatively unaffected by the thickness of the piezoelectric layer.

However, there are some disadvantages to the 33 mode. First of all, finite element modeling demonstrates that the piezoelectric material under the electrodes is not active [47]. This means that the electrode width should be minimized, within the constraints imposed by the need for effective poling. Furthermore, the electric field is not uniform through the thickness of the piezoelectric layer [47, 49–51]. Thus, design of IDE electrode systems requires optimization of variables such as the dimensions of the electrode and the piezoelectric layer [51].

In contrast, for typical excitation levels, the generated voltage of a 31 mode energy harvester is typically too low for simple rectification circuits, when the thickness of the PZT layer is less than 0.5 μm. However, with the use of TBEs, increase in PZT thickness produces higher voltages, and yields higher currents than 33 mode devices. Thus, the selection of electrode configuration for energy harvesting should take into account not only anisotropy in the dielectric and piezoelectric properties associated with the domain configuration but also the film thickness.

2.4.2 Unimorph and Bimorph Structures

Three configurations of piezoelectric energy harvesters are shown in Figure 2.13 [52]. In general, they can be prepared with piezoelectrics on one or both sides of the elastic layer for unimorph or bimorph configurations, respectively.

Two types of bimorphs can be distinguished based on the electrical connection of the piezoelectric elements: series or parallel connection. During bending, the two PZT layers undergo stresses of the opposite sign (e.g. the top layer is in compression whereas the bottom layer is in tension). As a consequence, for the series configuration shown in Figure 2.13a, the two piezoceramic layers should be poled in opposite directions. Likewise, Figure 2.13b represents the parallel connection of the bimorph structure with the piezoceramic layers poled in the same direction along the thickness. In this case, on bending, the two piezoelectric layers induced opposite directions of the electric field [53].

The series connection yields twice the output voltage from the bimorph piezoelectric layer relative to the parallel connection [54]. One of the merits of the

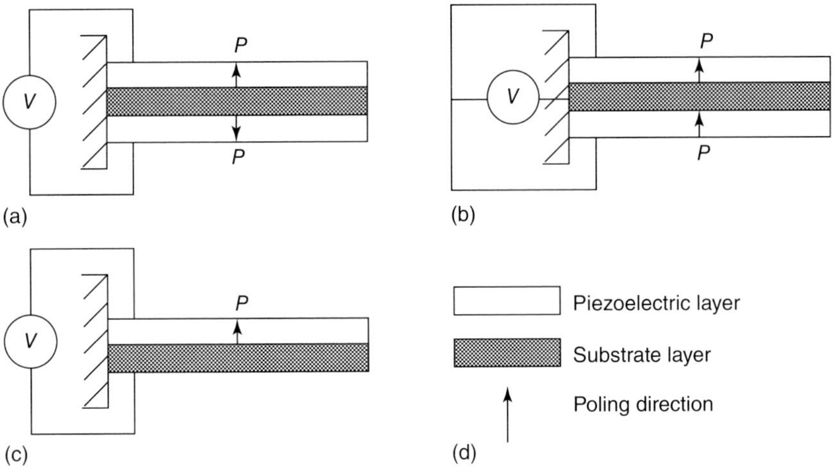

Figure 2.13 (a) Bimorph configuration in series connection, (b) bimorph configuration in parallel connection, (c) unimorph configuration, and (d) key to figure. *Source:* Yao et al. 2004 [52]. Reproduced with the permission of Elsevier.

bimorph structure is that high power density can be obtained even when the working area is limited. However, in the arena of MEMS, bimorph structures are hard to achieve by microfabrication processing [2].

2.4.3 Linear Piezoelectric Energy Harvesters

Linear (resonant based) piezoelectric generators are a simple approach for mechanical energy harvesting. These linear piezoelectric harvesters generate maximum power when the frequency of the device matches the ambient frequency, as described in Section 2.2.2. Common configurations are cantilever beams, as bending of the cantilever provides high strain; additional mass can be used to tune the resonant frequency (Figure 2.14).

Rectangular cantilever beams generally suffer from excessively high strains near the clamping point. The strain dramatically decreases in magnitude with distance away from the clamp. On the other hand, trapezoidal beam shapes improve the strain distribution along the length, and so help prevent overstrain at higher excitation [55–57]. Dhakar et al. proposed a new design using a bimorph piezoelectric cantilever connected to a polymer cantilever to improve the strain distribution along the beam for resonant-based energy harvesters operated at a low frequency [58].

2.4.4 Nonlinear Energy Harvesting

In contrast to linear oscillators, which typically have a narrowband response that decays rapidly if excited off their resonance frequency, nonlinear oscillators provide broadband performance. In general, nonlinear energy harvesting uses Duffing (hysteretic resonance), frequency up conversion, or bistable oscillator designs using magnetic force and special mechanical structures.

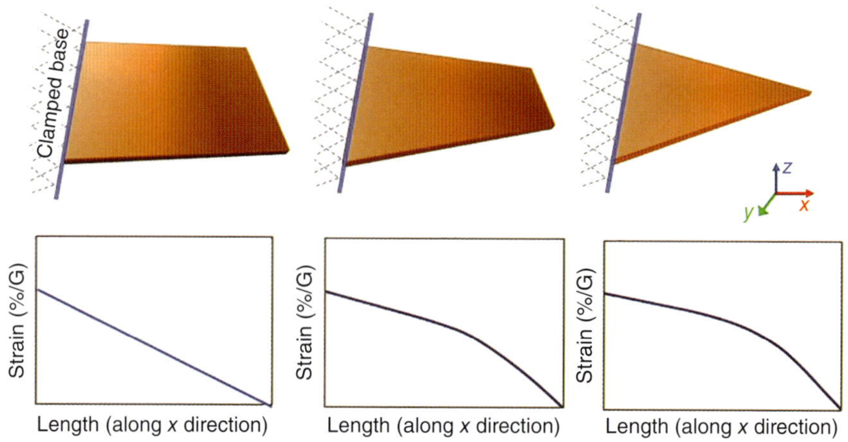

Figure 2.14 Three types of cantilever beams for piezoelectric energy harvesters and strain distribution along the beam.

(i) *Duffing resonance (nonlinear hysteretic resonance)*: Duffing oscillators are characterized by either softening (when the force decreases with the tip deflection) or hardening (when the force increases with the tip deflection) of the system stiffness [58]. Near the resonance frequency, i.e. $\approx\omega_n$, the result of hardening/softening is a broad resonance peak with three branches: the large resonant branch, B_r; the nonresonant branch, B_n; and the unstable branch (dashed lines), as shown in Figure 2.15. When δ is negative, i.e. softening nonlinearity, the deflection follows the nonresonant branch up to the saddle-node (*s.n*) point as the excitation frequency is increased toward ω_n. Then it shifts up to the resonant branch along the dashed line and follows the upper branch on increasing the frequency. On the other hand, when the excitation frequency is decreased toward resonance, the behavior follows the upper branch up to the higher *s.n* point and then falls to the lower branch B_n [59].

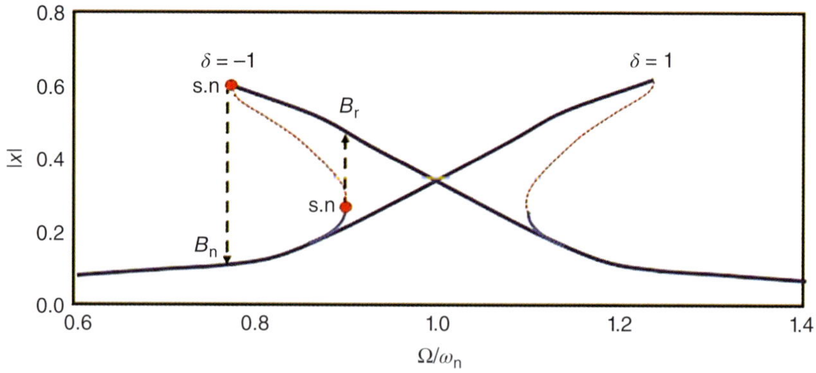

Figure 2.15 Frequency response of an oscillator with stiffness nonlinearity. ($|x|$ is the steady-state amplitude, Ω is the input excitation frequency, ω_n is the resonant frequency of the system, and δ is the coefficient of the cubic nonlinearity).

For instance, Hajati et al. demonstrated a MEMS piezoelectric bridge Duffing oscillator with a central proof mass. This device achieved a wide bandwidth due to nonlinearities in the stiffness caused by a net stretching with increasing acceleration amplitude [60]. Leland et al. demonstrated that compressive axial stress in the bimorph PZT ceramic beam induced using a small steel vise and a micrometer drive for loading reduces the transverse stiffness, and tunes the resonance frequency [61].

(ii) *Frequency up conversion*: Frequency up conversion is a good strategy to excite a high-frequency piezoelectric transducer from a low-frequency vibration output. In such a device, a large acceleration induced in a low-frequency resonator is transferred to a high-frequency resonator through a catch-and-release or plucking mechanism. The high-frequency output is produced by the piezoelectric elements. The main advantage of this method is that it allows more efficient scavenging from low-frequency mechanical energy. Pillatsch et al. suggested a rotational piezoelectric beam-plucking energy harvester with a magnet mass for frequency up conversion in wearable applications [62]. The maximum output power of this device was 7 μW during running, although the performance was reported to degrade quickly on operation [62].

(iii) *Bistable oscillator*: Nonlinear bistable oscillators typically use different permanent magnet configurations to induce a magnetoelastic oscillation in a piezoelectric material with a widened bandwidth of resonance frequency. Bistable behavior occurs when there are two shallow potential wells. For a small stroke per forcing period, the device oscillates within one well. When the force is increased, the oscillator can be excited over the saddle point to oscillate between wells, as depicted in Figure 2.16 [63].

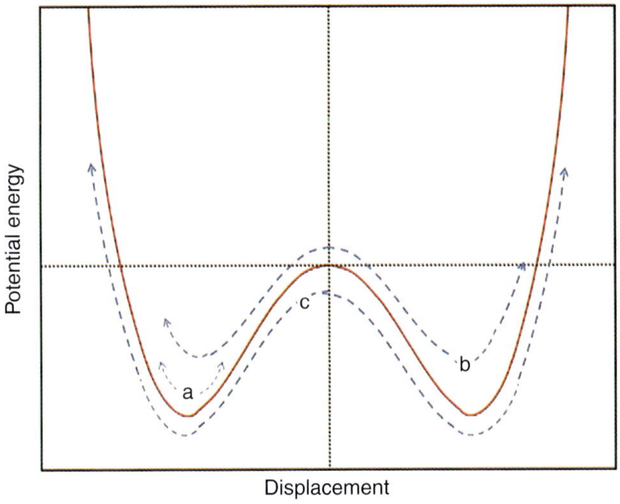

Figure 2.16 Potential as a function of position for a bistable resonator: (a) intrawell oscillations, (b) chaotic interwell vibrations, and (c) interwell oscillation. *Source:* Harne and Wang 2013 [63]. Reproduced with the permission of IOP.

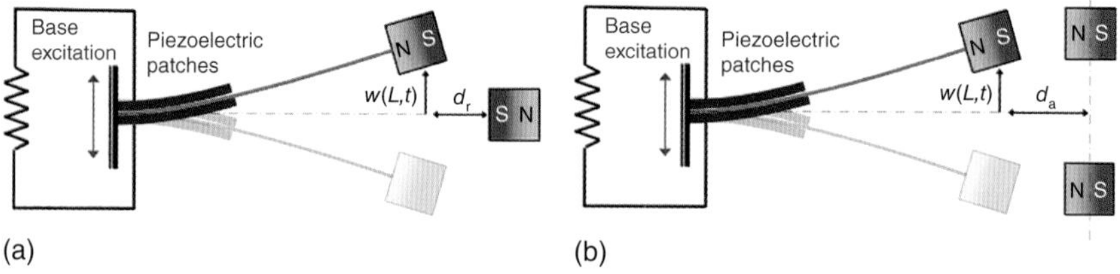

Figure 2.17 (a) Bistable cantilever type of resonator using magnetic repulsion and (b) magnetic attraction.

Figure 2.17 shows the concept of a bistable piezoelectric energy harvester with a permanent magnet attached to the tip of the cantilever beam (or on an adjacent platform) designed to induce repulsive or attractive magnetic forces, respectively. Under base excitation, as indicated by the black arrow, a repulsive magnetic force switches the cantilever beam between two stable states at a critical distance (d_r) between beam tip and external magnets in Figure 2.17a. Figure 2.17b shows a bistable harvester using attraction force by two separated magnets positioned at two stable states.

2.5 Review of Piezoelectric Thin Films on Metal Substrate (Foils)

Many types of MEMS piezoelectric energy harvesters are fabricated using Si-based micro-processing. However, Si-based MEMS energy harvesters face mechanical challenges such as weak fracture strength or rigidity. In contrast, base metal substrates, especially foils, are attractive for MEMS piezoelectric harvesters due to their small stiffness and high fracture strength [64]. In this case, the metal foil can be used as the mechanical elastic layer.

Secondly, the large TEC of metals such as copper, nickel, or stainless steel (as compared with that of PZT) produces compressive stress in plane during cooling down from the crystallization temperature to the Curie temperature. As described earlier, compressive stress in the PZT films can be used to reduce the permittivity and enhance the FoM for energy harvesting.

Additionally, metal foils are easily cut to shape. However, one challenge in the growth of oxide films such as PZT on metal foils is the thermodynamic incompatibility between the metal substrate and the oxide film. Several reports have been published regarding PZT films coated onto base metal substrates with buffer layers ($LaNiO_3$, Pt) by sputtering or CSD, particularly in the case where crystallization was conducted at low oxygen partial pressures [65, 66]. Sometimes, passivation layers such as SiO_2 or HfO_2 are added below the buffer layer electrode to minimize oxidation of non-noble-metal foil layers. Other papers have demonstrated PZT film transfer from oxide substrates to flexible metal foil using a laser lift-off process [67, 68].

Recently, a number of papers have demonstrated that both surfaces of the metal foil can be coated to produce bimorphs that double the energy output for the same

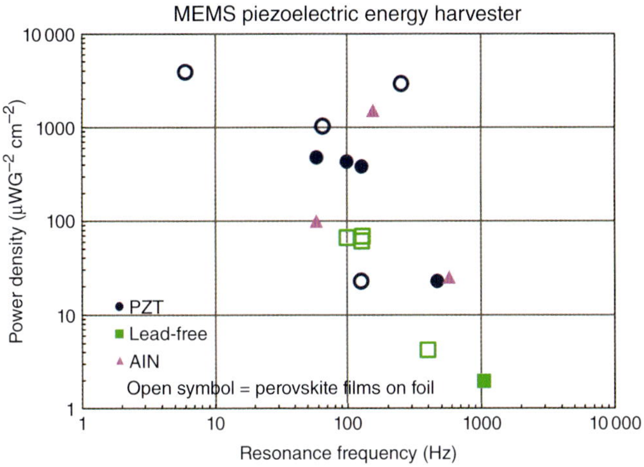

Figure 2.18 Comparison of a number of MEMS piezoelectric energy harvesters. Those shown in open symbols correspond to piezoelectrics on metal foils.

excitation and device geometry [69–71]. In this case, the two film layers can be poled to produce outputs either in parallel or in series, depending on the wiring. This allows an additional design degree of freedom that enables the user to choose whether to increase the output current or the output voltage. The bimorph geometry also helps minimize curling of the foils by balancing the stresses on the two surfaces.

A comparison of resonant harvesters made on metal foils with those on more brittle passive elastic layers such as Si, Si_xN_y, or SiO_2 is shown in Figure 2.18. With all other factors being held equal, in general perovskite piezoelectric layers on the higher TEC metal foils produce 2–5 times more power than comparable films on elastic layers that provide tensile stresses.

2.6 Conclusions

This chapter reviewed the field of mechanical energy harvesting, with an emphasis on material choices for piezoelectric micro-electromechanical systems for this application. It is shown that the key materials figures of merit are low permittivity and high piezoelectric $e_{31,f}$ coefficient. This is most easily achieved in high piezoelectric coefficient materials when they are placed under biaxial compressive stresses. Use of metal foils as substrates allows for robust mechanical harvesting devices, as well as efficient electromechanical conversion.

References

1 Priya, S. (2007). Advances in energy harvesting using low profile piezoelectric transducers. *Journal of Electroceramics* 19: 165–182.
2 Kim, S.-G., Priya, S., and Kanno, I. (2012). Piezoelectric MEMS for energy harvesting. *MRS Bulletin* 37: 1039–1050.

3 Cook-Chennault, K.A., Thambi, N., and Sastry, A.M. (2008). Powering MEMS portable devices – a review of non-regenerative and regenerative power supply systems with special emphasis on piezoelectric energy harvesting systems. *Smart Materials and Structures* 17 (4): 043001.
4 Beeby, S.P., Tudor, M.J., and White, N.M. (2006). Energy harvesting vibration sources for microsystems applications. *Measurement Science and Technology* 17: R175–R195.
5 Roundy, S.J. (2003). Energy scavenging for wireless sensor node with a focus on vibration to electricity conversion. PhD thesis. The University of California, Berkeley.
6 Williams, C.B. and Yates, R.B. (1996). Analysis of a micro-electric generator for microsystems. *Sensors Actuators A* 52: 8–11.
7 Roundy, S., Wright, P.K., and Rabaey, J. (2003). A study of low level vibrations as a power source for wireless sensor nodes. *Computer Communications* 26: 1131–1144.
8 Dutoit, N., Wardle, B., and Kim, S.-G. (2005). Design considerations for MEMS-scale piezoelectric mechanical vibration energy harvesters. *Integrated Ferroelectrics* 71 (1): 121–160.
9 Yun, J., Reynolds, S.N., Reynolds, M.S., and Abowed, G.D. (2011). Design and performance of an optimal inertial power harvester for human-powered devices. *IEEE Transactions on Mobile Computing* 10 (5): 669–683.
10 Mitcheson, P.D., Yeatman, E.M., Rao, G.K. et al. (2008). Energy harvesting from human and machine motion for wireless electronic devices. *Proceedings of the IEEE* 96 (9): 1457–1486.
11 Amirtharajah, R. and Chandrakasan, A.P. (1998). Self-powered signal processing using vibration-based power generation. *IEEE Journal of Solid-State Circuits* 33 (5): 687–695.
12 Trolier-McKinstry, S. (2008). Crystal chemistry of piezoelectric materials (Chapter 3). In: *Piezoelectric and Acoustic Materials for Transducer Applications* (ed. A. Safari and E.K. Akdogan), 39–56. Springer Science & Business Media.
13 Heywang, W., Lubitz, K., and Wersing, W. (ed.) (2008). *Piezoelectricity: Evolution and Future of a Technology*, vol. 114. Springer Science in Materials Science.
14 Roundy, S. (2005). On the effectiveness of vibration-based energy harvesting. *Journal of Intelligent Material Systems and Structures* 16: 809–823.
15 Yeager, C.B. and Trolier-McKinstry, S. (2012). Epitaxial $Pb(Zr_x,Ti_{1-x})O_3$ (0.3<x<0.63) films on (100)MgO substrates for energy harvesting applications. *Journal of Applied Physics* 112: 074107.
16 Trolier-McKinstry, S. and Muralt, P. (2004). Thin film piezoelectrics for MEMS. *Journal of Electroceramics* 12: 7–17.
17 Muralt, P. (2000). PZT thin films for microsensors and actuators: Where do we stand? *IEEE Transactions on Ultrasonics, Ferroelectrics and Frequency Control* 47 (4): 903–915.
18 Baek, S.H., Park, J., Kim, D.M. et al. (2011). Giant piezoelectricity on Si for hyperactive MEMS. *Science* 334: 958–961.

19 Ledermann, N., Muralt, P., Baborowski, J. et al. (2003). {100}-Textured, piezoelectric Pb(Zrx,Ti$_{1-x}$)O$_3$ thin films for MEMS: integration, deposition and properties. *Sensors and Actuators A: Physical* 105 (2): 162–170.
20 Priya, S. (2010). Criterion for material selection in design of bulk piezoelectric energy harvesters. *IEEE Transactions on Ultrasonics, Ferroelectrics, and Frequency Control* 57 (12): 2610–2612.
21 Lovinger, A.J. (1983). Ferroelectric polymers. *Science* 220: 1115–1121.
22 Trolier-McKinstry, S., Griggio, F., Yaeger, C. et al. (2011). Designing piezoelectric films for micro electromechanical systems. *IEEE Transactions on Ultrasonics, Ferroelectrics, and Frequency Control* 58 (9): 1782–1792.
23 Yeager, C.B. (2015). PZT thin films for piezoelectric MEMS mechanical energy harvesting. PhD Thesis. The Pennsylvania State University.
24 Mina, I.G., Kim, H., Kim, I. et al. (2007). High frequency piezoelectric MEMS ultrasound transducers. *IEEE Transactions on Ultrasonics, Ferroelectrics, and Frequency Control* 54 (12): 2422–2430.
25 Akiyama, M., Umeda, K., Honda, A., and Nagase, T. (2013). Influence of scandium concentration on power generation figure of merit of scandium aluminum nitride thin films. *Applied Physics Letters* 102 (2): 021915.
26 Matloub, R., Hadad, M., Mazzalai, A. et al. (2013). Piezoelectric $Al_{1-x}Sc_xN$ thin films: a semiconductor compatible solution for mechanical energy harvesting and sensors. *Applied Physical Letters* 102: 152903.
27 Ujimoto, K., Yoshimura, T., Ashida, A., and Fujimura, N. (2012). Direct piezoelectric properties of (100) and (111) $BiFeO_3$ epitaxial thin films. *Applied Physics Letters* 100 (10): 102901.
28 Won, S.S., Lee, J., Venugopal, V. et al. (2016). Lead-free Mn-doped $(K_{0.5},Na_{0.5})NbO_3$ piezoelectric thin films for MEMS-based vibrational energy harvester applications. *Applied Physics Letters* 108 (23): 232908.
29 Calame, F. and Muralt, P. (2007). Growth and properties of gradient free sol-gel lead zirconate titanate thin films. *Applied Physics Letters* 90 (6): 062907.
30 Yeager, C.B., Ehara, Y., Oshima, N. et al. (2014). Dependence of $e_{31,f}$ on polar axis texture for tetragonal $Pb(Zr_x,Ti_{1-x})O_3$ thin films. *Journal of Applied Physics* 116 (10): 104907.
31 Wasa, K., Matsushima, T., Adachi, H. et al. (2012). Thin-film piezoelectric materials for a better energy harvesting MEMS. *Journal of Microelectromechanical Systems* 12 (2): 451–457.
32 Kanno, I., Ichida, T., Adachi, K. et al. (2012). Power-generation performance of lead-free $(K,Na)NbO_3$ piezoelectric thin-film energy harvesters. *Sensors and Actuators A: Physical* 179: 132–136.
33 Jaffe, B., Cook, W.R., and Jaffe, H. (1971). *Piezoelectric Ceramics*. New York: Academic.
34 Tuttle, B.A., Voight, J.A., Goodnow, D.C. et al. (1993). Highly oriented, chemically prepared $Pb(Zr,Ti)O_3$ thin films. *Journal of the American Ceramic Society* 76 (6): 1537–1544.
35 Roundy, S., and Trolier-McKinstry, S. (2018). Materials and approaches for on-body energy harvesting. *MRS Bulletin* 43 (3): 206–213.

36 Muralt, P. (2000). Ferroelectric thin films for micro-sensors and actuators: a review. *Journal of Micromechanics and Microengineering* 10 (2): 136–146.
37 Cho, C.R. (1999). Heteroepitaxial growth and switching behaviors of PZT (53/47) films on $LaNiO_3$-deposited $LaAlO_3$ and $SrTiO_3$ substrates. *Materials Science and Engineering: B* 64 (2): 113–117.
38 Yokoyama, S., Honda, Y., Morioka, H. et al. (2005). Dependence of electrical properties of epitaxial $Pb(Zr,Ti)O_3$ thick films on crystal orientation and Zr/(Zr+ Ti) ratio. *Journal of Applied Physics* 98 (9): 094106.
39 Zhang, Q. and Whatmore, R.W. (2001). Sol-gel PZT and Mn-doped PZT thin films for pyroelectric applications. *Journal of Physics D: Applied Physics* 34 (15): 2296–2301.
40 Childambaram, N., Mazzalai, A., Balma, D., and Muralt, P. (2013). Comparison of lead zirconate titanate thin films for microelectromechanical energy harvester with interdigitated and parallel plate electrodes. *IEEE Transactions on Ultrasonics, Ferroelectrics, and Frequency Control* 60 (8): 1564–1571.
41 Cook, W.R., Berlincourt, D.A., and Scholz, F.J. (1963). Thermal expansion and pyroelectricity in lead titanate zirconate and barium titanate. *Journal of Applied Physics* 34 (5): 1392–1398.
42 Okada, Y. and Tokumaru, Y. (1984). Precise determination of lattice parameter and thermal expansion coefficient of silicon between 300 and 1500 K. *Journal of Applied Physics* 56 (2): 314–320.
43 Kollie, T.G. (1977). Measurement of the thermal-expansion coefficient of nickel from 300 to 1000 K and determination of the power-law constants near the Curie temperature. *Physical Review B* 16 (1): 4872–4882.
44 Brennecka, G.L., Huebner, W., Tuttle, B.A., and Clem, P.G. (2004). Use of stress to produce highly oriented tetragonal lead zirconate titanate (PZT 40/60) thin films and resulting electrical properties. *Journal of the American Ceramic Society* 87 (8): 1459–1465.
45 Han, G., Ryu, J., Yoon, W.-H. et al. (2011). Stress-controlled $Pb(Zr_{0.52}Ti_{0.48})O_3$ thick films by thermal expansion mismatch between substrate and $Pb(Zr_{0.52}Ti_{0.48})O_3$ film. *Journal of Applied Physics* 110: 124101.
46 Yeo, H.G. and Trolier-McKinstry, S. (2014). {001} Oriented piezoelectric films prepared by chemical solution deposition on Ni foils. *Journal of Applied Physics* 116 (1): 014105.
47 Kim, S.B., Park, H., Kim, S.H. et al. (2013). Comparison of MEMS PZT cantilevers based on and modes for vibration energy harvesting. *Journal of Microelectromechanical Systems* 22 (1): 26–33.
48 Anton, S.R. and Sodano, H.A. (2007). A review of power harvesting using piezoelectric materials (2003-2006). *Smart Materials Structures* 16 (3): R1–R21.
49 Kim, M., Kim, S.H., and Hong, S. (2013). Materials and devices for MEMS piezoelectric energy harvesting. In: *Advances in Energy Harvesting Methods*, 417–435. New York: Springer.
50 Childambaram, N., Balma, D., Nigon, R. et al. (2015). Converse mode piezoelectric coefficient for lead zirconate titanate thin film with interdigitated electrode. *Journal of Micromechanics Microengineering* 25: 045016.

51 Kim. M., Dugundji, J., Wardle, B.L. (2015). Effect of electrode configurations on piezoelectric vibration energy harvesting performance. *Smart Materials and Structures* 24: 045026.
52 Yao, L.Q., Zhang, J.G., Lu, L., and Lai, M.O. (2004). Nonlinear dynamic characteristics of piezoelectric bending actuators under strong applied electric field. *Sensors and Actuators A: Physical* 115 (1): 168–175.
53 Erturk, A. and Inman, D.J. (2009). An experimentally validated bimorph cantilever model for piezoelectric energy harvesting from base excitations. *Smart Materials and Structures* 18: 025009.
54 Safari, A. and Akdogan, E.K. (2008). Piezoelectric MEMS: Materials and Devices (Chapter 20). In: *Piezoelectric and Acoustic Materials for Transducer Applications*, 413–430. Springer Science & Business Media.
55 Calio, R., Rongala, U.B., Camboni, D. et al. (2014). Piezoelectric energy harvesting solutions. *Sensor* 14: 4755–4790.
56 Mateu, L. and Moll, F. (2005). Optimum piezoelectric bending beam structures for energy harvesting using shoe inserts. *Journal of Intelligent Material Systems and Structures* 16: 835–845.
57 Goldschmidtboeing, F. and Woias, P. (2008). Characterization of different beam shapes for piezoelectric energy harvesting. *Journal of Micromechanics and Microengineering* 18: 104013.
58 Dhakar, L., Liu, H., Tay, F.E.H., and Lee, C. (2013). A new energy harvester design for high power output at low frequencies. *Sensors and Actuators A* 199: 344–352.
59 Daqaq, M.F., Masana, R., Erturk, A., and Quinn, D.D. (2014). On the role of nonlinearities in vibratory energy harvesting: a critical review and discussion. *Applied Mechanics Reviews* 66: 040801.
60 Hajati, A. and Kim, S.-G. (2011). Ultra-wide bandwidth piezoelectric energy harvesting. *Applied Physics Letters* 99 (8): 083105.
61 Leland, E.S. and Wright, P.K. (2006). Resonance tuning of piezoelectric vibration energy scavenging generators using compressive axial preload. *Smart Materials and Structures* 15: 1413–1420.
62 Pillatsch, P., Yeatman, E.M., and Holmes, A.S. (2014). A piezoelectric frequency up-converting energy harvester with rotating proof mass for human body applications. *Sensors and Actuators A* 206: 178–185.
63 Harne, R.L. and Wang, K.W. (2013). A review of the recent research on vibration energy harvesting via bistable systems. *Smart Materials and Structures* 22: 023001.
64 Kanda, K., Kanno, I., Kotera, H., and Wasa, K. (2009). Simple fabrication of metal-based piezoelectric MEMS by direct deposition of $Pb(Zr,Ti)O_3$ thin films on titanium substrates. *Journal of Microelectromechanical Systems* 18 (3): 610–615.
65 Kingon, A.I. and Srinivasan, S. (2005). Lead zirconate titanate thin films directly on copper electrodes for ferroelectric, dielectric and piezoelectric applications. *Nature Materials* 4 (3): 233–237.
66 Losego, M.D., Jimison, L.H., Ihlefeld, J.F., and Maria, J.P. (2005). Ferroelectric response from lead zirconate titanate thin film prepared directly on low-resistivity copper substrate. *Applied Physics Letters* 86: 172906.

67 Morimoto, K., Kanno, I., Wasa, K., and Kotera, H. (2010). High-efficiency piezoelectric energy harvesters of c-axis oriented epitaxial PZT films transferred onto stainless steel cantilever. *Sensors and Actuators A: Physical* 163 (1): 428–432.

68 Suwa, E., Tsujiura, Y., Kurokawa, F. et al. (2016). Fabrication of high-efficiency piezoelectric energy harvesters of epitaxial $Pb(Zr,Ti)O_3$ thin films by laser lift-off. *Energy Harvesting and Systems* 3 (1): 61–67.

69 Yeo, H.G., Ma, X., Rahn, C., and Trolier-McKinstry, S. (2016). Efficient piezoelectric energy harvesters utilizing (001) textured bimorph PZT films on flexible metal foils. *Advanced Functional Materials* 26 (32): 5940–5946.

70 Tsujiura, Y., Suwa, E., Nishi, T. et al. (2017). Airflow energy harvester of piezoelectric thin-film bimorph using self-excited vibration. *Sensors and Actuators A: Physical* 261: 295–301.

71 He, X.Y., Zhang, Y., Ding, A.L. et al. (2008). Fabrication and characteristics of PZT thick films on Pt/Ti foil substrates for piezoelectric vibrators. *Journal of Electroceramics* 21: 871–874.

72 Li, W.J., Ho, T.C.H., Chan, G.M.H. et al. (2000). Infrared signal transmission by a laser-micromachined vibration-induced power generator. *Proceedings of the 43rd IEEE Midwest Symposium on Circuits and Systems*, vol. 1, Aug 236–239.

73 Ching, N.N.H., Chan, G.M.H., Li, W.J. et al. (2000). PCB integrated micro-generator for wireless systems. *Proceedings of the International Symposium on Smart Structures*, Oct 19–21.

74 Williams, C.B., Shearwood, S., Harradine, M.A. et al. (2001). Development of an electromagnetic micro-generator. *IEE Proceedings: Circuits, Devices and Systems* 148 (6): 337–342.

75 El-hami, M., Glynne-Jones, P., White, N.M. et al. (2001). Design and fabrication of a new vibration-based electromechanical power generator. *Sensors Actuators A Physical* 92 (1): 335–342.

76 Ching, N.N.H., Wong, H.Y., Li, W.J. et al. (2002). A laser-micromachined multi-modal resonating power transducer for wireless sensing systems. *Sensors Actuators A Physical* 97: 685–690.

77 Mizuno, M. and Chetwynd, D.G. (2003). Investigation of a resonance microgenerator. *Journal of Micromechanics and Microengineering* 13 (2): 209–216.

78 Lee, J.M.H., Yuen, S.C., Li, W.J., and Leong, P.H.W. (2003). Development of an AA size energy transducer with micro resonators. *Proceedings of the International Symposium on Circuits and Systems* 876–879.

79 Glynne-Jones, P., Tudor, M.J., Beeby, S.P., and White, N.M. (2004). An electromagnetic vibration-powered generator for intelligent sensor systems. *Sensors Actuators A Physical* 110: 344–349.

80 Beeby, S.P., Tudor, M.J., Torah, R.N. et al. (2006). Macro and micro scale electromagnetic kinetic energy harvesting generators. DTIP MEMS MOEMS. Stresa, Italy, p. 6.

81 Huang, W.-S., Tzeng, K.-E., Cheng, M.-C., and Huang, R.-S. (2007). A silicon MEMS micro power generator for wearable micro devices. *Journal of the Chinese Institute of Engineers* 30 (1): 133–140.
82 Perpetuum Limited, PMG17 datasheet, Jan. 7, 2008. [Online]. www .perpetuum.co.uk/resource/PMG17-100_dsheet.pdf.

3

Ferroelectric Photovoltaics

Akash Bhatnagar

Centre for Innovation Competence ZIK SiLi-nano®, Light for High-voltage Photovoltaics, Karl-Freiherr-von-Fritsch-Straße 3, D-06120 Halle (Saale), Germany

3.1 Introduction

The ever-growing demand for green energy cannot be understated. In this regard, photovoltaic effect, which involves conversion of light to electricity, has been considered as one of the most promising pathways. The field of photovoltaics is largely dominated by the use of semiconducting-materials-based devices. The photovoltaic device primarily involves a *pn*-junction. A space charge region (SCR) develops across this region between *p*- and *n*-type materials. The associated electric field within the region assists in the separation of photo-generated carriers and prevents them from recombination. The maximum efficiency obtainable from the *pn*-junction devices is defined by the Shockley–Quisser limit [1]. The limit was defined by assuming that the photons with energies higher than, and equal to the band gap of materials, will have the same effect, while photons with energies less than the band gaps will have no contribution. The efficiency (η) of the *pn*-junction-based photovoltaic cell can be then described with the following expression [1, 2]:

$$\eta = \frac{J_{sc} V_{oc} \mathrm{FF}}{P_{in}}$$

where J_{sc} is the short-circuit current/photovoltaic current, V_{oc} is the voltage measured in the open circuit conditions, P_{in} is the power of the incident illumination, and FF is the fill factor, which represents the ratio between actual power and ideal power obtained from the product $J_{sc}V_{oc}$. The maximum η, keeping in perspective the law of detailed balancing, has been calculated to be around 30%. This value has been calculated for a band gap of 1.1 eV and under the assumption that all the recombination processes are radiative in nature [1]. The innate limitation of the photovoltaic energy conversion in semiconductors has triggered the pursuit to find new materials that have a fundamentally different mechanism for charge separation.

In this context, the photovoltaic effect in ferroelectric materials, i.e. *photoferroics*, has attracted a lot of attention. Ferroelectric materials have been known

Ferroelectric Materials for Energy Applications, First Edition.
Edited by Haitao Huang and James F. Scott.

Table 3.1 Band gap values of Si and some of the well-known ferroelectric materials.

Material	Si	$BaTiO_3$	$LiNbO_3$	$KNbO_3$	SbSI	$PbTiO_3$	$Pb(Zr,Ti)O_3$	$BiFeO_3$
Band gap (eV)	1.1 [1]	3.28 [3]	3.78 [4]	3.1–3.3 [5, 6]	~2 [7]	~3.6 [8]	~3.5 [9]	2.2–2.7 [10, 11]

for decades for their functional properties. Ferroelectric materials can exist in different states of polarizations and can be switched from one state to the other by the application of an electric field. Along with their piezoelectric properties, ferroelectrics find a wide array of applications in the areas of sensing, actuation, and memory devices. The observation of a photovoltaic effect in these materials in the early 1950s was in a way a revelation that presented opportunities to study the interplay between ferroic and photovoltaic properties. The photovoltaic effect in these materials did not demand a junction, nor any form of chemical gradient. Other photo-assisted effects were also observed, such as photoconductivity and photorefraction. With the underlying advantages over semiconductors, research was directed toward exploring the origins and several mechanisms were considered. However, the inability to match the efficiencies obtained in conventional photovoltaics and band gaps that are near the ultra violet range of solar energy spectra (see Table 3.1) largely suppressed the interest in the field till the early 2000s.

In the recent past, development in thin-film technology for synthesis of oxides thin films not only boosted the research in ferroelectrics but also led to a revival of photoferroics. Thin-film engineering allowed epitaxial and pseudomorphic growth of ferroelectrics on a variety of substrates. Such kind of fabrication laid the ground to utilize the extremely malleable perovskite structure, in which most of the well-known ferroelectrics crystallize. The perovskite structure allows customization of optical properties such as band gap, absorption, and ferroelectric characteristics such as polarization and domain variants. The observation of photovoltaic effect in a multiferroic material such as $BiFeO_3$ (BFO) further motivated research in the direction of smart photovoltaic materials that are capable of much more than mere energy conversion. The objective of this chapter will be to provide the reader with some basic understanding about ferroelectric photovoltaics and the related processes. The chapter intends to bring the reader abreast with some of the instrumental work that has been done to understand and unravel the origin of this effect.

3.2 Historical Background

Photovoltaic (PV) effect in ferroelectric materials has stirred a lot of interest not only in the field of ferroics but also in the field of semiconductor photovoltaics. The effect is often referred to as *abnormal* or *anomalous photovoltaic (APV) effect*. Such a reference can be attributed to the unconventional characteristics of this effect when compared to the extensively investigated semiconductor-based photovoltaic effect, which forms the platform for solar cells. The underlying mechanism behind the PV effect in semiconductors involves the use of a *pn*-junction. The *pn*-junction forms a SCR that assists in keeping the charge carriers generated under illumination away from recombination. The *abnormal* part about the ferroelectric photovoltaic effect is the absence of any kind of junction to separate the charge carriers. Instead, the non-centrosymmetric character of the material is at the core of the PV effect and therefore is referred to as bulk photovoltaic effect (BPV). Another *abnormal* feature is the value of open circuit voltage that can be achieved. In the *pn*-junction-based PV effect the V_{oc} is largely limited by the built-in potential of the SCR. The electric field across the SCR is further dependent upon the band gap of the material itself, while in the case of ferroelectric PV effect, the V_{oc} could be several times higher than the band gap.

The first report about the photovoltaic effect in ferroelectric materials dates back to 1956 in the case of $BaTiO_3$ (BTO) [12]. Even though the work was largely focused on pyroelectric properties in BTO single crystals, there was a section stating the observation of a small current upon illumination when the crystal was just above the Curie temperature. In the subsequent years, the effect of light on ferroelectric single crystals of $LiNbO_3$ and $LiTaO_3$ was investigated. The work of Ashkin [13] and Chen [14], from 1966 and 1969, respectively, observed local changes in the indices of refraction in $LiNbO_3$ single crystals upon illumination. This effect was often referred to as "optical damage or inhomogeneity" and was partially stable for a few hours after switching off the illumination. The observation led to the ideas of local change in structure owing to the movement of either Nb or Li atom, which will vary the polarizability and thus the refractive index. The other idea was based on the effect, also observed in wide band gap semiconductors, involving development of SCR under illumination. The drifting away of charge careers under illumination from the center of the beam to the periphery and getting trapped resulted in the SCR. The reason for this drift was attributed to an uncompensated field within the ferroelectric material [14]. The electric field within the SCR was considered to be strong enough to change the refractive index via the electro-optic effect.

The phenomenon of "optical damage" was also measured in $BaTiO_3$ single crystals. The measurements apparently questioned the need for having doping centers for the manifestation of the BPV effect [15]. The underlying origin was still considered to be the drift of carriers to non-illuminated regions, but another theory proposed by Johnston was discussed to explain the drift [16]. The theory assumed that local variation in polarization is allowed in materials with pyroelectric crystal symmetry upon illumination. These variations could arise from ionization and, or, filling of certain traps. The electric field due to the variable or gradient of polarization was considered sufficient to drift the carriers to non-illuminated regions. Under steady-state conditions, gradient of polarization is compensated by the SCR created by the migration of carriers. The SCR via the electro-optic effect was able to change the local index of refraction. The role of pyroelectric effect, i.e. change of polarization upon heating, was ruled out as the effect of "optical damage" did not increase when the temperatures were raised to around the Curie temperature.

The work from Glass et al. from 1974 further shed some light on the origin of the photoinduced effects in ferroelectric materials using $LiNbO_3$ single crystals

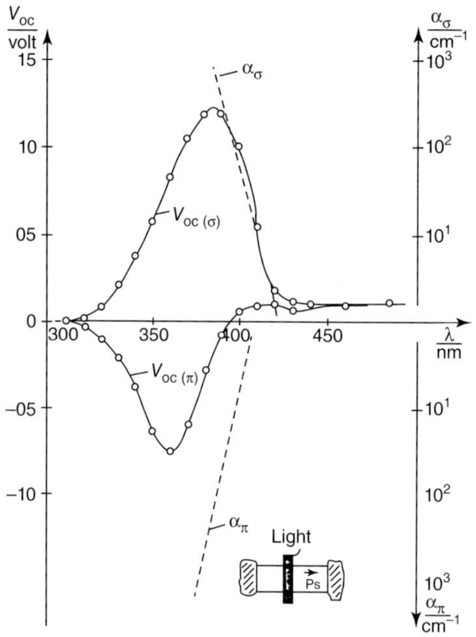

Figure 3.1 Open circuit voltage measured while varying the wavelength with the light polarization perpendicular (σ) and parallel (π) to the direction of spontaneous polarization (P_s) in the crystal. The dotted lines depict the corresponding absorption coefficients measured in the two geometries. *Source:* Koch et al. 1975 [18]. Reprinted with permission of Elsevier.

doped with iron oxide (Fe_2O_3) and copper oxide (CuO) [17]. It was probably the first work to utilize the sub-band-gap states to analyze the photovoltaic effect. The excitation source was chosen so that the carriers could be mostly excited from the impurity to the conduction band, which in the case of $LiNbO_3$ is formed by the *d*-orbitals of Nb. It was proposed that the distance between the Fe^{2+} impurity and Nb is different along the polar axis ($\pm c$). Hence the probability of charge transfer upon excitation with light will be also different, and this difference in the probabilities can potentially explain the observation of a photovoltaic current under light.

In the work of Koch on $BaTiO_3$ single crystals, focusing not on the optical damage but the PV effect, it was observed that above-band-gap photovoltages only appear when the samples were in the ferroelectric states and not otherwise [18]. Additionally, the influence of polarization direction of light with respect to the *c*-axis of the material was investigated. Intriguing observations were made about the direction and magnitude of photovoltages. The polarization direction of the illumination was irrelevant when the excitation energy was below the band gap. However, the wavelength-resolved measurements revealed a change in the sign of photovoltage when the excitation energy was around the band gap and the polarization of light was perpendicular to the *c*-axis as shown in Figure 3.1.

These results could not have been explained within the framework put forward by Johnston or Chen that utilized the macroscopic variations in polarization. Also, the work questioned the pre-requisite of sub-band-gap states for the manifestation of the photovoltaic effect, which was demonstrated in the work of Glass on $LiNbO_3$. The work instead addressed the issue taking into account the thermally induced lattice polarization fluctuations in perovskite-based ferroelectrics. The underlying phenomenology was developed by Wempel to understand the electron–phonon scattering in perovskite ferroelectric materials [19]. The fluctuations in polarization were considered "needle like" with an associated electric field. Similar fluctuations could also be produced upon illumination. The resultant needles will be oriented in the same direction due to symmetry. The photoexcited electrons will align themselves with the electric field associated with the needles, resulting in a steady-state photocurrent [18]. The polarization direction dependence of the photocurrent and voltage was explained by considering that the valence band is formed with p-orbitals of the oxygen ions. However, there are two distinct positions of oxygen ions; one is aligned along the *c*-direction and the other located within the lateral planes. Depending upon the direction of polarization of light, different oxygen ions can be excited resulting in changes in the charge distribution or needle-like fluctuations. The electric field of the respective needle will then drive the photocurrent.

The work led to further analysis of the theoretical background of the effect by neglecting the need for any intentionally induced sub-band-gap states. Chanussot in 1976 proposed an explanation that considered band–band transitions accompanied with coupling between excited electronic states with the lattice [20]. The photocurrent will then be due to relaxation of the system from the Frank–Condon state following excitation. The photocurrent comprised of different components, namely, ionic current and electronic charge transfer current arising from excitation and recombination processes. However, there will be a net current or photocurrent flowing only if the path for excitation is different from that of recombination. Even though this theory did not consider explicitly the presence of any intentionally induced states, the desired difference in the paths of excitation and recombination implied the necessity for levels or trap states. The directional properties of the effect, which were observed within the work of Koch, were explained by assuming that the Frank–Condon relaxation proceeds via the coupling of the electronic state with optic modes having a component along the *c*-axis.

The experimental work from Glass [21] described above presented the BPV in the following form:

$$j = k(\omega)\alpha(w)I_0 \tag{3.1}$$

where ω denotes the frequency, I_0 is the intensity of the radiation, α is the absorption coefficient of the crystal, and k is the glass coefficient. However, the tensorial and directional nature of the BPV effect was put forward by Belinicher et al. [22] while considering the Taylors expansion of the current generated in a medium under the influence of an electric field (E) associated with light. The electric field can be described by using the wave vector (q), frequency (ω), polarization vector (e), and amplitude (E_0) as

$$E\,(q,t) = eE_0e^{i(qr-\omega t)} + e^{*}E_0e^{-i(qr-\omega t)} \tag{3.2}$$

The expansion to describe the current density is given by

$$j_i = \sigma_{ij}^{\mathrm{d}}E_j + \sigma_{ijl}E_jE_l + \sigma_{ijlm}^{\mathrm{ph}}E_je_le_m^{*}I + \chi_{ijlm}q_je_le_m^{*}I + \beta_{ijl}e_je_m^{*}I + \cdots \tag{3.3}$$

wherein I denotes the intensity of light obtained by $|E_0|^2$. The expansion coefficients are the tensors that describe the electrical properties of the material, namely, σ^{d}_{ij} denotes the tensor describing the dark conductivity of the material or the Ohm's law, σ_{ijl} indicates the quadratic electrical conductivity, $\sigma^{\mathrm{ph}}_{ijlm}$ represents the anisotropic photoconductivity, and χ_{ijlm} represents the light entrainment effect. The term β_{ijl} is the term to describe the BPV tensor. It must be mentioned here that the current density, electric field, polarization, and wave vectors are polar vectors and change their signs upon spatial inversion. Therefore, in the case of a medium with a center of symmetry, $\beta_{ijl} = 0$. Hence, it can be implied that BPV is possible only in materials that lack center of symmetry.

The tensorial nature of the BPV was further validated by the results of Fridkin and Magomadov in 1979 in their work on Fe-doped $LiNbO_3$ [23]. Photovoltaic current was measured along different crystallographic directions while varying the angle that the plane of light polarization made with the crystallographic directions of the crystal (Figure 3.2). The results exhibited an angular dependency of the photocurrent, which is seldom observed in semiconductor-based photovoltaics. Baltz and Kraut further continued the work by basing their calculation on the quadratic response theory proposed by Belinicher [24]. Their work was crucial in proving that BPV can also exist in a pure crystal without any added impurities [25]. Calculations revealed that BPV is determined by the non-diagonal elements of the current operator. The virtual transitions, i.e. transitions occurring from and to the conduction and valence band, to other bands constitute the non-diagonal elements. Therefore, it was implied that while a pure non-centrosymmetric materials will exhibit BPV effect, the virtual transitions contribute extensively to the effect. The interference between the two types of transitions, namely, band-to-band and virtual, has a directional bias associated with it due to the non-centrosymmetric structure of the material. As result, a net and persistent photocurrent could be drawn from these materials under illumination. Such a phenomenology has been often referred to as *shift* mechanism as the carriers (electrons and holes) undergo a shift in real space [26–28].

Under open circuit condition, i.e. when no current is allowed to pass through the crystals, the photocurrent will charge the capacitance in the crystals resulting in an electric field. The electric field across the crystal will be then proportional to the photovoltaic current and inversely proportional to the conductivity of the crystal and is presented as [21]

$$V_{\mathrm{oc}} = \frac{J}{\sigma^{\mathrm{ph}} + \sigma^{\mathrm{d}}} l \tag{3.4}$$

The work of Krätzig further continued the research in $LiNbO_3$ doped with Fe_2O_3 with a focus on the correlation between the photoconductivity and resultant electric fields [29]. Similar observations were made by Günter in the case of $KNbO_3$ crystals, wherein the crystals were doped with Fe centers. However, in this case it was observed that the high dark conductivity of the crystals limited the values of the V_{oc} that can be achieved. These observation further emphasized the validity of the relation given in Eq. (3.4) [30].

It is rather intriguing that a similar photovoltaic effect has been also observed in wide-band-gap semiconductors such as ZnS [31, 32], CdTe, and PbS [33]. In all these systems, open circuit voltages were higher than the respective

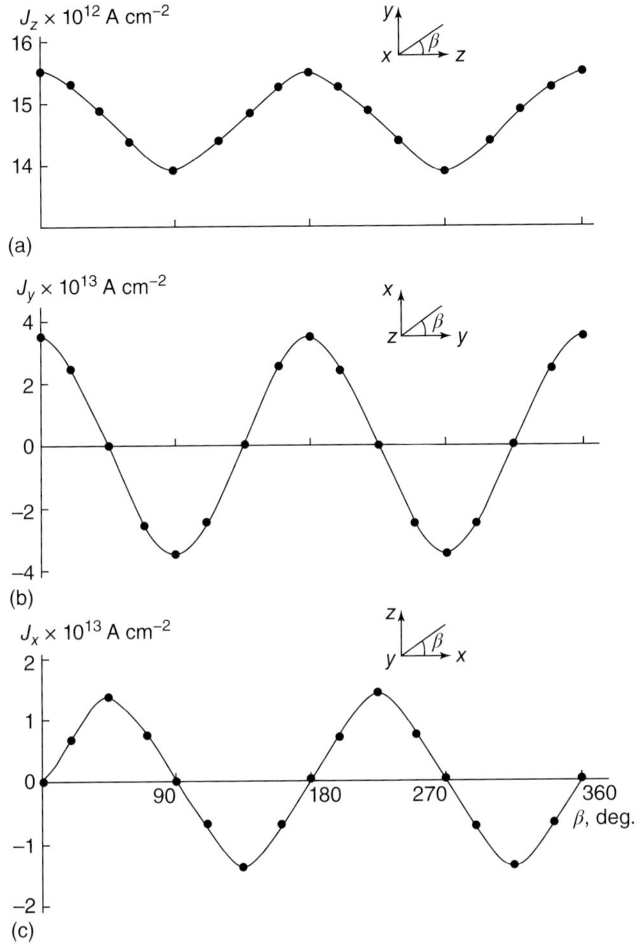

Figure 3.2 Photovoltaic current measured along the direction of polarization (J_z) and along the other two crystal axes (J_x and J_y). The current was measured as a function of the angle (β) which the plane of linearly polarized light makes with the crystal axis [23].

band gaps energies. Additionally, the V_{oc} also underwent a reversal in sign at certain wavelengths of illumination. At least in the case of ZnS, experiments demonstrated that only the samples with stacking faults, involving changes in structure from hexagon to cubic, exhibit the abnormal photovoltaic effect. Based on the observations it was proposed that the cubic region owing to smaller band gap will absorb more than the hexagonal region. This differential absorption was attributed with the generation of a potential gradient that could result in a photovoltaic current and a corresponding photovoltage [32]. In general, the observation of abnormal photovoltaic effect in wide-band-gap semiconductors has been mostly explained by assuming multiple junctions connected in series with each other. However, such an explanation was never applicable to the effect observed in ferroelectric materials.

3.2.1 Recent Studies

Much of the recent interest in the field of photo-ferroelectrics was triggered by the work of Choi et al. [34]. The work mostly focused on the diode-like conduction behavior in BFO single crystals that can be switched by reversing the direction of polarization. The other part of the work presented the photovoltaic effect in the crystals with the associated current flowing along the direction of polarization. Interestingly, the photocurrent was observed with excitation energies that were below the reported band gap of $BiFeO_3$ [11]. The study therefore provided direct evidence of the manifestation of BPV in $BiFeO_3$. This was further validated by the angular dependency of the PV current on the plane of polarization of light (Figure 3.3).

In the subsequent work of Yang et al. thin films of $BiFeO_3$ with well-oriented domains and corresponding domain walls were used for the investigation of the photovoltaic effect [35]. Above-band-gap open circuit voltages were observed under white illumination. Even though the characteristics of the observed effect largely resembled those of BPV, domain walls were proposed to be at the origin of the effect. The underlying mechanism was based on a previous work that established the presence of a potential step across the domain walls [36]. The potential was calculated to be arising from the component of the polarization perpendicular to the domain wall (PPDW). The corresponding charge density led to an electric field dipole. It was assumed that carriers are generated under light within the domain walls and separated under the influence of the potential. As a result, the carriers accumulated on either side of domain walls contribute towards the buildup of charges on either side (Figure 3.4). On the contrary, the carriers generated within the bulk were expected to recombine quickly in the absence of any

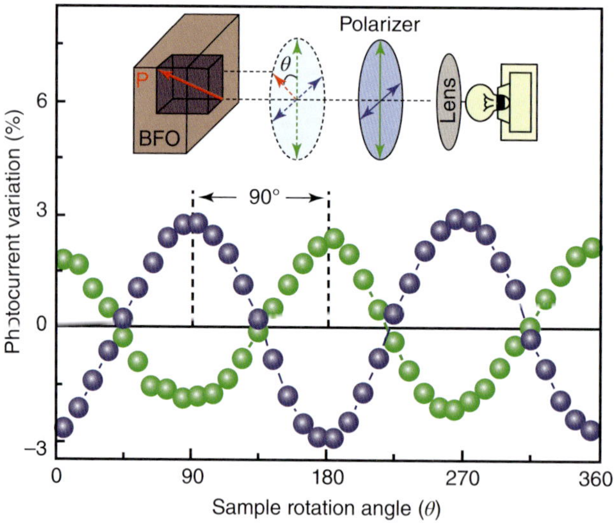

Figure 3.3 Variation of the photocurrent as the sample is rotated with respect to the electric field of the linearly polarized light. The polarization direction of the crystal is indicated as P. *Source:* Choi et al. 2009 [34]. Reprinted with permission of AAAS.

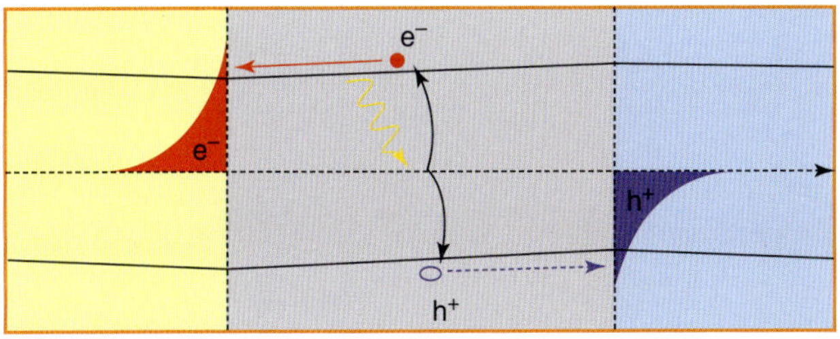

Figure 3.4 Schematic depicting the separation of electron–hole pair generated within a domain wall upon illumination. *Source:* Yang et al. 2010 [35]. Reprinted by permission from Macmillan Publishers Limited: Nature Nanotechnology.

effective electric field. The analogy with a *pn*-junction-based photovoltaic device was apparent with each and every domain wall acting like a *pn*-junction. In open circuit conditions, the charge built on either side of domain walls cumulatively will add up to result in an above-band-gap open circuit voltage.

Even though $LiNbO_3$ and BFO have the same bulk symmetry, any comparison with the BPV effect, similar to that observed in $LiNbO_3$, was *a priori* ruled. This was justified by considering that dopant levels have been more or less a perquisite for the observation of BPV effect in $LiNbO_3$ in most of the previous studies. Additionally, while $LiNbO_3$ undergoes a rhombohedral–rhombohedral phase transition and thus comprises of only 180° domain walls, BFO undergoes an orthorhombic–rhombohedral phase transition resulting in the formation of ferroelastic domains and corresponding 71°, 109° domain walls. This fact was utilized to support the role of domain walls in the observed PV effect. The work shed new light on the photovoltaic effect, which traditionally was always measured in pure and single domain ferroelectric samples. The manifestation of this effect in a multiferroic material such as BFO with periodically arranged domains also opened the doors to new pathways for novel devices. However, there were some issues that remained unresolved. For instance, the 71° domain walls were shown to have a smaller potential step across in comparison to the 109° domain walls [36]. Assuming the validity of the model, samples with 109° domain walls must exhibit comparable or much higher V_{oc}s. But there were no experimental reports to support this argument. The other issue was pertaining to the volume fraction of the domain walls. It was rather difficult to perceive that the charge carriers generated in a small volume fraction of domain walls can be responsible for an effect that spans across few tens of micrometers of measurement gap. These issues further motivated the research in the field to better understand the underlying origin.

In one of the subsequent works by Alexe and Hesse [37], an atomic force microscope (AFM) was used in conjunction with a light source and a conductive tip to investigate the local photoelectrical properties. In one of the measurement geometries, the tip was used as a moving electrode with the other electrode stationary. The setup allowed to scan over the surface of the sample while mapping the generation of photocurrents and also the piezo response. Hence, it was possible to visualize the domains and domain walls, and map the originating photocurrents. The measurements revealed uniform generation of photocurrents, irrespective of the position of the tip, i.e. whether it is positioned at the domain or at the domain wall. The results largely questioned the assumption made in the work of Yang et al. [35] regarding much higher recombination rates of photocarriers within the domains than at the domain walls. Surprisingly, the photocurrent collected with one electrode as the tip was largely similar to the photocurrent collected with two conventional stationary electrodes. This indicated much higher current density when the current was collected with a tip, thus enhancing the overall quantum efficiency (Figure 3.5).

Similar measurements were performed on BFO single crystals to measure the generation and recombination lifetimes, albeit with a pulsed light source [38]. The method was similar to the photoinduced transient spectroscopy (PITS) which is a rather conventional technique to characterize the lifetimes of photocarriers and activity of trap states [39]. The use of a conducting tip as an electrode allowed scanning of the surface while measuring the generation and recombination rates. The energy of the chosen light source was high enough to overcome the band gap energy of BFO. While the light was pulsed at high frequency rates, the PV current arising from the sample was followed by an evidently longer response time. The gradual rise and fall of the PV current suggested the activity of trap levels capable of capturing and reemitting the carriers, thereby increasing the rise and fall times [40]. As the tip was scanned across the domains and domain walls, the calculated recombination rates were not substantially different. On the contrary, the domain walls seemed to have a considerably lower generation rate, which was against the assumption made in the work of Yang et al.

In the work from Moubah et al. [41], the aspect of oxygen vacancy migration in connection to the photovoltaic effect was discussed. To keep the system simple, single domain crystals were used for the investigations. The samples in

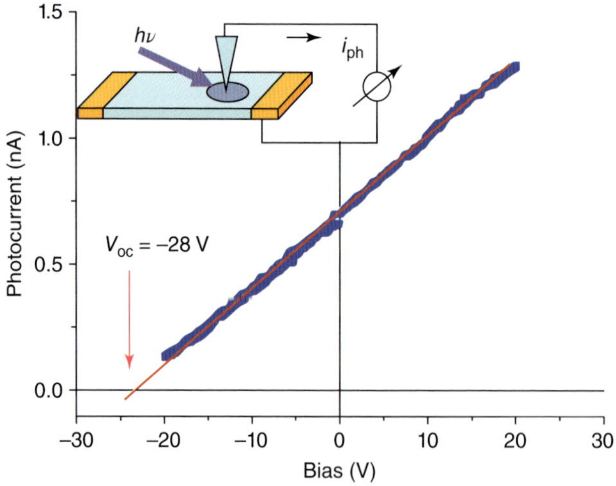

Figure 3.5 Current–voltage characteristics measured with a conductive tip as an electrode to collect photo-generated charge carriers within the area of illumination. *Source:* Alexe and Hesse 2011 [37]. Reprinted by permission from Macmillan Publishers Limited: Nature Communications.

the virgin states did not exhibit any kind of photovoltaic effect. Interestingly, a systematic treatment of the sample involving controlled illumination and application of electric field resulted in a drastic change in the transport properties and observation of photovoltaic effect. Electric fields much below the coercive field were applied while subjecting the sample to rather intense illumination intensities. An overall enhancement in the photo- and dark conductivities was observed in the subsequent measurements. Also, a substantial PV current was observed. The switchable character of the PV current was achieved by switching the direction of the electric field, albeit without changing the state of polarization. These results largely questioned the theories proposed in the previous works wherein the direction of polarization was considered crucial in determining the direction of PV current.

Electron spin resonance measurements revealed the generation of excessive oxygen vacancies under light. These vacancies could have been created under light when the electrons are removed from the $2p$ states of oxygen to populate the Fe $3d$ states that constitute the conduction band. This charge transfer, potentially, can cause some of the oxygen ions to transform into a nonstoichiometric state. The vacancies moved across the measurement gap under a bias voltage toward the negative electrode. While in the virgin sample, much higher field would have been required to cause these migrations, intense light illumination apparently mediated this phenomenon at lower fields. These electro migrations resulted in a doping gradient across the gap in conjunction with local doping around the electrodes. The carriers generated under light will have a tendency to get trapped at the surface or in defects such as the vacancies. When the electric field is removed, the trapped chargers will rearrange to produce a non-negligible internal field that will eventually decay with time. However, this field was proposed to be strong enough to produce a PV current when the sample was illuminated again. The diode-like behavior obtained in the current–voltage characterization and time-dependent decay of PV current supported this theory. The presented results certainly provided an interesting insight into the possible role of electronic entities in the PV effect. However, one of the key aspects of APV effect, i.e. above-band-gap V_{oc}, was not discussed and commented upon. It is well established now that oxygen vacancies form a level within the band gap, and thus can provide certain asymmetry in the generation and recombination processes [10, 11]. Therefore, one might speculate their role in the realization of PV effect.

The role of domain walls was investigated in a more systematic way by Bhatnagar et al. [42]. The work included two systems of BFO, i.e. thin films with 109° and 71° domain walls. This was somehow missing in the previous work where either single crystals (in a polydomain state) or thin films with only 71° domain walls were used. Results acquired at room temperature from samples with 109° domain walls were largely in accordance with the results from Yang et al. Above-band-gap V_{oc}s were measured only in the geometry where the electrodes were parallel to the domain walls or PLDW, while only a photoconductive behavior was observed in the other geometry where the electrodes were PPDWs. The results were rather contradictory for the sample with 71° domain walls for which above-band-gap V_{oc}s were obtained in both the geometries. To investigate this discrepancy, measurements were conducted at lower temperatures (Figure 3.6).

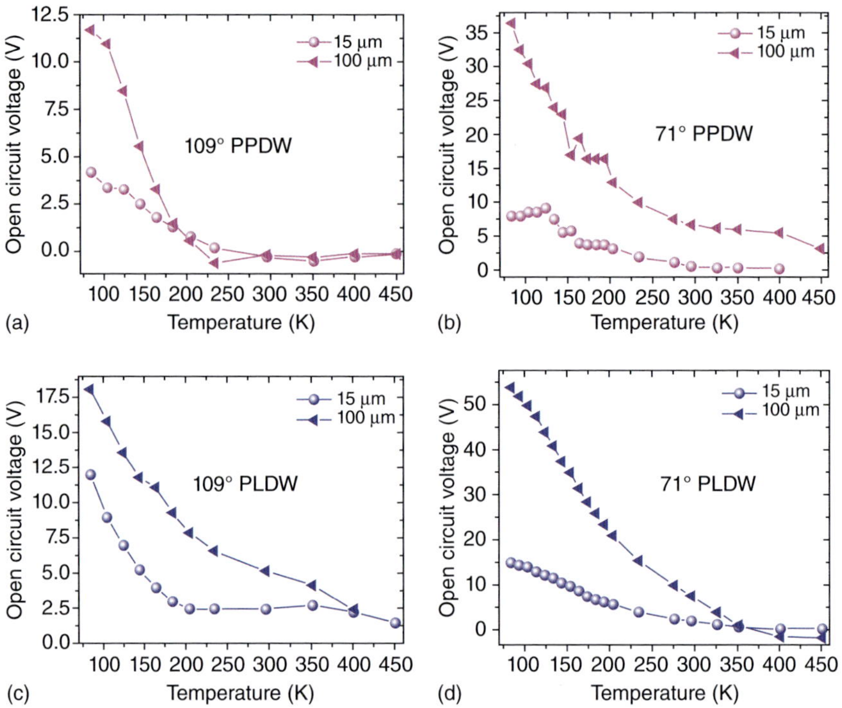

Figure 3.6 Temperature dependence of the open circuit voltage for 109° domain walls in (a) perpendicular and (c) parallel geometry, and for 71° domain walls in (b) perpendicular and (d) parallel geometry. *Source:* Bhatnagar et al. 2013 [42]. Reproduced with permission of Springer Nature.

Interestingly, higher V_{oc}s were obtained in both the systems and for both the geometries. In particular, a sample with 71° domain walls exhibited an overall higher increment in V_{oc} in comparison to the 109° domain wall sample. The photoconductivity of the samples was extracted in conjunction, which showed an exponentially decaying behavior as the temperature was lowered from room temperature. A similar behavior was observed for the dark conductivities. Assuming a uniform generation of photovoltaic current at all temperatures, the inverse relation between the V_{oc} and conductivity behavior matched well with the expression given in Eq. (3.4). This was already a good indication for the validity of the BPV phenomenon in these samples. The domain walls have been demonstrated to be more conductive in comparison to the bulk via micro- and macroscopic measurements. In this context, an equivalent circuit was proposed in which the overall conductivity of the gap is determined by the conductivity of the bulk (σ_{bulk}) and domain walls (σ_{dw}). In the PLDW configurations the conductivity of the gap is given by $(\sigma_{bulk})^{-1} + (\sigma_{dw})^{-1}$, and in the PPDW case the conductivities simply add up to give $\sigma_{bulk} + \sigma_{dw}$. Since the overall conductivity of the PPDW gap is higher than in the case of PLDW, the measured V_{oc}s were always lower in the PPDW case.

To further support this argument, experiments were performed to measure the photovoltaic current while varying the angle made by the plane of linearly polarized light with the direction of current flow. The PV current exhibited an angular dependency, which confirmed that the underlying mechanism was the BPV effect. However, there were certain unresolved issues. The current did not have simple sine and cosine dependency on the angle, as was observed in the previous work on $LiNbO_3$ where the current was measured along the principal crystallographic directions. There were certain offsets that were unaccounted. This issue was resolved by taking into account the orientation of two variants of domains existing in the samples, with respect to the measurement geometry [26]. It allowed to transform the material coordinate system into the laboratory/measurement coordinate system, and as a result the corresponding offsets were calculated.

The work from Belinicher [22] and Kraut and Baltz [24, 25] made the tensorial nature of bulk photoeffect rather obvious, which was further validated in the experimental work from Glass et al. [17], Fridkin and Magomadov [23], and Koch et al. [18]. But, surprisingly, there have not been much work dedicated to further exploit this aspect in thin films. This is despite the excellent advancement in thin-film growth techniques that allows engineering of films in the desired orientation. One of the reasons could be the lack of suitable substrates that can accommodate the inherent unit cell structure of the materials. The work from Basu et al. [43] attempted to investigate the PV effect in [111]-oriented films grown on $SrTiO_3$ (STO) substrates. Owing to the $R3c$ symmetry of the BFO structure, the direction of polarization is along [111]. The work mostly focused on the photoconductive and photoabsorption response from the thin-film samples. The work from Ji et al. [44] also utilized [111]-oriented films grown directly on the [111] STO substrates. In both of these cases, only the β_{22} component of the BPV tensor is active with no observation of above-band-gap V_{oc}. In the case of BFO samples with 71° and 109° domain walls, the other tensor components, namely, β_{33}, β_{31}, β_{22}, and β_{15}, are also relevant. The activity of these components can either enhance or diminish the overall effect. At least in the case of sample with 71° domain walls, and in a certain geometry, the components add up to result in a substantial PV effect. Therefore, thin films growth can provide an exciting playground to exploit the tensorial aspect of this effect. This is further possible by switching the domains in the in-plane direction. The switching may progress via ferroelectric or ferroelastic pathway [45]. The resultant variation in the domain configuration will cause different BPV tensor components to gain or lose their relevance. Such experiments, therefore, present more degrees of freedom to modulate the PV effect in these materials, or in other words, more ways to utilize the effect for application in actuation and detection mechanisms.

Additionally, most of the work has been focused till now on measuring the samples in the out-of-plane (OOP) geometry with a bottom and top electrode. The OOP measurements have been pursued to utilize the polarization as a controlling parameter for the PV effect, although the shift current theory explained it otherwise. It is not surprising that above-band-gap voltages have never been achieved in such geometries. The expression in Eq. (3.4) clearly underlines the significance of the width of the measurement gap. In the OOP geometry, the maximum thickness that can be obtained while maintaining the epitaxial relation with the substrates is largely limited to few hundreds of nanometers. On the contrary, measurements in the in-plane geometry have, under appropriate conditions, resulted in higher V_{oc}s due to larger widths of the gap.

3.3 Modulation of the Effect

3.3.1 Polarization

In the recent past, the first hints about the role of polarization in the switchability of the PV effect came from the work of Choi et al. [34] The diode-like behavior, analogous to a *pn*-junction, implied the possibility of a PV effect. This was validated with the observation of a PV current, and that too with a light of energy (2.33 eV) that was lower than the reported band gap of 2.8 eV [11]. The dependence of the diode behavior on polarization direction also implied a similar mechanism for the PV effect. An angular dependency of the PV current was also observed in relation to the angle that the electric field vector of linearly polarized light made with the ferroelectric polarization. Although such a behavior is also a footprint of the BPV, the proposed explanation considered the origin of the PV effect to be asymmetric polarization induced upon intense light illumination. The asymmetric induced polarization can result in rectification behavior such as the PV effect, as has been previously reported in potassium dihydrogen phosphate [46].

In an earlier work from Pintilie et al. [47] the contribution of the polarization and the electronic effects to the PV mechanism was further emphasized. $Pb(Zr,Ti)O_3$ (PZT) films with varying ratio of Zr/Ti were used in a capacitor geometry. Electrode interfaces that were symmetric in character were attempted by using $SrRuO_3$ (SRO) at the top and the bottom, with the bottom SRO grown epitaxially with pulsed laser deposition and the top one sputtered. Interesting results were obtained since the PV current/short circuit current remained high in the case of tetragonal films made from PZT 40/60 and PZT 20/80, while a drop was measured in the case of rhombohedral structured films. On the contrary, the band gap was found to be increasing as the Zr content was increased, and thus the *rhombohedrality*. This result was also recently confirmed in the calculations reported by Wang et al. based on density functional theory (DFT) wherein the rhombohedral-perovskite phase of materials always exhibits a higher band gap than the corresponding tetragonal phase [48].

The polarization control of the PV current was demonstrated by measuring the spectral distribution before and after switching the polarization state. While the spectral distribution did change its sign when measured immediately after polarization switch, it reverted back to its original form after five days. Correspondingly, this was associated with the back switching of the polarization. Additionally, a perfect overlap was found between the dynamic hysteresis loop of polarization and the value of the PV current. Some of the voltages applied during the polarization loop measurement were applied separately and the PV current was subsequently recorded. The resultant values of the PV current were plotted and the shape matched with the polarization loop. However, it must be noted that the PV current values were normalized in the positive and negative directions, and therefore a true dependence of the magnitude of PV current on the state of polarization was somehow ambiguous. These two observations apparently proved the utility of the PV current in reading the polarization state in ferroelectric materials for a nondestructive readout. The PV current generation was explained by the presence of the Schottky barriers formed at the interface between the ferroelectric layer and the metal electrodes. In a symmetric structure, with a non-ferroelectric material sandwiched between electrodes of the same material, the built-in fields associated with the electrode interfaces will be cancelled and as a result there will be no net field within the material. However, the presence of a ferroelectric material with a directional polarization disturbs this symmetry, resulting in a net field within the material. This net field was proposed to be the driving force behind the separation of electron–hole pairs that are generated under appropriate illumination. The field will drive the electrons and holes to an electrode of opposite polarity and this movement will constitute the PV current. The direction of this current could be conveniently switched by flipping the polarization, which varies the built-in potential at the electrodes. The treatment of a ferroelectric material such as a wide-band-gap semiconductor and the interface with the electrode such as a Schottky barrier vastly assisted in modeling the transport behavior in ferroelectric materials when incorporated in a capacitor geometry.

Similar conclusions were made in the work of Lee et al. that focused on the switchable diode-like behavior in ferroelectrics [49]. It was demonstrated that the magnitude and direction of the PV current can be successfully modulated by sequential variation of the polarization state. The model proposed took into account the presence of Schottky contacts at the top and bottom of the ferroelectric film (BFO/metal interface). The PV effect was henceforth attributed to the built-in field near the contact regions. The model was adequate to explain the switchable diode behavior with and without light, previously reported in the work of Cheong and coworkers [34], but the angular dependency of the PV current was not considered [49].

Guo et al. [50] utilized the polarization control of PV effect wherein the BFO capacitors were fabricated with $La_{0.7}Sr_{0.3}MnO_3$ as the bottom electrode and Pt/Fe forming the top electrode. The polarization state of the BFO film was detected by reading the sign of the open circuit voltage obtained from the current/voltage characteristics.

To prove the utility of such a mechanism for the purpose of polarization readout, a 16-cell memory prototype with cross bar architecture was fabricated by the authors (Figure 3.7). The bottom electrode, $La_{0.7}Sr_{0.3}MnO_3$, was patterned in a stripe form by photolithography and etching. A BFO layer was subsequently deposited on top of $La_{0.7}Sr_{0.3}MnO_3$, followed by patterning of Pt/Fe in a stripe shape to form the top electrode.

Although in the capacitor geometry a constant photocurrent was generated without any bias voltage, above-band-gap V_{oc}s were not observed. One of the reasons could be the relatively smaller measurement gaps. As a result, the interfacial properties at the electrode are dominant. Moreover, in such small measurement

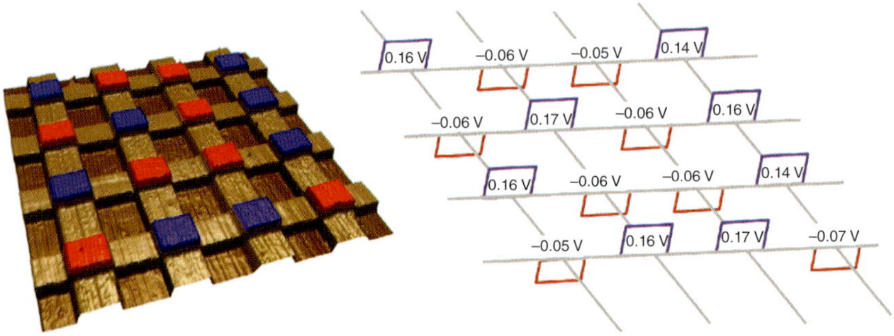

(a) (b)

Figure 3.7 (a) Topography of a 16-cell device with the state of polarization depicted by blue (up polarization) and red (down polarization) color. (b) V_{oc} measured across each cell. *Source:* Guo et al. 2013 [50]. Reprinted by permission from Macmillan Publishers Limited: Nature Communications.

gaps crystal lattice defects such as oxygen vacancies also become much more relevant in the transport mechanisms. To circumvent these issues, planar geometry was used in the case of $Pb_{0.97}La_{0.03}(Zr_{0.52}Ti_{0.48})O_3$ doped with 0.5% WO_3 [51]. Interdigital electrodes were patterned with the gap between the electrodes around 10 and 25 μm. While the gap of 10 μm resulted in a V_{oc} of 4.4 V, the 25-μm gap resulted in V_{oc}s in excess of 7 V. The V_{oc} was reversed to opposite sign by switching the direction of polarization. This was however not completely feasible in the large gap since the required poling fields can be rather high. The dependency of the V_{oc} on the electrode gap clearly indicated the dominant BPV mechanism.

While much work has been devoted to understand the role of polarization on PV effect, there have been also some experimental investigations to unravel the role of light on polarization-related characteristics. One of the most important parameters to test the feasibility of the material for commercial applications is the long-term usability. In the case of ferroelectric materials this usability is defined by the fatigue properties, i.e. a gradual decrement in the switchable polarization value upon subjecting the sample to repeated polarization reversal cycles. The origin of fatigue has been often related to pinning of domains due to charge trapping at internal domain walls. The polarization discontinuity at the domain walls could be one of the driving factors for the charge trapping effect. Illuminating the sample with above-band-gap light, and under a bias voltage, has been demonstrated as the pathway to restore the original state of the sample. Warren et al. in their work on Pt/PZT/Pt validated this role of illumination by first driving the sample to a condition of fatigue and then restoring it to nearly an original state by illuminating with UV light under a bias of 15 V [52]. The photoexcited carriers apparently recombine with the trapped charges, thus unpinning the domains, while the bias voltage assists in realignment of the domains. The results were largely corroborated by the subsequent work of Kholkin et al. in PZT films where the piezoelectric coefficients and the permittivity were clearly affected by the UV light [9].

3.3.2 Electrodes

As was discussed in the previous sections, metal electrodes tend to form Schottky-like contact with the ferroelectrics. Therefore, materials used for the electrodes are rather crucial as they can affect the barrier height at the interface, and thus photoconduction. This is even more relevant in the scenarios when the measurement is done across the film thickness and the measurement gap is limited to a few hundreds of nanometers. The height of the barriers is approximately equal to the difference between the work functions of the two materials. The use of a ferroelectric in the place of a conventional semiconductor provides additional degrees of freedom to tune the barrier height and width.

The work from Blom et al. first proposed the idea of a fundamentally different mechanism to vary the Schottky barrier by polarization of ferroelectric material [53]. The device of interest comprised of $PbTiO_3$ with $LaSrCoO_3$ as the bottom electrode and Au as the top electrode. Different conduction mechanisms, i.e. Ohmic, space charge limited, and Schottky, were observed as the bias voltage was varied. In conventional Schottky contact with a semiconductor, it is assumed that the dielectric constant (ε) of the material does not vary with the electric field. This assumption was considered invalid in the case of ferroelectric semiconductor wherein the ionic part of ε evidently changes with the applied electric field.

Such a scenario was further investigated in detail by Pintilie et al. [54, 55] The model proposed, based on the typical metal–semiconductor rectifying contacts, took into perspective the observation of higher sensitivity of current measurements to the electrode interface. On the contrary, the capacitance measurements are more sensitive to the charge density within the SCR. In contrast to the approach adopted by Blom et al., ε was considered as independent from the electric field.

To provide an exhaustive account on the role of electrodes, a wide range of materials were used as top electrodes while the bottom electrode was kept unchanged (SRO) [56]. The materials for top electrodes included Ag, Pd, Al, Au, V, and Cr (Figure 3.8). For each structure with a different top electrode, spectral distributions were measured for the two polarization states. For some top electrodes, an expected response, i.e. a change in the sign of photocurrent upon a change in the direction of polarization was observed. But, in the case of Al, Cr, and Pt, the sign of photo-response was independent of the direction of polarization. In the case of Cu with the up-polarization state, the photo-response underwent a change in the sign with wavelength. Imprint within the ferroelectric film, i.e. shift of the polarization loop along the axis of electric field due to the presence of an internal field, was proposed as a possible reason. However, polarization measurements revealed only small imprints in the case of Al and Cr [57]. In another report, the use of Cu was demonstrated to obtain completely symmetric polarization loops with no traces of imprint. The open question was whether such small imprint or internal electric field is sufficient to maintain the sign of photocurrent even upon reversal of polarization. The observation was critical in proving the role of factors other than just polarization, such as work function of electrode, quantum efficiency, and width of the depleted region, in determining the sign of photocurrent. Additionally, the observation

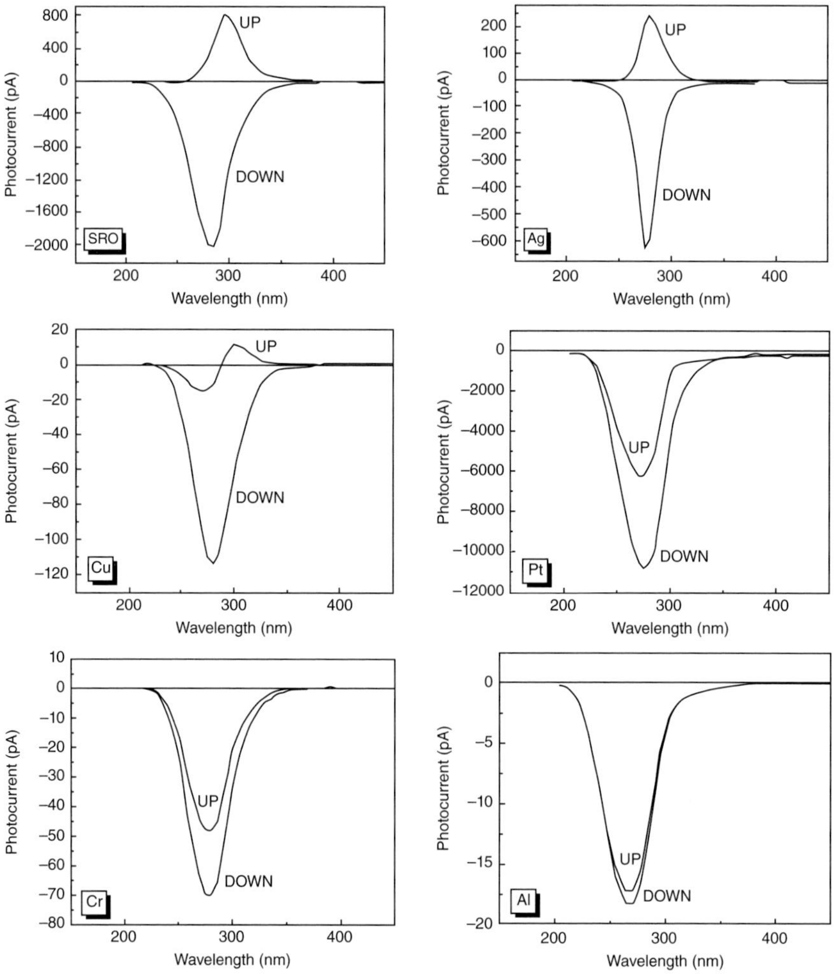

Figure 3.8 Spectral distribution of photovoltaic current measured with different materials as top electrode and for different orientations (up and down) of polarization. *Source:* Pintilie et al. 2011 [56]. Reprinted with the permission of AIP Publishing.

also indirectly implied the role of BPV, which has no dependency on the state of polarization [27].

To explain the above observations and the influence of other factors, the photocurrent generated in the film was considered to be different around the top and bottom electrodes:

$$I_{\text{top}} = A\emptyset n_{\text{ph}} T_{\text{top}} \left[1 - \exp(-\alpha W_{\text{top}})\right] q$$

$$I_{\text{SRO}} = A\emptyset n_{\text{ph}} T_{\text{top}} \exp\left[-\alpha(d - W_{\text{SRO}})\right] \left[1 - \exp(-\alpha W_{\text{top}})\right] q$$

where A is the area of the top electrodes, $\emptyset$ is the quantum efficiency, n_{ph} is the number of incident photons, T_{top} is the transparency of the top electrode, W_{top} is the width of the depleted region at the interface with top electrode, and W_{SRO} is the width of the depleted region around the bottom SRO electrode. The overall photocurrent generated within the bulk was assumed to be a sum of the two currents. The width of the depleted region in the ferroelectric-based Schottky junctions is given by [54, 55]

$$W = \sqrt{\frac{2\varepsilon_o\varepsilon_{st}(V + V'_{bi})}{qN_{eff}}}$$

Here, V_{bi} is given by

$$V'_{bi} = V_{bi} + \frac{P}{\varepsilon_o\varepsilon_{st}}\delta$$

where ε_o is the permittivity of the vacuum, ε_{st} is the permittivity of the material at low frequencies, V is the applied bias voltage, V'_{bi} is the built-in potential at the interfaces due to the presence of a ferroelectric, N_{eff} is the effective density of charge in the depleted region, P is the ferroelectric polarization, V_{bi} is the built-in potential at the interface in the case of a normal semiconductor, and δ is the distance between the polarization charge and the electrode. The origin of the distance δ arises from the assumption of treating the ferroelectric polarization as an infinite sheet of charges positioned at a certain distance away from the interface.

3.3.3 Band Gap Engineering

The large band gap in ferroelectric oxide materials apparently inhibits their photovoltaic efficiency as only 8–20% of the solar spectrum can be utilized. The band gap characteristics in perovskite (ABO_3)-structured ferroelectrics arise due to the A—O and B—O bonds. The excitation process across the band gap largely comprises of charge transfer that occurs from the 2*p* states of the oxygen to the metallic d states of the transition metal ion B. The large difference in the electronegativity between the oxygen and transition metal ion is at the origin of the large band gap [58, 59].

The dependence of the band gap on the A—O and B—O bonds provides certain degrees of freedom to tune the band gap. In this regard B-site doping can significantly affect the band structure and its properties [60]. Additionally, doping also can assist in the manifestation of the BPV effect, as was reported in the case of $LiNbO_3$ single crystals doped with Fe impurities [21]. On the theoretical front, different possibilities have been explored that either consider substitution of the B-site ions or formation of layered structures with the dopants.

In the case of $PbTiO_3$, Bennett et al. [61] calculated that replacing one Ti in every three with a metal from group 10 (e.g. Ni, Pd, Pt) can assist in lowering the band gap, while maintaining the polarization values. However, the presence of an accompanying oxygen vacancy is necessary to stabilize the structure. The resultant band gap values were calculated to be reduced by as much as 2 eV. An essential requirement was the presence of an oxygen vacancy within the usual oxygen octahedra cage to stabilize the structure. The vacancy resulted in substantial displacements of the cations, thus allowing larger values of polarization.

The lowering of the band gap was attributed to the development of new states in the electronic structure of HOMO (highest occupied molecular orbital) and LUMO (lowest unoccupied molecular orbital). In $PbTiO_3$, the HOMO is found to be mostly localized around the O, with its $2p$ states under the influence of $3d$ states of Ti. When doped with a group 10 metal, and oxygen vacancy, the metal d states tend to extend into the band gap, causing a reduction in the band gap.

The idea to replace the B-site ion was further utilized in the experimental work from Grinberg et al. [59] Two different metal ions were used at the B-site. One ion was utilized for maintaining the ferroelectricity, and the other, in conjunction with a vacancy, for reducing the band gap. $KNbO_3$ was used as the base ferroelectric since Nb-containing ferroelectrics have been demonstrated to accommodate oxygen vacancy. A single phase oxide solution was synthesized with the composition $[KNbO_3]_{1-x}[BaNi_{1/2}Nb_{1/2}O_{3-\delta}]_x$ (KBNNO). The B-site position was henceforth taken by either the Nb ion or a combination of Ni^{2+} and a vacancy. The DFT calculations, and also the experimental investigations, revealed a band gap of around 1.46 eV. The change, in comparison to the band gap of un-doped KNO, is about 150%. Analysis of the electronic band structure demonstrated that the bottom of the conduction band comprises of Nb $4d$ states. The maximum of the valence band consists of hybridized Ni $3d$ and O $2p$ states, rather than only O $2p$. Therefore, the Ni $3d$ states were considered responsible for the depreciation in the band gap values.

A different approach was adopted in the work of Zhang et al. [62], where reducing the coordination of oxygen was considered as a plausible route to lower the band gap energy. The reduction in coordination was achieved by utilizing the tetrahedral arrangement with four oxygens surrounding the iron ion, rather than the usual octahedral arrangement with six oxygen ions. A new multiferroic material was proposed with the chemical formula $KBiFe_2O_5$ with alternating blocks of the Fe tetrahedra and $(K, Bi)O_2$. The resultant band gap was found to be around 1.6 eV.

In conventional ferroelectrics such as BTO and PTO, the ferroelectricity and band gap properties are controlled by the B-site positioned cation. In this regard, $BiFeO_3$ can be considered unconventional as the A-site cation Bi is mostly responsible for the ferroelectricity, while the electronic structure and anti-ferromagnetism maintain their dependency upon the B-site cation Fe3+ [63, 64]. This provides a little more flexibility in tuning the electronic structure while maintaining the ferroelectricity. In a recent work from Nechache et al. [65] such a possibility was explored. The work involved use of Bi_2FeCrO_6 (BFCO) synthesized by mixing $BiFeO_3$ and $BiCrO_3$ in equal ratios to form a target for pulsed laser deposition. The fabricated films demonstrated a double perovskite structure with two different cations alternating at the B-site, Fe and Cr. Bi^{3+} ion at the A-site, as in the case of $BiFeO_3$, is mostly responsible for the ferroelectricity. The band gap (E_g) was shown to be controlled by the interaction between the Fe and Cr states. In an earlier work from Baettig and Spaldin [66] on BFCO, *ab initio* calculation predicted the conduction band to comprise of empty Fe $3d$ states and the valence band comprising of hybridized Cr $3d$ and O $2p$ states. This calculation already provided the first hints about how the ordering of Fe and Cr can be utilized to vary the band gap. In the work of Nechache et al. this theory was clearly demonstrated as the grown films exhibited different band gaps based on the cation ordering. The ordering was measured as a function of ratio R between the intensity of superlattice peaks and the primary (111) peak arising from the perovskite. The width of superlattice peaks, D, was another parameter to gauge the extent of ordering. Apparently, the films with high ratio R were found to be much more absorbing at higher wavelengths (lower energies), proving the lowering of the band gap. The parameters R and D were observed to be highly dependent upon the growth kinetics. The rate and temperature of growth were crucial in determining the values of R and D, and thus the E_g. This feature allowed the authors to engineer photovoltaic cells consisting of layers with different band gaps. The topmost layer was grown with the highest band gap to achieve maximum absorption while the bottom layer had the lowest band gap to obtain an overall maximum efficiency. The cell demonstrated an efficiency of 8%, the maximum ever achieved from a ferroelectric photovoltaic cell.

Strain engineering has been considered as yet another pathway to tune the band gap of perovskite-structured ferroelectric materials, as was demonstrated in the work of Wang et al. [48] Strain can be effectively used as a tool to cause further off-centering of B-site-positioned ion and distort the BO_6 octahedra, thereby affecting the electronic structure. On the contrary, the role of A-site ion has been demonstrated to be much more subsidized in determining the band gap values. The basis of the work was the density functional calculations that indicated substantial disparity in the band gaps arising from different phases of perovskites of same compositions. The rhombohedral or R phase always exhibited higher band gap values than the tetragonal or T phase (Figure 3.9). Additionally, an increase in the valence of the B-site ion, such as from +3 to +6 found in WO_3, had an incremental effect on E_g as well.

The primary reason behind such phenomenon was found to be the shift of the minima of the conduction band. The top of the valence band was consistent in both the phases for all the materials investigated. It was proposed that as the materials transform to the R- phase, the B-site cations undergo further off centering. As a result, the difference between the short and long bond lengths within the

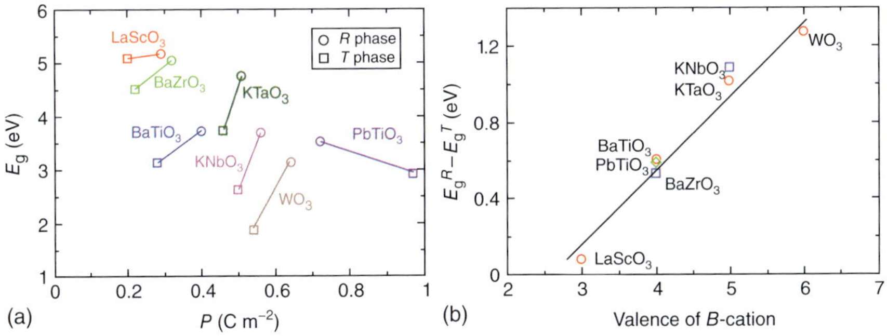

Figure 3.9 (a) Calculated values of band gap as a function of polarization for different materials in their respective tetragonal and rhombohedral phases. (b) Difference in the band gap between tetragonal and rhombohedral phases as a function of the *B*-site cation valence. *Source*: Wang et al. 2014 [48]. Reprinted with the permission of AIP Publishing.

BO_6 octahedral becomes much more apparent, with the B—O—B angles deviating from the usual 180°. The antibonding character induced at the gamma point causes the bottom of conduction band to shift further up. The calculations also explained the higher polarization values for the respective R-phases.

Similar predictions were made about the band gap in $BiFeO_3$ in the work of Tütüncü and Srivastava [67] The *ab initio* calculations revealed a direct band gap of as low as 0.55 eV in tetragonal BFO. However, experimental data from Chen et al. obtained from BFO films under high compressive strain demonstrated an opposite scenario [68]. BFO films were grown on orthorhombic and (110)-oriented $YAlO_3$ substrates. The mismatch between the in-plane lattice parameters caused the BFO film to grown in a tetragonal structure with an elongated OOP lattice. The optical properties were determined with ellipsometry and absorption techniques. The resultant band gap in the tetragonal films was calculated to be around 3.1 eV, which is approximately 0.4 eV higher than the band gap of rhombohedral films. The discrepancy between the measured and predicted E_g values most probably arose from the under- and overestimation of the movement of O-2*p* and Fe-3*d* orbitals, respectively. For comparison, in the work of Wang et al., the top of the valence band or position of O-2*p* orbitals remained unchanged for both the phases, and only the movement of B-site ion associated *d*-orbitals caused a variation in the E_g.

Himcinschi et al. reported similar results about the E_g values in tetragonal-like BFO films [69]. In addition to thinner films (29 nm), thicker films (118 nm) were also synthesized. The absorption measurements provided E_g value of 3.1 eV for thin and 3.05 eV for thick films (Figure 3.10). Raman spectra were measured and compared with relaxed and rhombohedral BFO films grown on $TbScO_3$. The spectra from tetragonal-like BFO presented new modes from which the strongest were at 263, 367, 481, and 691 cm^{-1}. In particular, the mode around 691 cm^{-1}

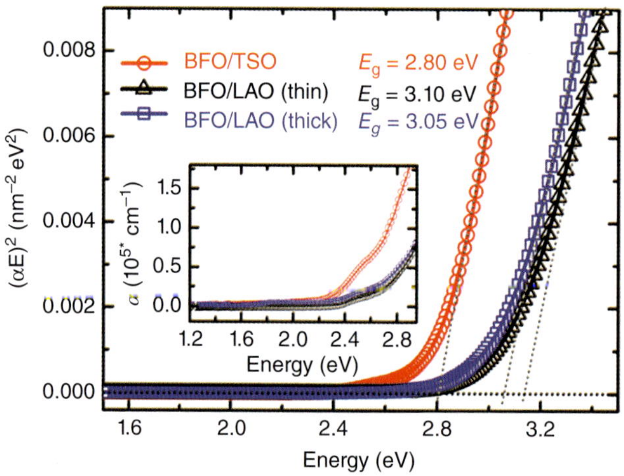

Figure 3.10 Absorption measured as function of photon energy. The linear extrapolation provides the values for direct optical band gap in $BiFeO_3$ thin films grown in relaxed and strained phases. The inset shows the absorption coefficients determined by numerical inversion. *Source:* Himcinschi et al. 2015 [69]. Reprinted the permission of AIP Publishing.

is considered as a footprint of high tetragonality [70, 71]. The mode has also been attributed to a five-oxygen coordination of Fe instead of six in an octahedral arrangement [72]. Such a change could potentially result in entirely different electronic properties. Surprisingly, all of these new modes were present in both the thin and thick films, indicating the persistence of strain even in thicker films.

Indeed, the electronic properties measured in tetragonal-like BFO films were found to be vastly different from the relaxed rhombohedral phase films. Bhatnagar et al. provided a direct evidence in this regard by measuring the spectral distributions from the two types of samples [73]. The photocurrent was measured from the samples as the wavelength of the illumination was scanned from high to low, and back to high values. The estimated E_g values were in agreement with the absorption measurements. In the case of rhombohedral films, the responses in both directions of wavelength scan were similar. On the contrary, the response was much different in the case of tetragonal samples as the photocurrent was enhanced by a factor of 3 within certain intervals of the wavelength. In particular, the photocurrent was enhanced around the energies of band gap transitions and between 600 and 675 nm. Such an increment in the photo-response explicitly suggests the activity of sub-band-gap states.

The activity of such states within the band gap can often result in exotic properties. In the case of wide-band-gap semiconductors such as CdS, and more recently in STO/$LaAlO_3$ interface and in pure STO, the defect levels are at the origin of persistent photoconductivity (PPC) [74, 75]. The phenomenon involves first illuminating the material with appropriate energy, as result of which a high conductivity state is achieved. The material stays in this state even after removal of the illumination. Eventually, there is a gradual decay with time to a high resistance state. Bhatnagar et al. reported such an effect in the strained films of BFO [76]. To examine the origin of the effect the authors utilized thermally stimulated current to prove the role of sub-band-gap states. The sample was charged at room temperature with above-band-gap energy and then quickly quenched to low temperature to freeze the state of electronic arrangement. As the sample was heated back to room temperature, a clear peak was observed in the current measurement. Analysis of the peak provided information about the position of a level at around 0.26 eV near an allowed band edge. Although such levels have been also reported in rhombohedral films, they have not resulted in any kind of PPC behavior. This implies that a change of structure can vary the capture cross section of these levels, and thus their activity.

In the case of single crystals of BFO, the sub-band-gap levels have been effectively used for even tuning the BPV effect. This was recently demonstrated in the work from Yang et al. wherein crystals from different batches of processing and from different groups (from Viret and Schmidt group) were used [77]. The spectral distribution and current–voltage characteristics were measured in all the samples. Distinctive features were observed in the spectral response within the energy range of 470–700 nm in all the samples. The area under these features was calculated and plotted against the open circuit voltages that were extracted from the current–voltage curves (Figure 3.11). A direct correlation between the two parameters means that the activity of these levels is crucial to achieve an above-band-gap V_{oc}. This observation is also consistent with theoretical

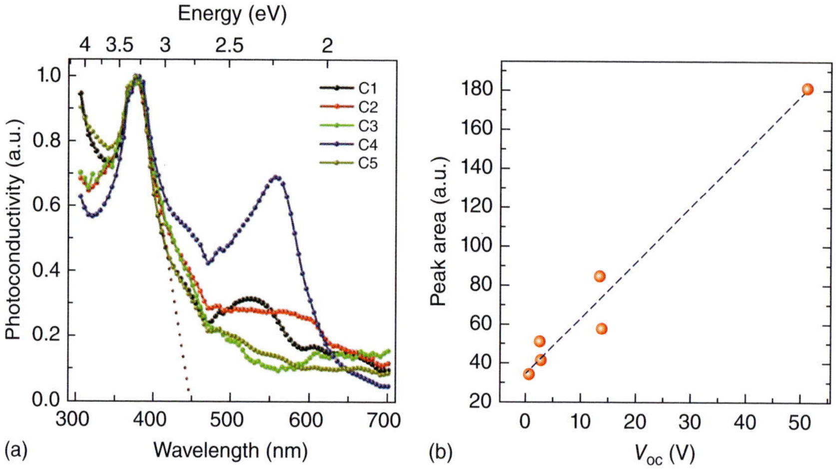

Figure 3.11 (a) Spectral distribution of photoconductive current measured from crystals grown in different batches. Crystals C3 and C5 were synthesized in the group of Viret and Schmid, respectively. (b) Open circuit voltage plotted as a function of the area measured under the peaks positioned around 470–700 nm in (a). *Source:* Yang et al. 2015 [77]. Reprinted with permission of John Wiley & Sons.

predictions for the manifestation of BPV effect wherein the paths of carrier recombination and generation must be different [20, 22, 25]. Such a possibility can be facilitated by the presence of sub-band-gap states.

3.3.4 Photo-mechanical Coupling

The change in the dimensions of a material upon illumination is often referred to as photostriction. The phenomenon has been reported in polar and nonpolar semiconductors and different polymers. However, the underlying origin of this phenomenon varies from material to material. The observation of this effect in ferroelectric and multiferroic materials opened the door to pathways for new functionalities. The possibility to combine this effect with other properties of ferroelectric and multiferroic materials can form the core for realization of several devices, ranging from sensing to actuation.

The effect was first observed in ferroelectric materials by Tatsuzaki et al. [78] in the case of SbSI, which has been also investigated for its non-trivial photoconductive properties [79]. The work demonstrated a change in the length of the single crystal along the polar axis upon illumination, albeit under a bias DC voltage (Figure 3.12). The induced strain in the crystals was observed to have gradual settling time, which indicated a charge-assisted mechanism. The strain behavior was measured as a function of temperature. The measurement clearly demonstrated a correlation with the ferroelectric Curie temperature (T_c) since the strain underwent a discontinuity as the sample was heated. The strain was found to be positive below T_c and negative above T_c. To rule out the strain

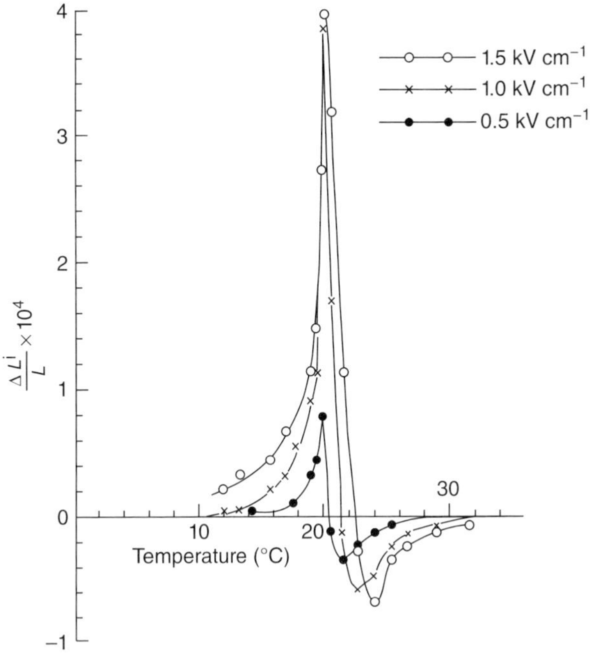

Figure 3.12 Photostriction in SbSI measured by analyzing the behavior of $\Delta L/L$ with temperature. *Source:* Tatsuzaki et al. 1966 [78]. Copyright (1966). Reprinted with permission of American Physical Society.

arising from piezoelectricity, the authors also measured the sample without illumination. Interestingly, an increase in the applied bias across the sample caused an overall increment in the effect.

Photostriction was also observed in the ceramic samples of PZT. The work from Brody used PZT samples to engineer an optomechanical bimorph actuator [80]. The device comprised of two plates of PZT that were cemented together. The plates were pre-polarized in parallel but opposite directions. The ends of the plates were electroded to develop an electric field. Only one plate/face of the device was illuminated. Upon illumination, the device bent in an arc away from the illumination. The proposed reason behind deflection of the device was based on the BPV. The illumination on one side of the device resulted in a photocurrent that flows opposite to the direction of polarization. The current flow charges the electrodes and generates an electric field. The field in one plate is parallel to the polarization and is opposite in the other. As a result, and due to inverse piezoelectric effect, one plate undergoes expansion while the other experiences contraction. Since the plates were cemented together, the strain was absorbed within the device by bending one end of the device in an arc.

The utilization of BPV in explaining the photostriction in ferroelectric materials was further extended in the work of Uchino and Poosanaas [81]. The work made use of La-doped PZT ceramics with the composition $Pb_{1-x}La_x(ZrTi_{1-y})_{1-x/4}O_3$. The performance of the sample was evaluated by the product of photovoltage and piezoelectric coefficient. An empirical formula was proposed to define the photostriction in the form

$$x_{\mathrm{ph}} = d_{33}E_{\mathrm{ph}}\left(1 - \exp\left(-\frac{t}{RC}\right)\right)$$

with x_{ph} defining the photostriction, d_{33} the piezoelectric coefficient, E_{ph} the photovoltage arising due to the BPV effect, and RC being the product of the resistance and capacitance. It is apparent from the expression that higher photostriction can be obtained with either higher d_{33} and E_{ph} or with lower permittivity.

While the photostriction measured in much of the early work dealt with strain occurring on a macroscopic scale, recent work on $PbTiO_3$ by Daranciang et al. revealed changes at an atomic scale [82]. This was possible by using femtosecond time-resolved X-ray scattering measurements. Thin films of $PbTiO_3$ were used for the study. The films were illuminated with 400 nm femtosecond pulses of laser and the omega scans around (003) reflections were analyzed for strain (Figure 3.13). Intriguing observations were made regarding the positions of peaks. The time-dependent rocking curves revealed a slight movement of the peak toward higher 2θ angles that indicated a decrement in the tetragonality and OOP lattice parameter. This movement of the peak was measured over a timescale of 5 ps. However, a contrary behavior was observed in the subsequent period of time. The peak symmetrically and gradually (within a time scale of 10 ps) shifted toward lower 2θ angles, indicating an increment in the tetragonality. Eventually, the peak relaxes to its initial position before illumination over a timescale of nanoseconds.

A two-step mechanism was proposed to explain such a behavior. In the first instance, when the sample is impinged with photons, there is a coherent shift of the free carriers due to the BPV effect. The associated current flows along the direction of polarization toward the interface with the substrate. This causes an increase in the depolarization field, resulting in a contraction of the unit cell. In the second step, the photo-generated carriers play a dominant role and have a tendency to flow opposite to the direction of depolarization. In doing so, the depolarization field is screened and the lattice undergoes expansion. As the carriers eventually recombine, the lattice relaxes to its original state. The role of depolarization field in the short-lived photostriction was validated when similar

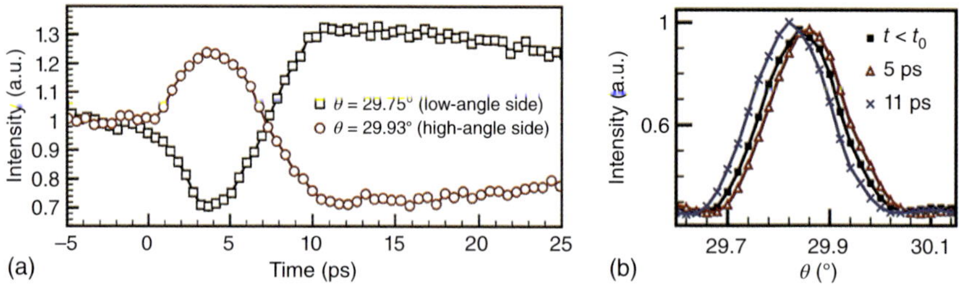

Figure 3.13 (a) Time scans measured at the low- and high-angle sides of the (003) reflection of PTO depicting the movement of peak toward lower angles with time. (b) Rocking curves (ω-scans) measured around (003) reflection at the time intervals shown in (a). *Source:* Daranciang et al. 2012 [82]. Copyright (2012). Reprinted with permissionof American Physical Society.

measurements were done above the Curie temperature, when the depolarizing field is nonexistent. No shift of the peak was observed within the resolution of the setup, which apparently implied the relevance of depolarization field.

Photostriction was observed in a multiferroic material, $BiFeO_3$, for the first time by Kundys et al. [83]. Although BFO has been known for its ferroelectric properties since 1960s from the work of Smolenski and Bokov [84], its photo-ferroelectric properties, i.e. BPV and photostriction, attracted attention only recently. The work from Kundys et al. involved the use of a single domain BFO single crystal. The change in the length was measured along the [010] direction while the sample was illuminated uniformly. To utilize the magnetic degrees of freedom available in a multiferroic material, the authors also measured photostriction under magnetic fields of up to 1.5 T. However, the magnetic field was observed to have a negative effect on the photostriction. The measured strain dropped by 30% upon the application of a magnetic field. The result is rather surprising as other studies have not reported any measurable effect on the polarization or in other words, in the ferroelectric character, with these values of magnetic fields. Therefore, the idea of photostriction being a net result of BPV effect and inverse piezoelectric effect needs further validation and investigations.

In all of the work described above, above-band-gap energy was utilized for photostriction. That only made it necessary to understand and draw a correlation between the amount of photo-generated carriers and the extent of photostriction. Precisely this issue was addressed in the work of Wen et al. [85] where time-resolved X-ray methods and transient optical absorption were used. The structural change in the samples was analyzed by using the (002) reflection from BFO as the reference. The change in the peak position and intensity was measured with pulsed X-rays that were time synchronized with the excitation source, which was laser light of 400 nm wavelength. In a separate measurement, the time-resolved transient optical absorption was measured from the samples using the same excitation source.

The excitation caused a left shift of the (002) peak of BFO, indicating an increase in the lattice parameters. The increment had a decaying character that lasted over several nanoseconds. The absorption measurement revealed similar decaying character albeit on a shorter timescale. The discrepancy in the timescales was explained by slower strain relaxation, which is achieved via thermal transport. Nevertheless, a direct and linear correlation was observed between the induced strain and the difference in optical densities. The difference in optical densities (ΔOD) is a measure of the optically excited carriers. It is rather interesting that a maxima in the ΔOD was observed at around 540 nm or 2.29 eV, which is well below the reported band gap of BFO of around 2.7 eV. This explicitly suggests that excitation via sub-band-gap states can also assist in photostriction. The authors also attempted similar measurements with a lower energy light of 800 nm but it did not result in any observable change in the lattice.

The wavelength dependence of photostriction was investigated in the work of Kundy et al. on BFO single crystals [86]. A systematic study was done that involved using light emitting diodes of different wavelength that were shone on the sample for the same amount of time (Figure 3.14). The variation in the

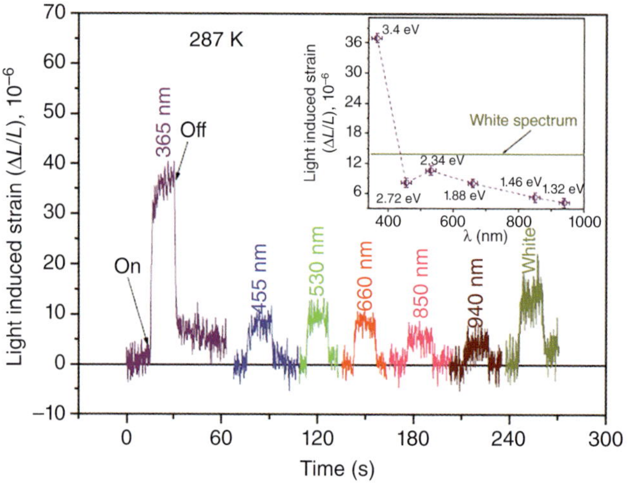

Figure 3.14 Photostriction measured in the $BiFeO_3$ crystal as a function of time, under illumination of different wavelengths. The inset compares the photostriction measured at different wavelengths and white light for the same duration of time. *Source:* Kundys et al. 2012 [86]. Copyright (2012). Reprinted with permission of American Physical Society.

resultant photostriction apparently proved that band–band excitation is not a necessary criterion for photostriction.

A higher photostriction was observed at a wavelength of 530 nm than at 455 nm, which corresponds to the reported band gap. This behavior also supports the results from Wen et al. mentioned above. Additionally, much higher photostriction was measured at around 365 nm. One of the reasons could be that the carriers excited to higher states than the edges of the band gap eventually lose their energy via electron–phonon interaction, which results in a larger variation in lattice. However, the timescales for such relaxation mechanisms, as proposed by Wen et al. [85], can be in the range of picoseconds, and not several seconds. The large photostriction observed at 365 nm could be therefore much more complicated, involving virtual transitions via sub-band-gap states.

3.4 Summary and Outlook

The fundamentally different mechanism for charge separation associated with the BPV indeed provides new pathways for the fabrication of efficient photovoltaic modules. The fact that the BPV effect does not require any artificial junction and is observed in bare materials is an apparent benefit. The perovskite structure, in which the effect has been mostly observed, provides multiple degrees of freedom to manipulate the effect in terms of efficiency and utility. Although a lot of work has been done to unravel the origin and is related to application, when compared with the state of knowledge available in semiconductors, the field of BPV can still be considered in preliminary stages. Crucial information pertaining to mobility, drift, diffusion, and lifetime of carriers is still missing and requires further research. The results can, and will, greatly assist in determining appropriate device architectures to harness the maximum output. One of the primary reasons for the scarcity of information in this regard is the inherent insulating character of the ferroelectric materials. However, recent advancements in cryogenic techniques, with the option of illumination, will certainly allow overcoming these hurdles.

Thin-film technologies have made it possible to engineer a variety of structure-related characteristics such as termination at the surface, type of domain wall, strained state, and systematic/sequential doping. In this regard, the significance of scanning probe methods is much more relevant as they allow localized measurement around topological entities such as domain walls and vary the local state of strain. This can be further extended to tune the local non-centrosymmetric character via flexoelectric effects [87]. The method can be also utilized in identifying the contribution from different kinds of domain walls, i.e. ferroelastic and ferroelectric. Measurement of efficiencies around these local features might lead to non-trivial observations.

The recent reports about efficiencies approaching 8% [65] truly illustrate the potential of the effect to evolve into a meaningful alternative. New avenues must be explored that aim for an increment in the power efficiency. The tensorial nature of the BPV till now has not been sufficiently utilized. The bulk photovoltaic tensor is largely analogous to the piezoelectric tensor of the ferroelectric material. The similarity is sufficiently compelling to channel research toward material compositions that are around the morphotropic phase boundary. One of the reasons for losses in photovoltaic modules is associated with the band offsets at the interfaces with the electrodes. Polarization control in ferroelectrics has been successfully demonstrated as an effective tool to vary the band alignment with the electrodes. This aspect must be further explored to obtain Ohmic contacts. The non-centrosymmetric nature of material is at the core of photovoltaic effect in ferroelectrics. Attempts therefore must be made to directly address this aspect. Antiferroelectrics, which exhibit ferroelectricity under appropriate fields, have been recently reported to show above-band-gap photovoltages [88]. Relaxor-type ferroelectrics have a rather broad dielectric response near Curie temperature due to nano-polar regions. However, with stoichiometric substitutions at A- and B-sites, giant piezoelectric coefficients have been achieved. Such an approach is worthy to be attempted for the photovoltaic effect. In addition, parallel efforts must be directed towards obtaining better interfaces with semiconductors. The semiconducting layer will be useful in absorbing a majority of the solar spectra, and the ferroelectric for charge separation.

References

1 Shockley, W. and Queisser, H.J. (1961). Detailed balance limit of efficiency of p–n junction solar cells. *Journal of Applied Physics* 32 (3): 510–519.

2 Sze, S.M. (1981). *Physics of semiconductor devices*. Wiley.

3 Wemple, S.H. (1970). Polarization fluctuations and the optical-absorption edge in $BaTiO_3$. *Physical Review B: Condensed Matter* 2 (7): 2679–2689.
4 Redfield, D. and Burke, W.J. (1974). Optical absorption edge of $LiNbO_3$. *Journal of Applied Physics* 45 (10): 4566–4571.
5 Kam, K., Henkel, J.H., and Hwang, H. (1978). Band structure and spontaneous polarization of ferroelectric $LiNbO_3$. *Journal of Chemical Physics* 69 (5): 1949–1951.
6 Kaifu, Y., Komatsu, T., Hirano, T., and Inoguchi, T. (1967). Temperature dependence of the absorption edge in ferroelectric $KNbO_3$ crystals. *Journal of the Physical Society of Japan* 23: 903–903.
7 Nitsche, R. and Merz, W.J. (1960). Photoconduction in ternary V-VI-VII compounds. *Journal of Physics and Chemistry of Solids* 13 (1–2): 154–155.
8 Bao, D., Yao, X., Shinozaki, K., and Mizutani, N. (2003). Crystallization and optical properties of sol–gel-derived $PbTiO_3$ thin films. *Thin Solid Films* 36: 2141–2145.
9 Kholkin, A.L., Iakovlev, S.O., and Baptista, J.L. (2001). Direct effect of illumination on ferroelectric properties of lead zirconate titanate thin films. *Applied Physics Letters* 79 (13): 2055–2057.
10 Hauser, A.J., Zhang, J., Mier, L. et al. (2008). Characterization of electronic structure and defect states of thin epitaxial $BiFeO_3$ films by UV-visible absorption and cathodoluminescence spectroscopies. *Applied Physics Letters* 92 (22): 222901.
11 Clark, S.J. and Robertson, J. (2007). Band gap and Schottky barrier heights of multiferroic $BiFeO_3$. *Applied Physics Letters* 90 (13): 1–4.
12 Chynoweth, A.G. (1956). Surface space-charge layers in Barium Titanate. *Physical Review* 102 (3): 705–714.
13 Ashkin, A., Boyd, G.D., Dziedzic, J.M. et al. (1966). Optically-induced refractive index in homogeneities in $LiNbO_3$ and $LiTaO_3$. *Applied Physics Letters* 9 (1): 72–74.
14 Chen, F.S. (1969). Optically induced change of refractive indices in $LiNbO_3$ and $LiTaO_3$. *Journal of Applied Physics* 40 (8): 3389–3396.
15 Townsend, R.L. and LaMacchia, J.T. (1970). Optically induced refractive index changes in $BaTiO_3$. *Journal of Applied Physics* 41 (13): 5188–5192.
16 Johnston, W.D.J. (1970). Optical index damage in $LiNbO_3$ and other pyroelectric insulators. *Journal of Applied Physics* 41 (8): 3279–3285.
17 Glass, A.M., Von Der Linde, D., and Negran, T.J. (1974). Highvoltage bulk photovoltaic effect and the photorefractive process in $LiNbO_3$. *Applied Physics Letters* 25 (4): 233–235.
18 Koch, W.T.H., Munser, R., Ruppel, W., and Würfel, P. (1975). Bulk photovoltaic effect in $BaTiO_3$. *Solid State Communications* 17: 847–850.
19 Wemple, S.H., Didomenico, M., and Jayaraman, A. (1969). Electron scattering in perovskite-oxide ferroelectric semiconductors. *Physical Review* 180 (2): 547–556.
20 Chanussot, G. and Glass, A.M. (1976). A bulk photovoltaic effect due to electron-phonon coupling in polar crystals. *Physics Letters A* 59 (5): 405–407.

21 Glass, A.M., von der Linde, D., Auston, D.H., and Negran, T.J. (1975). Excited state of polarization, bulk photovoltaic effect and the photorefractive effect in electrically polarized media. *Journal of Electronic Materials* 4 (5): 915–943.
22 Belinicher, V.I., Malinovskii, V.K., and Sturman, B.I. (1977). Photogalvanic effect in a crystal with polar axis. *Soviet Physics – JETP* 46 (2): 362–366.
23 Fridkin, V.M. and Magomadov, R.M. (1979). Anomalous photovoltaic effect in $LiNbO_3$: Fe in polarized light. *Pis'ma v zhurnal eksperimental'noi i teoreticheskoi fiziki* 30 (11): 723–726.
24 Kraut, W. and von Baltz, R. (1978). Theory of the anomalous bulk photovoltaic effect in ferroelectrics. *Physica Status Solidi A Applications and Material Science* 89: 419–429.
25 Von Baltz, R. (1981). The bulk photovoltaic effect in ferro- and piezoelectric materials. *Ferroelectrics* 35 (February): 131–136.
26 Young, S.M., Zheng, F., and Rappe, A.M. (2012). First-principles calculation of the bulk photovoltaic effect in bismuth ferrite. *Physical Review Letters* 109 (23): 1–5.
27 Tan, L.Z., Zheng, F., Young, S.M. et al. (2016). Shift current bulk photovoltaic effect in polar materials – hybrid and oxide perovskites and beyond. *NPJ Computer Materials* 2.
28 Young, S. (2016). Bulk photovoltaic effect for next-generation light-sensitive devices. *SPIE Newsroom* doi: 10.1117/2.1201610.006641.
29 Krätzig, E. and Kurz, H. (1976). Photo-induced currents and voltages in $LiNbO_3$. *Ferroelectrics* 13: 295–296.
30 Günter, P. (1978). Photovoltages, photocurrents and photorefractive effects in $KNbO_3$:Fe. *Ferroelectrics* 22: 671–674.
31 Lempicki, A. (1959). Anomalous photovoltaic effect in ZnS single crystals. *Physical Review* 113 (5): 1204–1209.
32 Neumark, G.F. (1962). Theory of the anomalous photovoltaic effect of ZnS. *Physical Review* 125 (3): 838–845.
33 Pensak, L. (1958). High-voltage photovoltaic effect. *Physical Review* 109 (2): 601.
34 Choi, T., Lee, S., Choi, Y.J. et al. (2009). Switchable ferroelectric diode and photovoltaic effect in $BiFeO_3$. *Science* 324 (5923): 63–66.
35 Yang, S.Y., Seidel, J., Byrnes, S.J. et al. (2010). Above-bandgap voltages from ferroelectric photovoltaic devices. *Nature Nanotechnology* 5 (2): 143–147.
36 Seidel, J., Martin, L.W., He, Q. et al. (2009). Conduction at domain walls in oxide multiferroics. *Nature Materials* 8 (3): 229–234.
37 Alexe, M. and Hesse, D. (2011). Tip-enhanced photovoltaic effects in bismuth ferrite. *Nature Communications* 2: 256.
38 Alexe, M. (2012). Local mapping of generation and recombination lifetime in $BiFeO_3$ single crystals by scanning probe photo- induced transient spectroscopy. *Nano Letters* 12: 3–6.
39 Coffey, D.C. and Ginger, D.S. (2006). Time-resolved electrostatic force microscopy of polymer solar cells. *Nature Materials* 5 (9): 735–740.
40 Rose, A. (1955). Recombination processes in insulators and semiconductors. *Physical Review* 97 (2): 322–333.

41 Moubah, R., Rousseau, O., Colson, D. et al. (2012). Photoelectric effects in single domain $BiFeO_3$ crystals. *Advanced Functional Materials* 22 (22): 4814–4818.

42 Bhatnagar, A., Roy Chaudhuri, A., Heon Kim, Y. et al. (2013). Role of domain walls in the abnormal photovoltaic effect in $BiFeO_3$. *Nature Communications* 4 (May): 2835.

43 Basu, S.R., Martin, L.W., Chu, Y.H. et al. (2008). Photoconductivity in $BiFeO_3$ thin films. *Applied Physics Letters* 92 (9): 10–13.

44 Ji, W., Yao, K., and Liang, Y.C. (2011). Evidence of bulk photovoltaic effect and large tensor coefficient in ferroelectric $BiFeO_3$ thin films. *Physical Review B: Condensed Matter* 84 (9): 1–5.

45 Balke, N., Choudhury, S., Jesse, S. et al. (2009). Deterministic control of ferroelastic switching in multiferroic materials. *Nature Nanotechnology* 4 (12): 868–875.

46 Bass, M., Franken, P.A., Ward, J.F., and Weinreich, G. (1962). Optic rectification. *Physical Review Letters* 9 (11): 446–448.

47 Pintilie, L., Vrejoiu, I., Le Rhun, G., and Alexe, M. (2007). Short-circuit photocurrent in epitaxial lead zirconate-titanate thin films. *Journal of Applied Physics* 101 (6).

48 Wang, F., Grinberg, I., and Rappe, A.M. (2014). Band gap engineering strategy via polarization rotation in perovskite ferroelectrics. *Applied Physics Letters* 104 (15): 152903.

49 Lee, D., Baek, S.H., Kim, T.H. et al. (2011). Polarity control of carrier injection at ferroelectric/metal interfaces for electrically switchable diode and photovoltaic effects. *Physical Review B: Condensed Matter* 84 (12): 125305.

50 Guo, R., You, L., Zhou, Y. et al. (2013). Non-volatile memory based on the ferroelectric photovoltaic effect. *Nature Communications* 4 (May): 1–5.

51 Yao, K., Gan, B.K., Chen, M., and Shannigrahi, S. (2005). Large photo-induced voltage in a ferroelectric thin film with in-plane polarization. *Applied Physics Letters* 87 (21): 1–3.

52 Warren, W.L., Dimos, D., Tuttle, B.A. et al. (1994). Electronic domain pinning in $Pb(Zr,Ti)O_3$ thin films and its role in fatigue. *Applied Physics Letters* 65 (8): 1018–1020.

53 Blom, P.W.M., Wolf, R.M., Cillessen, J.F.M., and Krijn, M.P.C.M. (1994). Ferroelectric Schottky diode. *Physical Review Letters* 73 (15): 2107–2111.

54 Pintilie, L. and Alexe, M. (2005). Metal-ferroelectric-metal heterostructures with Schottky contacts. I. Influence of the ferroelectric properties. *Journal of Applied Physics* 98 (12): 124103.

55 Pintilie, L., Boerasu, I., Gomes, M.J.M. et al. (2005). Metal-ferroelectric-metal structures with Schottky contacts. II. Analysis of the experimental current-voltage and capacitance-voltage characteristics of $Pb(Zr,Ti)O_3$ thin films. *Journal of Applied Physics* 98 (12): 124104.

56 Pintilie, L., Dragoi, C., and Pintilie, I. (2011). Interface controlled photovoltaic effect in epitaxial $Pb(Zr,Ti)O_3$ films with tetragonal structure. *Journal of Applied Physics* 110 (4): 44105.

57 Pintilie, L., Vrejoiu, I., Hesse, D., and Alexe, M. (2008). The influence of the top-contact metal on the ferroelectric properties of epitaxial ferroelectric $Pb(Zr_{0.2}Ti_{0.8})O_3$ thin films. *Journal of Applied Physics* 104 (11): 114101.
58 Cohen, R.E. (1992). Origin of ferroelectricity in perovskite oxides. *Nature* 358: 136–138.
59 Grinberg, I., West, D.V., Torres, M. et al. (2013). Perovskite oxides for visible-light-absorbing ferroelectric and photovoltaic materials. *Nature* 503 (7477): 509–512.
60 Yang, C.-H., Kan, D., Takeuchi, I. et al. (2012). Doping $BiFeO_3$: approaches and enhanced functionality. *Physical Chemistry Chemical Physics* 14 (46): 15953–15962.
61 Bennett, J.W., Grinberg, I., and Rappe, A.M. (2008). New highly polar semiconductor ferroelectrics through d^8 cation-O vacancy substitution into $PbTiO_3$: a theoretical study. *Journal of the American Chemical Society* 130 (51): 17409–17412.
62 Zhang, G., Wu, H., Li, G. et al. (2013). New high T(c) multiferroics $KBiFe_2O_5$ with narrow band gap and promising photovoltaic effect. *Scientific Reports* 3: 1265.
63 Cohen, R. and Krakauer, H. (1992). Electronic structure studies of the differences in ferroelectric behavior of $BaTiO_3$ and $PbTiO_3$. *Ferroelectrics* 136 (January 2013): 65–83.
64 Seshadri, R. and Hill, N.A. (2001). Visualizing the role of Bi 6s "Lone Pairs" in the off-center distortion in ferromagnetic $BiMnO_3$. *Chemistry of Materials* 13: 2892–2899.
65 Nechache, R., Harnagea, C., Li, S. et al. (2014). Bandgap tuning of multiferroic oxide solar cells. *Nature Photonics* 61 (1): 61–67.
66 Baettig, P. and Spaldin, N.A. (2005). Ab initio prediction of a multiferroic with large polarization and magnetization. *Applied Physics Letters* 86 (1): 12505–12501.
67 Tütüncü, H.M. and Srivastava, G.P. (2008). Electronic structure and lattice dynamical properties of different tetragonal phases of $BiFeO_3$. *Physical Review B: Condensed Matter* 78 (23): 1–10.
68 Chen, P., Podraza, N.J., Xu, X.S. et al. (2010). Optical properties of quasi-tetragonal $BiFeO_3$ thin films. *Applied Physics Letters* 96 (13): 2008–2011.
69 Himcinschi, C., Bhatnagar, A., Talkenberger, A. et al. (2015). Optical properties of epitaxial $BiFeO_3$ thin films grown on $LaAlO_3$. *Applied Physics Letters* 106 (1): 012908.
70 Hermet, P., Goffinet, M., Kreisel, J., and Ghosez, P. (2007). Raman and infrared spectra of multiferroic bismuth ferrite from first principles. *Physical Review B: Condensed Matter* 75 (22): 3–6.
71 Kreisel, J., Jadhav, P., Chaix-Pluchery, O. et al. (2011). A phase transition close to room temperature in $BiFeO_3$ thin films. *Journal of Physics: Condensed Matter* 23 (34): 342202.
72 Hatt, A.J., Spaldin, N.A., and Ederer, C. (2010). Strain-induced isosymmetric phase transition in $BiFeO_3$. *Physical Review B: Condensed Matter* 81 (5): 1–5.

73 Bhatnagar, A., Kim, Y.H., Hesse, D., and Alexe, M. (2014). Sub-band level-assisted photoconduction in epitaxial $BiFeO_3$ films. *Applied Physics Letters* 105 (12): 12–16.
74 Tebano, A., Fabbri, E., Pergolesi, D. et al. (2012). Room-temperature giant persistent photoconductivity in $SrTiO_3/LaAlO_3$ heterostructures. *ACS Nano* 6 (2): 1278–1283.
75 Tarun, M.C., Selim, F.A., and McCluskey, M.D. (2013). Persistent photoconductivity in strontium titanate. *Physical Review Letters* 111 (18): 187403.
76 Bhatnagar, A., Kim, Y.H., Hesse, D., and Alexe, M. (2014). Persistent photoconductivity in strained epitaxial $BiFeO_3$ thin films. *Nano Letters* 14 (9): 5224–5228.
77 Yang, M., Bhatnagar, A., and Alexe, M. (2015). Electronic origin and tailoring of photovoltaic effect in $BiFeO_3$ single crystals. *Advanced Electronic Materials* 1 (11): 1–5.
78 Tatsuzaki, I., Itoh, K., Ueda, S., and Shindo, Y. (1966). Strain along C-axis of SbSI caused by illumination in DC electric field. *Physical Review Letters* 17 (4): 198.
79 Fridkin, V.M., Grekov, A.A., Ionov, P.V. et al. (1974). Photoconductivity in certain ferroelectrics. *Ferroelectrics* 8: 433–435.
80 Brody, P.S. (1983). Optomechanical bi-morph actuator. *Ferroelectrics* 50 (1): 27–32.
81 Poosanaas, P. and Uchino, K. (1999). Photostrictive effect in lanthanum-modified lead zirconate titanate ceramics near the morphotropic phase boundary. *Materials Chemistry and Physics* 61 (1): 36–41.
82 Daranciang, D., Highland, M.J., Wen, H. et al. (2012). Ultrafast photovoltaic response in ferroelectric nanolayers. *Physical Review Letters* 108 (8): 1–6.
83 Kundys, B., Viret, M., Colson, D., and Kundys, D.O. (2010). Light-induced size changes in $BiFeO_3$ crystals. *Nature Materials* 9 (10): 803–805.
84 Smolenskii, G.A. and Bokov, V.A. (1964). Coexistence of magnetic and electric ordering in crystals. *Journal of Applied Physics* 35 (3): 915–918.
85 Wen, H., Chen, P., Cosgriff, M.P. et al. (2013). Electronic origin of ultrafast photoinduced strain in $BiFeO_3$. *Physical Review Letters* 110 (3): 1–5.
86 Kundys, B., Viret, M., Meny, C. et al. (2012). Wavelength dependence of photoinduced deformation in $BiFeO_3$. *Physical Review B: Condensed Matter* 85 (9): 1–4.
87 Catalan, G., Lubk, A., Vlooswijk, A.H.G. et al. (2011). Flexoelectric rotation of polarization in ferroelectric thin films. *Nature Materials* 10 (12): 963–967.
88 Perez-Tomas, A., Lira-Cantu, M., and Catalan, G. (2016). Above-bandgap photovoltages in antiferroelectrics. *Advanced Materials* 28 (43): 9644–9647.

4

Organic–Inorganic Hybrid Perovskites for Solar Energy Conversion

Peng You and Feng Yan

The Hong Kong Polytechnic University, Department of Applied Physics, Hung Hom, Kowloon, Hong Kong

4.1 Introduction

Solar cells based on organic–inorganic hybrid perovskite materials have attracted enormous attention during the past few years. PSC was selected as one of the major scientific breakthroughs in 2013. Since the first report of the material used in solar cells in 2009 [1], power conversion efficiencies (PCEs) of perovskite solar cell (PSC) have now reached a certified value of 22.1% [2], with an unprecedented fast growth rate. PSCs have attained comparable photovoltaic performance to cadmium telluride (CdTe) and copper indium gallium diselenide (CIGS) solar cells (22.1% for CdTe, and 22.6% for CIGS), and are approaching that of monocrystalline silicon solar cells (25.3% for single-junction single crystal Si cells without concentrators) [2]. The semiconducting organic–inorganic hybrid perovskites have shown great potential for photovoltaic applications, which are likely be attributed to their excellent optoelectronic properties, ease of processing, and low fabrication costs.

The hybrid perovskites possess several outstanding optoelectronic properties, which make them ideal materials for photovoltaic applications. The bandgap of the most often used methylammonium lead iodide perovskite ($CH_3NH_3PbI_3$, abbreviated as $MAPbI_3$) is nearly ideal for light absorption and it can be continuously tuned from 1.17 to 2.3 eV through compositional engineering. The optical absorption coefficient of $MAPbI_3$ is higher than that of many other photoactive materials such as Si, CdTe, and amorphous silicon; thus the film thickness of the active layer can be reduced to only several hundred nanometers. In contrast to organic photovoltaic (OPV) materials, the low exciton binding energy of the hybrid perovskites allows the spontaneous dissociation of excitons into free carriers upon light absorption. Moreover, the high electron and hole mobilities as well as the long carrier lifetimes result in long diffusion lengths; thus charge carriers can be freely transported across the whole perovskite film before recombination.

In addition to the outstanding optoelectronic properties, the processing of PSCs has shown excellent flexibility. High-quality perovskite films can be easily fabricated through either solution or vapor-phase deposition, which will be

Ferroelectric Materials for Energy Applications, First Edition.
Edited by Haitao Huang and James F. Scott.

discussed in the following sections. Many of these deposition methods are quite compatible with industrial large-scale production techniques, which strengthens the potential for the commercialization of PSCs. A large number of research groups all over the world, especially those with experience in dye-sensitized solar cells (DSSCs) and OPVs, have been attracted to work on this hot research topic, because of its ease of processing. As a result, this research area has experienced tremendous development during the past few years.

Besides, all the components of the hybrid perovskites are quite cheap materials including some organic compounds (e.g. methylammonium (MA) and formamidinium (FA)) and inorganic elements such as lead and iodine. Thus, the cost of raw materials of PSCs is dramatically decreased compared with other photovoltaic technologies. Also, the processing cost of PSCs is quite low because of the facile fabrication conditions.

Despite demonstrations of high efficiencies, PSCs have also been suffering from some severe problems such as long-term stability, current-voltage ($I-V$) hysteresis, and toxicity of raw materials (lead), which hindered their large-scale commercial applications. How to address these issues has been identified as the main challenge for the future research work. In this chapter, we first summarize briefly the fundamental properties of the hybrid perovskite materials. The state-of-the-art deposition methods of the perovskite films are presented in the following part. We then review the development process of organic–inorganic hybrid PSCs. The working mechanisms and device architectures of PSCs are also described to give a better understanding of high-performance PSCs. Finally, some crucial challenges such as long-term stability, $I-V$ hysteresis, and toxicity of raw materials are discussed, as well as an outlook toward the further improvement of PSCs.

4.2 Fundamental Properties of Hybrid Perovskites

4.2.1 Crystal Structures

The organic–inorganic hybrid perovskites generally have a chemical formula of ABX_3 (where A is MA or FA, MA = $CH_3NH_3^+$, FA = $CH_3(NH_2)_2^+$; B is Pb^+ or Sn^+; X is Cl^-, Br^-, or I^-). In a standard cubic crystal structure, the body-centered cation B has 6 nearest X anions, while the cation A located at vertex positions possesses 12 nearest X anions, as shown in Figure 4.1a. In order to maintain a stable crystal structure, it is generally accepted that the tolerance factor t ($t = (r_A + r_X)/[2^{1/2}(r_B + r_X)]$, where r_A, r_B, r_X represent the effective ionic radii for the ions at A, B, and X sites, respectively) should be in the range between 0.813 and 1.107 [5]. Thus, for the compositional change of perovskites, the tolerance factor should be carefully considered according to the ion radius of the cations and anions. It is interesting that the hybrid perovskites undergo structural phase transitions upon temperature change because of the rotation of A-site cations at high temperatures [5, 6]. For example, $MAPbI_3$ perovskite with a stabilized orthorhombic structure at low temperatures will have an orthorhombic to tetragonal phase transition at around 162 K. With further increase of the temperature to 327 K, another phase transition from tetragonal

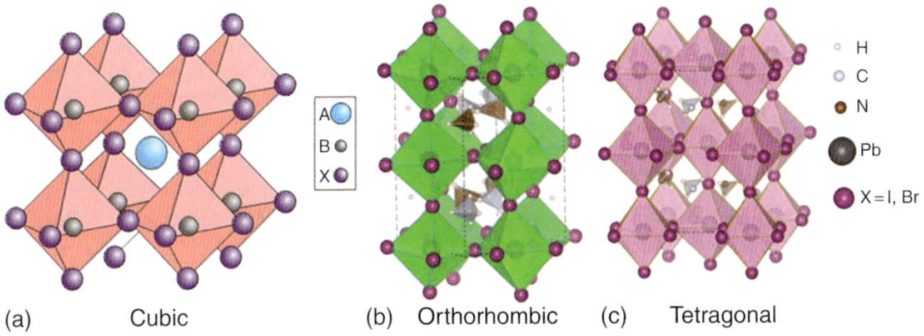

Figure 4.1 Schematic diagram of the crystal structures of organic–inorganic hybrid perovskites. *Source:* Reprinted with permission from Ref. [3, 4].

to cubic will occur. The schematic crystal structures of these three phases are shown in Figure 4.1. A similar phase transition occurs at lower temperatures for $MAPbBr_3$ and $FAPbI_3$, which gives them better thermal stability [5, 6].

4.2.2 Optical Properties

The hybrid perovskite materials have been confirmed to be exceptionally good at absorbing photons. Being direct bandgap semiconductors, the hybrid perovskites possess extremely high absorption coefficients, with reported values in the order of $10^5\ cm^{-1}$ [7]. Thus, a perovskite film ~400 nm thick is generally sufficient to absorb the most sunlight, which greatly reduces the material costs [8, 9]. In contrast to OPV materials, the exciton binding energy of organic–inorganic hybrid perovskites was demonstrated to be only 16 meV at low temperatures, and this falls to even smaller values of only a few millielectronvolts at room temperature, which explains their excellent device performance as being due to spontaneous free-charge separation upon light illumination [10].

The color of the hybrid halide perovskites is tunable through compositional engineering, covering the entire visible range and extending to the near infrared region. Seok and coworkers studied the chemically tuned organic–inorganic hybrid perovskite materials based on $MAPb(I_{1-x}Br_x)_3$, which shows a bandgap covering almost the entire visible spectrum, enabling the realization of colorful solar cells [11]. Their work represents one step toward the realization of colorful solar cells with low-cost, high-efficiency, and long-term stability. The color-tuned PSCs have been proved to be a promising candidate for building integrated photovoltaics (BIPVs).

The hybrid perovskites also show strong photoluminescence (PL). The PL spectra are different for various organometal halide perovskite materials. In most reports at room temperature, the $MAPbI_3$ perovskite has a PL emission peak at around 775 nm, while $FAPbI_3$ has a peak at around 800 nm [12, 13]. Compared with lead-based perovskites, tin-based perovskites, such as $MASnI_3$, which has a smaller bandgap, shows a broader PL peak at ~980 nm [14]. The PL peaks of $(MA)_x(FA)_{1-x}PbI_3$ ($x = 0$–1) films can be continuously tuned by changing the molar ratio between MA and FA cations, ranging from 776 to 803 nm [13]. The gradual shift in PL emission indicates the formation of a solid solution of MA and FA in the perovskite lattice [13]. Similarly, a gradual shift in PL peaks is also observed in $CH_3NH_3Pb(I_{1-x}Br_x)_3$ films ($x = 0$–1) by tuning the iodide–bromide ratio [15].

4.2.3 Charge Transport Properties

One of the most important features of all the semiconducting materials is the charge transport property. The outstanding performance of organometal halide perovskites is partly attributed to their excellent charge transport properties. The small photocarrier effective masses, high carrier mobilities, long carrier lifetimes, long diffusion lengths, and low trap densities ensure that charge carriers can be freely transported across the entire perovskite film (normally several hundred nanometers) before recombination [16–22].

Originally in 2013, the electron and hole diffusion lengths of solution-processed $MAPbI_3$ perovskite films were calculated to be around 100 nm (the carrier lifetime is only several nanoseconds) based on the results of transient optical spectroscopy and time-resolved PL-quenching measurements [19]. This value of carrier diffusion length at that time was actually already much higher than that of traditional low-temperature solution-processed light absorbers (typically ~10 nm) [19]. It was also found that the diffusion lengths of $MAPbI_{3-x}Cl_x$ mixed halide perovskites are larger than 1 μm, which is an order of magnitude greater than the triiodide perovskite absorbers [18]. Owing to the much longer carrier diffusion lengths of mixed halide perovskites than the light absorption depth (100–200 nm), the mesoporous TiO_2 structure is no longer necessary so that planar structure is adoptable [18]. The charge-carrier mobilities of solution-processed organometal trihalide perovskite films were first estimated to be around 10–20 $cm^2\ V^{-1}\ s^{-1}$, which is rather high for solution-processed materials [21, 22].

The successful growth of large organometal trihalide perovskite single crystals in 2015 greatly promoted our understanding about the bulk electrical properties of this material [17, 20, 23]. Huang's group reported that the diffusion length in the solution-grown $MAPbI_3$ single crystals can exceed 175 μm under 1 sun illumination, and is larger than 3 μm under weak light condition, which has far exceeded the diffusion length obtained in organometal perovskite thin films [20]. In addition, the carrier mobility they measured in $MAPbI_3$ single crystals was as high as 164 $cm^2\ V^{-1}\ s^{-1}$, and the trap density was calculated to be as low as $3.6 \times 10^{10}\ cm^{-3}$ [20]. Cooke and coworkers even reported a record high mobility of 800 $cm^2\ V^{-1}\ s^{-1}$ on a single crystal lead halide perovskite through optical pump-multi-terahertz measurements [23].

4.2.4 Compositional Engineering and Bandgap Tuning

The dominating perovskites ($MAPbI_3$) possess a bandgap around 1.5 eV, which is extremely close to the "ideal" single-junction bandgap (1.1–1.4 eV) [11, 24, 25]. Depending on the halide used, the bandgap can be continuously tuned from ~1.5 eV (pure I) to ~3.2 eV (pure Cl) [11, 25, 26]. Even smaller bandgaps can be achieved by using different organic cations or inorganic cations, covering the entire visible spectral range and expanding to the near infrared region [8, 26].

The bandgap of $MAPbI_3$ perovskite was estimated to be ~1.5 eV from ultraviolet photoelectron spectroscopy (UPS), UV–visible, and incident photon conversion efficiency (IPCE) spectra. However, the ~1.5 eV bandgap is not enough to cover the whole visible and the near-infrared region, which is necessary to get ultrahigh PCE. Therefore, bandgap tuning is generally considered to be a good way to extend the absorption to longer wavelengths without sacrificing the absorption coefficient.

Replacing MA with other organic cations (e.g. FA, 5-AVA) or inorganic cations (e.g. Cs, Rb) is one approach to realize the bandgap tuning, because changing the organic cations at A sites of AMX_3 perovskites can tune the bandgap by modifying the metal–halide–metal bond length and angle without affecting the valence band maximum [27]. Partial replacement of MA with FA is an effective way to extend the absorption to longer wavelengths and enhance the thermal stability [25]. With the combination of FA, MA, and cesium (Cs), the resulting triple-cation perovskites are less sensitive to processing conditions, thermally more stable, and show higher PCEs with better reproducibility [28, 29]. Besides, the bandgap of the hybrid perovskites can also be continuously tuned from 1.5 to 2.3 eV by introducing mixed halides (I and Br) [11]. The introduction of bromide element also enhances the stability of the perovskites in ambient conditions.

In addition to altering the organic cations and the halide anions, another approach for tuning the bandgap is the direct change or substitution of the metal cations. For instance, the bandgap of $MASn_xPb_{1-x}I_3$ perovskites can be tuned from 1.3 to 1.55 eV by varying the molar ratio of lead and tin [30]. Owing to concerns about the toxicity of lead, lead-free perovskites such as $MASnX_3$ have attracted increasing attention recently. Compared with the lead-based perovskite, the relatively lower band gap of $MASnI_3$ (~1.3 eV) allows light absorption over a much broader range with an increased current density. However, both the efficiency and the stability of the Sn-based devices are still not comparable to that of the lead-based cells at this moment.

4.3 Synthesis of Hybrid Perovskite Crystals

Much effort has been devoted by researchers to obtain high-quality organometal halide perovskite films for solar cell applications. However, the synthesis of high-quality large perovskite single crystals as well as a variety of nanostructures are still of great importance to the fundamental research about the physical properties of these materials and a better understanding of the crystal growth process.

4.3.1 Bulk Crystal Growth

The growth of organometal halide perovskite bulk crystals is generally similar to the traditional solution-based crystal-growth methods. In most cases, the oversaturated perovskite precursor solution with organic halide salts and inorganic lead halide salts is slowly cooled to evaporate the solvents and form the hybrid perovskite single crystals. So far, many researchers have reported the successful synthesis of large hybrid perovskite single crystals up to a centimeter scale [17, 20, 31–33].

Bakr and coworkers reported an antisolvent vapor-assisted crystallization (AVC) process to grow high-quality, millimeter-sized $MAPbBr_3$ and $MAPbI_3$

single crystals with exceptionally low trap densities (10^9–10^{10} cm^{-3}) and charge-carrier diffusion lengths (>10 μm) [17]. In this method, the antisolvent dichloromethane (DCM) is slowly diffused into the perovskite precursor solution to promote the growth of sizable perovskite crystals of high quality. This concept of antisolvent method has also been widely used to fabricate high-quality perovskite thin films through a fast-crystallization process. Bakr and coworkers also developed an inverse temperature crystallization (ITC) method to grow $MAPbBr_3$ and $MAPbI_3$ crystals at a rate much faster than the previously reported methods [31]. This method is based on the phenomenon that the solubility of $MAPbBr_3$ or $MAPbI_3$ in dimethyl formamide (DMF) drops dramatically with the increase of solution temperature from room temperature to 80 °C. This inverse solubility phenomenon was utilized to grow $MAPbX_3$ crystals rapidly from hot precursor solutions by carefully controlling the solution temperature and concentrations.

By using a top-seeded solution-growth (TSSG) method, Huang and coworkers grew large-sized $MAPbI_3$ single crystals (the largest size is around 10 mm) with electron–hole diffusion lengths larger than 175 μm under 1 sun illumination [20]. In this process, small $MAPbI_3$ crystals at the bottom of the solution in a container were used to help maintain a saturated solution while the cooler top half of the solution was supersaturated. Thus, a crystal seed put in the top region of the solution can grow into larger size by consumption of small crystals in the bottom. Recently, Liu et al. also developed a facile method to prepare inch-sized $MAPbX_3$ (X = Cl, Br, I) perovskite crystals (the largest one is 71 mm × 54 mm × 39 mm) based on a seed-induced heterogeneous nucleation mechanism [32].

4.3.2 Nanocrystal Synthesis

In the early work on synthesis of hybrid perovskite nanocrystals, mesoporous TiO_2 or Al_2O_3 films were normally used as the template to control the growth of the perovskite nanocrystals [34, 35]. In these reports, the regular perovskite precursor solution for solar cell fabrication was used and no surfactant ligands were adopted; thus the crystal size was actually controlled by the pore size of these mesoporous scaffold.

Later, a series of hybrid perovskite nanostructures with different sizes and morphologies were synthesized through various solution processes without using templates. Jin and coworkers successfully fabricated single crystalline nanowires, nanorods, nanobelts, and nanoplates of $MAPbI_3$ and $MAPbBr_3$ perovskites through a dissolution–recrystallization conversion from lead halide films coated on substrates to perovskites [36]. These perovskite nanocrystals show stronger room-temperature photoluminescence and longer carrier lifetimes than the bulk crystals and thin films. Galian and coworkers reported the non-template synthesis of monodispersed 6-nm-sized hybrid $MAPbBr_3$ perovskite nanoparticles for the first time [37]. The medium-sized chain alkyl ammonium bromide was used to react with $PbBr_2$ in the presence of oleic acid and octadecene. These nanoparticles can be dispersed and kept stable in a variety of solvents for more than three months. By using similar methods mentioned above, all inorganic lead halide nanocrystals (e.g. $CsPbX_3$) can also be obtained from inexpensive commercial raw materials [38]. Through compositional modulations, the emission spectra are widely tunable from 410 to 700 nm, which makes them appealing for many optoelectronic applications.

In addition to the three-dimensional organic–inorganic hybrid perovskites, the two-dimensional counterparts are also fairly appealing for their special novel features and wide applications. Two-dimensional $MAPbX_3$ (X = Cl, Br, or I) perovskite nanosheets with thickness of single unit cell were synthesized by Bao and coworkers using a hybrid method that combines the solution and vapor-phase processes [39]. The high-quality two-dimensional nanocrystals exhibit high photoluminescence with broad wavelength tunability. Yang's group reported the direct growth of atomically thin single-crystalline two-dimensional hybrid $(C_4H_9NH_3)_2PbBr_4$ perovskites from solution-phase growth [40]. Much larger organic cations are needed for the fabrication of this kind of layered two-dimensional perovskites.

4.4 Deposition Methods of Perovskite Films

In order to prepare high-efficiency PSCs, the fabrication of high-quality perovskite films is the most important process. So far, a variety of deposition methods have been developed by researchers. The most widely adopted deposition methods for perovskite films can be categorized into three groups: one-step solution process, two-step solution, process and vapor-phase deposition.

4.4.1 One-Step Solution Process

For standard one-step solution method (as shown in Figure 4.2a), the perovskite raw materials including lead sources (PbI_2, $PbCl_2$, etc.) and organic sources (methylammonium iodide (MAI), formamidinium iodide (FAI), etc.) are first dissolved in appropriate solvents such as DMF, gamma-butyrolactone (GBL), or dimethyl sulfoxide (DMSO), and the solution is then spin-coated on the substrate. A subsequent thermal annealing process (mostly around 100 °C) for a proper period of time is needed in most cases for the complete crystallization of the perovskite film.

One-step solution process was first used for the fabrication of $MAPbI_{3-x}Cl_x$ perovskite films from a mixed solution of $PbCl_2$ and MAI (with a 1 : 3 molar ratio) in DMF [43]. The annealing process of the $MAPbI_{3-x}Cl_x$ films is extremely sensitive to temperature, and thus the film coverage and morphology are difficult to control. In addition, the annealing atmosphere is just as important since the crystallization process is susceptible to the solvent atmosphere in the glovebox or moisture in air. Yang and coworkers have reported that the thermal annealing process for $MAPbI_{3-x}Cl_x$ film under proper humidity in air can greatly improve the film morphology and carrier lifetime [44, 45].

In the case of $MAPbI_3$-based perovskites, it is difficult to obtain a high-quality and uniform perovskite film through a simple one-step spin-coating process owing to the film shrinkage phenomenon, which is related to the simultaneous

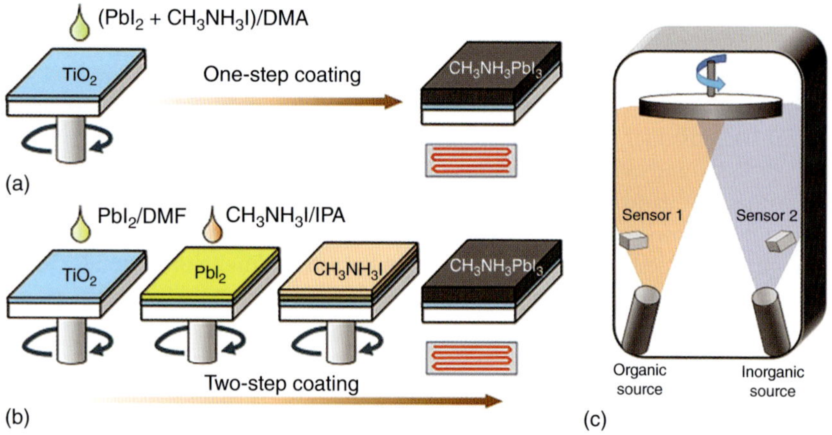

Figure 4.2 Schematic illustration of the one-step (a) and two-step (b) spin-coating procedures for the fabrication of perovskite films. *Source:* Reprinted with permission from Ref. [41]. (c) Dual-source thermal evaporation system for depositing the perovskite absorbers. *Source:* Liu et al. 2013 [42]. Reprinted with permission of Springer Nature.

solvent evaporation and perovskite crystallization. Generally, two strategies are used to address this problem, i.e. slow crystallization and fast crystallization. Interestingly, it was demonstrated that some additives (such as hydroiodic acid (HI) [24, 46] and MACl [47]) can significantly improve the morphology and surface coverage of perovskite films by retarding the crystallization process. However, the slow-crystallization annealing process often takes a relatively long time, which will prolong the preparation period.

Recently, an antisolvent rapid crystallization process has been widely adopted by researchers. Cheng and coworkers reported that the dripping of chlorobenzene during the one-step spin-coating process could induce fast crystallization, and thus very flat and uniform perovskite films with large grain size could be obtained [48]. The role of chlorobenzene is to rapidly reduce the solubility of the perovskites in the mixed solvents, thereby promoting the rapid precipitation and growth of perovskite grains. Soek's group also developed a novel solvent-engineering technique for high-performance PSCs [25, 49]. The adoption of a mixed solvent of GBL and DMSO followed by toluene dripping could lead to quite uniform and dense perovskite layers through the formation of a PbI_2-MAI-DMSO intermediate phase. The solar cells fabricated with this method exhibited a PCE of 16.2%, which was further improved to 17.9% by delicate compositional engineering. To date, this one-step antisolvent fast crystallization process has already been widely adopted for fabricating PSCs with PCEs higher than 20% [28, 29, 50].

4.4.2 Two-Step Solution Process

For the two-step solution process in very early stages [51], PbI_2 solution in DMF is spin-coated on the substrate to form a PbI_2 layer and the film is kept at a low temperature (e.g. 70 °C) for a certain time. After drying, the film is immersed into a solution of MAI in 2-propanol (e.g. 10 mg ml^{-1}) for the appropriate time and is then rinsed in pure 2-propanol to wash away the residual MAI. Normally, a further thermal annealing process is also required for crystallization. Gratzel and coworkers first adopted this two-step solution process and successfully achieved a certified efficiency of 14.1% [51].

Later, a thermal annealing-induced interdiffusion method was developed by Huang's group to fabricate pin-hole free perovskite films [52]. As is illustrated in Figure 4.2b, the stacked PbI_2/MAI bilayer structure will convert to perovskites upon thermal annealing. Many groups later adopted this approach for the fabrication of PSCs with relatively high efficiencies [53–56].

4.4.3 Vapor-Phase Deposition

Thermal evaporation, which is often used to deposit metal electrodes, can also be adopted to fabricate perovskite films. Figure 4.2c shows a dual-source thermal evaporation system that was first used for the depositing of uniform flat perovskite absorbers [42]. $PbCl_2$ and MAI (molar ratio, 4 : 1) were evaporated simultaneously from two separate ceramic crucibles in an evaporation chamber at 10^{-5} mbar. The crystallization started immediately after the evaporation, judged according to the color change. Further thermal annealing was needed to get fully crystallized perovskite films. Compared with solution-processed perovskite films, the vapor-deposited ones were extremely uniform, and the surfaces were quite flat with no pin holes or voids. The solar cell based on the vapor-deposited perovskite film demonstrated an efficiency higher than 15%. The deposition rates of the two precursor materials are crucial for the fabrication of high-quality perovskite films. However, it is often difficult to precisely control the molar ratio of the two precursor materials. To solve this problem, a layer-by-layer growth approach was developed to fabricate uniform $MAPbI_{3-x}Cl_x$ films with the desired thickness for high performance and stable PSCs [57]. Briefly, a thin layer of $PbCl_2$ (~50 nm) was thermally evaporated on the substrate, and then dipped in the 2-propanol solution of MAI, so that it was transferred into the perovskite. This process was repeated several times to get a desirable final thickness. PSCs fabricated through this approach showed PCEs up to 15%. However, the thermal deposition process requires high vacuum and expensive equipment, which is energy consuming and is incompatible with large-scale mass production. What is more, the growth condition and mechanisms of vapor deposition are relatively complicated, which hinders the large-scale application of this technique.

4.5 Efficiency Roadmap of Perovskite Solar Cells

From the best research cell efficiencies chart provided by the National Renewable Energy Laboratory (NREL) [2], it is obvious that the efficiency of PSCs has been growing at an unprecedented rate, which is faster than that of all the other types of solar cells. Some of the landmark efficiencies during its development

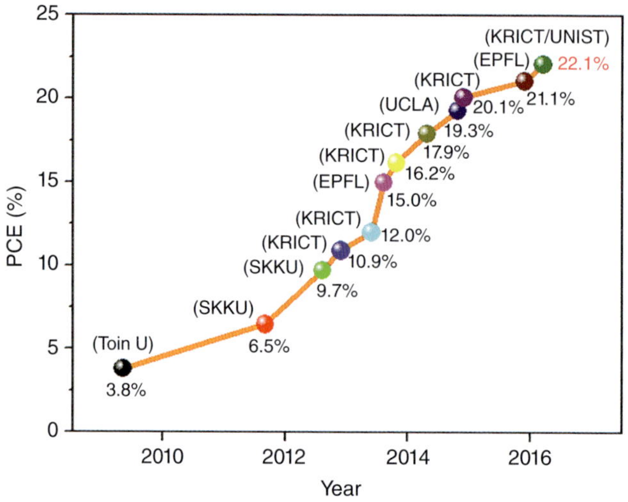

Figure 4.3 Efficiency roadmap for organic–inorganic hybrid PSCs [1, 2, 25, 28, 34, 43, 44, 49, 51, 53, 58, 59].

process are exhibited in Figure 4.3 and Table 4.1, based on published papers or certified efficiencies [1, 2, 25, 28, 34, 43, 44, 49, 51, 53, 58, 59]. In 2009, Miyasaka and coworkers reported the photovoltaic function of the organic–inorganic lead halide perovskite materials for the first time [1]. The perovskite nanocrystalline particles ($MAPbBr_3$ and $MAPbI_3$) self-organized on TiO_2 surface were adopted as visible-light sensitizers in the DSSCs with liquid electrolyte. A conversion efficiency of 3.8% on an $MAPbI_3$-based cell was obtained. The efficiency was actually much lower than that of the conventional DSSCs (~11%) at that time, and the perovskite compounds showed very poor stability in the liquid electrolytes. Two years later, Park and coworkers fabricated a perovskite quantum-dot-sensitized solar cell with a PCE of approximately 6.5% by using 2- to 3-nm-sized $MAPbI_3$ nanocrystals [34]. However, the lifetime of the perovskite quantum-dot-sensitized solar cell was only about 10 min due to the easy dissolution of the perovskite nanocrystals in the liquid electrolyte. Thus, little attention had been paid to the hybrid perovskite materials until the successful development of the first all-solid-state mesoscopic solar cells with impressive PCE of 9.7% by Park's group in 2012 [58]. The use of a solid hole transport material (spiro-MeOTAD) dramatically improved the efficiency as well as the device stability compared with the former electrolyte-based perovskite-sensitized cells. This improvement triggered a quick research explosion on PSCs. Since then, hybrid halide perovskites have attracted enormous attention worldwide. Soon after, in the same year, Snaith and coworkers pushed the efficiency of PSCs to 10.9% with dramatically improved V_{oc} (0.98 V) by coating a new mixed-halide perovskite ($MAPbI_2Cl$) on the surfaces of a mesostructured insulating Al_2O_3 scaffold, termed as "meso-superstructured solar cell" (MSSC) [43]. This iodide–chloride mixed-halide perovskite was found to be more stable

Table 4.1 Summary of the device information corresponding to efficiencies shown in Figure 4.3. (mp = mesoporous).

Device structure	J_{sc} (mA cm^{-2})	V_{oc} (V)	FF	PCE (%)	Year	References
mp-TiO_2/$MAPbI_3$/ liquid electrolyte	11.0	0.61	0.57	3.8	2009	[1]
mp-TiO_2/$MAPbI_3$-QDs/ liquid electrolyte	15.82	0.706	0.586	6.5	2011	[34]
mp-TiO_2/$MAPbI_3$/ spiro-MeOTAD	17.0	0.888	0.62	9.7	2012	[58]
mp-Al_2O_3/$MAPbI_2Cl$/ spiro-MeOTAD	17.8	0.98	0.63	10.9	2012	[43]
mp-TiO_2/$MAPbI_3$/PTAA	16.5	0.997	0.727	12.0	2013	[59]
mp-TiO_2/$MAPbI_3$/ spiro-MeOTAD	20.0	0.993	0.73	15.0	2013	[2, 51]
mp-TiO_2/$MAPb(I_{1-x}Br_x)_3$/ PTAA	19.64	1.11	0.742	16.2	2013	[2, 49]
mp-TiO_2/$(FAPbI_3)_{1-x}$ $(MAPbBr_3)_x$/PTAA	21.84	1.114	0.736	17.9	2014	[2, 25]
Y:TiO_2/$MAPbI_{3-x}Cl_x$/ spiro-MeOTAD	22.75	1.13	0.75	19.3	2014	[44]
mp-TiO_2/$(FAPbI_3)_{1-x}$ $(MAPbBr_3)_x$/PTAA	24.7	1.06	0.775	20.1	2015	[2, 53]
mp-TiO_2/$Cs_x(MA_{0.17}FA_{0.83})_{(1-x)}$ $Pb(I_{0.83}Br_{0.17})_3$/spiro-MeOTAD	23.5	1.147	0.785	21.1	2015	[2, 28]
Unknown				22.1	2016	[2]

in air and possessed better electron diffusion property than the previously reported pure iodide equivalent.

In 2013, Gratzel and Seok et al. introduced a layered sandwich-type device structure, comprising of a 3D-nanocomposites of mesoporous TiO_2 scaffold (infiltrated with perovskite material), a dense well-crystallized $MAPbI_3$ perovskite capping layer, and a polymeric hole transport layer. By comparison of several different kinds of hole transport materials, they found that poly(triarylamine) (PTAA) provided the maximum efficiency of 12.0% under standard air mass (AM) 1.5 light illumination, with a substantially increased fill factor (FF) (72.7%). Later in 2013, Gratzel and coworkers achieved a PCE of 15.0% (the certified efficiency was 14.1%) with improved perovskite morphology through a sequential deposition route (also called two-step method) [51], which was widely adopted by other researchers ever after. This technique for the fabrication of perovskite material increased the reproducibility compared to the previously employed one-step route, and provided opportunities for high-performance PSCs. A close PCE of 15.4% with an open-circuit voltage of 1.07 V was illustrated by Snaith and coworkers by using a sequential vapor-deposited $MAPbI_{3-x}Cl_x$ perovskite layer as the absorbing layer [42]. It was proved that the mesostructured TiO_2 scaffold was not necessary and a simple planar heterojunction thin-film structure can also achieve high photovoltaic performance, although this device structure tends to show more significant hysteresis phenomenon in $I-V$ characteristics.

Soon afterwards, Seok's research team reported a certified PCE of 16.2% without $I-V$ hysteresis by using $MAPb(I_{1-x}Br_x)_3$ as the absorbing layer and PTAA as a hole transport layer [2, 49]. A solvent-engineering technique was developed to obtain extremely dense and uniform perovskite films. Here, a mixed solvent of GBL and DMSO was used to prepare the perovskite precursor solution. The better morphology was actually attributed to the formation of the intermediate phase PbI_2-MAI(Br)-DMSO film when non-dissolving solvent (such as toluene) was drop-casted during the spin-coating process. For this kind of bilayer architecture, the thickness ratio of the continuous perovskite upper layer and the mesoporous TiO_2 layer (also infiltrated with perovskites) is the key factor for the brilliant performance. This was further increased to a confirmed PCE of 17.9%, also by Soek's group in early 2014 [2, 25]. A mixed composition of $(FAPbI_3)_{1-x}(MAPbBr_3)_x$ was used as light absorber. According to their report, the incorporation of $MAPbBr_3$ into $FAPbI_3$ could stabilize the perovskite phase and improve the efficiency. Besides, Yang and coworkers also reported a high efficiency of 19.3% in a planar device geometry by means of delicate interface engineering throughout the entire device [44].

In late 2014, Seok and colleagues reported an exciting PCE of 20.1% [2, 53], which was the first one that surpassed the 20% mark. The perovskite film was fabricated essentially through a two-step solution process. Unlike the conventional two-step method, the PbI_2 film was replaced by a PbI_2(DMSO) complex film, which will interact with FAI (MABr) in 2-propanol through an intramolecular exchange process (IEP). Negligible volume expansion occurred during the IEP process and high-quality perovskite film was obtained. High-efficiency PSCs with average efficiency of over 19% can be obtained through this technique.

In late 2015, Saliba et al. reported a higher efficiency of 21.1% by employing a triple-cation perovskite composition (FA/MA/Cs) [2, 28], which was thermally more stable, and more robust to the fabrication conditions. This provided a novel strategy for the further improvement of PSCs. Recently, in early 2016, a new record efficiency of 22.1% was developed by Korea Research Institute of Chemical Technology (KRICT) and Ulsan National Institute of Science and Technology (UNIST) [2]. Details about the solar cell are still unknown. Further improvement of the efficiency of PSCs is on the way.

4.6 Working Mechanism and Device Architectures of Perovskite Solar Cells

The operation of a solar cell requires four basic processes, i.e. light absorption, charge separation, charge transport, and charge collection. For light absorbers with different optoelectronic properties, diverse device structures are selected. For example, an n–i–p (or p–i–n) structure is required if the light-absorbing

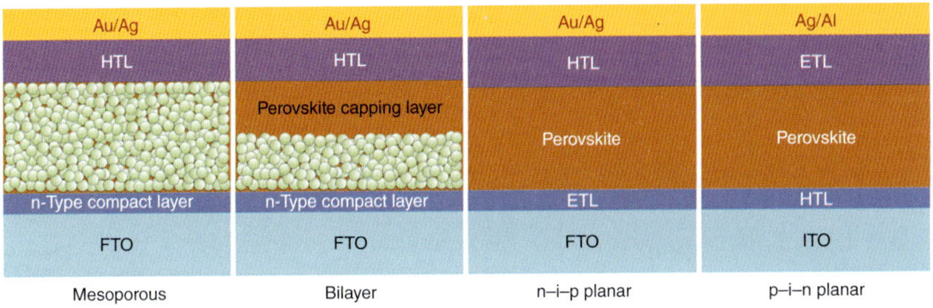

Figure 4.4 Schematic illustration of the mesoporous, bilayer, n–i–p planar and p–i–n planar device architectures of PSCs. HTL represents the hole transport layer; ETL represents the electron transport layer.

material is an intrinsic semiconductor, while only a p–n structure is needed when the light harvester itself has a p-type (or n-type) property. Thus, the organic–inorganic hybrid perovskite materials can be applied to either p–i–n or p–n structures owing to their long-range balanced ambipolar charge transport nature [19].

The original PSCs employed a mesoporous structure with a thick mesoporous metal-oxide layer (mp-TiO_2 or mp-Al_2O_3) as scaffold [1, 34, 43, 51, 58], which was evolved from the typical DSSCs. In this structure, as shown in Figure 4.4a, the mesoporous scaffold is filled with perovskites and sandwiched between a compact electron transport layer (mostly TiO_2) and a hole transport layer. The mesoporous structure is thought to enhance the charge collection efficiency by decreasing the carrier transport distance. In the very early stage, the thickness of the mesoporous layer was usually larger than 500 nm (or even several micro meters), in order to absorb sufficient light [1, 34, 43, 51, 58]. But because the crystal growth of perovskites is confined by the dimensions of the pores, a great amount of the material is present in disordered or amorphous phases [60], which will definitely deteriorate the device performance.

Interestingly, when the mesoporous layer thickness decreases to less than 300 nm, a dense and uniform perovskite capping layer with large grains will form on top of the porous layer [25, 28, 49, 53], which is called a bilayer structure (shown in Figure 4.4b). The complete pore filling together with the superior crystallinity of the continuous perovskite capping layer can assure high charge collection and transport efficiencies. As a result, PSCs based on the bilayer structure have demonstrated much higher efficiencies than those based on the traditional mesoporous structure [25, 28, 49, 53]. Actually, the state-of-the-art PSCs are usually based on this device configuration [28, 29, 50, 53, 61].

Further development of the device architectures brings out the n–i–p or p–i–n planar configurations (as shown in Figure 4.4c,d), which are similar to OPVs. In planar structures, the mesoporous layer is omitted and the perovskite layer is sandwiched between two opposite charge transport layers (hole transport layer and electron transport layer), which simplifies the preparation process and makes it more convenient for low-temperature fabrication. To date, the best n–i–p planar devices have efficiencies around 19% with different electron transport layers (TiO_2, SnO_2, etc.) [44, 62, 63]. Recently, the p–i–n planar devices have also demonstrated efficiencies higher than 18% due to the employment of advanced film preparation methods [46, 54, 55, 64, 65].

4.7 Key Challenges of Perovskite Solar Cells

Despite its excellent properties and performances, the hybrid PSC is currently still a fragile new star in the photovoltaic community. There are a variety of challenging issues to be addressed before its commercial applications.

4.7.1 Long-Term Stability

Compared with the enormous breakthrough in device efficiencies of PSCs, study on the long-term stability has been lagging far behind. However, the long-term stability is just as important as efficiency when it comes to practical application. High efficiencies are meaningless if the solar cells are unstable.

Some representative stabilities are listed in Table 4.2. In 2011, the 6.5% efficient perovskite quantum-dot-sensitized solar cell fabricated by Park's group [34] suffered an 80% degradation in just 10 min, because the liquid electrolyte tends to dissolve the perovskite quantum dot gradually. By 2012, the device stability was improved substantially from minutes to over 500 hours thanks to the employment of all-solid-state hole transport material (spiro-MeOTAD), which was a revolutionary step for long-term stability. Moreover, the PSC based on the two-step sequential process, developed by Gratzel and coworkers, could also maintain more than 80% of its initial PCE after a period of 500 hours [51]. In order to overcome the ultraviolet light instability of mesoporous TiO_2 film, Snaith and coworkers developed TiO_2-free cells with Al_2O_3 scaffold, which delivered stable photocurrent for over 1000 hours under continuous full spectrum sunlight [66]. Similarly, a polymeric hole transport material poly[2,5-bis(2-decyldodecyl)pyrrolo[3,4-c]pyrrole-1,4(2H,5H)-dione-(E)-1,2-di(2,20-bithiophen-5-yl) ethene] (PDPPDBTE) was used by Park and coworkers to replace spiro-MeOTAD [67]. The stability of the PDPPDBTE cells was improved remarkably, and the initial efficiency showed almost no degradation after long-term aging test for over 1000 hours. The superior stability of the PDPPDBTE cell may be attributed to its hydrophobic properties, which prevented water penetration into the perovskite layer.

In 2014, Han and coworkers reported a novel hole-conductor-free mesoporous device structure with considerably high stability [68]. They employed a mesoporous TiO_2/ZrO_2/carbon triple-layer structure, infiltrated with perovskite by drop-casting from perovskite solution, so that no hole transport layer is required. The cell achieved a certified PCE of 12.8% and was stable for over 1008 hours in air under AM 1.5 simulated sunlight.

Besides, some recent work on device stability indicated that the use of inorganic charge transport layers was quite effective to maintain long-term stability [69, 70]. For example, Han and coworkers adopted two heavily doped inorganic charge extraction layers in the planar structure of fluorine-doped tin

Table 4.2 The stability test of some representative PSCs.

Device structure	Initial PCE	Degradation percentage	Stability test condition	Test time (h)	Year	References
mp-TiO_2/$MAPbI_3$-QDs/liquid electrolyte/Pt	6.5%	~80%	Under continued irradiation	0.17	2011	[34]
mp-TiO_2/$MAPbI_3$/spiro-MeOTAD/Au[a)]	~7%	Improved for about 14%	In air, RT, 100 mW cm^{-2}	500	2012	[58]
mp-TiO_2/$MAPbI_3$/spiro-MeOTAD/Au[b)]	~9%	~20%	45 °C, at maximum power point, 100 mW cm^{-2}	500	2013	[51]
mp-Al_2O_3/$MAPbI_3$/spiro-MeOTAD/Au[b)]	~11%	~47%	40 °C, at open-circuit condition, 76.5 mW cm^{-2}	1000	2013	[66]
mp-TiO_2/$MAPbI_3$/PDPPDBTE/Au[a)]	~8.5%	Almost no degradation	In air, RT, $H \sim 20\%$	1000	2014	[67]
mp-TiO_2/ZrO_2/(5-AVA)$_x$ $MA_{1-x}PbI_3$/C[a)]	~9.5%	Almost no degradation	In air, RT, 100 mW cm^{-2}	1008	2014	[68]
NiMgLiO/$MAPbI_3$/PCBM/Ti(Nb)O_x/Ag[b)]	~14.6% (large area)	~10%	In air, RT, at short-circuit condition, 100 mW cm^{-2}	1000	2015	[69]
NiO_x/$MAPbI_3$/ZnO/Al[a)]	~14.6%	~10%	In air, RT, H =30–50%	1440	2015	[70]

RT, room temperature; *H*, humidity.
a) Without encapsulation.
b) Encapsulated.

oxide (FTO)/NiMgLiO/$MAPbI_3$/[6,6]-phenyl-C71-butyric acid methyl ester (PCBM)/Ti(Nb)O_x/Ag [69]. The robust nature of these inorganic layers allowed for the long-term stability (maintaining over 90% of the initial efficiency after long light soaking for over 1000 hours). Yang and coworkers fabricated PSCs with p-type NiO_x and n-type ZnO nanoparticles as hole and electron transport materials respectively [70]. About 90% of the original efficiency could be retained after 60 days of storage without encapsulation. The dense and robust ZnO layer was not only able to isolate the perovskite layer and Al electrode, but also may serve as an excellent diffusion barrier against moisture.

Some recent studies have demonstrated that adding a small quantity of inorganic cesium (Cs) into the perovskite recipes can produce efficient solar cells that are far more stable exposed in ambient air. Park and colleagues reported that by partial substitution of FA with Cs in $FAPbI_3$ perovskite (i.e. $FA_{0.9}Cs_{0.1}PbI_3$), the photo and moisture stability along with photovoltaic performance of the PSCs were found to be substantially improved due to the enhanced interaction between FA and iodide and the reduced trap density [71]. Graetzel and coworkers reported perovskite cells with a mixture of MA, FA, and cesium cations that had a stabilized PCE of over 21% (18% after 250 hours under operational conditions) [28]. It shows that the triple-cation perovskite compositions are less sensitive to processing conditions, contain less phase impurities, and are thermally much more stable.

Although some improvements have been achieved in the stability of PSCs, the lifetime of the current state-of-the-art PSC is still far from the requirements of practical applications. Not only the intrinsic stability of organic–inorganic hybrid perovskite materials but also the influence of charge transport layers and metal electrodes remains the main research focus for long-term device stability. Only when stable perovskite material systems are well combined with optimized device architectures, PSCs with high performance and good long-term stability can be achieved in the future.

4.7.2 *I–V* Hysteresis

The anomalous hysteresis issue in the current–voltage ($I-V$) curves of PSCs was first reported by Snaith et al. [72]. It refers to the large discrepancies among the $I-V$ curves in different bias sweeping directions or scan rates [73]. The presence of $I-V$ hysteresis hinders the determination of the accurate PCEs of PSCs and the commercial application. Many researchers have devoted themselves to the study of the $I-V$ hysteresis in PSCs. However, the origins of this phenomenon are still controversial. Now, some possible mechanisms have been proposed to explain the origins of $I-V$ hysteresis, such as defect-trapping effect, ferroelectric effect, and ion migration [72, 74–77].

The perovskite absorbers may have a large defect density within or near the surface of the material [72]. These defects could act as traps for electrons and holes in devices under working conditions. Huang and coworkers claimed that the trap states on the surface and grain boundaries of the perovskite films in the planar inverted solar cell are the origin of $I-V$ hysteresis [78]. They found that the fullerene layers deposited on top of perovskite films can effectively passivate these charge trap states and eliminate the hysteresis, which seems to be a simple and effective method to eliminate the hysteresis effect.

Some researchers have proposed the ferroelectric effect as one possible origin for the $I-V$ hysteresis in PSCs [72, 74, 77]. It is believed that the inverse ferroelectric polarization field would reduce the net built-in electric field and hinder the charge separation. Miyasaka and coworkers demonstrated that the reverse bias poling (from −0.3 to −1.1 V) of $MAPbI_3$ perovskite layers before the $I-V$ test will cause stronger hysteresis depending on the device structure [77]. This phenomenon is considered as the effect of ferroelectric remanent polarization in the perovskite film on the photocurrent, which is more obvious in planar structure devices. However, a recent report by Wang and coworkers concluded that $MAPbI_3$ perovskite is not ferroelectric at room temperature, according to polarization–electric field hysteresis measurements and piezoresponse force microscopy [79].

Recently, ion migration has been suggested to be a more important factor contributing to the $I-V$ hysteresis behavior. Huettner and coworkers studied the iodine migration and the effect on hysteresis in planar PSCs ($FTO/TiO_2/MAPbI_{3-x}Cl_x$/spiro-MeOTAD/Ag) [80]. They indicated that the iodide migration under applied bias voltage will shift down the work function of TiO_2 and shift up that of spiro-MeOTAD and lead to interfacial barriers at the interfaces, which is the origin of the hysteresis. Moehl and coworkers have investigated the origin of the hysteresis phenomenon with a combination of experimental and computational approaches [75]. The computationally determined activation energies for halide ion (or vacancy) migration match very well with the experimentally determined values involved in hysteresis, suggesting that ion migration is the cause of the observed hysteresis behavior in PSCs.

4.7.3 Toxicity of Raw Materials

Since most organic–inorganic hybrid PSCs are lead based, there are several concerns about the possible toxicity issues associated with the lead content in terms of commercialization and real applications. It is believed that the organic–inorganic lead halide perovskites are highly sensitive to moisture or water, and may be released into the environment after decomposition. However, the amount of Pb diffusion to the environment in the case of module failure was estimated to be far from catastrophic for the environment, because of the small amount of lead halide perovskite material in the ultrathin PSCs. Compared with other lead emission sources, such as mining and fossil fuels, the potential lead pollution from perovskite solar panels is insignificant.

Owing to the potential threat of the lead element in PSCs to the environment or human health, lead-free perovskites are considered a good solution for this problem. Another metal element, tin (Sn), which is more environment friendly, has been proposed as an alternative to Pb. Actually, the lead-free $MASnBr_{3-x}I_x$ perovskite was already synthesized by Weber et al. early in the 1970s. Unfortunately, the Sn-based PSCs (completely lead free) have shown much lower photovoltaic performance compared with the lead-based counterparts, with PCEs of only around 6% [14, 81, 82]. It is difficult to obtain full-covered and uniform Sn-based perovskite films due to the rapid crystallization of Sn-based perovskite at room temperature. What is more, these Sn-based perovskites are unstable in air because the Sn^{2+} in the films are prone to be oxidized into Sn^{4+} by oxygen in air, which is detrimental to device performance. More research is needed in order to find lead-free metal halide perovskites with higher photovoltaic performance and better stability.

4.8 Summary and Perspectives

In summary, the organic–inorganic metal halide PSCs have experienced a rapid development in the past few years and have revolutionized the prospects of new-generation photovoltaics. They have redefined solar cell research due to their high PCEs, easy processing, and low material and fabrication cost. The organic–inorganic hybrid perovskites have shown great potential for a variety of optoelectronic applications and not just for photovoltaic devices. During the past few years, many approaches have been successfully developed to produce high-quality perovskites and devices with great performances. Besides the continued advances in perovskite device performance, further development of perovskites will be promoted by more fundamental investigations including theoretical calculations and various experimental characterizations. It is of great importance to have a deeper understanding about the fundamental properties related to the hybrid perovskite materials.

Despite a great speed of development, the organic–inorganic hybrid PSCs are still in their early stages of laboratory research. In order to realize industrial production in the future, PSCs must have a substantially higher solar module performance with much lower manufacturing costs than the other existing solar technologies. Thus, there is still much room for improvement. Also, some challenging issues, such as large-area module, reproducibility, long-term stability, $I-V$ hysteresis, and toxicity of lead, should be carefully addressed through more research before their real application.

References

1 Kojima, A., Teshima, K., Shirai, Y., and Miyasaka, T. (2009). Organometal halid perovskites as visible-light sensitizers for photovoltaic cells. *Journal of American Chemical Society* 131: 6050–6051.

2 NREL Chart (2016). https://www.nrel.gov/pv/assets/images/efficiency-chart.png.

3 Feng, J. and Xiao, B. (2014). Crystal structures, optical properties, and effective mass tensors of $CH_3NH_3PbX_3$ (X = I and Br) phases predicted from HSE06 (vol 5, pg 1278, 2014). *Journal of Physical Chemistry Letters* 5 (10): 1719–1720.

4 Green, M.A., Ho-Baillie, A., and Snaith, H.J. (2014). The emergence of perovskite solar cells. *Nature Photonics* 8: 506–514.

5 Kim, H.S., Im, S.H., and Park, N.G. (2014). Organolead halide perovskite: new horizons in solar cell research. *Journal of Physical Chemistry C* 118 (11): 5615–5625.

6 Kawamura, Y., Mashiyama, H., and Hasebe, K. (2002). Structural study on cubic-tetragonal transition of $CH_3NH_3PbI_3$. *Journal of the Physical Society of Japan* 71 (7): 1694–1697.

7 Yakunin, S., Sytnyk, M., Kriegner, D. et al. (2015). Detection of X-ray photons by solution-processed lead halide perovskites. *Nature Photonics* 9 (7): 444–U44.

8 Yin, W.J., Shi, T.T., and Yan, Y.F. (2014). Unique properties of halide perovskites as possible origins of the superior solar cell performance. *Advanced Materials* 26 (27): 4653.

9 De Wolf, S., Holovsky, J., Moon, S.J. et al. (2014). Organometallic halide perovskites: sharp optical absorption edge and its relation to photovoltaic performance. *Journal of Physical Chemistry Letters* 5 (6): 1035–1039.
10 Miyata, A., Mitioglu, A., Plochocka, P. et al. (2015). Direct measurement of the exciton binding energy and effective masses for charge carriers in organic–inorganic tri-halide perovskites. *Nature Physics* 11 (7): 582–587.
11 Noh, J.H., Im, S.H., Heo, J.H. et al. (2013). Chemical management for colorful, efficient, and stable inorganic-organic hybrid nanostructured solar cells. *Nano Letters* 13 (4): 1764–1769.
12 Stoumpos, C.C., Malliakas, C.D., and Kanatzidis, M.G. (2013). Semiconducting tin and lead iodide perovskites with organic cations: phase transitions, high mobilities, and near-infrared photoluminescent properties. *Inorganic Chemistry* 52 (15): 9019–9038.
13 Pellet, N., Gao, P., Gregori, G. et al. (2014). Mixed-organic-cation perovskite photovoltaics for enhanced solar-light harvesting. *Angewandte Chemie-International Edition* 53 (12): 3151–3157.
14 Noel, N.K., Stranks, S.D., Abate, A. et al. (2014). Lead-free organic-inorganic tin halide perovskites for photovoltaic applications. *Energy & Environmental Science* 7 (9): 3061–3068.
15 Sadhanala, A., Deschler, F., Thomas, T.H. et al. (2014). Preparation of single-phase films of $CH_3NH_3Pb(I_{1-x},Br_x)(3)$ with sharp optical band edges. *Journal of Physical Chemistry Letters* 5 (15): 2501–2505.
16 Giorgi, G., Fujisawa, J.I., Segawa, H. et al. (2013). Small photocarrier effective masses featuring ambipolar transport in methylammonium lead iodide perovskite: a density functional analysis. *Journal of Physical Chemistry Letters* 4 (24): 4213–4216.
17 Shi, D., Adinolfi, V., Comin, R. et al. (2015). Low trap-state density and long carrier diffusion in organolead trihalide perovskite single crystals. *Science* 347 (6221): 519–522.
18 Stranks, S.D., Eperon, G.E., Grancini, G. et al. (2013). Electron-hole diffusion lengths exceeding 1 micrometer in an organometal trihalide perovskite absorber. *Science* 342: 341–344.
19 Xing, G.C., Mathews, N., Sun, S.Y. et al. (2013). Long-range balanced electron- and hole-transport lengths in organic-inorganic $CH_3NH_3PbI_3$. *Science* 342 (6156): 344–347.
20 Dong, Q.F., Fang, Y.J., Shao, Y.C. et al. (2015). Electron-hole diffusion lengths $> 175\,\mu m$ in solution-grown $CH_3NH_3PbI_3$ single crystals. *Science* 347 (6225): 967–970.
21 Leijtens, T., Stranks, S.D., Eperon, G.E. et al. (2014). Electronic properties of meso-superstructured and planar organometal halide perovskite films: charge trapping, photodoping, and carrier mobility. *ACS Nano* 8 (7): 7147–7155.
22 Wehrenfennig, C., Eperon, G.E., Johnston, M.B. et al. (2014). High charge carrier mobilities and lifetimes in organolead trihalide perovskites. *Advanced Materials* 26 (10): 1584–1589.
23 Valverde-Chavez, D.A., Ponseca, C.S., Stoumpos, C.C. et al. (2015). Intrinsic femtosecond charge generation dynamics in single crystal $CH_3NH_3PbI_3$. *Energy & Environmental Science* 8 (12): 3700–3707.

24 Eperon, G.E., Stranks, S.D., Menelaou, C. et al. (2014). Formamidinium lead trihalide: a broadly tunable perovskite for efficient planar heterojunction solar cells. *Energy & Environmental Science* 7 (3): 982–988.
25 Jeon, N.J., Noh, J.H., Yang, W.S. et al. (2015). Compositional engineering of perovskite materials for high-performance solar cells. *Nature* 517 (7535): 476–480.
26 Amat, A., Mosconi, E., Ronca, E. et al. (2014). Cation-induced band-gap tuning in organohalide perovskites: interplay of spin-orbit coupling and octahedra tilting. *Nano Letters* 14 (6): 3608–3616.
27 Geng, W., Zhang, L., Zhang, Y.N. et al. (2014). First-principles study of lead iodide perovskite tetragonal and orthorhombic phases for photovoltaics. *Journal of Physical Chemistry C* 118 (34): 19565–19571.
28 Saliba, M., Matsui, T., Seo, J.-Y. et al. (2016). Cesium-containing triple cation perovskite solar cells: improved stability, reproducibility and high efficiency. *Energy & Environmental Science* 9: 1989–1997.
29 Jacobsson, T.J., Correa-Baen, J.P., Pazoki, M. et al. (2016). Exploration of the compositional space for mixed lead halogen perovskites for high efficiency solar cells. *Energy & Environmental Science* 9 (5): 1706–1724.
30 Hao, F., Stoumpos, C.C., Chang, R.P.H., and Kanatzidis, M.G. (2014). Anomalous band gap behavior in mixed Sn and Pb perovskites enables broadening of absorption spectrum in solar cells. *Journal of the American Chemical Society* 136 (22): 8094–8099.
31 Saidaminov, M.I., Abdelhady, A.L., Murali, B. et al. (2015). High-quality bulk hybrid perovskite single crystals within minutes by inverse temperature crystallization. *Nature Communications* 6.
32 Liu, Y.C., Yang, Z., Cui, D. et al. (2015). Two-inch-sized perovskite $CH_3NH_3PbX_3$ (X = Cl, Br, I) crystals: growth and characterization. *Advanced Materials* 27 (35): 5176–5183.
33 Dang, Y.Y., Zhou, Y.A., Liu, X.L. et al. (2016). Formation of hybrid perovskite tin iodide single crystals by top-seeded solution growth. *Angewandte Chemie-International Edition* 55 (10): 3447–3450.
34 Im, J.H., Lee, C.R., Lee, J.W. et al. (2011). 6.5% efficient perovskite quantum-dot-sensitized solar cell. *Nanoscale* 3 (10): 4088–4093.
35 Kojima, A., Ikegami, M., Teshima, K., and Miyasaka, T. (2012). Highly luminescent lead bromide perovskite nanoparticles synthesized with porous alumina media. *Chemistry Letters* 41 (4): 397–399.
36 Fu, Y.P., Meng, F., Rowley, M.B. et al. (2015). Solution growth of single crystal methylammonium lead halide perovskite nanostructures for optoelectronic and photovoltaic applications. *Journal of the American Chemical Society* 137 (17): 5810–5818.
37 Schmidt, L.C., Pertegas, A., Gonzalez-Carrero, S. et al. (2014). Nontemplate synthesis of $CH_3NH_3PbBr_3$ perovskite nanoparticles. *Journal of the American Chemical Society* 136 (3): 850–853.
38 Protesescu, L., Yakunin, S., Bodnarchuk, M.I. et al. (2015). Nanocrystals of cesium lead halide perovskites ($CsPbX_3$, X = Cl, Br, and I): novel optoelectronic materials showing bright emission with wide color gamut. *Nano Letters* 15 (6): 3692–3696.

39 Liu, J.Y., Xue, Y.Z., Wang, Z.Y. et al. (2016). Two-dimensional CH(3)NH(3)PbI(3) perovskite: synthesis and optoelectronic application. *ACS Nano* 10 (3): 3536–3542.

40 Dou, L.T., Wong, A.B., Yu, Y. et al. (2015). Atomically thin two-dimensional organic-inorganic hybrid perovskites. *Science* 349 (6255): 1518–1521.

41 Im, J.H., Kim, H.S., and Park, N.G. (2014). Morphology-photovoltaic property correlation in perovskite solar cells: one-step versus two-step deposition of $CH_3NH_3PbI_3$. *Applied Materials* 2 (8): 081510.

42 Liu, M., Johnston, M.B., and Snaith, H.J. (2013). Efficient planar heterojunction perovskite solar cells by vapour deposition. *Nature* 501 (7467): 395–398.

43 Lee, M.M., Teuscher, J., Miyasaka, T. et al. (2012). Efficient hybrid solar cells based on meso-superstructured organometal halide perovskites. *Science* 338: 643–647.

44 Zhou, H., Chen, Q., Li, G. et al. (2014). Interface engineering of highly efficient perovskite solar cells. *Science* 345 (6196): 542–546.

45 You, J., Yang, Y.M., Hong, Z. et al. (2014). Moisture assisted perovskite film growth for high performance solar cells. *Applied Physics Letters* 105 (18): 183902.

46 Heo, J.H., Han, H.J., Kim, D. et al. (2015). Hysteresis-less inverted $CH_3NH_3PbI_3$ planar perovskite hybrid solar cells with 18.1% power conversion efficiency. *Energy & Environmental Science* 8: 1602–1608.

47 Zhao, Y.X. and Zhu, K. (2014). Efficient planar perovskite solar cells based on 1.8 eV band gap $CH_3NH_3PbI_2Br$ nanosheets via thermal decomposition. *Journal of the American Chemical Society* 136 (35): 12241–12244.

48 Xiao, M.D., Huang, F.Z., Huang, W.C. et al. (2014). A fast deposition-crystallization procedure for highly efficient lead iodide perovskite thin-film solar cells. *Angewandte Chemie-International Edition* 53 (37): 9898–9903.

49 Jeon, N.J., Noh, J.H., Kim, Y.C. et al. (2014). Solvent engineering for high-performance inorganic-organic hybrid perovskite solar cells. *Nature Materials* 13 (9): 897–903.

50 Saliba, M., Orlandi, S., Matsui, T. et al. (2016). A molecularly engineered hole-transporting material for efficient perovskite solar cells. *Nature Energy* 1: 1–7.

51 Burschka, J., Pellet, N., Moon, S.J. et al. (2013). Sequential deposition as a route to high-performance perovskite-sensitized solar cells. *Nature* 499 (7458): 316–319.

52 Xiao, Z.G., Bi, C., Shao, Y.C. et al. (2014). Efficient, high yield perovskite photovoltaic devices grown by interdiffusion of solution-processed precursor stacking layers. *Energy & Environmental Science* 7 (8): 2619–2623.

53 Yang, W.S., Noh, J.H., Jeon, N.J. et al. (2015). High-performance photovoltaic perovskite layers fabricated through intramolecular exchange. *Science* 348: 1234–1237.

54 Bi, C., Wang, Q., Shao, Y.C. et al. (2015). Non-wetting surface-driven high-aspect-ratio crystalline grain growth for efficient hybrid perovskite solar cells. *Nature Communications* 6.

55 Wu, C.G., Chiang, C.H., Tseng, Z.L. et al. (2015). High efficiency stable inverted perovskite solar cells without current hysteresis. *Energy & Environmental Science* 8 (9): 2725–2733.

56 Dong, Q.F., Yuan, Y.B., Shao, Y.C. et al. (2015). Abnormal crystal growth in $CH_3NH_3PbI_{3-x}Cl_x$ using a multi-cycle solution coating process. *Energy & Environmental Science* 8 (8): 2464–2470.

57 Chen, Y.H., Chen, T., and Dai, L.M. (2015). Layer-by-layer growth of $CH_3NH_3PbI_{3-x}Cl_x$ for highly efficient planar heterojunction perovskite solar cells. *Advanced Materials* 27 (6): 1053–1059.

58 Kim, H.-S., Lee, C.-R., Im, J.-H. et al. (2012). Lead iodide perovskite sensitized all-solid-state submicron thin film mesoscopic solar cell with efficiency exceeding 9%. *Scientific Reports* 2: 591.

59 Heo, J.H., Im, S.H., Noh, J.H. et al. (2013). Efficient inorganic–organic hybrid heterojunction solar cells containing perovskite compound and polymeric hole conductors. *Nature Photonics* 7: 486–491.

60 Choi, J.J., Yang, X.H., Norman, Z.M. et al. (2014). Structure of methylammonium lead iodide within mesoporous titanium dioxide: active material in high-performance perovskite solar cells. *Nano Letters* 14 (1): 127–133.

61 Bi, D., Tress, W., Dar, M.I. et al. (2016). Efficient luminescent solar cells based on tailored mixed-cation perovskites. *Science Advances* 2: e1501170.

62 Ke, W.J., Xiao, C.X., Wang, C.L. et al. (2016). Employing lead thiocyanate additive to reduce the hysteresis and boost the fill factor of planar perovskite solar cells. *Advanced Materials* 28 (26): 5214.

63 Seo, J.-Y., Matsui, T., Luo, J. et al. (2016). Ionic liquid control crystal growth to enhance planar perovskite solar cells efficiency. *Advanced Energy Materials* 1600767.

64 Nie, W.Y., Tsai, H.H., Asadpour, R. et al. (2015). High-efficiency solution-processed perovskite solar cells with millimeter-scale grains. *Science* 347 (6221): 522–525.

65 Wang, Q., Dong, Q., Li, T. et al. (2016). Thin insulating tunneling contacts for efficient and water-resistant perovskite solar cells. *Advanced Materials* 28 (31): 6734–6739.

66 Leijtens, T., Eperon, G.E., Pathak, S. et al. (2013). Overcoming ultraviolet light instability of sensitized TiO_2 with meso-superstructured organometal tri-halide perovskite solar cells. *Nature Communications* 4: 2885.

67 Kwon, Y.S., Lim, J., Yun, H.-J. et al. (2014). A diketopyrrolopyrrole-containing hole transporting conjugated polymer for use in efficient stable organic–inorganic hybrid solar cells based on a perovskite. *Energy & Environmental Science* 7 (4): 1454–1460.

68 Mei, A., Li, X., Liu, L. et al. (2014). A hole-conductor-free, fully printable mesoscopic perovskite solar cell with high stability. *Science* 345: 295–298.

69 Chen, W., Wu, Y., Yue, Y. et al. (2015). Efficient and stable large-area perovskite solar cells with inorganic charge extraction layers. *Science* 350: 944–948.

70 You, J., Meng, L., Song, T.-B. et al. (2016). Improved air stability of perovskite solar cells via solution-processed metal oxide transport layers. *Nature Nanotechnology* 11: 75–81.

71 Lee, J.W., Kim, D.H., Kim, H.S. et al. (2015). Formamidinium and cesium hybridization for photo- and moisture-stable perovskite solar cell. *Advanced Energy Materials* 5 (20): 1501310.

72 Snaith, H.J., Abate, A., Ball, J.M. et al. (2014). Anomalous hysteresis in perovskite solar cells. *Journal of Physical Chemistry Letters* 5 (9): 1511–1515.

73 Tress, W., Marinova, N., Moehl, T. et al. (2015). Understanding the rate-dependent J-V hysteresis, slow time component, and aging in $CH_3NH_3PbI_3$ perovskite solar cells: the role of a compensated electric field. *Energy & Environmental Science* 8 (3): 995–1004.

74 Wei, J., Zhao, Y.C., Li, H. et al. (2014). Hysteresis analysis based on the ferroelectric effect in hybrid perovskite solar cells. *Journal of Physical Chemistry Letters* 5 (21): 3937–3945.

75 Meloni, S., Moehl, T., Tress, W. et al. (2016). Ionic polarization-induced current-voltage hysteresis in $CH_3NH_3PbX_3$ perovskite solar cells. *Nature Communications* 7: 10334.

76 van Reenen, S., Kemerink, M., and Snaith, H.J. (2015). Modeling anomalous hysteresis in perovskite solar cells. *Journal of Physical Chemistry Letters* 6 (19): 3808–3814.

77 Chen, H.W., Sakai, N., Ikegami, M., and Miyasaka, T. (2015). Emergence of hysteresis and transient ferroelectric response in organo-lead halide perovskite solar cells (vol 6, pg 164, 2015). *Journal of Physical Chemistry Letters* 6 (6): 935–935.

78 Shao, Y.H., Xiao, Z.G., Bi, C. et al. (2014). Origin and elimination of photocurrent hysteresis by fullerene passivation in $CH_3NH_3PbI_3$ planar heterojunction solar cells. *Nature Communications* 5.

79 Fan, Z., Xiao, J.X., Sun, K. et al. (2015). Ferroelectricity of CH(3)NH(3)PbI(3) perovskite. *Journal of Physical Chemistry Letters* 6 (7): 1155–1161.

80 Li, C., Tscheuschner, S., Paulus, F. et al. (2016). Iodine migration and its effect on hysteresis in perovskite solar cells. *Advanced Materials* 28 (12): 2446–2454.

81 Liao, W., Zhao, D., Yu, Y. et al. (2016). Lead-free inverted planar formamidinium tin triiodide perovskite solar cells achieving power conversion efficiencies up to 6.22. *Advanced Materials* 28 (42): 9333–9340.

82 Hao, F., Stoumpos, C.C., Cao, D.H. et al. (2014). Lead-free solid-state organic-inorganic halide perovskite solar cells. *Nature Photonics* 8 (6): 489–494.

5

Dielectric Ceramics and Films for Electrical Energy Storage

Xihong Hao

Inner Mongolia University of Science and Technology, No. 7 Inner Mongolia Arding Street, Baotou, Inner Mongolia 014010, PR China

5.1 Introduction

Dielectric capacitors, an important type of key passive devices, are widely used in different electronic and electrical power systems, of which the volume and quality is occupied mainly by the capacitors [1]. Although capacitors can take various functions, such as coupling, decoupling, filtering, high-frequency vibration suppression, resonance, bypass, neutralization, timing, integration, differential, compensating, bootstrap, frequency division, and loader, their basic operating principle is the electrical energy charge and discharge [2]. Compared with other electrical storage devices, such as fuel cells, batteries, and super-capacitors, dielectric capacitors possess higher discharge speed, which makes them more likely be used in pulsed systems. However, dielectric capacitors usually have low energy-storage density. Thus, inspired by the devices developed toward miniaturization, lightweight, and integration, dielectric capacitors with high energy-storage performance, mainly including high energy-storage density, low loss, and good temperature stability, are desired. Apart from the geometric structure, the energy-storage performance of a capacitor is mainly determined by the selected dielectric materials. As a result, explorations of dielectrics with high energy-storage performance have been widely carried out all over the world in recent decades, and many outstanding achievements have been reported. Generally speaking, the potential candidates for electrical energy storage could be divided into two classes, namely, inorganic and organic dielectrics. It should be noted here that organic polymers as important potential candidates for electrical energy storage have been explored widely, due to their excellent electric field endurance (up to 8000 $kV\,cm^{-1}$), good flexibility, and low fabricating temperature. To date, quite a few polymers, including polypropylene (PP), polyester (PT), polycarbonate (PC), polyimide, aromatic polyurea, aromatic polythiourea (ArPTU), polyvinylidenefluoride (PVDF), and their related organic and inorganic composites, have been investigated for electrical energy storage, and some of them have been applied in commercial capacitors. Among these polymers, PVDF-based materials are the most studied ones due to the larger

Ferroelectric Materials for Energy Applications, First Edition.
Edited by Haitao Huang and James F. Scott.

dielectric constant caused by their ferroelectric nature, and have received much attention in recent years. Accordingly, many researched papers, reviews, and book chapters have been published recently [3–5]. In order to avoid repetition and due to the length limitation, a summary on the energy-storage behavior in polymer-based dielectrics is not covered in this chapter. Thus, a timely overall summary on the state-of-the-art progress on electrical energy-storage performance of inorganic dielectrics will be presented; emphasis is mainly placed on inorganic compounds for application in high-temperature capacitors.

In order to easily understand the reported results and complete the logicality, the basics on capacitors, including the physical principles for improving energy-storage properties, some important parameters, and the related testing methods, are first given in this chapter. Following this, the energy-storage performance in four kinds of inorganic compounds, namely, simple metal oxides, antiferroelectrics (AFEs), dielectric glass-ceramics, and relaxor ferroelectrics, which are believed to be the most promising candidates for next-generation high energy-storage capacitors at elevated temperatures, are reviewed in turn. It should be noted here that, compared with bulk ceramics, dielectrics in thin- and thick-film form usually display excellent electric field endurance, which is favorable to the improvement of the final energy-storage properties. Accordingly, investigations on dielectric capacitors in single layer and multilayer forms are receiving increasing attention, as they have more potential for application in some portable electronic devices. Therefore, apart from the bulk-ceramic form, the current state of energy-storage performance in thin- and thick-film forms is especially focused on. Lastly, some general future prospects are also provided from the existing theoretical and experimental results, in order to propel their application in practice.

5.2 Principles of Dielectric Capacitors for Electrical Energy Storage

5.2.1 The Basic Knowledge on Capacitors

A capacitor was historically known as an electric condenser, and it is a passive electrical component that stores electrical energy under an applied electric field. Although the physical form and construction of practical capacitors are varied widely, the basic unit contains at least two electrical conductors and a separating dielectric medium [6]. The conductor may be a foil, metal thin film, or electrolyte. The dielectric mediums are often gas, glass, paper, mica, polymer film, and ceramic, which are used to increase the effect of the capacitors. The effect of a capacitor is defined as capacitance, which is only determined by the physical dimension (geometry) of the conductors and the dielectric constant of the dielectrics. Taking the simplest parallel-plate capacitor as an example, as shown in Figure 5.1a, the capacitance of a capacitor constructed of two parallel plates filled with certain dielectrics is approximately equal to the following [7]:

$$C = \varepsilon_r \varepsilon_0 \frac{A}{d} \tag{5.1}$$

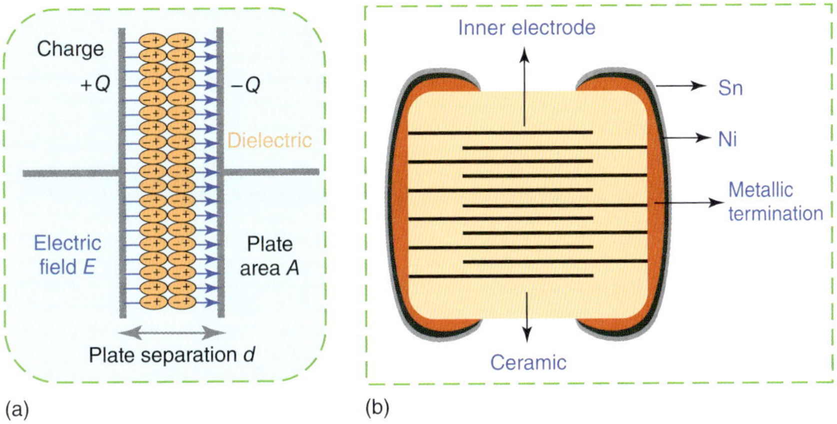

Figure 5.1 (a) The diagram of charge separation in parallel-plate capacitor under an electric field. (b) The diagram of a multilayer capacitor.

where C is the capacitance, A is the area of overlap of the two plates, ε_r is the relative dielectric constant, ε_0 is the vacuum electric constant ($\approx 8.85 \times 10^{-12}$ F m^{-1}), and d is the distance between the plates. The unit of capacitance is farad (F). Sometimes, the relative dielectric constant of the dielectrics is changed by the external bias, causing the capacitance to vary. In this case, capacitance is defined in terms of incremental change:

$$C = \frac{\mathrm{d}q}{\mathrm{d}v} \tag{5.2}$$

This model can also be used to make qualitative predictions for other device geometries. Obviously, the capacitance is directly proportional to the overlap area of the conductor plates and the relative dielectric constant of the dielectrics and inversely proportional to the separation distance between the plates. Thus, to increase the capacitance of a capacitor, there are only two methods. One is to select a proper dielectric medium with large dielectric constant, which is the main topic of this chapter. The other is to design the geometry with a large ratio of A/d, such as a multilayer ceramic capacitor (MLCC), as shown in Figure 5.1b. In fact, the multilayer capacitor is the most widely used form in electrical devices.

If an external voltage V is applied on the conductor plates, the charges between the conductors in a capacitor move, and at the same time electric polarization also occurs. This will result in positive and negative charges of equal amount accumulating on the two plates, respectively, which is the so-called charge process of the capacitor. The charge process ends when the electrical potential caused by the accumulated charge $\pm Q$ on both plates is equal to the external applied voltage V. Q/V is equal to the capacitance C of the capacitor. If the plates are uniformly charged over an area A, a charge density $\pm\rho = \pm Q/A$ exists on the surface. Assuming that the length and width of the plates are much greater than their separation d, the electric field near the center of the capacitor is uniform with a magnitude $E = \rho/\varepsilon$. The voltage V is equal to the line integral of the electric field between the plates:

$$V = \int_0^d E\,\mathrm{d}x = \int_0^d \frac{\rho}{\varepsilon}\mathrm{d}x = \frac{\rho d}{\varepsilon} = \frac{Qd}{\varepsilon A} \tag{5.3}$$

During the charge process, the charges are moved between the conductor plates under an external bias, indicating that work must be done and that the electric energy is stored in the dielectrics at the same time. Hence, the amount of the stored energy U could be obtained from the following formula:

$$U = \int_0^Q V\,\mathrm{d}q = \int_0^Q \frac{q}{C}\mathrm{d}q = \frac{1}{2}\frac{Q^2}{C} = \frac{1}{2}CV^2 = \frac{1}{2}\frac{\varepsilon A}{d}(Ed)^2 \tag{5.4}$$

5.2.2 Some Important Parameters for Electrical Energy Storage

5.2.2.1 Energy-Storage Density

In order to evaluate the electrical energy-storage performance of a capacitor, the energy-storage density W as an important parameter is always studied. It is defined as the energy stored per unit volume of a dielectric under the application of an external electric field:

$$W = \frac{U}{Ad} = \frac{1}{2}\varepsilon E^2 = \frac{1}{2}\varepsilon_0\varepsilon_r E^2 \tag{5.5}$$

Evidently, the stored energy density W is proportional to the relative dielectric constant and the square of the applied electric field. Thus, finding an ideal dielectric material (ideal insulator) with giant relative dielectric constant and super-high electric field endurance is the only way for the fabrication of high energy-storage capacitors.

However, in practice, the ideal dielectric material does not exist. Under the function of external electric field, a leakage current is always detected, which leads to energy loss. Moreover, for ferroelectrics/antiferroelectrics, the electric-field-derived movement of domains also consumes energy. This means that only part of the stored energy could be released during the discharge process because some of it is lost during the charge and discharge process. For convenience, the released energy density is labeled as W_{rec} and the lost energy density is named as W_{loss}. Clearly, $W = W_{rec} + W_{loss}$. It should be noted that in most cases the relative dielectric constant is temperature and frequency dependent. As a result, the stored energy density W also depends on the temperature and frequency.

5.2.2.2 Energy Efficiency

Apart from the energy-storage density, energy efficiency is another parameter to evaluate the electrical energy-storage performance of a dielectric, which is defined as

$$\eta = \frac{W_{rec}}{W} = \frac{W_{rec}}{W_{rec} + W_{loss}} \tag{5.6}$$

Thus, in order to realize high energy efficiency, large discharge energy density and small loss energy density should be obtained at the same time. Similar to the energy density, the energy efficiency is also a temperature- and frequency-dependent parameter.

5.2.2.3 Breakdown Strength (BDS)

From Eq. (5.5), it can be seen that a high external electric field E is favorable to improving the energy-storage density of a dielectric. However, as mentioned above, the dielectric in practice is not an ideal insulator. This means that an electric breakdown cannot be avoided under an applied electric field. The critical field E_c is defined as the breakdown strength (BDS), which is determined by the chemical composition, grain size, pore, and interface. Apart from the material condition, the BDS value is also strongly affected by the temperature, time, and frequency of the field. Based on the breakdown process, three mechanisms are currently accepted, namely, electrical breakdown, thermal breakdown, and partial discharging breakdown. In order to reduce or eliminate the partial discharge breakdown, the capacitors are often placed in silicon oil during the measurement. Irrespective of whether it is the electrical breakdown or the thermal breakdown, they are closely related to the leakage current at the applied electric field. Thus, the conductive performance of the dielectric is also extensively investigated for energy storage. Currently, some conductive mechanisms are also proposed, such as Ohmic behavior, space-charge-limited current (SCLC) and Poole–Frenkel (PE) emission from the bulk effect, and Schottky emission (SE) and Fowler–Nordheim (FN) tunneling from the interface effect [8]. Because of the limit on the length, a detailed description on the abovementioned conductive mechanisms is not given here.

Weibull distribution is often used to analyze the BDS [9]. The average values of BDS can be described by

$$X_i = \ln(E_i) \tag{5.7}$$

$$Y_i = \ln(\ln(1/(1 - P_i))) \tag{5.8}$$

$$P_i = i/(n + 1) \tag{5.9}$$

where X_i and Y_i are two parameters in Weibull distribution function, and E_i is the specific breakdown voltage of each specimen in the experiment. The samples are arranged in ascending order of BDS values, so that $E_1 \leq E_2 \leq E_3 \leq \cdots \leq E_i \leq \cdots \leq E_n$. P_i is the probability for dielectric breakdown, n is the sum of specimens of each sample, and i is the serial number of the specimen. According to the Weibull distribution, X_i and Y_i have a linear relationship. The average BDS value can be extracted from the points where the fitting lines intersect with the horizontal line through $Y_i = 0$. In order to obtain an accurate average BDS value, the number of specimens should not be less than eight, and should be as high as possible.

For a given dielectric material under certain operating conditions, the maximum energy-storage density W_{max} is equal to $\frac{1}{2}\varepsilon_0\varepsilon_r E_c^2$. Generally, a dielectric material with smaller dielectric constant illustrates a higher BDS value. For the bulk ceramics (≥100 μm) prepared by the conventional solid-state sintering technique, their BDS values are often less than 200 kV cm^{-1}, which is mainly caused by the low relative density of about 95–98%. On the other hand, a denser structure for a dielectric material in thin-film form (≤1 μm) can be easily obtained presently. Thus, ceramic thin films often display excellent electric field endurance with a BDS value of above 1000 kV cm^{-1}. Therefore, compared with the bulk ceramic, their thin-film counterparts usually possess enhanced energy-storage performance.

5.2.2.4 Thermal Stability

For the practical application of capacitors, a good thermal stability is always desired, which is defined as the ratio of maximum capacitance change in the temperature range of interest with respect to the capacitance at 0 °C, namely, $\Delta C/C_0$. The capacitors prepared with different dielectric materials often show quite different thermal stability, and are applied for different purposes. The capacitors are divided in three classes, based on the thermal stability of the ceramic material used.

Class 1: Ceramic capacitors are composed of paraelectric materials with low relative dielectric constant (6–200) and low loss ($\tan\theta \leq 0.15\%$), such as titanium dioxide (TiO_2) modified by zinc, zirconium, niobium, tantalum, and strontium. This kind of capacitors usually have a low capacitance, but display good temperature, electric field, and frequency-dependent stability, and are mainly used in resonant circuits.

Class 2: Ceramic capacitors are made of ferroelectric materials such as barium titanate ($BaTiO_3$) added with aluminum oxide, magnesium silicate, and aluminum silicate. This kind of dielectric material usually has a large relative dielectric constant (≥200) and moderate loss. Thus, the obtained capacitors show high capacitance values in small size, but display nonlinear behavior as a function of temperature, applied electric field, and frequency. They are used for bypass, coupling, and decoupling in frequency-discriminating circuits. Generally, class 2 capacitors are labeled according to the change in capacitance over the temperature range. The most widely used classification is based on the EIA RS-198 standard and uses a three-digit code [10]. The first character is a letter that gives the low-end operating temperature, the second gives the high-end operating temperature, and the final character gives capacitance change over that temperature range, as shown in Table 5.1. For instance, a Z5U capacitor will operate from +10 to +85 °C with a capacitance change of at most +22% to −56%, and an X7R capacitor will operate from −55 to +125 °C with a capacitance change of at most ±15%.

Class 3: Ceramic capacitors, also named barrier or semiconductive capacitors, are usually from doped ferroelectric materials such as $BaTiO_3$, which have very

Table 5.1 Code system according to EIA RS-198 for class 2 capacitors.

Lower temperature	Upper temperature	Change of capacitance over the temperature range (%)
$X = -55$ °C (−67 °F)	$4 = +65$ °C (+149 °F)	$P = \pm 10$
$Y = -30$ °C (−22 °F)	$5 = +85$ °C (+185 °F)	$R = \pm 15$
$Z = +10$ °C (+50 °F)	$6 = +105$ °C (+221 °F)	$S = \pm 22$
	$7 = +125$ °C (+257 °F)	$T = +22/-33$
	$8 = +150$ °C (+302 °F)	$U = +22/-56$
	$9 = +200$ °C (+392 °F)	$V = +22/-82$

large relative dielectric constants up to 50 000. This kind of capacitors show worse thermal, electric field, and frequency-dependent stability, accompanied by very high loss and age over time, but can realize a large capacitance of about 100 μF.

5.2.2.5 Power Density

As compared with other energy-storage components, a unique character of ceramic capacitors is huge power density, which means that the stored energy could be released in a very short time, up to several nanoseconds, in proper circuits. The power density P is defined as the energy released per second per volume:

$$P = \frac{\mathrm{d}U}{Ad\,\mathrm{d}t} = \frac{1}{2Ad}\frac{\mathrm{d}CV^2}{\mathrm{d}t} = \frac{CV(t)}{Ad}\frac{\mathrm{d}V(t)}{\mathrm{d}t} \tag{5.10}$$

In most cases, the power density P in libraries is calculated by monitoring the current (i) through an LCR circuit (as mentioned in the following part), according to the following formulas:

$$i(t) = C\frac{\mathrm{d}V}{\mathrm{d}t} \tag{5.11}$$

$$V(t) = \frac{1}{C}\int_0^{\infty} i(t)\mathrm{d}t \tag{5.12}$$

$$P = V(t) \cdot i(t) \tag{5.13}$$

5.2.2.6 Service Life

A long service life is necessary for the application of capacitors in practice, which means a maximum number of repeated charge–discharge cycles (N). Under the long-term function of electric field and temperature, the electrical properties of the dielectric ceramics, especially for the ferroelectrics, often illustrate deterioration, which in turn leads to declined energy-storage performance of the capacitors. Comparatively, ferroelectric/antiferroelectric-based dielectric capacitors usually show relative short service life, mainly caused by the polarization fatigue, but their charge–discharge cycle numbers can still reach 10^6.

5.2.3 Measurement Techniques of Energy Density

From the above section, it can be seen that although there are several parameters to evaluate the energy-storage performance, energy-storage density is the most important one. Presently, there are three methods employed to obtain the energy-storage density. Thus, in this section, the detailed descriptions on these methods will be presented.

5.2.3.1 Polarization-Based Method

This method is to obtain the energy density from polarization–electric field (P–E) loops. Physically, it is well known that the charge density (Q/A) on the conductor plate of a capacitor is equal to the electrical displacement D ($D = \varepsilon_0\varepsilon_r E$). Thus, combining with Eq. (5.4), the energy-storage density W could be also expressed as follows:

$$W = \frac{U}{Ad} = \frac{\int_0^Q V\,\mathrm{d}q}{Ad} = \int_0^{D_{\max}} E\,\mathrm{d}D \tag{5.14}$$

where E is the external applied electrical field and equal to V/d; other variables are defined as before. For dielectrics with a high permittivity, the electrical displacement D is very close to their electrical polarization P. As a result, formula (5.14) can be rewritten as

$$W = \int_0^{P_{\max}} E\,\mathrm{d}P \tag{5.15}$$

Evidently, based on formula (5.15), the W value of the dielectrics can be easily obtained by numerical integration of the area between the polarization and the curves of the P–E loops. As shown by the P–E loop in the first quadrant in Figure 5.2, as the electric field increases from zero to the maximum E_{max}, the polarization also increases to its maximum P_{max}. As a result, the electrical energy with a density W is charged in the dielectrics, represented by the overall area of green and red. When the electric field decreases from E_{max} to zero, the polarization also declines from P_{max} to P_{rem}. Accordingly, some charged energy, named as the recoverable electrical energy density W_{rec}, is then discharged, represented by the green area in Figure 5.2. The un-discharged energy, indicated by the red area enclosed by the loops, is the energy loss density W_{loss}, which is exhausted during the charge–discharge process. Clearly, based on the P–E loops, the energy density including W, W_{rec}, and W_{loss} could be easily obtained by numerical integration. This is also a widely used method to study energy-storage currently.

From the above description, it can be concluded that the large energy density as well as high energy efficiency can be realized in dielectrics with large ($P_{max} - P_{rem}$) value and high BDS. Accordingly, five kinds of dielectrics are believed to have the most potential for high electrical energy-storage application: (i) paraelectric-like metal oxides with high BDS, (ii) AFEs with large P_{max} and $P_{rem} = 0$, (iii) relaxor ferroelectrics with large ($P_{max} - P_{rem}$) value, (iv) glass-ceramic ferroelectric with high BDS, and (v) polymers with high BDS.

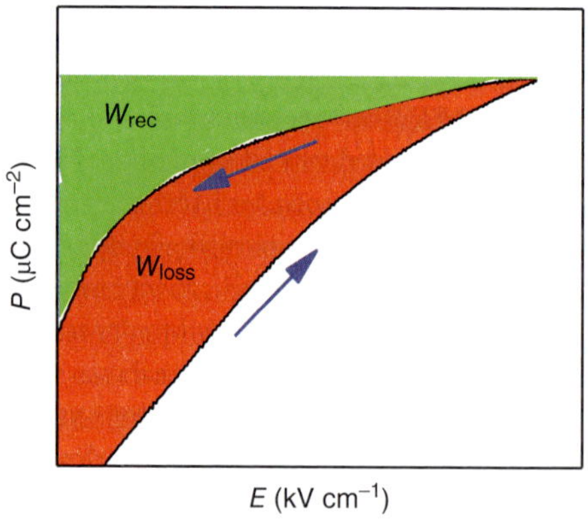

Figure 5.2 Diagram of the charge–discharge process from P–E measurements.

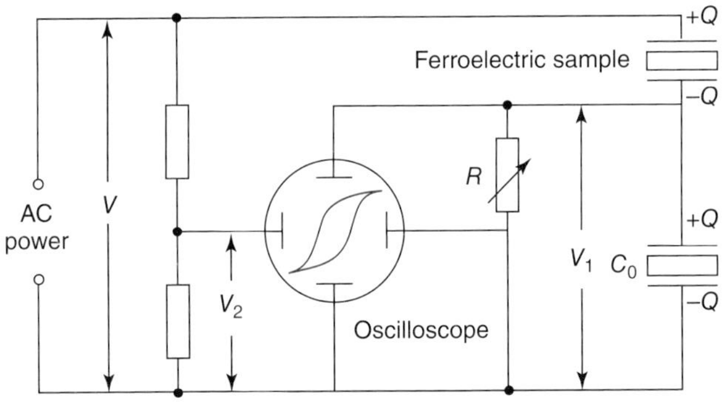

Figure 5.3 Schematic circuit of the Sawyer–Tower circuit for measuring the *P–E* loops. *Source:* Li et al. 2014 [11]. Reproduced with the permission of John Wiley & Sons.

The key to employing this method to obtain the energy-storage density is dependent on accurately measuring the *P–E* loops and avoiding the effect of leakage current. Currently, the measurement of *P–E* loops is based on the Sawyer–Tower circuit, as shown in Figure 5.3 [11]. As an AC voltage is imposed on an electroded sample, the quantity plotted on the horizontal axis is proportional to the field across the crystal. A linear capacitor C_0 is connected in series with the sample. The voltage across C_0 is proportional to the polarization (P) of the ferroelectric sample. In fact, dielectric displacement (D) and polarization have a relationship

$$D = \varepsilon_0 E + P \tag{5.16}$$

where D is the charge density collected by C_0. Compared to the larger value of P, the contribution by $\varepsilon_0 E$ can be omitted. Therefore, the obtained D is considered as P in practice. Currently, the commercial apparatus for *P–E* measurement are mainly from *aixACCT* in Germany and *Radiant* in the Unites States.

5.2.3.2 Indirect Calculated Method

As given in formula (5.5), it can be seen that W is equal to $^1/_2\varepsilon_0\varepsilon_r E^2$. Therefore, for the linear dielectrics (electric-field-independent dielectric constant) such as glass, metal oxides, and glass-ceramics, the energy-storage density is easily given by measuring their dielectric constant. Accordingly, the maximum density can be estimated after detecting its BDS value. From this method, the only overall energy-storage density W can be roughly given; the recoverable and loss energy density cannot be obtained. Moreover, at high electric fields, the variation of dielectric constant is difficult to avoid, which leads to an inaccurate result. Thus, this method is seldom used in practice.

5.2.3.3 Direct Charge–Discharge Method

Although the above two methods can estimate the energy-storage performance in some degree, they cannot give the real discharged energy density and the discharge time of the dielectrics in the circuit. Thus, in order to realize this aim, a specially designed home-made measurement circuit was proposed by Dong,

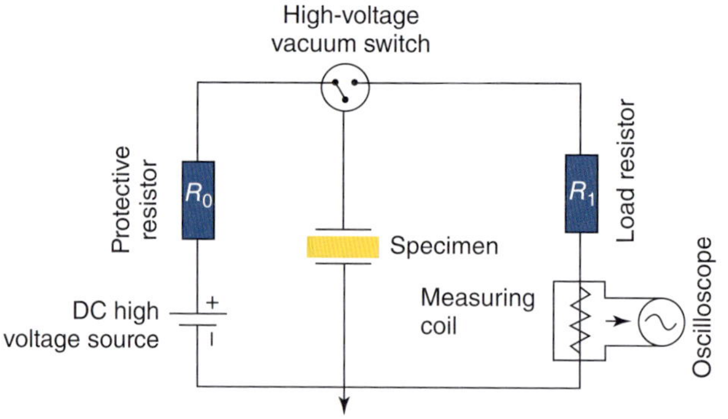

Figure 5.4 Diagram of the charge–discharge measurement circuit. *Source:* Xiao et al. 2016 [12]. Reproduced with the permission of Elsevier.

Zhai, and Xu's groups in China, respectively. Figure 5.4 gives the diagram of the charge–discharge measurement circuit [12]. In this case, the sample capacitor is first charged by an external bias, and thus the electric energy is stored in the dielectric. Then, the charged capacitor is connected to the load R by a switch, to form a discharged circuit (LCR), mainly constituted by the protective resistor, load resistor, sample capacitor, and monitor coil. Accordingly, part of the stored energy is discharged, accompanied by the transient current formed in the closed circuit, which is monitored by the coil. Based on the $I(t)$–t curve, as shown in Figure 5.5, the recoverable energy density W_{rec} can be calculated by the following formula:

$$W_{\text{rec}} = \frac{R}{Ad} \int I^2(t)\text{d}t \tag{5.17}$$

where R is the resistance of the load, t is the discharge time, and Ad is the volume of the capacitor.

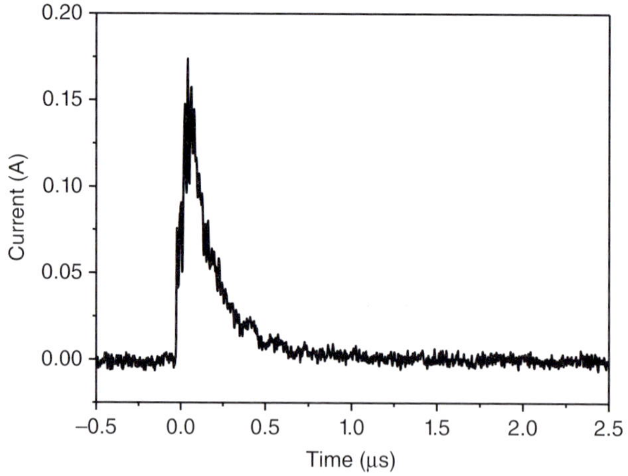

Figure 5.5 The discharged $I(t)$–t curve under different charged electric fields.

Another advantage of this method is that the discharge time of the capacitor in the circuit can be obtained directly from the $I(t)$–t curve, which is meaningful for the investigation on pulsed dielectric capacitors. In fact, the discharge time (T) of a capacitor in a specific circuit can be expressed as

$$T = \frac{2\pi}{\omega} = \frac{2\pi}{\sqrt{\frac{1}{LC} - \left(\frac{R}{2L}\right)^{-2}}} \tag{5.18}$$

Apparently, the discharge time T is determined by R, L, and C in the circuit. If the values of R and L are constant, the T value is only dependent on C. Thus, the discharge time can be electric field dependent for the capacitor from a nonlinear dielectric medium, while it is constant for the capacitor from a linear dielectric medium.

Finally, it should be noted that the results obtained from the above three methods are often slightly different, especially at high electric fields. This is due to the different measurement processes.

5.3 The Energy-Storage Performance in Paraelectric-Like Metal Oxides

In theory, the paraelectric-like metal oxides should have great potential for high energy-storage application owing to their linear dielectric properties and high BDS value. For convenience, paraelectric-like metal oxides are divided into two subclasses, namely, simple metal oxides such as TiO_2 and multi-metal oxides such as $SrTiO_3$. Recently, some attention has also been paid to the study of the energy-storage performance of paraelectric-like metal oxides. Thus, in the following section, a brief summary is given for the two kinds of paraelectric-like metal oxides, respectively.

5.3.1 Simple Metal Oxides

Simple metal oxides such as TiO_2, ZrO_2, and Al_2O_3 are good candidates for electrical energy-storage application, because of their large forbidden bandwidth E_g and moderate dielectric constant. This means that a quite high BDS value can be obtained in these oxides. However, limited by the fabrication procedure and the purity of raw materials, high-quality pure ceramics of these metal oxides are still hard to obtain presently, which leads to a decreased BDS value and large dielectric loss. As a result, no outstanding energy-storage performance in these metal oxides is obtained currently, and only a few tentative results are reported. A concise summary on the energy-storage performance of some typical metal oxides is presented in this section.

5.3.1.1 TiO_2

TiO_2 is a widely used material, which can exist in anatase, rutile, and brookite phases at room temperature. TiO_2 in the anatase and rutile phases is obtained comparatively more easily at room temperature, and has an E_g value of 3.2 eV and relative dielectric constant of about 80–110.

By selecting nano-sized TiO_2 (40 nm) and proper dispersants, thick ceramic films with a density of above 95% were fabricated through tape-casting method [13]. It was found that the dispersants had a strong effect on the dielectric property of the films. By using AD1 + PE as the dispersant, a larger BDS value of 1402 kV cm^{-1} was achieved in the final films, which led to an improved recoverable energy-storage density W_{rec} of 14 J cm^{-3}. Quite recently, it was reported that a colossal relative dielectric constant ($\geq 10^4$) could be achieved in co-doped TiO_2 ceramics in a broad temperature range, which may lead to good energy-storage performance [14, 15].

5.3.1.2 ZrO_2

ZrO_2 possesses an abundant temperature-dependent phase structure. It exists in the monoclinic phase below 1170 °C, turns into the tetragonal phase in the temperature range of 1170–2370 °C, and into the cubic phase above 2370 °C. Depending on its phase structure, its E_g value varies between 5.0 and 7.0 eV.

A superthin ZrO_2 was deposited on silicon substrates successfully, and TiN was selected as the bottom and top electrodes to form paralleled capacitors [16]. It is interesting that this thin film showed a slim double P–E loop. Accordingly, a W_{rec} value of 37 J cm^{-3} and an efficiency of 51% were achieved at room temperature in these films. Moreover, a giant W_{rec} value of 937 J cm^{-3} was obtained in the 3D-structured ZrO_2 thin films.

5.3.1.3 Al_2O_3

Al_2O_3 can also exist in different phases. It has a trigonal structure and possesses a large E_g value of about 6.2 eV.

By using the atomic layer deposition (ALD) method, Al_2O_3-based superthin films including pure Al_2O_3, a nanolaminate based on Al_2O_3 and ZrO_2, and a nanolaminate based on Al_2O_3 and TiO_2 were deposited on ITO substrates [17]. An improved relative dielectric constant and BDS value were obtained in the nanolaminate films. As a result, maximum W_{rec} values of 50 and 60 J cm^{-3} were obtained in Al_2O_3 and Al_2O_3/TiO_2 nanolaminate films, respectively.

Although little work has been carried out on the electrical energy storage of simple metal oxides to date, the reported results still indicate that this kind of dielectrics, especially in thin-film form, have the potential for application in high energy-storage capacitors. It could be predicated that with the help of advanced material fabrication techniques, further success can be achieved in simple metal oxide-based dielectric capacitors.

5.3.2 Multi-metal Oxides

Compared with the simple metal oxides, paraelectric-like metal oxides usually possess a larger dielectric constant above 100, but their BDS is slightly lower. Therefore, this kind of oxides were investigated to enhance the electric endurance and at the same time maintain or increase their dielectric constant. However, because of length limitation, only two typical oxides, namely, $SrTiO_3$

and $Bi_{1.5}Zn_{0.9}Nb_{1.5}O_{6.9}$ are presented here to illustrate the current state of energy-storage performance in the paraelectric-like multi-metal oxides.

5.3.2.1 $SrTiO_3$

$SrTiO_3$ is a typical paraelectric-like oxide with perovskite structure, which is widely used in electric ceramics. Presently, two routes are employed to improve its energy-storage performance, which are ion doping for enhancing dielectric constant and glass addition for increasing BDS, respectively.

For example, by doping 10–23.3 mol% Bi^{3+} into $SrTiO_3$ ceramics, higher dielectric constant, better frequency stability, larger resistivity, and lower dielectric loss were obtained, which was ascribed to the lattice change of $SrTiO_3$ [18]. Accordingly, the ceramics with 10 mol% Bi^{3+} had the highest recoverable energy-storage density of 1.63 J cm^{-3} and energy efficiency of 62% at 217 kV cm^{-1}. By addition of 15 vol.% SiO_2 and 1 mol% $BaCu(B_2O_5)$ sintering aids, respectively, the maximum recoverable energy-storage density of 1.16 J cm^{-3} at 361 kV cm^{-1} and 1.05 J cm^{-3} at 288 kV cm^{-1} were achieved in the dense $SrTiO_3$ ceramics with a reduced sintering temperature, which was owing to the improved BDS [19].

5.3.2.2 $Bi_{1.5}Zn_{0.9}Nb_{1.5}O_{6.9}$

$Bi_{1.5}Zn_{0.9}Nb_{1.5}O_{6.9}$ is a cubic pyrochlore at room temperature with a dielectric constant of 150–200, a low dielectric loss, and high BDS. Recently, a series of $Bi_{1.5}Zn_{0.9}Nb_{1.5}O_{6.9}$ thin films with a thickness of about 150 nm were prepared on Pt(111)/Pt/Ti/SiO_2/Si substrates via chemical solution deposition method by Michael et al. [20]. When an AC electric field with different frequencies between 10 Hz and 10 kHz was applied, the BDS value of the films varied between 4700 and 5200 kV cm^{-1}. At 10 kHz, a maximum recoverable energy-storage density of 60.8 J cm^{-3} was reached. As the environmental temperature increased to 200 °C, the BDS and dielectric loss gradually declined, indicating a reduction in energy-storage performance. Subsequently, by the addition Ta, the authors prepared a $Bi_{1.5}Zn_{0.9}Nb_{1.35}Ta_{0.15}O_{6.9}$ thin film. It was found that Ta had little effect on the dielectric constant and dielectric loss, but greatly improved BDS from 5100 to 6100 kV cm^{-1} at 10 kHz. As a result, a further increased recoverable energy-storage density of 66.9 J cm^{-3} was reached. Moreover, a huge power storage density of 850 MW cm^{-3} at 10 kHz and 45 MW cm^{-3} at 1 kHz were achieved in Ta-doped $Bi_{1.5}Zn_{0.9}Nb_{1.5}O_{6.9}$ thin films. These results indicated that $Bi_{1.5}Zn_{0.9}Nb_{1.5}O_{6.9}$-based thin films had a great potential for application in pulsed high energy-storage capacitors.

5.4 The Energy-Storage Performance in Antiferroelectrics

As a subclass of dielectrics, in the true AFEs, the spontaneous polarization directions of adjacent dipoles are opposite, could be induced to the same orientation as a function of electric field, and then be changed to the ferroelectric phase [21]. Thus, AFEs possess two distinct features. One is that the net macroscopic remnant polarization is zero. Another is that *P–E* curves under sufficient high electric

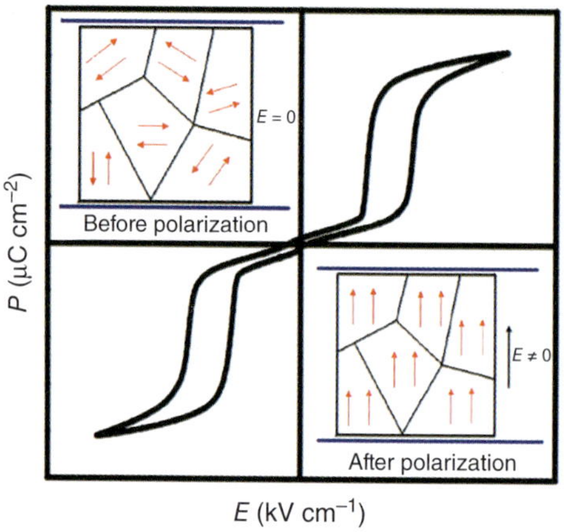

Figure 5.6 The diagram of the polarization process and *P–E* loops of AFEs.

field display double loops. The diagram for the electric-field-induced phase transition and the corresponding *P–E* result of AFEs are illustrated in Figure 5.6.

Because of their zero remnant polarization and large maximum polarization, AFEs also have the potential for application in high energy-storage capacitors. Moreover, AFEs possess the capacity for fast electric-field-induced phase transition, indicating a fast charge and discharge speed. Thus, much work has been carried out to find the potential AFEs for energy-storage applications. Currently, besides the typical $PbZrO_3$-based AFEs, $(Na_{0.5}Bi_{0.5})TiO_3$, $AgNbO_3$, and HfO_2-based AFEs are also widely investigated for this purpose.

5.4.1 $PbZrO_3$-Based Antiferroelectric

The notion of antiferroelectricity dates back to the early 1950s. At that time, Kittel proposed a theory of antiparallel ionic displacements in dielectrics using the antiferromagnetism scheme [22]. In parallel, $PbZrO_3$ with perovskite structure was the first AFE demonstrated by Sawaguchi and coauthors [23]. Its Curie point is about 230 °C. According to the X-ray pattern and detailed optical observations, $PbZrO_3$ at room temperature is orthorhombic with $a = 5.78$ A, $b = 11.74$ A, and $c = 8.20$ A. Each orthorhombic unit cell contains eight primitive cells, which have a tetragonal structure and spacing $a_0 = b_0 = 4.15$ A, $c_0 = 4.10$ A [23]. As for the origination of AFE in $PbZrO_3$-based materials, it was proved to be due to the antiparallel displacement of Pb ions in the plane perpendicular to the c axis in the orthorhombic unit cell. The displacement was calculated to be about 0.2 A. Thus, it could be concluded that the anti-polar direction of lead-based AFE is along the pseudocubic $\langle 110 \rangle$ plane, and that the polar direction of parallel dipoles is along the $\langle 111 \rangle$ plane of the rhombohedral structure after being induced into the FE phase [24, 25]. Accordingly, the final properties of AFE materials including the electrical energy-storage performance can be tailored by selecting the proper growth orientation.

To date, room temperature electric-field-induced phase transformation between AFE and FE has not been observed in pure $PbZrO_3$ bulk ceramics, and only at higher temperatures (above 210 °C) electric-field-induced double

P–E loops were obtained. This is caused by the higher critical field for the phase transition from AFE to FE, which is higher than its BDS values. Thus, in order to investigate $PbZrO_3$-based AFEs in detail and obtain stable double *P–E* loops in ceramics at room temperature, much efforts has been made to reduce the critical field by chemical substitution. The most studied doping ions are La^{3+}, Ba^{2+}, and Sr^{2+} at the A site and Sn^{4+}, Ti^{4+}, and Nb^{5+} at the B site [26–30]. Among all of these investigations, $(Pb,La)(Zr,Sn,Ti)O_3$ (PLZST) and $Pb(Zr,Sn,Ti)NbO_3$ (PNZST) are the most important systems and are explored extensively. The phase diagrams of $(Pb_{0.97}La_{0.02})(Zr,Sn,Ti)O_3$ and $Pb_{0.99}(Zr,Sn,Ti)_{0.98}Nb_{0.02}O_3$ were first obtained by Berlincourt successfully, and AFEs with orthorhombic and tetragonal structure at room temperature were obtained by composition control [31].

The history of the study on energy storage in AFEs could be traced back to 1961, when Jaffe predicated high energy-storage density in $PbZrO_3$ [28]. However, due to the high critical electric field for antiferroelectric–ferroelectric phase transition, which often was larger than the BDS of the bulk ceramics, the study on energy storage in AFEs attracted little attention for a long time. With the rapid development of advanced ceramic fabrication techniques during the last decades, good electric field endurance in $PbZrO_3$-based ceramics and thin and thick films was realized, and then their energy-storage performance was studied widely. Here, a detailed summary on $PbZrO_3$-based AFEs is given in the bulk ceramic, thin-film, and thick-film forms in turn. For convenience, we define the dielectrics with a thickness above 100 μm as bulk ceramics, below 1 μm as thin films, and in the range of 1–100 μm as thick films.

Owing to the moderate phase transition electric field, the study on energy-storage performance in AFE bulk ceramics was mainly focused on the composition-modified $PbZrO_3$ such as $(Pb_{0.97}La_{0.02})(Zr,Sn,Ti)O_3$ and $Pb_{0.99}(Zr,Sn,Ti)_{0.98}Nb_{0.02}O_3$ systems. The reported results indicated that chemical composition has a strong effect on the phase transition and polarization character of $PbZrO_3$-based AFEs, and thus on the energy-storage performance. For instance, by changing the ratio of Zr/Sn and Zr/Ti, a larger W_{rec} value of 4.426 J cm^{-3} was obtained in $Pb_{0.97}La_{0.02}(Zr_{0.90}Sn_{0.05}Ti_{0.05})O_3$ bulk ceramics with orthorhombic structure, accompanied by a moderate energy efficiency of 61.2% [32]. It was also reported that by controlling the ratio of Zr/Ti, a maximum recoverable density of 4.2 J cm^{-3} with an efficiency of 82% was achieved in $Pb_{0.97}La_{0.02}(Zr_{0.5}Sn_{0.44}Ti_{06})O_3$ ceramics with tetragonal structure, and good thermal stability was observed in a wide temperature range [33]. Apart from *B* site substitution, the *A* site elements also could be used to tailor the energy-storage performance of $PbZrO_3$-based AFEs. By introducing Sm, a moderate recoverable energy-storage density of 1.743 J cm^{-3} with good frequency-dependent stability in the range of 1–10 Hz was achieved in $Pb_{9.85}Sm_{0.01}(Zr_{0.64}Sn_{0.28}Ti_{0.08})O_3$ ceramics [34]. Similarly, it was found that Ba, Y, La, and Sr addition with varied content also has an obvious effect on the electrical energy-storage performance including thermal and frequency-dependent stability. As a typical example, $(Pb_{0.995}Sr_{0.015}La_{0.02})(Zr_{0.75}Sn_{0.195}Ti_{0.055})O_3$ with orthorhombic structure showed a recoverable density of 5.56 J cm^{-3}, which is the highest value so far for the dielectric ceramics prepared by a conventional sintering technique, and very small energy density variation (<12%) in the range of 30–90 °C [35].

Besides, by controlling phase transition and polarization character by selecting proper chemical composition to optimize the electrical energy-storage performance of $PbZrO_3$-based AFE bulk ceramics, a few fabrication techniques such as addition of glass powder and new annealing methods were also employed to realize this aim by improving their BDS values. For example, when 4 wt% $CdO–Bi_2O_3–PbO–ZnO–Al_2O_3–B_2O_3–SiO_2$ low-softening-point glass powders were added into $Pb_{0.97}La_{0.02}(Zr_{0.56}Sn_{0.35}Ti_{0.09})O_3$, a large recoverable energy density of 3.3 J cm^{-3} with an efficiency of 80% was achieved after being annealed at 1130 °C, which was due to the improved BDS value of 175 kV cm^{-1}, in contrast to that (117 kV cm^{-1}) for the glass-free ceramics [36]. Very recently, Jiang's group successfully improved the BDS value by using the hot-press sintering (HPS) and spark plasma sintering (SPS) methods. For instance, SPS ceramics possessed a considerable BDS value of 306 kV cm^{-1} and a high recoverable density of 6.40 J cm^{-3}, which is the highest value obtained in dielectric ceramics to date [37].

Generally speaking, dielectric materials in film form usually show an increased BDS value, as compared with their bulk-ceramic counterpart, which makes the study on $PbZrO_3$-based AFEs with large critical electric fields of antiferroelectric–ferroelectric phase transition possible. Thus, it could be predicted that enhanced energy-storage performance must be realized in AFE films. First, a moderate energy-storage density with a value of 5–7 J cm^{-3} was reported in $(Pb_{0.97}La_{0.02})(Zr,Sn,Ti)O_3$ and $Pb_{0.99}(Zr,Sn,Ti)_{0.98}Nb_{0.02}O_3$ thin films on Pt-buffered silicon substrates by Xu et al. in 2000 [38]. Subsequently, the energy-storage performance in La, Sr, Pr, Eu, and Ba-doped $PbZrO_3$ thin films were investigated by different authors. A large recoverable energy-storage density of 14–40 J cm^{-3} was achieved, accompanied by a moderate energy efficiency of about 60% and an acceptable thermal stability. For instance, a larger recoverable energy density J_{reco} of 14.5 J cm^{-3} at 900 kV cm^{-1} at room temperature was obtained in Sr-doped $PbZrO_3$ films, corresponding to a higher energy efficiency η of 78% [39]. An enhanced fatigue endurance and stable temperature-dependent energy-storage performance were also observed in the Sr-doped $PbZrO_3$ thin films. Similar to the AFE bulk ceramics, the composition dependence of energy-storage behaviors in $(Pb,La)(Zr,Sn,Ti)O_3$ AFE thin films with orthorhombic structure was studied by Mirshekarloo et al. [40]. It was found that all the samples shared a similar J_{reco} value above 10 J cm^{-3} at room temperature at the same applied electric field. It is well known that excess lead content has a deep influence on the final properties of lead-based films. Therefore, the lead-content-dependent energy-storage performance in thin films was also investigated. It was found that W_{rec} value in 400-nm-thick $Pb_{0.97}La_{0.02}(Zr_{0.97}Ti_{0.03})TiO_3$ films at 1200 kV cm^{-1} was remarkably improved from 3.3 to 11.7 J cm^{-3} by 20 mol% excess lead addition [41].

For the dielectrics in thin-film form, apart from chemical composition, the electrodes, thickness, interface, strain, and orientation also have strong effects on their final properties. Very dense and uniform $(Pb_{0.92}La_{0.08})(Zr_{0.95}Ti_{0.05})O_3$ AFE thin films (about 1000 nm) were successfully fabricated on $LaNiO_3$-buffered

Ni foil substrates via a chemical solution deposition method by Ma et al. [42]. Excellent breakdown endurance (above $3500\,kV\,cm^{-1}$) can be realized in this film, and thus a considerably large W_{rec} value of $37\,J\,cm^{-3}$ was achieved. A series of dense and uniform $(Pb_{0.96}La_{0.04})(Zr_{0.98}Ti_{0.02})O_3$ thin films with a thickness range of 150–850 nm were deposited on $(La_{2/3}Sr_{1/3})MnO_3/SrTiO_3$ substrates [43]. All the films had a similar W_{rec} value of about $20\,J\,cm^{-3}$, while their energy loss density declined from 15.2 to $7.4\,J\,cm^{-3}$ as the thickness increased, which led to a significant improvement in energy efficiency from 56.8% to 74.1%. Pure $PbZrO_3$ thin films of 300 nm with (100), (110), and (111)-preferred orientation were deposited on $SrTiO_3$ single crystal substrates with different orientations [44]. In this case, in-plant compressive strain existed in (110)- and (111)-oriented films, which stabilized the AFE phase, while a contrary effect developed in (100)-oriented films. As a result, the critical field for the antiferroelectric–ferroelectric phase transition improved by a $120\,kV\,cm^{-1}$ increment, and hence an increment of the recoverable energy-storage density by $5.3\,J\,cm^{-3}$ at $700\,kV\,cm^{-1}$ was also observed in (110)- and (111)-oriented films, as shown in Figure 5.7. To stress the function of interface, $PbZrO_3$ AFE and $0.88BaTiO_3$–$0.12Bi(Mg_{0.5}Ti_{0.5})O_3$ ferroelectric layers with a multilayered structure were deposited on Pt-butterflied silicon substrates [45]. A quite high recoverable energy-storage density of $33\,J\,cm^{-3}$ with an efficiency of about 70% was obtained in the layer-by-layer annealed thin films. Recently, it has also been found that the formation of local field in the dielectric matrix could enhance polarization and thus improve the electrical energy-storage density. For example, by adding the appropriate content of α-Fe_2O_3 nanoparticles in $PbZrO_3$ thin films, a maximum polarization of $78\,\mu C\,cm^{-2}$ and a recoverable energy density as high as $17.4\,J\,cm^{-3}$ were achieved at $600\,kV\,cm^{-1}$, which are higher by 69.6% and 65.7% compared to the pure thin films, respectively [46].

Although, $PbZrO_3$-based AFEs in thin-film form have a large energy-storage density, the overall stored energy is usually lower because of thickness limitation, which in some cases could not meet the practical requirement. Thus, the energy-storage properties in thick films with a thickness of 1–100 μm

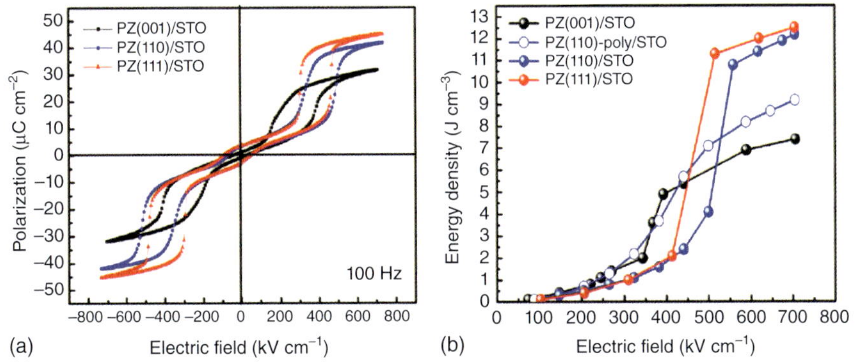

Figure 5.7 (a) *P–E* loops and (b) electric field-dependent recoverable of $PbZrO_3$ thin films with different orientation. *Source:* Ge et al. 2014 [44]. Reproduced with the permission of AIP.

may be more interesting for practical application, and some attempts have been made. In 1971, a moderate recoverable energy density of 2.1 J cm^{-3} was achieved in 65-μm-thick $PbZrO_3$ AFE thick films with a small amount of low-softening-point SiO_2–Bi_2O_3 glass by the screen-printing method [47]. Following this work, Yang' group investigated the energy-storage performance in 50 μm-thick $Pb_{0.97}La_{0.02}(Zr_{0.92}Sn_{0.05}Ti_{0.03})O_3$ AFE films added with $0.8PbO$–$0.2B_2O_3$ glass, which were also prepared by the screen-printing method. The largest recordable energy-storage density with a value of 7.4 J cm^{-3} measured at 475 kV cm^{-1} was obtained in the 3-wt%-glass-added films at room temperature, which was 1.75 times higher than that of the pure films (4.4 J cm^{-3}) [48]. Meanwhile, the energy-storage efficiency was also increased by the added glass. Although, as compared with bulk ceramic, the energy-storage performance was improved in the screen-printed thick films, the larger W_{rec} value was only about 7 J cm^{-3}, which was still caused by the pores-contained microstructure. It could be concluded that if a uniform and dense microstructure can be realized in the materials, a great improvement can be achieved.

Therefore, more recently, the energy-storage properties in the chemical-solution-derived AFE thick films with a thickness of 1–10 μm have attracted our interest, and some attempts have been made. We investigated the effects of heat-treatment process on energy-storage performance of 3.3-μm $Pb_{0.97}La_{0.02}(Zr_{0.98}Ti_{0.02})O_3$ AFE thick films deposited on $LaNiO_3$/Si(100) substrates from a polyvinylpyrrolidone (PVP)-modified chemical solution, in which each wet layer was first dried at 300 °C, and then pyrolyzed at higher temperatures of 600, 650, or 700 °C, respectively. It was found that on increasing the pyrolysis temperature of the single layer, the crystallized films displayed a more uniform and dense surface microstructure [49]. As a result, the dielectric properties, AFE characterization, and energy-storage performance were remarkable improved for the AFE films, as shown in Figure 5.8. A maximum recoverable energy-storage density of 58.1 J cm^{-3} and an energy-storage efficiency of 37.3% were obtained in the films pyrolyzed at 700 °C for every layer.

Based on this result, a series of trials that included chemical composition, design, orientation, and grain size control were carried out to optimize the electrical energy-storage performance in this kind of thick films and some interesting results were obtained. For instance, 1.8-μm-$(Pb_{0.97}La_{0.02})(Zr_{0.95-x}Sn_xTi_{0.05})O_3$ AFE thick films with orthorhombic ($x = 0.05$ and 0.25) and tetragonal ($x = 0.40$) structure were deposited on platinum-buffered silicon substrates using a chemical solution method [50]. With increasing x value, the dielectric constant and critical electric breakdown field of the thick films gradually increased, while their saturated polarizations decreased. As shown in Figure 5.9, their maximum recoverable energy-storage density increased for the thick films with larger x values. A huge recoverable energy-storage density of 56 J cm^{-3} was obtained at its BDS in AFE thick films with $x = 0.40$. Moreover, good temperature-dependent stability of the energy storage was obtained in all the films from 20 to 120 °C. Because of the negligible effect of strain in thick films, their orientation-dependent dielectric properties showed a different tendency, in contrast to thin films. $(Pb_{0.97}La_{0.02})(Zr_{0.73}Sn_{0.22}Ti_{0.05})O_3$ (PLZST) AFE films of 1.5 μm thickness with (100), (110), and (111) orientation were deposited on

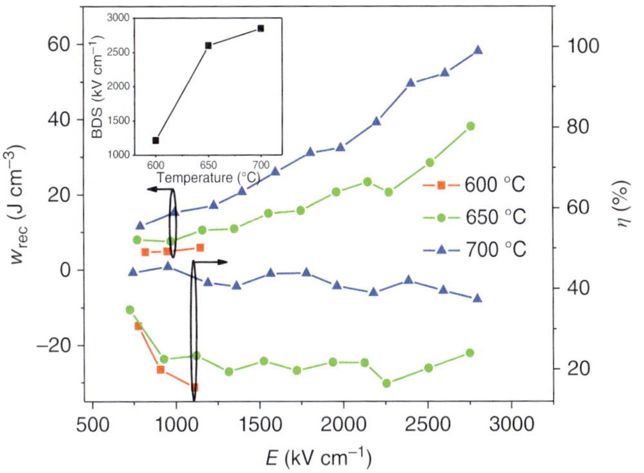

Figure 5.8 Electric-field-dependent energy-storage density and energy-storage efficiency curves of the $Pb_{0.97}La_{0.02}(Zr_{0.98}Ti_{0.02})O_3$ thick films measured at room temperature and up to their critical breakdown field. The insert shows their room-temperature breakdown field. *Source:* Wang et al. 2012 [49]. Reproduced with the permission of AIP.

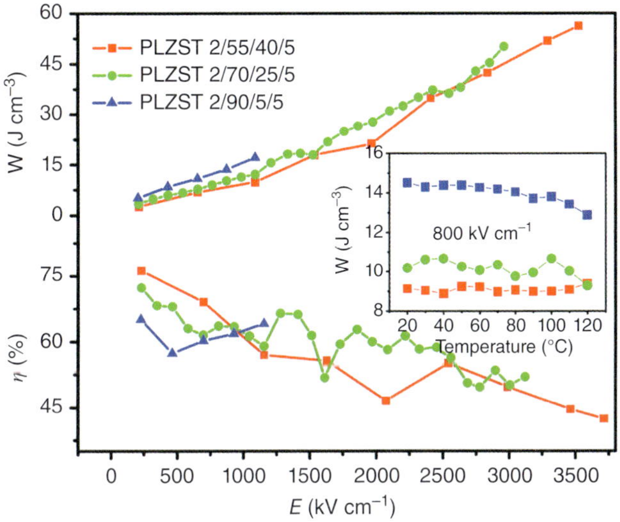

Figure 5.9 Electric-field dependence of recoverable energy-storage density and energy-storage efficiency of the PLZST AFE thick films with $x = 0.05$, 0.25, and 0.40. The inset gives the corresponding temperature dependence of the recoverable energy-storage density measured at 800 kV cm^{-1}. *Source:* Hao et al. 2013 [50]. Reproduced with the permission of AIP.

Pt(111), $LaNiO_3$(110)/Pt(111), and $LaNiO_3$(100)/Pt(111) bottom electrodes, respectively [51]. The corresponding maximum polarizations at 900 kV cm^{-1} for the PLZST(111), PLZST(110), and PLZST(100) films are 77.3, 66.2, and 59.1 μC cm^{-2}, respectively. The schematic diagram of the primitive cells in AFE and FE phases with different orientations is shown in Figure 5.10. The polar

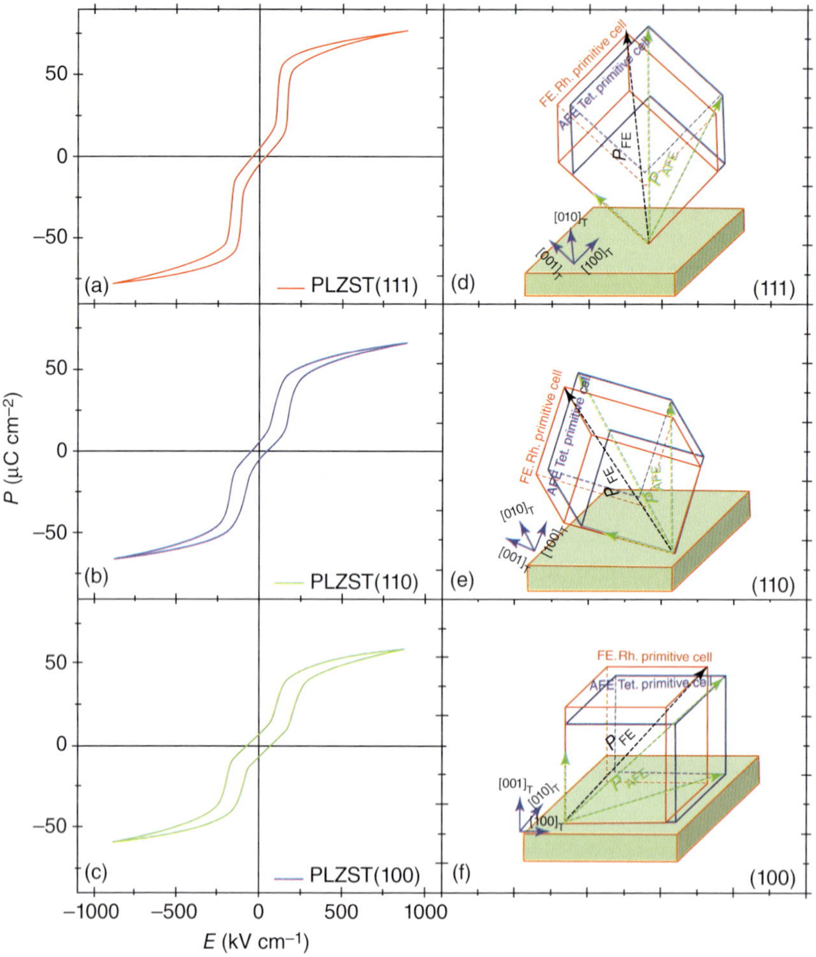

Figure 5.10 (a–c) Room temperature *P–E* loops of the PLZST(111), PLZST(110), and PLZST(100) AFE thick films at 1 kHz. (d–f) The corresponding schematic diagram of the primitive cells. *Source:* Zhao et al. 2016 [51]. Reproduced with the permission of Elsevier.

direction of the antiparallel dipoles is along the [110] direction of the tetragonal primitive cell. When the AFE phase changes into the FE phase under applied electric field, the tetragonal primitive cell becomes rhombohedral with polar directions along the [111] direction. Therefore, the PLZST(111) film possesses the highest P_{max}, followed by the PLZST(110), and then the PLZST(100). As a result, an enhanced energy-storage density of 13.5 J cm^{-3} at 900 kV cm^{-1} was obtained in the PLZST thick film with (111) orientation due to high polarization. By controlling the annealing time, 1.5-μm-thick $Pb_{0.85}Ba_{0.10}La_{0.10}(Zr_{0.90}Ti_{0.10})O_3$ AFE films with average grain size of 0.19, 0.40, and 0.59 μm were fabricated [52]. As the grain size increased, their BDS values decreased from 1882 to 1556 kV cm^{-1}. Accordingly, a maximum recoverable energy-storage density of

33.6 J cm^{-3} and an efficiency of 73% were obtained in the film with a grain size of 0.19 μm.

Apart from higher energy-storage density, AFEs also possess a high discharge speed, which is an important parameter for their practical application. It was reported by Pan et al. that the ferroelectric–antiferroelectric switching time for (Pb,La)(Zr,Sn,Ti)O_3 bulk ceramics first gradually increased with increase in applied electric field and then held constant with an order of 1–2 μs and that samples with larger switching field also had a larger switching speed [53]. The electric field dependence of switching speed was supposed to be related with the stability of the induced ferroelectric phase. It was believed that the stability of the induced ferroelectric phase increased with increasing applied electric field, and thus the energy barrier for the recovery of the AFE phase increased with increasing applied field. As a result, the switching time also increased. When the applied field is high enough to complete the antiferroelectric–ferroelectric transition, a further increase in the applied field would not increase the stability of the induced ferroelectric phase and thus the switching time levels off with increasing applied field. In recent studies, a higher charge releasing speed was observed by Dong's group [54]. It is found that more than 80% stored electric charge can be released in 65 ns for the ceramic with a composition of $Pb_{0.925}La_{0.05}(Zr_{0.42}Sn_{0.40}Ti_{0.18})O_3$ and that this AFE ceramic capacitor can withstand 2000 times of charge–discharge cycling with no significant degradation of properties under 35 kV cm^{-1}, indicating a good fatigue endurance, as shown in Figure 5.11. Similar results were also reported by Xu's group recently [55].

The ferroelectric–antiferroelectric switching speeds in AFE thin films were also investigated by several groups. A typical work was reported by Bharadwaja, in which the electric field, temperature, and thickness dependence of ferroelectric–antiferroelectric switching process were systemically studied in pure $PbZrO_3$ thin films. The results indicated that 60% of the stored energy in

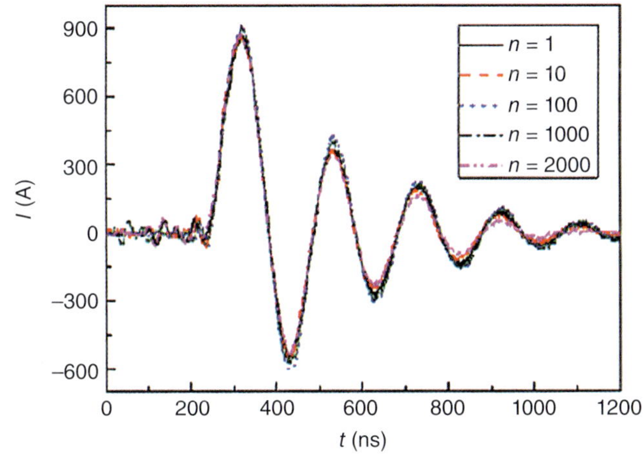

Figure 5.11 Discharged I–t curves of the $Pb_{0.925}La_{0.05}(Zr_{0.42}Sn_{0.40}Ti_{0.18})O_3$ ceramics capacitor after 1, 10, 100, 1000, and 2000 charge–discharge cycling under 3.5 kV mm^{-1}. *Source:* Zhang et al. 2010 [54]. Reproduced with the permission of John Wiley & Sons.

the 550-nm-thick films could be released in less than 7 ns at room temperature and at an applied electric field of 200 kV cm^{-1} [56]. With further increase in the applied field, the switching time was slightly varied and the net released charge was increased, whereas with increasing temperature, the switching time was reduced to 6 ns and the released charge was found to decrease. For the $PbZrO_3$ films with three different thicknesses of 250, 380, and 600 nm, at a fixed applied field of 200 kV cm^{-1}, the maximum released charge increased with the increase of the thickness, but the switching time had no obvious variation.

In summary, due to comparatively low BDS value for bulk ceramics, the maximum recoverable energy-storage density in $PbZrO_3$-based ceramics is still less than 7 J cm^{-3}, which usually is about 2–4 J cm^{-3}. On the contrary, because of their good electric field endurance, an improved recoverable energy-storage density even as high as about 60 J cm^{-3}, accompanied by acceptable energy efficiency of 60–80% and good thermal stability, could be achieved in this kind of AFE thin and thick films, by properly controlling their chemical composition and microstructure. Moreover, both bulk ceramic and film AFEs have a quite fast discharge speed in the time range of several nanoseconds to several microseconds. It should be noted that apart from the dielectric materials, the discharge speed is also dependent on the electric circuit conditions, such as R and L.

5.4.2 $(Na_{0.5}Bi_{0.5})TiO_3$-Based Antiferroelectric

Although a good electric energy-storage performance could be obtained in lead-based AFEs, the toxicity of lead to humans and the environment limits their application in practice. In fact, the use of lead-containing materials has been restricted by some countries and international organizations. Thus, the search for lead-free materials, including lead-free AFEs, has been attracting increasing attention recently. Among the newly exploited lead-free AFEs, $(Na_{0.5}Bi_{0.5})TiO_3$-based AFEs are believed to be the most potential candidates.

$Bi_{0.5}Na_{0.5}TiO_3$ (BNT) with rhombohedral $R3c$ structure at room temperature, which was discovered by Smolenskii in the 1960s, is a typical perovskite-type ferroelectric with a large coercive field (E_c = 73 kV cm^{-1}) and remnant polarization (P_r = 38 μC cm^{-2}) [57, 58]. It exists in the ferroelectric phase up to 200 °C and paraelectric phase above ~320 °C. In the paraelectric regime, there is a ferroelastic–paraelastic transition at ~520 °C. The intermediate phase between 200 and 320 °C was suggested to be AFE based on the presence of double P–E loops. However, its origin is still in dispute. It was initially considered in BNT that rhombohedral ferroelectric phase transits to tetragonal AFE phase at the depolarization temperature (T_d), and the AFE phase exists stably between T_d and T_c (or T_m, where the AFE phase transits to paraelectric phase). However, there is no structural evidence from TEM and neutron diffraction. Yet another view holds that BNT ceramics are only a relaxor, which is widely accepted by researchers. In a wide temperature range, rhombohedral $R3c$ and tetragonal $P4bm$ nanoregions merge in a cubic matrix, which is justified by *in situ* TEM.

Regardless of the origin of its AFE nature, the double *P–E* loops indicate that this kind of material is suitable for high energy-storage applications. Thus, recently, quite a few works have been carried out mainly to decrease the T_d at room temperature and then to achieve acceptable energy-storage density. Generally speaking, to realize this aim, some other compounds, such as $BaTiO_3$, $SrTiO_3$, $(K_{0.5}Na_{0.5})NbO_3$, $(K_{0.5}Bi_{0.5})TiO_3$, and $CaZrO_3$, were introduced into the BNT matrix to form a binary or ternary solid solution with the perovskite structure [59]. The reported results indicated that obtaining an ideal AFE-like *P–E* loop with large maximum polarization and zero remnant polarization at room temperature is still not easy in this kind of dielectrics. As a result, the recoverable energy-storage density at room temperature was about 1 J cm^{-3}, and the maximum value was only 1.58 J cm^{-3} to date in NBT-based bulk ceramics, as shown in Table 5.2. The corresponding energy efficiency is not very high and is usually less than 70%. The comparatively low dielectric property and inferior energy-storage performance were mainly caused by the low BDS value and worse polarization, because of the violation of Bi^{3+} and Na^+.

The energy-storage performance in NBT-based thick films was studied in the author's laboratory. With a small addition of $Bi_2O_3–Li_2O$ glass, $(1-x)$NBT $-x$BT and $(1-x)$NBT $-x$ST films with a thickness of about 20 μm were prepared on the Ag–Pd (95/5)-buffered alumina substrates by the screen-printing method [77]. A maximum recoverable energy-storage density of 2.7 J cm^{-3} with efficiency of 45% was achieved at 600 kV cm^{-1} in 0.70NBT-0.30ST thick films, which showed a good thermal stability in a wide temperature range from 20 to 200 °C. Subsequently, 1.5-μm $(1-x)$NBT $-x$ST ($x = 0$, 0.05, 0.10, and 0.15) thick films were successfully fabricated on $LaNiO_3$/Si(100) substrate by using a PVP-modified chemical solution route [78]. It was found that the addition of ST enhanced BDS and improved the difference between the maximum polarization (P_{max}) and the remnant polarization (P_r) of the thick films. A maximum recoverable energy density of 36.1 J cm^{-3} and the corresponding efficiency of 40.8% were obtained in a 0.095NBT-0.05ST thick film, which also displayed good energy-storage stability in the temperature range from room temperature to 80 °C. Moreover, under an external electric field of 500 kV cm^{-1}, the thick film showed a low leakage current density of 1.93×10^{-5} A cm^{-2} at room temperature. Because of their good electric field endurance, a highly recoverable energy-storage density could be achieved in NBT-based thick films, but the energy efficiency was quite low, less than 50%, which induced high energy loss at elevated electric fields. In fact, the *P–E* loops at high electric fields of NBT-based thick films always displayed ferroelectric-like curves.

Interestingly, recently, a giant energy-storage effect (137 and 154 J cm^{-3} for (100) and (111)-oriented thin films, respectively) with high efficiency above 90%, excellent anti-fatigue property (10^8 cycles), and thermal stability (variation below 2.6% at 2180 kV cm^{-1} from room temperature to 110 °C) has been achieved in the AFE/FE-coexisting $(Bi_{0.5}Na_{0.5})_{0.9118}La_{0.02}Ba_{0.0582}(Ti_{0.97}Zr_{0.03})O_3$ epitaxial thin films with a thickness 350 nm, prepared by PLD method [79]. In this case, the thin films show a very slim loop with AFE-like character. This result indicated that high-quality film fabrication is the key to obtaining excellent energy-storage performance in NBT-based AFE materials.

Table 5.2 The reported energy-storage performance of NBT-based bulk ceramics at room temperature.

Composition	W_{rec} (J cm^{-3})	E (kV cm^{-1})	η (%)	Sintering method	References
11 mol%-Mn-doped 0.7[0.94NBT-0.06BT]-0.3ST	1.06	95	79.1	Conventional sintering method	[60]
0.89NBT-0.06BT-0.05KNN	0.90	99	—	Two-step sintering	[61]
5 wt% MgO-doped 0.475NBT-0.525BCTZ	1.37	179.1	61.4	Microwave sintering method	[62]
0.84NBT-0.16BKT	1.2	100	—	Conventional sintering method	[63]
0.97[0.94NBT-0.06BT]-0.03CZ	0.7	70	—	Conventional sintering method	[64]
$[(Bi_{1/2}Na_{1/2})_{0.94}Ba_{0.06}]La_{0.98}Zr_{0.02}TiO_3$	1.58	84.3	—	Conventional sintering method	[65]
$(Bi_{0.5}Na_{0.5})_{0.94}Ba_{0.06}Ti_{0.96}(Al_{0.5}Nb_{0.5})_{0.04}O_3$	0.70	70	70.4	Conventional sintering method	[66]
0.96[0.94NBT-0.06BT]-0.06KN	0.89	100	—	Conventional sintering method	[67]
0.90[0.92NBT-0.08BT]-0.10NN	0.71	70	—	Conventional sintering method	[68]
0.90[0.92NBT-0.08BT]-0.10BMT	0.77	70	54.5	Conventional sintering method	[69]
[0.950.97NBT-0.07BT]-0.05KN	1.72	168	—	Conventional sintering method	[70]
0.75[0.80NBT-0.20KBT]-0.25ST	0.97	—	—	Two-step sintering	[71]
3 mol%-BaO–B_2O_3–SiO_2–0.95NBT–0.05BT	0.68	94.6	71	Conventional sintering method	[72]
0.85[0.96NBT-0.04BT]-0.15SZ	1.32	155	58	Conventional sintering method	[73]
0.80[0.92NBT-0.08BT]-0.20NBN	1.36	136	73.9	Conventional sintering method	[74]
0.89NBT-0.06BT-0.05KNN	0.59	56	—	Conventional sintering method	[75]
0.90NBT-0.10KN	1.17	104	—	Conventional sintering method	[76]

5.4.3 $AgNbO_3$-Based Antiferroelectric

Similar to NBT ceramics, $AgNbO_3$ also has several temperature-dependent phase transitions. With the temperature increasing, it first changes from the orthorhombic to the tetragonal phase at 183 °C, and then into the cubic phase at 251 °C. For the pure $AgNbO_3$ bulk ceramics, an ideal double *P–E* loop could be achieved at room temperature; its saturated polarization is about 52 μC cm^{-2} and the critical transition field of AFE to ferroelectric phase is about 110 kV cm^{-1} [80]. Although $AgNbO_3$ is a good candidate for electric energy-storage application, only a few works have been reported. This is due to two reasons. One is that the price of Ag raw material is relatively high. The other reason is that Ag_2O decomposes above 200 °C, which leads to a significant deterioration in the electrical properties of $AgNbO_3$.

Recently, Xu and Li's groups, respectively, successfully prepared high-quality pure $AgNbO_3$ bulk ceramics by the conventional solid-state reaction method under annealing in O_2 atmosphere [81, 82]. Similar results were obtained. It was found that $AgNbO_3$ ceramics showed a characteristic AFE double *P–E* loop at an electric field of 130 kV cm^{-1}, with a peak recoverable energy-storage density of 1.6 J cm^{-3} at 140 kV cm^{-1}. In addition, the incorporation of 0.1–0.3 wt% MnO_2 into $AgNbO_3$ can further increase W_{rec}, exceeding 2.3 J cm^{-3} at 150 kV cm^{-1} by the reduction of the remnant polarization, which is due to the enhanced AFE stability induced by the addition of MnO_2. Of particular importance is the fact that the 0.1 wt% MnO_2-doped $AgNbO_3$ ceramics were found to possess good thermal stability with $W_{rec} = 2.5–2.9$ J cm^{-3} over a temperature range of 20–180 °C at 150 kV cm^{-1} and 1 Hz.

5.4.4 HfO_2-Based Antiferroelectric

HfO_2 has various polymorphs in the bulk phase depending on the temperature and pressure, which transits from the monoclinic to the tetragonal phase at 1700 °C, and to the cubic phase at 2500 °C. It is an electrical insulator with an energy gap of 5.3–5.7 eV, and its relative dielectric constant ε_r is about 17 at room temperature [83].

The ferroelectricity and antiferroelectricity in HfO_2-based superthin films were first discovered in 2007 by Böscke et al., but their origins are still unclear [83]. It was reported that Y-, Al-, Gd-, Sr-, and La-doped HfO_2 films usually displayed ferroelectric properties, while Si- and Zr-doped films showed AFE properties. The study on the electrical energy-storage performance of HfO_2-based AFE was mainly focused on Zr-doped superthin films.

AFE $Hf_xZr_{1-x}O_2$ $(0.1 < x < 0.4)$ films with a thickness of 7.7–29 nm were deposited on a TiN-electroded SiO_2/Si substrate via thermal ALD at 280 °C [16]. All the thin $Hf_xZr_{1-x}O_2$ films with a thickness of 9.2 nm could have a large recoverable energy-storage density higher than 30 J cm^{-3}, and a maximum value of 46 J cm^{-3} was achieved in 9.2-nm-thick $Hf_{0.3}Zr_{0.7}O_2$ films at the applied electric field of 4.5 MV cm^{-1}. The recoverable energy-storage density of $Hf_{0.3}Zr_{0.7}O_2$ films decreased with an increase in the film's thickness, likely due to the increased formation of the m-phase, but was still as high as ~21 J cm^{-3}

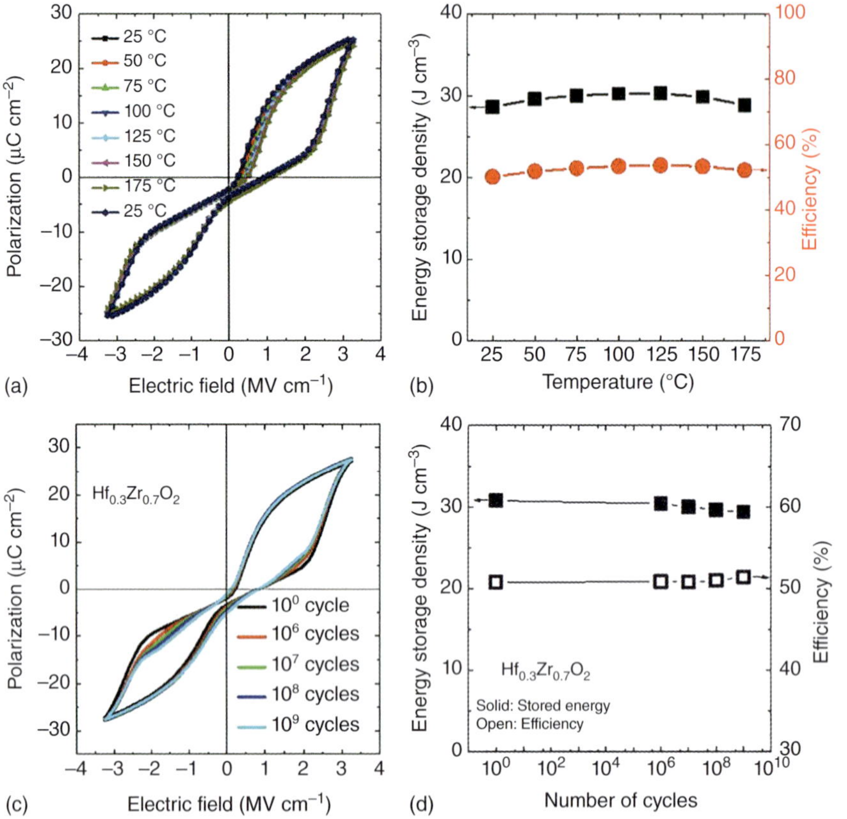

Figure 5.12 (a) *P–E* hysteresis curves and (b) energy-storage density as a function of the electric field of the 9.2-nm-thick $Hf_{0.3}Zr_{0.7}O_2$ capacitor at various temperatures (25–175 °C). (c) *P–E* hysteresis curves and (d) energy-storage density as a function of the electric field of the 9.2-nm-thick $Hf_{0.3}Zr_{0.7}O_2$ capacitor after various electric field cycling procedures with a pulse height of 3.26 MV cm^{-1} at the cycling frequency of 500 kHz. *Source:* Park et al. 2014 [16]. Reproduced with the permission of John Wiley & Sons.

for the maximum thickness of 29 nm. Acceptable energy efficiency of about 55% was also obtained in 9.2-nm-thick $Hf_{0.3}Zr_{0.7}O_2$ films. A good thermal stability of energy-storage performance was achieved in $Hf_{0.3}Zr_{0.7}O_2$ films in a wide temperature range from 25 to 175 °C. Moreover, the large recoverable energy-storage density decreased by only about 4.5% even after 10^9 times of electric field cycling at a switching electric field of 3.26 MV cm^{-1}. The results are illustrated in Figure 5.12 [16].

5.5 Energy-Storage Performance in Glass-Ceramic Ferroelectrics

Generically, high BDS up to 10 MV cm^{-1} could be realized in glass because of the pore-free microstructure. Thus, it is supposed that good energy-storage performance should be achieved in the composition of glass-ceramic ferroelectrics by employing the advantages of high BDS of glass and large dielectric constant of ferroelectrics. Recently, much work on the energy-storage performance in glass-ceramic ferroelectrics has been reported [84]. Glass-ceramics are usually synthesized from two different methods: compositing method and body-crystallization method (including melting–quenching technique and sol–gel route). In the ensuing section, the energy-storage performance in the glass-ceramic ferroelectrics prepared using the two methods is first summarized, and then the interface effect on their energy-storage performance is highlighted.

5.5.1 Glass-Ceramic Ferroelectrics Prepared by Compositing Method

In this case, the desired glass (or low melting point oxide) and ceramic powders are first prepared, respectively; then they are mixed at the determined ratio and pressed into certain shapes; finally, the compositions are sintered at a certain temperature to form the product. Glass-ceramics synthesized in this way usually possess ideal chemical compositions, but some pores often exist. Generally, with increase in glass content, the grain size in the glass-ceramic ferroelectrics declines and uniformity increases, which is favorable to improving BDS, while the dielectric constant decreases. Therefore, the optimized energy-storage performance could be obtained by properly controlling the glass content.

Currently, investigations on the energy-storage performance in glass-ceramic ferroelectrics prepared by this method are mainly focused on $BaTiO_3$-based system, while the recoverable energy-storage density obtained is usually less than 1 J cm^{-3}. Here, some typical examples are illustrated. $Ba_{0.4}Sr_{0.6}TiO_3$ ceramics containing 5 vol.% $BaO-SiO_2-B_2O_3-Al_2O_3-ZrO_2-SrO$ glass displayed a recoverable energy density of 0.89 J cm^{-3}, which was 2.4 times higher than that of the pure ceramics (0.37 J cm^{-3}) [85]. A recoverable energy density of 0.45 J cm^{-3} with an energy efficiency of 88.2% was achieved at its BDS in $Ba_{0.4}Sr_{0.6}Zr_{0.15}Ti_{0.85}O_3$ ceramics with 5 wt% $SrO-B_2O_3-SiO_2$ glass addition [86]. Similarly, 5 wt% $MgO-CaO-Al_2O_3-SiO_2$ glass-added $Ba_{0.95}Sr_{0.05}Zr_{0.02}Ti_{0.8}O_3$ bulk ceramics showed a recoverable energy density of 0.42 J cm^{-3} and an energy efficiency of 82.8% at its critical breakdown field of 140 kV cm^{-1} [87]. These results indicated that the improvement of energy-storage performance for the mixture by simply mixing the glass and ceramic is quite finite.

Quite recently, a new technique was employed to fabricate glass-ceramic ferroelectrics by the compositing method, in which the fine ferroelectric ceramic grain was coated with glass to form the core–shell structure and to modify the surface of the grain while maintaining a fine and uniform microstructure, as shown in the diagram given in Figure 5.13. Excellent energy-storage performance was achieved in $Ba_{0.4}Sr_{0.6}TiO_3$ particles coated with Al_2O_3 and SiO_2 oxide. Under the optimized composition of $3Al_2O_3$:SiO_2, a maximum calculated stored energy density of 5.09 J cm^{-3} was achieved, whereas it was only 0.24 J cm^{-3} for the uncoated ceramics [88].

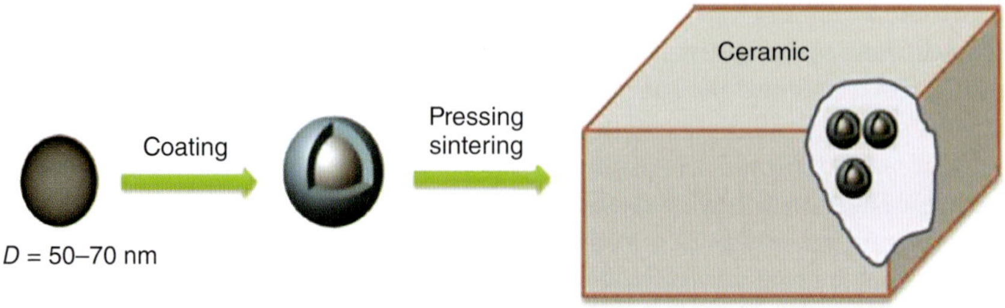

Figure 5.13 Schematic diagram of technological processes of the glass-coated ferroelectric ceramic grain and ceramics. *Source:* Wang et al. 2013 [88]. Reproduced with the permission of Springer Nature.

5.5.2 Glass-ceramic Prepared by Body-crystallization Method

For this method, an amorphous phase with the desired chemical elements is first fabricated; then the amorphous particles are partly crystallized at a certain temperature to obtain the final glass-ceramic. In this case, a pore-free microstructure can be easily obtained in the glass-ceramics, but their composition is difficult to be carefully controlled. Currently, many glass-ceramic ferroelectrics have been fabricated via this method and their energy-storage performance studied. According to the chemical composition, three kinds of glass-ceramic ferroelectrics were widely studied, as illustrated in this section.

5.5.2.1 Lead-Containing Glass-ceramic

Du's group studied the effect of crystallizing temperature on the energy-storage performance of the Na_2O–PbO–Nb_2O_5–SiO_2 system [89]. $Pb_2Nb_2O_7$ and $NaNbO_3$ were detected in the crystallized glass-ceramics, and the content of $NaNbO_3$ increased as the crystallized temperature increased from 700 to 900 °C. Moreover, as the crystallizing temperature increased, the dielectric constant increased, but the BDS decreased. At 850 °C, a maximum stored energy density of about 17 J cm^{-3} was obtained. Subsequently, the author studied the effect of crystallization time on the discharged energy properties in a Sr–PbO–Na_2O–Nb_2O_5–SiO_2 system, which was crystalized at 900 °C for 1, 10, 100, and 1000 minutes, respectively [90]. XRD result indicated that regardless of the crystallization time, both ferroelectric (Sr,Pb)Nb_2O_6 and AFE $NaNbO_3$ phases were formed. However, the grain size increased as the crystallization time increased, which led to improved dielectric constant of the final glass-ceramics. Accordingly, the recoverable energy-storage density was increased from 0.18 to 0.28 J cm^{-3}, but the energy efficiency declined from 96.3% to 89.8%.

5.5.2.2 $BaTiO_3$-Based Glass-ceramic

Gorzkowski et al. studied the energy-storage properties of BaO–SrO–TiO_2–Al_2O_3–SiO_2 glass-ceramics with (Ba,Sr)TiO_3 ferroelectric phase [91]. A relative dielectric constant of 1000 and BDS of 800 kV cm^{-1} were achieved; however, the recoverable energy-storage density from *P–E* results was only 0.3–0.9 J cm^{-3}, because of the large hysteresis loss. Following this, systemic work on BaO–SrO–TiO_2–Al_2O_3–SiO_2 system was carried out by Zhang's group [92]. It was found that sintering temperature had a deep influence on the phase composition and the dielectric properties. (Ba,Sr)TiO_3 ferroelectric phase formed after annealing at 830 °C, and the degree of crystallinity increased with increasing annealing temperature from 830 to 950 °C. A maximum J_{reco} of 0.18 J cm^{-3} was obtained at 280 kV cm^{-1}.

Moreover, it was reported that MnO_2, AlF_3, La_2O_3, and so on were efficient additives to optimize the microstructure and energy-storage properties of $BaTiO_3$-based glass-ceramics. For instance, Zhai's group prepared (Ba,Sr)TiO_3 glass-ceramics doped with different La contents from the Ba–SrO–TiO_2–Al_2O_3–SiO_2 glass system [93]. The results revealed that the La additive had little influence on the dielectric constant but significantly changed the microstructure of the glass-ceramics, which led to an improved BDS. An optimized energy-storage density of 3.18 J cm^{-3} was achieved in the glass-ceramics with 1.0 wt% La_2O_3 content, which was 2.56 times higher than that in pure (Ba,Sr)TiO_3 glass-ceramics. Recently, it was also found that crystallization route had a strong effect on the energy-storage performance of glass-ceramics. (Ba,Sr)TiO_3 glass-ceramics were fabricated via conventional sintering and microwave sintering [94]. It was found that the microwave process had little impact on the crystallinity (about 90 wt%), but it increased the crystallization of $SrAl_4O_7$. In addition, microwave crystallization can increase the dielectric constant from 33 to 38 and enhance BDS from 561.3 to 791.4 kV cm^{-1}. It was believed that the improved BDS was primarily ascribed to a trade-off between the growth in grain size and the inexistence of dendrites, as demonstrated in Figure 5.14. As a result, the energy-storage density was increased from 0.47 to 1.05 J cm^{-3}.

5.5.2.3 Nb-Containing Glass-ceramic

Comparatively, Nb-containing glass-ceramics were the most studied for high energy-storage application. The substrates in crystalline phase usually are $NaNbO_3$, $Ba_2NaNb_5O_{15}$, $Ba_2KNb_5O_{15}$, $Sr_2KNb_5O_{15}$, $Ba_{0.5}Ba_{0.5}Nb_2O_6$, and so on. Similarly, the studies also focused on the effects of glass compositions, nucleating agent additives, crystallization temperature and time,

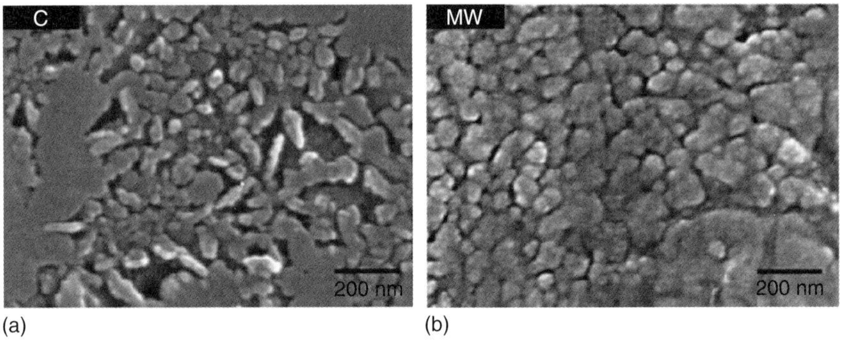

Figure 5.14 FE-SEM micrographs of (Ba,Sr)TiO_3 glass-ceramics crystallized by conventional sintering (a) and microwave sintering (b). *Source:* Wang et al. 2013 [94]. Reproduced with the permission of Cambridge University Press.

sintering method, etc. on the microstructure, dielectric properties, and energy-storage performance. For example, ferroelectric glass-ceramics of the $SrO{-}BaO{-}Nb_2O_5{-}CaO{-}SiO_2{-}B_2O_3$ system ($Sr_{0.5}Ba_{0.5}Nb_2O_6$ phase) with different nucleating agents Cr_2O_3, BaF_2, ZrO_2, CaF_2 and CeO_2 were prepared by conventional melt casting followed by controlled crystallization processing [95]. It was found that the addition of different nucleating agents decreased the crystallization temperatures, accelerated the secondary phase (SiO_2) elimination, and promoted the uniform distribution of grain size. The good development of microstructure promoted by the nucleating agents resulted in an improvement in the dielectric constant as well as energy-storage density remarkably. The sample with CeO_2 as the optimum nucleating agent achieved a relative dielectric constant of 301, BDS of 622 kV cm^{-1}, and the highest stored energy density of 5.15 J cm^{-3}. For simplicity, the current reported results on the energy-storage performance of Nb-containing glass-ceramics are summarized in Table 5.3.

5.5.3 Interface Effect-Related Energy-Storage Performance

Although high stored energy density could be achieved in glass-ceramics, the recoverable energy-storage density is often not as large as expected. This means that much of the stored energy could not be discharged. It was generally believed to be caused by the interfacial polarization between the amorphous phase and the crystalline phase, which had a long relaxation process. Recently, Zhai's group studied the correlation between energy conversion efficiency and interface polarization according to the discharge performance and impedance spectroscopy as a function of annealing temperatures in the $BaO{-}Na_2O{-}Nb_2O_5{-}SiO_2$ glass ceramic system [105]. The experimental results showed that energy conversion efficiency strongly depended on the interface polarization. Because of the significant difference in dielectric constant and conductivity between the ferroelectric phase and the glass matrix, interfacial polarization occurred under the influence of applied electric field; as a result, large amounts of charges accumulated at the interface. Thus, the charges at the interface cannot be eliminated during the discharging process because of the long relaxation process of interfacial polarization. As a result, some charges stored in electrodes were trapped to offset the electric field caused by the interfacial polarization charges in the discharge process. As shown in Figure 5.15a, without interfacial polarization, there appeared to be no significant gain in the charge and discharge process for the dielectric capacitor. All charges stored in capacitor electrodes could be completely released in the discharge process. As shown in Figure 5.15b, in the case of low interfacial polarization, the majority of charges stored in capacitor electrodes could be released, so that only a small amount of charges were trapped in the discharge circle, which indicated high energy conversion efficiency. With increasing interfacial polarization, more charges could be stored in the charging process. However, under the influence of electric field caused by the interfacial polarization charges, more charges were trapped in the discharge process, as shown in Figure 5.15c. Consequently, lower discharged charges and lower energy efficiency were obtained.

Table 5.3 The reported energy-storage performance of Nb-contained glass-ceramics.

Glass system	Nucleating agent	Crystalline phase	W (J cm^{-3})	W_{rec} (J cm^{-3})	BDS (kV cm^{-1})	η (%)	Sintering method	References
$BaO–Na_2O–Nb_2O_5–SiO_2$	Sm_2O_3	$NaNbO_3/Ba_2NaNb_5O_{15}$	0.541	0.368	—	68	Conventional sintering	[96]
$BaO–Na_2O–Nb_2O_5–SiO_2$	La_2O_3	$NaNbO_3/Ba_2NaNb_5O_{15}$	—	1.2	280	—	Conventional sintering	[97]
$BaO–Na_2O–Nb_2O_5–SiO_2$	La_2O_3	$NaNbO_3/Ba_2NaNb_5O_{15}$	8.4	—	1375	—	Conventional sintering	[98]
$BaO–K_2O–Nb_2O_5–SiO_2$	—	$KNb_2O_6/Ba_2KNb_5O_{15}$	14.58	—	2382	—	Conventional sintering	[99]
$K_2O–SrO–Nb_2O_5–B_2O_3–Al_2O_3–SiO_2$	—	$AlNbO_4/KSr_2Nb_5O_{15}$	8.99	—	1410.81	—	Microwave sintering	[100]
$SrO–BaO–Nb_2O_5–B_2O_3$	—	$SrNb_2O_6/Sr_{0.5}Ba_{0.5}Nb_2O_6$	5.71	—	1050	—	Conventional sintering	[101]
$SrO–BaO–Nb_2O_5–B_2O_3$	Gd_2O_3	$Sr_{0.5}Ba_{0.5}Nb_2O_6$	6.94	—	1075	—	Conventional sintering	[102]
$SrO–BaO–Nb_2O_5–CaO–SiO_2–B_2O_3$	CeO_2	$Sr_{0.5}Ba_{0.5}Nb_2O_6$	5.15	—	622	—	Conventional sintering	[103]
$Al_2O_3–SiO_2–SrO–BaO–Nb_2O_5$	ZnO	$Sr_{0.5}Ba_{0.5}Nb_2O_6$	6.0	—	950	—	Conventional sintering	[104]

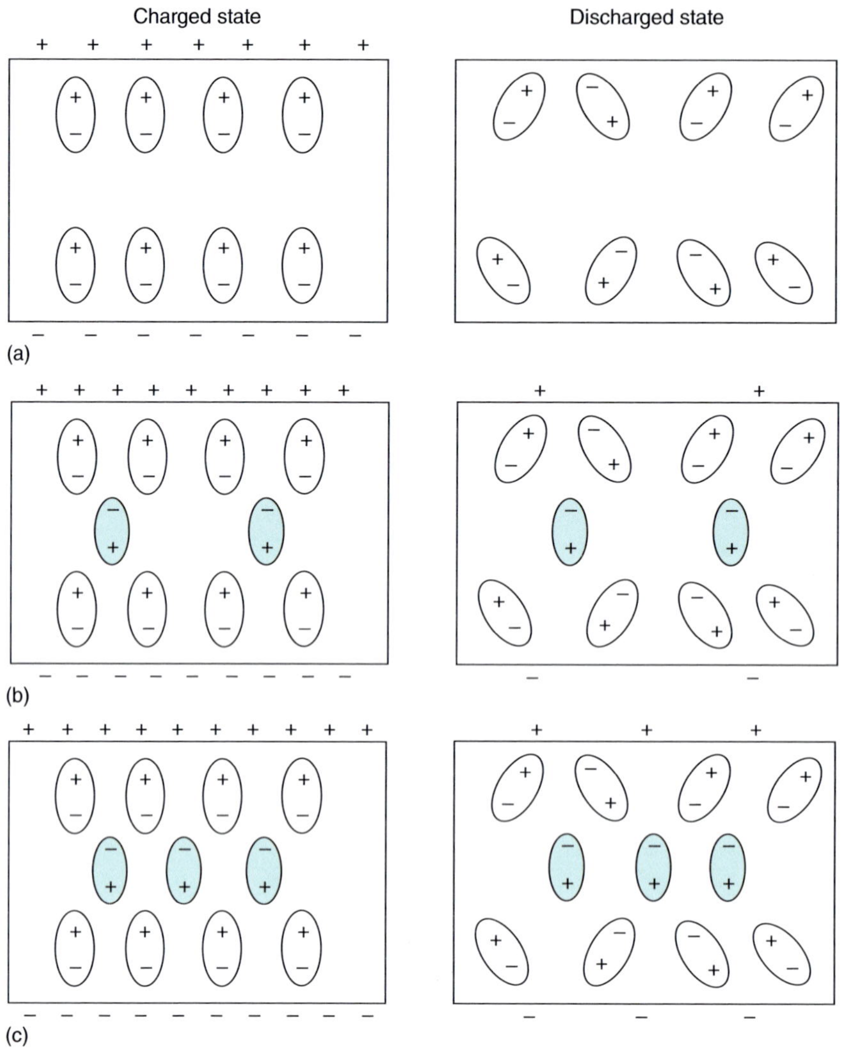

Figure 5.15 Schematic of charge and discharge mechanism under an electric field in a dielectric capacitor: (a) without interfacial polarization, (b) with low interfacial polarization, and (c) with high interfacial polarization. *Source:* Xue et al. 2015 [105]. Reproduced with the permission of AIP.

Apart from energy efficiency, interfacial polarization also has a strong effect on the BDS of glass-ceramics. Zhang's group studied the correlation between the dielectric breakdown performance and interface polarization by the measurement of the dielectric BDS and impedance spectroscopy as a function of sintering temperatures in a $BaO–SrO–TiO_2–Al_2O_3–SiO_2$ glass ceramic system [106]. The sintering-temperature-dependent BDS and activation energy E_a are given in Figure 5.16. Clearly, BDS and E_a showed opposing trends of variation with temperature. The activation energy, corresponding to the relaxation of space charge,

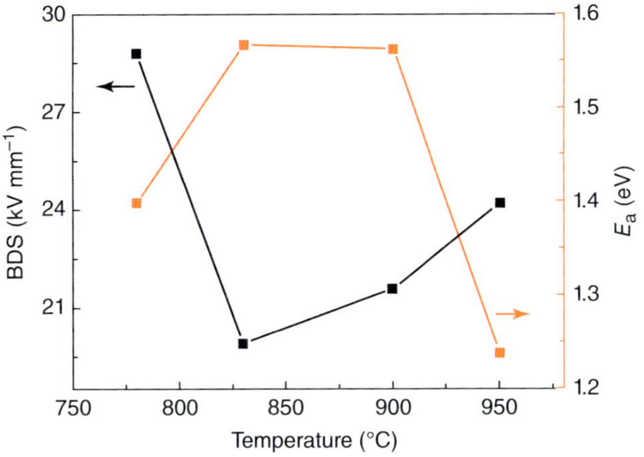

Figure 5.16 BDS and activation energy E_a as a function of sintering temperature for the glass ceramics. *Source:* Huang et al. 2010 [106]. Reproduced with the permission of AIP.

indicated the effective traps of energy or the interfacial mobility in the defects where polarizability was modified and where charge and energy localization could occur. A lower value of E_a represented better charge spreading behavior. The sample sintered at 780 °C, which is lower than its crystallization temperature contained either a significant volume fraction of the amorphous or the intermediate phase, indicating high homogeneity, low defect, and low interface number. As a result, large BDS and small E_a were obtained in this case. For the sample sintered at 830 °C, the major phase (Ba,Sr)TiO_3 and secondary phase (Ba,Sr)$Al_2Si_2O_8$ with smaller grain size were formed in the glass matrix, which led to the production of a large amount of interfaces between the crystallized phases and the residual glass. Thus, a drastic increase in accumulation of charges at the interfaces occurred owing to the big difference in their dielectric constants. Accordingly, the BDS decreased, and at the same time E_a increased. For the sample sintered at further higher temperature of 900–950 °C, both the volume of the crystallized phase and the grain size increased, which led to a reduction of the interfaces. Therefore, BDS increased and E_a decreased. Moreover, in this case, new interfaces with small difference in dielectric constant were also formed between the major crystallized phase and the secondary crystallized phase, which also led to a reduction of E_a. In summary, properly controlling the interface polarization is a key to optimizing the energy-storage performance of glass-ceramics.

5.6 Energy-Storage Performance in Relaxor Ferroelectrics

Different from the normal ferroelectric, under the function of external electric field, the relaxor ferroelectrics typically show a slim *P*–*E* loop with a large dielectric constant, small remnant polarization, and a coercive field, which was believed to originate from the nanometer-sized polar regions [107, 108]. Another general feature of relaxor ferroelectrics is the existence of a quite wide peak in the temperature-dependent dielectric constant. Moreover, the temperature T_m corresponding to the peak of the dielectric constant is shifted to higher temperatures, as the measuring frequency increases. These characteristics offer relaxor ferroelectrics the potential for application in high energy storage. Quite a few types of relaxor ferroelectrics, such as $BaTiO_3$-, $PbTiO_3$-, $BiFeO_3$-based systems, were investigated recently worldwide for electrical energy storage. Thus, the next section presents a timely summary to highlight the achievements in relaxor ferroelectrics.

5.6.1 PLZT Relaxor Ferroelectrics

The phase diagram of $(Pb_{1-3x/2}La_x)(Zr_{1-y}Ti_y)O_3$ (PLZT) was developed by Haertling. As shown in Figure 5.17, by properly controlling their chemical composition, relaxor ferroelectric phase cloud be obtained [109]. Currently, the investigation on the electrical energy storage of relaxor PLZT is mainly carried out in the film form. Much attention was also paid to this kind of relaxor films by the author. We successfully fabricated 1-μm PLZT 9/65/35 relaxor ferroelectric films on $Pt(111)/Pt/Ti/SiO_2/Si$ substrates by sol–gel method. A considerable capacitance density of $925\,nF\,cm^{-2}$ at 1 MHz and high BDS of $2177\,kV\,cm^{-1}$ were realized in thin films [110]. As shown in Figure 5.18, this relaxor film also possessed a slim *P–E* loop with a large difference between the saturated polarization and the remnant polarization, which was $49.1\,\mu C\,cm^{-2}$ at $1580\,kV\,cm^{-1}$. As the external field increased from 200 to $2040\,kV\,cm^{-1}$, the recoverable energy-storage density at 1 kHz was raised from 1.7 to $28.7\,J\,cm^{-3}$. However, the energy efficiency showed good stability and varied between 52.4% and 59.4% in the measurement range. It was also found that the relaxor thin film possessed a good temperature- and frequency-dependent stability. Moreover, a good fatigue against performance and a low leakage current density with an order of $10^{-6}\,A\,cm^{-2}$ were also detected in this film.

Following this, in order to optimize the electrical energy-storage performance of PLZT relaxor films, both Mn-doped and chemical compositionally graded films were deposited on $LaNiO_3/Si(100)$ substrates via the same

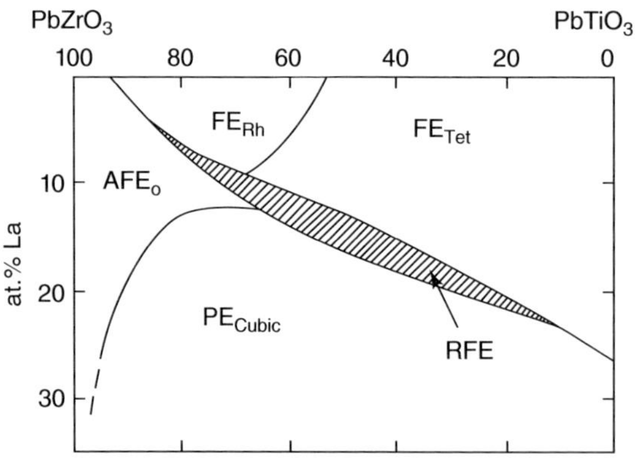

Figure 5.17 Room temperature phase diagram of PLZT solution. *Source:* Haertling and Land 1971 [109]. Reproduced with the permission of John Wiley & Sons.

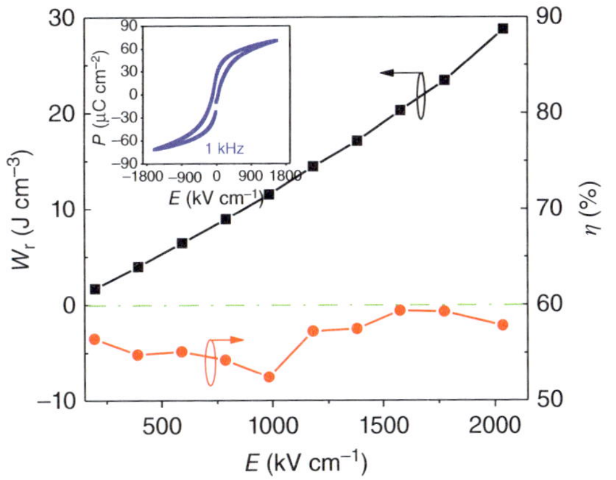

Figure 5.18 The electric field dependence of energy-storage density and energy-storage efficiency of $(Pb_{0.91}La_{0.09})(Zr_{0.65}Ti_{0.35})O_3$ thin film measured at room temperature and 1 kHz. The inset shows the *P–E* loop of the thin film measured at room temperature and at 1 kHz. *Source:* Hao et al. 2012 [110]. Reproduced with the permission of AIP.

method [111, 112]. The thickness of the films was about 1.5 μm in both cases. An obvious improvement in dielectric properties and energy-storage performance was realized in Mn-doped films. The dielectric constant, $P_{max} - P_{rem}$, and BDS attained their maximum values in 1-mol% Mn-doped PLZT films, which were 1378, 58.9 μC cm^{-2} and 1679 kV cm^{-1}, respectively. As a result, the recoverable energy density of 1-mol% Mn-doped films at 1185 kV cm^{-1} increased by 60% in contrast to un-doped PLZT films. At critical breakdown fields, a huge recoverable energy density of 30.8 J cm^{-3} was achieved in 1-mol% Mn-doped films, as shown in Figure 5.19a. Similarly, it was found that the dielectric properties and the energy-storage performance of PLZT relaxor films were also strongly affected by the composition graded sequence. For convenience, the films with La content varying from 8-mol% at the substrate interface to 10-mol% at the top surface were named "upgraded" films, whereas films with opposite gradient were called "downgraded" films. Compared with the single-composition films, the up- and downgraded films showed higher dielectric constant, and larger difference between the maximum polarization and the remnant polarization. As a result,

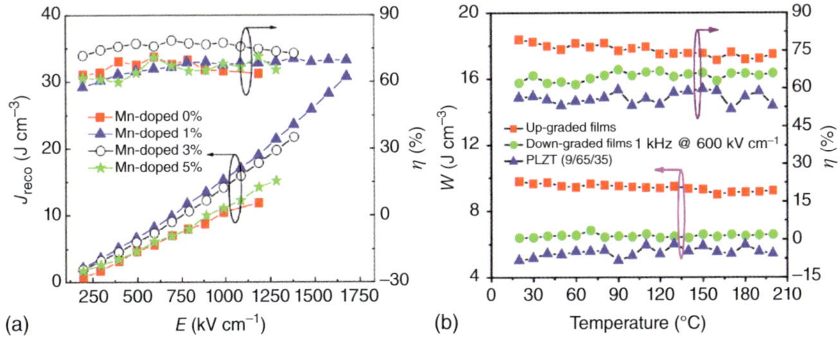

Figure 5.19 (a) Electric field-dependent energy-storage performance of PLZT films doped with 0, 1, 3, and 5 mol% Mn. *Source:* Liu et al. 2013 [111]. Reproduced with the permission of AIP. (b) Temperature-dependent energy-storage performance of composition-graded PLZT films. *Source:* Zhang et al. 2013 [112]. Reproduced with the permission of AIP.

improved energy-storage performance was realized in the compositionally graded films. At a measurement field of 600 kV cm^{-1} and at 1 kHz, the largest recoverable energy-storage density of 12.4 J cm^{-3} was obtained in the upgraded films, which was 75% higher than that of the single-composition films, as shown in Figure 5.19b. Moreover, a higher energy-storage efficiency above 75% was also realized in the upgraded films in a wide temperature range. These results indicated that PLZT relaxor ferroelectric films are promising candidates for high energy storage.

Much work on the electrical energy storage of PLZT relaxor films was also reported by Argone National Laboratory. It was found that Zr/Ti ratios had a strong effect on the dielectric properties and energy-storage performance [113]. The highest recoverable energy density of about 30 J cm^{-3} and the corresponding energy efficiency of about 78% were achieved in PLZT 8/52/48 thin films at 2180 kV cm^{-1}. Following this, the authors studied the effect of bottom electrodes (Pt(111)/Pt/Ti/SiO_2/Si and $LaNiO_3$/Ni) on PLZT 8/52/48 films with a thickness of about 3 μm [114]. PLZT 8/52/48 films deposited on $LaNiO_3$/Ni showed a dense polycrystalline phase without apparent defects or secondary phase, which led to a greater BDS of 1600 kV cm^{-1}. A recoverable energy-storage density of 22 J cm^{-3} and energy efficiency of 77% were achieved in the films with $LaNiO_3$/Ni substrates. PLZT film-on-foil (Ni) capacitor could be easily embedded into a printed wire board (PWB), which was meaningful for its practical application.

5.6.2 $BaTiO_3$-Based Relaxor Ferroelectrics

It is well known that pure $BaTiO_3$ is a normal ferroelectric with a Curie point of 120 °C. As the temperature increases, $BaTiO_3$ experiences a phase transition from rhombohedral ferroelectric to orthorhombic ferroelectric at −80 °C, to tetragonal ferroelectric at −5 °C, and then to cubic paraelectric at 120 °C. The relaxor state could be induced in $BaTiO_3$ by both isovalent and/or heterovalent ionic substitutions. In the first case, the relaxor state was observed in (Ba,Sr)TiO_3, Ba(Ti,Sn)O_3, and Ba(Zr,Ti)O_3-based compounds. In the second case, the relaxor behavior was detected in $BaTiO_3$–$BiScO_3$, $BaTiO_3$–$BiYbO_3$, Bi($Mg_{0.5}Ti_{0.5}$)O_3, $BaTiO_3$–Bi($Zn_{0.5}Ti_{0.5}$)O_3, and $BaTiO_3$–Bi($Mg_{2/3}Nb_{1/3}$)O_3 solid solutions, and so on. For convenience, the energy-storage performance of the two types of $BaTiO_3$-based relaxor ferroelectrics are summarized, respectively.

For the first type of $BaTiO_3$-based relaxor ferroelectrics, the study on the electrical energy-storage performance covered bulk ceramics, thick films, and thin films, and mainly focused on effects of the chemical composition, grain size, oxide doping, sintering temperature, and method. The latest report is summarized in Table 5.4 for comparison. Here, only the authors' preferred works are presented. By tuning the chemical compositions, the optimized recoverable energy-storage density of 0.68 J cm^{-3} and energy efficiency of 72% were achieved in 0.85Ba($Zr_{0.2}Ti_{0.8}$)O_3–0.15($Ba_{0.70}Ca_{0.30}$)TiO_3 bulk ceramics [116]. By controlling the grain size in the range of 0.5–5.6 μm, the energy-storage performance of ($Ba_{0.4}Sr_{0.6}$)TiO_3 ceramics changed greatly [115]. As the grain size increased, both BDS and recoverable energy-storage density decreased. A maximum density of 1.28 J cm^{-3} was achieved in ($Ba_{0.4}Sr_{0.6}$)TiO_3 ceramics

Table 5.4 The reported energy-storage performance of $BaTiO_3$-based relaxor ferroelectrics.

Composition	Form	W_{rec} ($J\,cm^{-3}$)	BDS ($kV\,cm^{-1}$)	η (%)	Fabrication method	References
$(Ba_{0.4}Sr_{0.6})TiO_3$	Ceramics	1.28–0.60	243–114	—	Solid-state reaction + Conventional sintering	[115]
$0.85Ba(Zr_{0.2}Ti_{0.8})O_3$–$0.15(Ba_{0.70}Ca_{0.30})TiO_3$	Ceramics	0.68	170	72	Solid-state reaction + Conventional sintering	[116]
$0.90Ba(Zr_{0.2}Ti_{0.8})O_3$–$0.10(Ba_{0.70}Ca_{0.30})TiO_3$	Ceramics	0.60	73.16	—	Solid-state reaction + Conventional sintering	[117]
$BaTiO_3$–$(Sr_{0.865}Bi_{0.09})TiO_3$	Ceramics	0.06	—	—	Solid-state reaction + Conventional sintering	[118]
$(Ba_{0.4}Sr_{0.6})TiO_3$	Ceramics	1.15	180	82	Solid-state reaction + Microwave sintering	[119]
95wt%$Ba_{0.4}Sr_{0.6}TiO_3$–5wt%MgO	Ceramics	1.50	300	88.5	Solid-state reaction + Conventional sintering	[120]
70wt%$Ba_{0.4}Sr_{0.6}TiO_3$–30wt%MgO	Ceramics	1.14	331	—	Solid-state reaction + Conventional sintering	[121]
$BaTiO_3/Ba_{0.3}Sr_{0.7}TiO_3$	Thin films	12.24	—	59	Pulsed laser deposition method	[122]
$Ba_{0.7}Ca_{0.3}TiO_3/BaZr_{0.2}Ti_{0.8}O_3$	Thin films	26.4–52.4	3000–4500	81–60	Magnetron sputtering method	[123]
$BaTiO_3/SrTiO_3$	Thick films	1.16	281	—	Tape-casting method	[124]
$BaTiO_3$–$Bi(Zn_{0.5}Ti_{0.5})TiO_3$	MLCC	2.8	330	—	Tape-casting method	[125]
0.91 $BaTiO_3$–$0.09BiYbO_3$	Ceramics	0.71	93	82.6	Solid-state reaction + Conventional sintering	[126]
$0.7BaTiO_3$–$0.3BiScO_3$	Ceramics	2.3	225	—	Solid-state reaction + Conventional sintering	[127]
$0.7BaTiO_3$–$0.3BiScO_3$	Thick films	6.1	730	—	Tape-casting method	[127]
$0.9BaTiO_3$–$0.1Bi(Mg_{2/3}Ti_{1/3})O_3$	Ceramics	1.13	143.5	96	Solid-state reaction + Conventional sintering	[128]
$0.88BaTiO_3$–$1.2Bi(Mg_{0.5}Ti_{0.5})O_3$	Thin film	37	1900	—	Chemical solution deposition method	[129]

with a grain size of 0.5 μm at a BDS of 243 kV cm^{-1}. By proper addition of oxides, such as MgO, ZnO, and Gd_2O_3, improved electrical energy-storage performance was also realized in $BaTiO_3$-based relaxor ferroelectrics. For instance, a high BDS of 300 kV cm^{-1}, highest recoverable energy-storage density of 1.50 J cm^{-3}, and high energy efficiency of 88.5% were obtained in 95 wt%$Ba_{0.4}Sr_{0.6}TiO_3$–5 wt%MgO ceramic [120]. Owing to their excellent electric field endurance, much improved electrical energy-storage performance was achieved in $BaTiO_3$-based relaxor ferroelectric films. By using pulsed laser deposition method, 600-nm-thick $BaTiO_3/Ba_{0.3}Sr_{0.7}TiO_3$ superlattices with a constant modulation period of 8 nm were grown on MgO(001) substrates, and a high recoverable energy-storage density of 12.24 J cm^{-3} was reported in this film [122]. A further improved energy-storage performance was obtained in 100-nm-thick $(Ba_{0.7}Ca_{0.3}TiO_3/BaZr_{0.2}Ti_{0.8}O_3)_N$ ($N = 2$, 4, 8) superlattices that were deposited on Nb:$SrTiO_3$ substrates by magnetron sputtering technique. N is the number of repetitions of $(Ba_{0.7}Ca_{0.3}TiO_3/BaZr_{0.2}Ti_{0.8}O_3)$ unit, and the thickness ratio of the two compounds was 3/7 [123]. It was found that BDS value was significantly improved by increasing the number of the interfaces in the superlattices, which effectively impedes the spread and growth of the electric trees, as revealed by numerical simulations based on finite element method. As a result, an ultrahigh recoverable energy-storage density of 52.4 J cm^{-3} with great energy efficiency of 72.3% was achieved at room temperature.

In the second kind of $BaTiO_3$-based relaxor ferroelectrics, $Bi(Me)O_3$ was usually composited to form $BaTiO_3$–$Bi(Me)O_3$ solid solution. In $Bi(Me)O_3$ compounds, Me can either be a single trivalent cation or two cations with an average of +3 valence, which occupy the octahedral sites. Inspired by the excellent relaxor behavior of $BaTiO_3$–$Bi(Me)O_3$ solid solution, increasing attention was paid to the investigation of their energy-storage performance and a few works were reported, which had great potential for application in MLCC. For example, in the $(1-x)BaTiO_3$–$xBi(Mg_{2/3}Nb_{1/3})O_3$ ($x = 0.05$–0.2) ceramics, a maximum recoverable energy-storage density of 1.13 J cm^{-3} was achieved for $x = 0.1$ at its BDS of 287.7 kV cm^{-1} [128]. Similarly, a large recoverable energy-storage density of 2.3 J cm^{-3} was obtained in $0.7BaTiO_3$–$0.3BiScO_3$ bulk ceramics at 225 kV cm^{-1} [127]. On decreasing the thickness to 15 μm, the recoverable energy-storage density increased to 6.1 J cm^{-3} at 730 kV cm^{-1}. Moreover, $0.7BaTiO_3$–$0.3BiScO_3$ also showed good thermal stability superior to the typical commercial XR7 capacitors. In the thin film case, the energy-storage performance in a $(1-x)BaTiO_3\_xBi(Mg_{0.5}Ti_{0.5})O_3$ ($x = 0.1$, 0.12, and 0.15) system was studied [129]. These thin films showed a nearly linear polarization response with high relative dielectric constant, exceeding 900, which is beneficial for high energy-density capacitors. An average BDS of 2.08 MV cm^{-1} and very low leakage current densities of the order of 10^{-8} A cm^{-2} even at temperatures of 200 °C were obtained for the films with $x = 0.12$, resulting in a recoverable energy-storage density as high as 37 J cm^{-3} at 1.9 MV cm^{-1}. Owing to the length limitation, some unmentioned works are given in Table 5.4 for comparison.

5.6.3 $PbTiO_3$-Based Relaxor Ferroelectrics

$PbTiO_3$ with tetragonal structure at room temperature is also a typical normal ferroelectric, whose Curie point is about 492 °C. By compositing with $Pb(Mg_{1/3}Nb_{2/3})O_3$ and $Bi(Me)O_3$, $PbTiO_3$-based relaxor ferroelectrics could be produced. For example, $(1-x)Pb(Mg_{1/3}Nb_{2/3})O_3$–$xPbTiO_3$ (PMN–PT) with compositions near the morphotropic phase boundary (MPB) ($0.30 < x < 0.35$) are one of the most studied relaxor ferroelectrics because of their excellent known ferroelectric, dielectric, and piezoelectric properties. Recently, some attention has also been paid to the energy-storage performance of $PbTiO_3$-based relaxor ferroelectrics. Here, some typical results are presented.

Kang and coauthors fabricated $0.24Pb(In_{1/2}Nb_{1/2})O_3$–$43Pb(Mg_{1/3}Nb_{2/3})O_3$–$33PbTiO_3$ films with a thickness of about 2.5 μm on Pt(111)/Pt/Ti/SiO_2/Si substrates by aerosol-disposition method [130]. The as-deposited film showed a recoverable energy-storage density of 3.5 J cm^{-3}, which increased to 4.1 J cm^{-3} after being annealed at 650 °C for one hour. However, the annealed films displayed an increased hysteresis loss, thus reducing the energy efficiency. Followed this work, we prepared $(1-x)Pb(Mg_{1/3}Nb_{2/3})O_3$–$xPbTiO_3$ ($x = 0.1$, 0.2, and 0.3) relaxor ferroelectric thin films with a thickness of about 400 nm by RF magnetron sputtering technique [131]. The film with $x = 0.2$ possessed the highest difference of 51 μC cm^{-2} between maximum polarization and remnant polarization, which led to a larger recoverable energy-storage density of 31 J cm^{-3} with energy efficiency of 64% at 2000 kV cm^{-1} at room temperature. Moreover, these films also showed an excellent stability of energy-storage performance at 20–200 °C and 100 Hz–10 kHz. These results indicated that PMN–PT relaxor ferroelectrics were ideal candidates for high energy storage.

Similarly to the $BaTiO_3$–$Bi(Me)O_3$ system, the energy-storage performance in $PbTiO_3$–$Bi(Me)O_3$ relaxor ferroelectrics was also investigated recently, focusing mainly on thin films prepared by the chemical solution method by Yue's group [132–136]. Huge recoverable energy-storage density of 46.7 J cm^{-3} at 2250 kV cm^{-1}, 45.1 J cm^{-3} at 2321 kV cm^{-1}, and 39.8 J cm^{-3} at 2167 kV cm^{-1} were obtained in $0.5PbTiO_3$–$0.5Bi(Ni_{0.5}Ti_{0.5})O_3$, $0.62PbTiO_3$–$0.38Bi(Ni_{0.5}Hf_{0.5})O_3$, and $0.60PbTiO_3$–$0.40Bi(Ni_{0.5}Zr_{0.5})O_3$ thin films, respectively. Similarly, a moderate energy efficiency of above 60% was realized in these thin films. In order to further improve the energy-storage performance of $PbTiO_3$–$Bi(Me)O_3$ relaxor ferroelectric films, the author fabricated $0.60PbTiO_3$–$0.40Bi(Ni_{0.5}Zr_{0.5})O_3$ thin films by inserting PbO seeding layers. It was found that highly (100)-oriented films were achieved by alternately introducing the PbO layers. As a result, a maximum recoverable energy-storage density of 56.1 J cm^{-3} at 2167 kV cm^{-1} was achieved, which was increased by 40.6% compared to the films without seeding layers.

5.6.4 $BiFeO_3$-Based Relaxor Ferroelectrics

$BiFeO_3$, as a typical multiferroic material, had been investigated for a long time for different purposes. It has a high Curie point of 830 °C and a large maximum polarization of 100 μC cm^{-2}. Thus, it should also have the potential for application

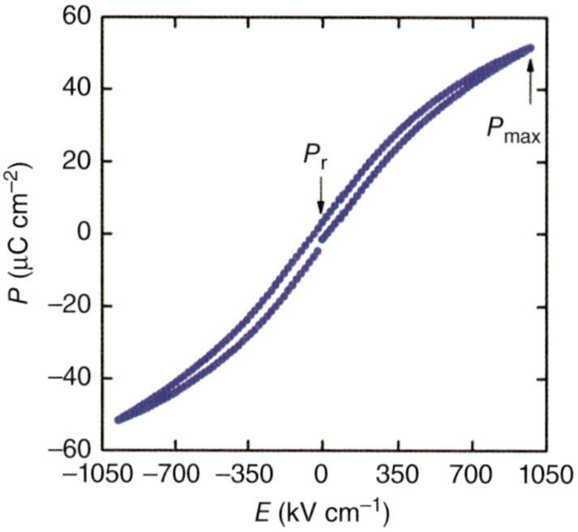

Figure 5.20 *P–E* loop of a $0.4BiFeO_3$–$0.6SrTiO_3$ thin film at 1 kHz at room temperature. *Source:* Correia et al. 2013 [137]. Reproduced with the permission of John Wiley & Sons.

in high energy-storage capacitors. However, serious leakage current issues have made this impossible. Recently, it was found that high-quality relaxor ferroelectric thin films could be formed from $BiFeO_3$–$SrTiO_3$ solid solution. Accordingly, some trial studies on the energy-storage performance of $BiFeO_3$–$SrTiO_3$ thin films were carried out, and outstanding results were achieved.

$0.4BiFeO_3$–$0.6SrTiO_3$ films of 400 nm thickness were deposited on $SiTiO_3$ substrates by pulsed laser deposition method [137]. A very slim *P–E* loop was obtained, as shown in Figure 5.20, which was characterized by a high maximum polarization of $52\,\mu C\,cm^{-2}$ and low remnant polarization of $3\,\mu C\,cm^{-2}$ at $972\,kV\,cm^{-1}$. Therefore, high recoverable energy-storage density of $18.6\,J\,cm^{-3}$ and energy efficiency of 85% were achieved. Moreover, these thin films showed a low temperature coefficient of capacitance of 11%. In order to further enhance the energy-storage density, a 0.5 mol%-Mn doped $0.4BiFeO_3$–$0.6SrTiO_3$ film with a thickness of 500 nm was deposited on Nb:$SrTiO_3$ substrates via pulsed laser deposition method. The thin film showed a (100)-epitaxial nature [138]. An ultrahigh recoverable energy-storage density of $51\,J\,cm^{-3}$ and a high energy efficiency of 64% were achieved in the Mn-doped film. At the same time, strong fatigue endurance after 2×10^7 cycles and good thermal stability in a wide range of −40 to 140 °C were also received. The excellent energy-storage performance of this film was attributed to its strong relaxor behavior, low leakage current ($1\times10^{-5}\,A\,cm^{-2}$ at $1000\,kV\,cm^{-1}$), and high BDS ($3600\,kV\,cm^{-1}$).

5.7 The General Future Prospects

The above-summarized results indicate that some outstanding achievements for electrical energy storage have been obtained in the dielectric materials, but there still exist many problems that need to be solved in order to propel the application of energy-storage capacitors in practice. So, in this section, we will present some key aspects to throw further light on the investigation in this area.

(1) From the environmental and human health point of view, lead-based dielectric materials have to be prevented in practical applications. Thus,

developing or exploring lead-free materials with good energy-storage performance should be received increasing attention.

(2) From the scientific research point of view, there still exists much physical mechanism (such as interface effect in glass-ceramics and in thin films) to be revealed, in order to optimize the electrical energy-storage performance. Therefore, an in-depth study on the physical mechanism should be continued.

(3) From the practical application point of view, the current investigation is mainly focused on the improvement of energy-storage density in the dielectrics, but their charge and discharge speed, fatigue behavior, temperature stability, and lifetime are often neglected. In fact, these parameters are also very important for their application in capacitors. So, these aspects have to be addressed in future studies.

(4) Form the device point of view, the low DBS for bulk ceramics and low volume for thin films often lead to low overall electrical energy storage. Accordingly, work to exploit MLCC with high energy-storage performance should be carried in the very near future.

References

1 Sarjeant, W.J., Zirnheld, J., and Macdougall, F.W. (1998). Capacitors. *IEEE Transactions on Plasma Science* 26 (5): 1368–1392.

2 Nishino, A. (1996). Capacitors: operating principles, current market and technical trends. *Journal of Power Sources* 60 (2): 137–147.

3 Huan, T.D., Boggs, S., Teyssedre, G. et al. (2016). Advanced polymeric dielectrics for high energy density applications. *Progress in Materials Science* 83: 236–269.

4 Prateek, Thakur, V.K., and Gupta, R.K. (2016). Recent progress on ferroelectric polymer-based nanocomposites for high energy density capacitors: synthesis, dielectric properties, and future aspects. *Chemical Reviews* 116 (7): 4260–4317.

5 Mannodi-Kanakkithodi, A., Treich, G.M., Huan, T.D. et al. (2016). Rational co-design of polymer dielectrics for energy storage. *Advanced Materials* 28 (30): 6277–6291.

6 Sarjeant, W.J., Clelland, I.W., and Price, R.A. (2001). Capacitive component for powder electronics. *Proceedings of the IEEE* 89 (6): 846–855.

7 Hao, X. (2013). A review on the dielectric materials for high energy-storage application. *Journal of Advanced Dielectrics* 3 (1): 1330001-1-14.

8 Sun, W., Zhou, Z., Luo, J. et al. (2017). Leakage current characteristics and Sm/Ti doping effect in $BiFeO_3$ thin films on silicon wafers. *Journal of Applied Physics* 121 (6): 064101-1-9.

9 Weibull, W., Stockholm, and Swenen (1951). A statistical distribution function of wide applicability. *Journal of Applied Mechanics* 13 (2): 293–297.

10 Pan, M.-J. and Randall, C.A. (2010). A brief introduction to ceramic capacitors. *IEEE Electrical Insulation Magazine* 26 (3): 44–50.

11 Li, J., Li, F., and Zhang, S. (2014). Decoding the fingerprint of ferroelectric loops: comprehension of the material properties and structures. *Journal of the American Ceramic Society* 97 (1): 1–27.
12 Xiao, S., Xiu, S., Bo, S., and Zhai, J. (2016). Microstructure evolution and energy storage properties of potassium strontium niobate boroaluminosilicate glass-ceramics by microwave crystallization. *Journal of the European Ceramic Society* 36 (16): 4071–4076.
13 Chao, S. and Dogan, F. (2011). Processing and dielectric properties of TiO_2 thick films for high-energy density capacitor applications. *International Journal of Applied Ceramic Technology* 8 (6): 1363–1373.
14 Dong, W., Wanbiao, H., Berlie, A. et al. (2015). Colossal dielectric behavior of Ga+Nb co-doped rutile TiO_2. *ACS Applied Materials & Interfaces* 7 (45): 25321–25325.
15 Gai, Z., Cheng, Z., Wang, X. et al. (2014). A colossal dielectric constant of an amorphous TiO_2:(Nb,In) film with low loss fabrication at room temperature. *Journal of Materials Chemistry C* 2 (33): 6790–6795.
16 Park, M.H., Kim, H.J., Jin Kim, Y. et al. (2014). Thin $Hf_xZr_{1-x}O_2$ films: a new lead-free system for electrostatic supercapacitors with large energy storage density and robust thermal stability. *Advanced Energy Materials* 4 (16): 1400610-1-7.
17 Spahr, H., Nowak, C., Hirschberg, F. et al. (2013). Enhancement of the maximum energy density in atomic layer deposited oxide based thin film capacitors. *Applied Physics Letters* 103 (4): 042907-1-6.
18 Zhang, G.-f., Cao, M., Hao, H., and Liu, H. Energy storage characteristics in $Sr_{(1-1.5x)}Bi_xTiO_3$ ceramics. *Ferroelectrics* 447 (1): 86–94.
19 Li, L., Xiaoxu, Y., Cai, H. et al. (2013). Preparation and dielectric properties of $BaCu(B_2O_5)$-doped $SrTiO_3$-based ceramics for energy storage. *Materials Science and Engineering B* 178 (20): 1509–1514.
20 Michael, E.K. and Trolier-Mackinstry, S. (2015). Bismuth pyrochlore thin films for dielectric energy storage. *Journal of Applied Physics* 118 (5): 054101-1-7.
21 Hao, X., Zhai, J., Kong, L., and Zhengkui, X. (2014). A comprehensive review on the progress of lead zirconate-based antiferroelectric materials. *Progress in Materials Science* 63: 1–57.
22 Kittel, C. (1951). Theory of antiferroelectric crystal. *Physics Review* 82 (5): 729–732.
23 Sawaguchi, E., Maniwa, H., and Hoshino, S. (1951). Antiferroelectric structure of lead zirconate. *Physics Review* 83 (5): 1078.
24 Jona, F., S.G., Mazzi, F., and Pepinsky, R. (1957). X-ray and neutron diffraction study of antiferroelectric lead zirconate, $PbZrO_3$. *Physical Review* 105 (3): 849–856.
25 Ayyub, P., Chattopadhyay, S., Pinto, R., and Multani, M.S. (1998). Ferroelectric behavior in thin films of antiferroelectric materials. *Physical Review B* 57 (10): R5559–R5562.
26 Shirane, G. (1952). Ferroelectricity and antiferroelectricity in ceramic $PbZrO_3$ containing Ba or Sr. *Physics Review* 86 (2): 219–227.

27 Frjtz, I.J. and Kect, J.D. (1978). Pressure–temperature phase diagrams for several modified lead zirconate ceramics. *Journal of Physics and Chemistry of Solids* 39 (11): 1163–1167.
28 Jaffe, B. (1961). Antiferroelectric ceramics with field-enforced transition: a new nonlinear circuit element. *Proceedings of the IRE* 49 (8): 1264–1267.
29 Berlincourt, D., Kruegger, H.H.A., and Jaffe, B. (1964). Stability of phases in modified lead zirconate with variation in pressure, electric field, temperature and composition. *Journal of Physics and Chemistry of Solids* 25 (7): 659–674.
30 Land, C.E. (1988). Bistable optical information storage using antiderroelectric-phase lead lanthanum zirconate titanate ceramics. *Journal of the American Ceramic Society* 71 (11): 905–908.
31 Berlincourt, D. (1966). Transducers using forced transition between ferroelectric and antiferroelectric state. *IEEE Transactions on Sonics and Ultrasonics* 13: 116–125.
32 Wang, J., Yang, T., Chen, S., and Xi, Y. (2014). Small hysteresis and high energy storage powder of antiferroelectric ceramics. *Functional Materials Letters* 7: −1. 1350064-1-4.
33 Wang, X., Shen, J., Yang, T. et al. (2016). High energy-storage performance and dielectric properties of antiferroelectric $(Pb_{0.97}La_{0.02})(Zr_{0.5}Sn_{0.5-x}Ti_x)O_3$ ceramics. *Journal of Alloys and Compounds* 655: 309–313.
34 Chen, Q.Z.J., Yinmei, L., Yang, T. et al. (2016). $(Pb,Sm)(Zr,Sn,Ti)O_3$ multifunctional ceramics with larger electric-field-induced strain and high-energy storage density. *Journal of the American Ceramic Society* 99 (12): 3853–3856.
35 Zhang, Q., Tong, H., Chen, J. et al. (2016). High recoverable energy density over a wide temperature range in Sr modified $(Pb,La)(Zr,Sn,Ti)O_3$ antiferroelectric ceramics with an orthorhombic phase. *Applied Physics Letters* 109 (26): 262901-1-4.
36 Chen, S., Yang, T., Wang, J., and Yao, X. (2013). Effects of glass additions on the dielectric properties and energy storage performance of $Pb_{0.97}La_{0.02}(Zr_{0.56}Sn_{0.35}Ti_{0.09})O_3$ antiferroelectrics ceramics. *Journal of Materials Science: Materials in Electronics* 24 (12): 4764–4768.
37 Zhang, L., Jiang, S., Fan, B., and Zhang, G. (2015). Enhanced energy storage performance in $(Pb_{0.858}Ba_{0.1}La_{0.02}Y_{0.08})(Zr_{0.65}Sn_{0.3}Ti_{0.05})O_3$-$(Pb_{0.97}La_{0.02})(Zr_{0.9}Sn_{0.05}Ti_{0.05})O_3$ anti-ferroelectric composite ceramics by spark plasma sintering. *Journal of Alloys and Compounds* 622: 162–165.
38 Xu, B., Pai, N.G., Moses, P., and Cross, L. (1998). Antiferroelectric thin films for depcoupling capacitor and microactuator applications. *Materials Research Society Symposium Proceedings* 493: 403–408.
39 Hao, X., Zhai, J., and Xi, Y. (2009). Improved energy storage performance and fatigue endurance of Sr-doped $PbZrO_3$ antiferroelectric thin films. *Journal of the American Ceramic Society* 92 (5): 1133–1135.
40 Mirshekarloo, M.S., Yao, K., and Sritharan, T. (2010). Large strain and high energy storage density in orthorhombic perovskite

$(Pb_{0.97}La_{0.02})(Zr_{1.x-y}Sn_xTi_y)O_3$ antiferroelectric thin films. *Applied Physics Letters* 97 (14): 142902-1-3.
41 Hao, X., Zhou, J., and An, S. (2011). Effects of PbO content on the dielectric properties and energy storage performance of $(Pb_{0.97}La_{0.02})(Zr_{0.97}Ti_{0.03})O_3$ antiferroelectric thin films. *Journal of the American Ceramic Society* 94 (6): 1647–1650.
42 Ma, B., Kwon, D.-K., Narayanan, M., and Balachandran, U. (2009). Dielectric properties and energy storage capability of antiferroelectric $Pb_{0.92}La_{0.08}Zr_{0.95}Ti_{0.03}O_3$ film-on foil capacitors. *Journal of Materials Research* 24 (9): 2993–2996.
43 Pan, H., Yi, Z., Shen, Y. et al. (2016). Thickness-dependent dielectric and energy storage properties of $(Pb_{0.96}La_{0.04})(Zr_{0.98}Ti_{0.02})O_3$ antiferroelectric thin films. *Journal of Applied Physics* 119 (12): 124106-1-5.
44 Ge, J., Remiens, D., Dong, X. et al. (2014). Enhancement of energy storage in epitaxial $PbZrO_3$ antiferroelectric films using strain engineering. *Applied Physics Letters* 105 (11): 112908-1-5.
45 Chen, T., Wang, J., Zhong, X. et al. (2014). High energy density capacitors based on $0.88BaTiO_3–0.12(Mg_{0.5}Ti_{0.5})O_3/PbZrO_3$ multilayered thin films. *Ceramics International* 40 (4): 5327–5332.
46 Sa, T., Cao, Z., Wang, Y., and Zhu, H. (2014). Enhancement of charge and energy storage in $PbZrO_3$ thin films by local field engineering. *Applied Physics Letters* 105 (4): 043902-1-5.
47 Burn, I. (1971). Field-enforced ferroelectricity in glass-bonded lead zirconate. *American Ceramic Society Bulletin* 50 (5): 501–505.
48 Chen, S., Yang, T., Wang, J., and Yao, X. (2013). Effects of glass additions on the energy storage performance of $(Pb_{0.97}La_{0.02})(Zr_{0.92}Sn_{0.05}Ti_{0.03})O_3$ antiferroelectric thick films. *Journal of Advanced Dielectrics* 3 (2): 1350012-1-5.
49 Wang, Y., Hao, X., Yang, J. et al. (2012). Fabrication and energy-storage performance of (Pb,La) $(Zr,Ti)O_3$ antiferroelectric thick films derived from polyvinylpyrrolidone-modified chemical solution. *Journal of Applied Physics* 112 (3): 034105-1-6.
50 Hao, X., Wang, Y., Le, Z. et al. (2013). Composition-dependent dielectric and energy-storage properties in $(Pb,La)(Zr,Sn,Ti)O_3$ antiferroelectric thick films. *Applied Physics Letters* 102 (16): 163903-1-4.
51 Zhao, Y., Gao, H., Hao, X., and Qi, Z. (2016). Orientation-dependent energy-storage performance and electrocaloric effect in PLZST antiferroelectric thick films. *Materials Research Bulletin* 84: 177–184.
52 Gao, H., Hao, X., Zhang, Q. et al. (2016). Giant electrocaloric effect and energy-storage properties in grain-size-engineered $Pb_{0.85}Ba_{0.05}La_{0.10}(Zr_{0.90}Ti_{0.10})O_3$ antiferroelectric thick films. *Journal of Materials Science: Materials in Electronics* 27 (10): 10309–10319.
53 Pan, W.Y., Dam, C.Q., Zhang, Q.M., and Cross, L.E. (1989). Large displacement transducers based on electric field forced phase transitions in the tetragonal $(Pb_{0.97}La_{0.02})(Ti,Zr,Sn)O_3$ family of ceramics. *Journal of Applied Physics* 66 (12): 6014–6023.

54 Zhang, H., Chen, X., Cao, F. et al. (2010). Charge–discharge properties of an antiferroelectric ceramics capacitor under different electric fields. *Journal of the American Ceramic Society* 93 (12): 4015–4017.

55 Ran, X., Zhuo, X., Feng, Y. et al. (2016). Nonlinear dielectric and properties of $Pb_{0.94}La_{0.04}[(Zr_{0.56}Sn_{0.44})_{0.84}Ti_{0.16}]O_3$ antiferroelectric ceramics. *Journal of Applied Physics* 120 (14): 144102-1-6.

56 Bharadwaja, S.S.N. and Krupanidhi, S.B. (2001). Backward switching phenomenon from field forced ferroelectric to antiferroelectric phases in antiferroelectric $PbZrO_3$ thin films. *Journal of Applied Physics* 89 (8): 4541–4547.

57 Tan, X., Ma, C., Frederick, J. et al. (2011). The antiferroelectric ↔ ferroelectric phase transition in lead-containing and lead-free perovskite ceramics. *Journal of the American Ceramic Society* 94 (12): 4091–4107.

58 Guo, Y., Liu, Y., Withers, R.L. et al. (2011). Large electric field-induced strain and antiferroelectric behavior in $(1-x)(Na_{0.5}Bi_{0.5})TiO_{3-x}BaTiO_3$ ceramics. *Chemistry of Materials* 23 (2): 219–228.

59 Qi, X., Liu, H., Song, Z. et al. (2016). A new energy-storage ceramic system based on $Bi_{0.5}Na_{0.5}TiO_3$ ternary solid solution. *Journal of Materials Science: Materials in Electronics* 27 (1): 322–329.

60 Cao, W., LIi, W., Feng, Y. et al. (2016). Defect dipole induced large recoverable strain and high energy-storage density in lead-free $Na_{0.5}Bi_{0.5}TiO_3$-based systems. *Applied Physics Letters* 108 (20): 202902-1-5.

61 Ding, J., Liu, Y., Lu, Y. et al. (2014). Enhanced energy-storage properties of $0.89Bi_{0.5}Na_{0.5}TiO_3-0.06BaTiO_3-0.05K_{0.5}Na_{0.5}NbO_3$ lead-free anti-ferroelectric ceramics by two-step sintering method. *Materials Letters* 114 (1): 107–110.

62 Yongping, P., Zhang, L., Yao, M. et al. (2017). Improved energy storage properties of microwave sintered 0.475BNT-0.525BCZT-$x$wt%MgO ceramics. *Materials Letters* 189: 232–235.

63 Hao, J., Xu, Z., Chu, R. et al. (2015). Enhanced energy-storage properties of $(1-x)[(1-y)(Bi_{0.5}Na_{0.5})TiO_3-y(Bi_{0.5}K_{0.5})TiO_3]-x(K_{0.5}Na_{0.5})NbO_3$ lead-free ceramics. *Solid State Communications* 204: 19–22.

64 Li, Q., Wang, J., Ma, Y. et al. (2016). Enhanced energy-storage performance and dielectric characterization of $0.94Bi_{0.5}Na_{0.5}TiO_3-0.06BaTiO_3$ modified by $CaZrO_3$. *Journal of Alloys and Compounds* 663: 701–707.

65 Wang, Y., Lv, Z., Xie, H., and Cao, J. (2014). High energy-storage properties of $[(Bi_{1/2}Na_{1/2})_{0.94}Ba_{0.06}]La_{(1-x)}Zr_xTiO_3$ lead-free anti-ferroelectric ceramics. *Ceramics International* 40 (3): 4323–4326.

66 Zhao, Y., Xu, J., Zhou, C. et al. (2015). High energy storage properties and dielectric behavior of $(Bi_{0.5}Na_{0.5})_{0.94}Ba_{0.06}Ti_{1-x}(Al_{0.5}Nb_{0.5})_xO_3$ lead-free ferroelectric ceramics. *Ceramics International* 42 (2): 2221–2226.

67 Wang, B., Luo, L., Jiang, X. et al. (2013). Energy-storage properties of $(1-x)Bi_{0.47}Na_{0.47}Ba_{0.06}TiO_3-xKNbO_3$ lead-free ceramics. *Journal of Alloys and Compounds* 585: 14–18.

68 Qi, X., Li, T., Huao, H. et al. (2015). Enhanced energy storage properties of $NaNbO_3$ modified $Bi_{0.5}Na_{0.5}TiO_3$ based ceramics. *Journal of the European Ceramic Society* 35 (2): 545–553.

69 Pan, C. and Chu, B. (2016). Improvement of dielectric and energy storage properties in $Bi(Mg_{1/2}Ti_{1/2})O_3$-modified $(Na_{1/2}Bi_{1/2})_{0.92}Ba_{0.08}TiO_3$ ceramics. *Journal of the European Ceramic Society* 36 (1): 81–88.

70 Qi, X., Xie, J., He, Z. et al. (2017). Energy-storage properties of $Bi_{0.5}Na_{0.5}TiO_3$–$BaTiO_3$–$KNbO_3$ ceramics fabricated by wet-chemical method. *Journal of the European Ceramic Society* 37 (1): 99–106.

71 Ye, J., Liu, Y., Yinong, L. et al. (2014). Enhanced energy-storage properties of $SrTiO_3$ doped $(Bi_{1/2}Na_{1/2})TiO_3$–$(Bi_{1/2}K_{1/2})TiO_3$ lead-free antiferroelectric ceramics. *Journal of Materials Science: Materials in Electronics* 25 (10): 4632–4637.

72 Feng, Q., Yuan, C., Liu, X. et al. (2015). Microstructure and energy-storage properties of $(1-x)(Na_{0.5}Bi_{0.5})TiO_3$–$xBaTiO_3$ with BaO–B_2O–SiO_2 additions. *Journal of Materials Science: Materials in Electronics* 26 (7): 5113–5119.

73 Zhou, X., Yuan, C., Li, Q. et al. (2016). Energy storage properties and electric electrical behavior of lead-free $(1-x)Ba_{0.04}Bi_{0.48}Na_{0.48}TiO_3$–$xSrZrO_3$ ceramics. *Journal of Materials Science: Materials in Electronics* 27 (4): 3948–3956.

74 Qi, X., Liu, H., Song, Z. et al. (2016). A new energy-storage ceramics system based on $Bi_{0.5}Na_{0.5}TiO_3$ ternary solid solution. *Journal of Materials Science: Materials in Electronics* 27 (1): 322–329.

75 Gao, F., Dong, X., Mao, C. et al. (2011). Energy-storage properties of $0.89Bi_{0.5}Na_{0.5}TiO_3$–$0.06BaTiO_3$–$0.05K_{0.5}Na_{0.5}NbO_3$ lead-free antiferroelectric ceramics. *Journal of the American Ceramic Society* 94 (12): 4382–4386.

76 Luo, L., Wang, B., Jiang, X., and Li, W. (2014). Energy storage properties of $(1-x)(Bi_{0.5}Na_{0.5})TiO_3$–$xKNbO_3$ lead-free ceramics. *Journal of Materials Science: Materials in Electronics* 49 (4): 1659–1665.

77 Le, Z. and Hao, X. (2014). Dielectric properties and energy-storage performances of $(1-x)(Na_{0.5}Bi_{0.5})TiO_3$–$xSrTiO_3$ thick films prepared by screen printing technique. *Journal of Alloys and Compounds* 586: 674–678.

78 Xu, Z., Hao, X., and An, S. (2015). Dielectric properties and energy-storage performance of $(Na_{0.5}Bi_{0.5})TiO_3$–$SrTiO_3$ thick films derived from polyvinylpyrrolidone-modified chemical solution. *Journal of Alloys and Compounds* 639: 387–392.

79 Peng, B., Qi, Z., Li, X. et al. (2015). Giant electric energy density in epitaxial lead-free thin films with coexistence of ferroelectrics and antiferroelectrics. *Advanced Electronic Materials* 1 (5): 1500052-1-7.

80 Desheng, F., Endo, M., Taniguchi, H. et al. (2007). $AgNbO_3$: a lead-free materials with large polarization and electromechanical response. *Applied Physics Letters* 90 (25): 252907-1-3.

81 Zhao, L., Liu, Q., Zhang, S., and Li, J. (2016). Lead-free $AgNbO_3$ anti-ferroelectric ceramics with an enhanced energy storage performance using MnO_2 modification. *Journal of Materials Chemistry C* 4 (36): 8380–8384.

82 Ye, T., Li, J., Zhang, H. et al. (2016). High energy density in silver niobate ceramics. *Journal of Materials Chemistry A* 4 (44): 17279–17287.

83 Park, M.H., Lee, Y.H., Kim, H.J. et al. (2015). Ferroelectricity and antiferroelectricity of doped thin HfO_2-based films. *Advanced Materials* 27 (11): 1811–1831.

84 Yadav, A.K. and Gautam, C. (2014). Dielectric behavior of perovskite glass ceramics. *Journal of Materials Science: Materials in Electronics* 25 (12): 5165–5187.

85 Zhang, Q., Wang, L., Luo, J. et al. (2009). Improved energy storage density in barium strontium titanate by addition of $BaO–SiO_2–B_2O_3$ glass. *Journal of the American Ceramic Society* 92 (8): 1871–1873.

86 Wu, T., Yongping, P., and Chen, K. (2013). Dielectric rrelaxation behavior and energy storage properties in $Ba_{0.4}Sr_{0.6}Zr_{0.15}Ti_{0.85}O_3$ ceramics with glass additives. *Ceramics International* 39 (6): 6787–6793.

87 Xu, Q., Zhan, D., Huang, D.-P. et al. (2013). Effect of $MgO–CaO–Al_2O_3–SiO_2$ glass additive on dielectric properties of $Ba_{0.95}Sr_{0.05}Zr_{0.2}Ti_{0.8}O_3$ ceramics. *Journal of Alloys and Compounds* 558: 77–83.

88 Wang, J., Chao, X., Bo, S., and Zhai, J. (2013). Enhancing energy storage density of $(Ba,Sr)TiO_3$ ceramics particles by coating with Al_2O_3 and SiO_2. *Journal of Materials Science: Materials in Electronics* 24 (9): 3309–3314.

89 Luo, J., Jun, D., Tang, Q., and Mao, C. (2008). Lead sodium niobate glass-ceramic dielectrics and internal electrode structure for high energy storage density capacitors. *IEEE Transactions on Electron Devices* 55 (12): 3549–3554.

90 Zhang, Q., Li, C., Liu, H. et al. (2014). Discharged energy properties of $SrO–PbO–Na_2O–Nb_2O_5–SiO_2$ glass-ceramics with different crystallization time. *Journal of the American Ceramic Society* 98 (2): 3660369.

91 Gorzkowski, E.P., Pan, M.-J., Bender, B., and Wu, C.C.M. (2007). Glass-ceramics of barium strontium titanate for high energy density capacitors. *Journal of Electroceramics* 18 (3–4): 269–276.

92 Zhang, Y., Huang, J., Ma, T. et al. (2011). Sintering temperature dependence of energy-storage properties in $(Ba,Sr)TiO_3$ glass-ceramics. *Journal of the American Ceramic Society* 94 (6): 1805–1810.

93 Zhang, W., Wang, J., Xue, S. et al. (2014). Effect of La_2O_3 additive on the dielectric properties of barium strontium titanate glass-ceramics. *Journal of Materials Science: Materials in Electronics* 25 (9): 4145–4149.

94 Wang, J., Tang, L., Shen, B., and Zhai, J. (2013). Effect of microwave processes on the energy-storage properties of barium strontium titanate glass ceramics. *Journal of Materials Research* 29 (2): 288–293.

95 Liu, X., Yongping, P., Li, P. et al. (2014). Influence of different nucleating agent additives on phase structure and ferroelectric properties of $SrO–BaO–Nb_2O_5–CaO–SiO_2–B_2O_3$ glass-ceramics. *Journal of Materials Science: Materials in Electronics* 25 (7): 3044–3051.

96 Yi, Z., Qiao, Y., Tian, Y. et al. (2017). Improvement in structural, dielectric and energy-storage properties of lead-free niobate glass-ceramic with Sm_2O_3. *Journal of the European Ceramic Society* 37 (3): 995–999.

97 Zhou, Y., Zhang, Q., Luo, J. et al. (2013). Structural optimization and improved discharged energy density for niobate glass-ceramics by La_2O_3 addition. *Journal of the American Ceramic Society* 96 (2): 372–375.

98 Xue, S., Liu, S., Zhang, W. et al. (2015). Improved discharged properties for (BaO, Na_2O)–Nb_2O_5–SiO_2 glass-ceramics by La_2O_3 addition. *Ceramics International* 41: S441–S446.

99 Wang, H., Liu, J., Zhai, J., and Shen, B. (2016). Ultra high energy-storage density in the barium potassium niobate-based glass-ceramics for energy-storage application. *Journal of the American Ceramic Society* 99 (9): 2909–2912.

100 Xiao, S., Xiu, S., Bo, S., and Zhai, J. (2016). Microstructure evolution and energy storage properties of potassium strontium niobate boroaluminosilicate glass-ceramics by microwave crystallization. *Journal of the European Ceramic Society* 36 (16): 4071–4076.

101 Yang, Y., Song, J., Chen, G. et al. (2015). Effect of crystallization temperature on the dielectric property and energy-density of SrO–BaO–Nb_2O_6–B_2O_3 glass-ceramics. *Journal of Non-Crystalline Solids* 410: 96–99.

102 Song, J. and Chen, G. (2014). Dielectric behavior and energy storage properties in BaO–SrO–Nb_2O_5–B_2O_3 system glass-ceramics with Gd_2O_3 addition. *Journal of Materials Science: Materials in Electronics* 25 (1): 349–354.

103 Liu, X., Yongping, P., Li, P. et al. (2014). Influence of different nucleating agent additives on phase structure and ferroelectric properties of SrO–BaO–Nb_2O_5–CaO–SiO_2–B_2O_3 glass-ceramics. *Journal of Materials Science: Materials in Electronics* 25 (7): 3044–3051.

104 Tang, L., Wang, W., Bo, S., and Kong, L.B. (2015). Crystallization and properties of strontium barium niobate-based glass-ceramics for energy-storage application. *Journal of Electronic Materials* 44 (1): 227–234.

105 Xue, S., Liu, S., Zhang, W. et al. (2015). Correlation of energy conversion efficiency and interface polarization in niobate glass-ceramics for energy-storage application. *Applied Physics Letters* 106 (16): 162903-1-4.

106 Huang, J., Zhang, Y., Ma, T. et al. (2010). Correlation between dielectric breakdown strength and interface polarization in barium strontium tatanate glass ceramics. *Applied Physics Letters* 96 (4): 042902-1-3.

107 Shvartsman, V.V. and Lupascu, D.C. (2012). Lead-free relaxor ferroelectrics. *Journal of the American Ceramic Society* 95 (1): 1–26.

108 Eric Cross, L. (1994). Relaxore ferroelectrics: an overview. *Ferroelectrics* 151 (1): 305–320.

109 Haertling, G.H. and Land, C.E. (1971). Hot-pressed $(Pb,La)(Zr,Ti)O_3$ ferroelectric ceramics for electrooptic applications. *Journal of the American Ceramic Society* 54 (1): 1–11.

110 Hao, X., Wang, Y., Yang, J. et al. (2012). High energy-storage performance in $Pb_{0.91}La_{0.09}(Ti_{0.65}Zr_{0.35})O_3$ relaxor ferroelectric thin films. *Journal of Applied Physics* 112 (11): 114111-1-6.

111 Liu, Y., Hao, X., and An, S. (2013). Significant enhancement of energy-storage performance of $(Pb_{0.91}La_{0.09})(Zr_{0.65}Ti_{0.35})O_3$ relaxor ferroelectric thin films by Mn doping. *Journal of Applied Physics* 114 (17): 174102-1-6.

112 Zhang, L., Hao, X., Yang, J. et al. (2013). Large enhancement of energy-storage properties of compositional graded $(Pb_{1-x}La_x)(Zr_{0.65}Ti_{0.35})O_3$ relaxor ferroelectric thick films. *Applied Physics Letters* 103 (11): 113902-1-3.

113 Zhongqiang, H., Ma, B., Liu, S. et al. (2014). Relaxor behavior and energy storage performance of ferroelectric PLZT thin films with different Zr/Ti ratios. *Ceramics International* 40 (1): 557–562.

114 Tong, S., Ma, M., Narayanan, M. et al. (2013). Lead lanthanum zirconate titanate ceramic thin films for energy storage. *ACS Applied Materials & Interfaces* 5 (4): 1474–1480.

115 Song, Z., Liu, H., Zhang, S. et al. (2014). Effect of grain size on the energy storage properties of $(Ba_{0.4}Sr_{0.6})TiO_3$ paraelectric ceramics. *Journal of the European Ceramic Society* 34 (5): 1209–1217.

116 Puli, V.S., Pradhan, D.K., Chrisey, D.B. et al. (2013). Structure, dielectric, ferroelectric, and energy density properties of $(1-x)$BZT$-x$BCT ceramic capacitors for energy storage applications. *Journal of Materials Science* 48 (5): 2151–2157.

117 Puli, V.S., Kumar, A., Chrisey, D.B. et al. (2011). Barium zirconate-titanate/barium clacimum-titanate ceramics via sol–gel process: novel high-energy-density capacitors. *Journal of Physics D: Applied Physics* 44 (39): 385403-1-10.

118 Zhang, Q., Zhang, Y., Wang, X. et al. (2012). Influence of sintering temperature on energy storage properties of $BaTiO_3-(Sr_1-_{1.5x}Bi_x)TiO_3$ ceramics. *Ceramics International* 38 (6): 4765–4770.

119 Song, Z., Zhang, S., Liu, H. et al. (2015). Improved energy storage properties accompanied by enhanced interface polarization in annealed microwave-sintering BST. *Journal of the American Ceramic Society* 98 (10): 3212–3222.

120 Hui Huang, Y., Yong Jun, W., Qiu, W.J. et al. (2015). Enhanced energy storage density of $Ba_{0.4}Sr_{0.6}TiO_3$–MgO composite prepared by spark plasma sintering. *Journal of the European Ceramic Society* 35 (5): 1469–1476.

121 Zhang, Q., Wang, L., Luo, J. et al. (2010). $Ba_{0.4}Sr_{0.6}TiO_3$/MgO composites with enhanced energy storage density and low dielectric loss for solid-state pules-forming line. *International Journal of Applied Ceramic Technology* 7 (S1): E124–E128.

122 Ortega, N., Kumar, A., Scott, J.F. et al. (2012). Relaxor-ferroelectric superlattices: high energy density capacitor. *Journal of Physics: Condensed Matter* 24 (44): 445901-1-8.

123 Sun, Z., Ma, C., Liu, M. et al. (2017). Ultrahigh energy storage performance of lead-free oxide multilayer film capacitors via interface engineering. *Advanced Materials* 29 (5): 1604427-1-6.

124 Chao, S. and Dogan, F. (2011). $BaTiO_3-SrTiO_3$ layered dielectrics for energy storage. *Materials Letters* 65 (6): 978–981.

125 Kumar, N., Ionin, A., Ansell, T. et al. (2015). Multilayer ceramic capacitors based on relaxor $BaTiO_3-Bi(Zn_{1/2}Ti_{1/2})O_3$ for temperature stable and high energy density capacitor application. *Applied Physics Letters* 106 (25): 202901-1-4.

126 Shen, Z., Wang, X., Luo, B., and Li, L. (2015). $BaTiO_3$–$BiYbO_3$ perovskite materials for storage applications. *Journal of Materials Chemistry A* 3 (35): 18146–18153.

127 Ogihara, H., Randall, C.A., and Trolier-Mckinstry, S. (2009). High-energy density capacitors utilizing $0.7BaTiO_3$–$0.3BiScO_3$ ceramics. *Journal of the American Ceramic Society* 92 (8): 1719–1724.

128 Wang, T., Li, J., Li, C. et al. (2015). Relaxor ferroelectric $BaTiO_3$–$Bi(Mg_{2/3}Nb_{1/3})O_3$ ceramics for storage application. *Journal of the American Ceramic Society* 98 (2): 559–566.

129 Kwon, D.-K. and Lee, M.H. (2012). Temperature-stable high-energy-density capacitors using complex perovskite thin films. *IEEE Transactions on Ultrasonics, Ferroelectrics, and Frequency Control* 59 (9): 1894–1899.

130 Kang, S.-B., Kim, H.S., Lee, J.G. et al. (2016). Dielectric properties of $Pb(In_{1/2}Nb_{1/2})O_3$–$Pb(Mg_{1/3}Nb_{2/3})O_3$–$PbTiO_3$ film by aerosol deposition for energy storage applications. *Ceramics International* 42(1: 1740–1745.

131 Wang, X., Zhang, L., Hao, X. et al. (2015). Dielectric properties and energy-storage performances $(1-x)Pb(Mg_{1/3}Nb_{2/3})O_3-xPbTiO_3$ relaxor ferroelectric thin films. *Journal of Materials Science: Materials in Electronics* 26 (12): 9583–9590.

132 Xie, Z., Peng, B., Zhang, J. et al. (2015). Highly (100)-oriented $Bi(Ni_{1/2}Hf_{1/2})O_3$–$PbTiO_3$ relaxor-ferroelectric films for integrated piezoelectric energy harvesting and storage system. *Journal of the American Ceramic Society* 98 (10): 2968–2971.

133 Xie, Z., Yue, Z., Peng, B. et al. (2015). Large enhancement of the recoverable energy storage density and piezoelectric response in relaxor-ferroelectric capacitors by utilizing the seeding layer engineering. *Applied Physics Letters* 106 (20): 202901-1-5.

134 Xie, Z., Peng, B., Zhang, J. et al. (2015). Effects of thermal anneal temperature on electrical properties and energy-storage density of $Bi(Ni_{1/2}Ti_{1/2})O_3$–$PbTiO_3$ thin films. *Ceramics International* 41: S206–S212.

135 Xie, Z., Peng, B., Meng, S. et al. (2013). High-energy-storage density capacitors of $Bi(Ni_{1/2}Ti_{1/2})O_3$–$PbTiO_3$ thin films with good temperature stability. *Journal of the American Ceramic Society* 96 (7): 2061–2064.

136 Xie, Z., Yue, Z., Ruehl, G. et al. (2014). $Bi(Ni_{1/2}Ti_{1/2})O_3$–$PbTiO_3$ relaxor-ferroelectric films for piezoelectric energy harvesting and electrostatic storage. *Applied Physics Letters* 104 (24): 243902-1-5.

137 Correia, T.M., Mamillen, M., Rokosz, M.K. et al. (2013). A lead-free and high-energy density ceramic for energy storage applications. *Journal of the American Ceramic Society* 96 (9): 2699–2702.

138 Pan, H., Yi, Z., Shen, Y. et al. (2017). $BiFeO_3$–$SrTiO_3$ thin films as a new lead-free relaxor ferroelectric capacitors with ultrahigh energy storage performance. *Journal of Materials Chemistry A* 5 (12): 5920–5926.

6

Ferroelectric Polymer Materials for Electric Energy Storage

Zhi-Min Dang[1], *Ming-Sheng Zheng*[2], *and Jun-Wei Zha*[2]

[1] *Tsinghua University, Department of Electrical Engineering, State Key Laboratory of Power System, Zhongguancun North Street, Beijing, 100084, PR China*
[2] *University of Science and Technology Beijing, School of Chemistry and Biological Engineering, Department of Polymer Science and Engineering, No. 30 Xueyuan Road, Beijing, 100083, PR China*

6.1 Introduction

Ferroelectric polymer materials are rapidly booming as novel materials for a broad range of advanced fields in recent years. It is well known that dielectric material is the core component of polymer-based capacitors, which are charge storage and energy storage devices commonly used in the electronics and electrical power systems [1]. Because of the increasing demand for energy storage technologies, great progress has been made to understand and develop new polymers materials with high energy density. In this chapter, based on the inherent dielectric properties, the use of ferroelectric polymers in electric energy storage application, especially the ferroelectric polyvinylidene fluoride (PVDF), PVDF copolymers, PVDF terpolymers, and their dielectric composites are thoroughly investigated. Polymers usually exhibit excellent electric breakdown strength, which is much higher than that of ceramics. However, the permittivity of polymers (mainly 2–5) is much lower than that of ceramics (over 1000), which is the main reason for the low energy density [2]. For example, the commercial polymer-based film capacitors are primarily made of biaxially oriented polypropylene (BOPP); however, the permittivity of BOPP is as low as 2.2, which results in poor energy density of about 2 J cm^{-3} [3–5]. Numerous other polymers also have been used including polyimides [6, 7], epoxy [8, 9], etc. to prepare nanocomposites. Apparently, the addition of high permittivity fillers provides enhanced dielectric properties [10]. However, these polymers have low permittivity in nature, which makes it harder to improve their permittivity. Normal elastomers also show some disadvantages, which cannot be ignored. For instance, polydimethylsiloxane (PDMS), a kind of silicon-based elastomer, is a flexible, transparent, and biocompatible polymer matrix [11]. However, the permittivity of PDMS is only 2.3, while the intrinsic loss tangent is as high as 0.13 at 100 Hz, respectively. In addition, its electric breakdown strength is too low (about 66 MV m^{-1}) [12]. Thus, due to their high permittivity, relatively

Ferroelectric Materials for Energy Applications, First Edition.
Edited by Haitao Huang and James F. Scott.

low dielectric loss, and large breakdown strength, the ferroelectric polymers, especially PVDF, its copolymers, and terpolymers serve as promising candidates for high energy density applications.

6.2 Energy Storage Theory

In this section, we give an overview of the basic concepts such as dielectric properties, breakdown strength, and energy density. As is wells known, the battery stores the electric energy as chemical energy, while the polymeric film capacitor turns the electric energy into physical changes instead, which result in much higher power density than with the battery. However, the low permittivity of polymer dielectrics seriously limits the development of polymer-based capacitors due to the diminutive energy density, which is at least one order of magnitude lower than that of batteries and electrochemical capacitors [13, 14]. The energy density of a polymeric film capacitor depends on the dielectrics we use, which can endure high electric field and impose restrictions on the long-range movement of charges [15, 16]. It is well known that dielectrics are broadly divided into linear and nonlinear dielectrics [17]. Polypropylene is a kind of classic linear dielectrics. Because linear dielectrics do not contain permanent dipoles, the dielectric displacement linearly changes following the electric field without remnant polarization. The electric displacement (D) for the linear dielectric can be calculated as in Eq. (6.1), for the relative permittivity is not influenced by the electric field:

$$D = \varepsilon_0 \varepsilon_r E \tag{6.1}$$

where ε_0 is the permittivity of free space (8.85×10^{-12} F m^{-1}), ε_r is the relative permittivity of the dielectric, and E represents the applied electric field. The energy density of dielectrics can be defined as the amount of energy stored in the unit volume or mass, and it can be described as in Eq. (6.2):

$$W = \int E \, dD \tag{6.2}$$

where W is the energy density, which is usually transformed into J cm^{-3}. Because the permittivity of linear dielectrics is independent of the external electric field, according to Eq. (6.1), the electric energy is expressed as follows:

$$W = \frac{1}{2} \varepsilon_0 \varepsilon_r E^2 \tag{6.3}$$

Thus, the energy density of linear dielectric can be affected by two parameters: the relative permittivity (ε_r) and the applied electric field (E). This equation also shows that the energy density is directly proportional to ε_r, and has a square relationship with the applied electric field, so promoting breakdown strength is a good choice for the purpose of high energy storage. The discharged energy density can be counted from the electric displacement–electric field loops (D–E loops) through the integral of the shaded gray area. As shown in Figure 6.1a, it is the D–E loop of a linear dielectric, and the area of the gray shadow represents the value of energy density. However, the permittivity of ferroelectric materials will not be a constant when a different electric field is applied. As shown in

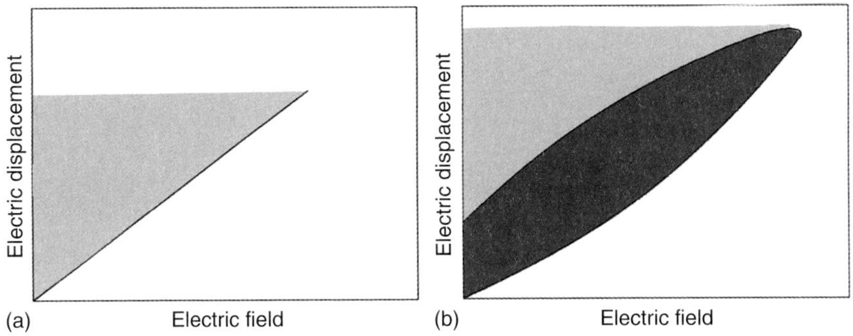

Figure 6.1 The D–E loops of (a) linear dielectrics and (b) ferroelectric dielectrics.

Figure 6.1b, it is clear that the electric displacement varies significantly with the electric field [19].

The ferroelectric property of nonlinear materials such as PVDF shows a significant hysteresis loop and large remnant polarization. The polarization can exist even when the applied electric field on the medium is removed, which is called the remnant polarization. This is from the presence of ferroelectric domains, and grows during the application of electric field and remains even when the electric field is zero due to the strong interdomain coupling [20, 21]. For the ferroelectric dielectrics, the discharged energy density does not equal the charged energy density due to the remarkable remnant polarization. As shown in Figure 6.1b, the area of gray shadow represents the discharged energy density, while the dark area is the wasted energy density that cannot be discharged because of the remnant polarization. As we can see, the remnant polarization associated with the irreversible dipoles shows a negative impact on the discharged energy density. For example, the total charged energy density (including the gray and dark area of Figure 6.1b) of PVDF is over $8\,\mathrm{J\,cm^{-3}}$; however, the discharged energy density is as low as $4.5\,\mathrm{J\,cm^{-3}}$, which is still inadequate for high energy density capacitor applications due to its obvious higher cost than polypropylene [22].

Therefore, the charge–discharge efficiency is also an indicator that can be used to evaluate the properties of energy storage, and this parameter represents the efficiency of energy conversion for capacitors. The efficiency can be shown as in Eq. (6.4):

$$\eta = \frac{W_1}{W_1 + W_2} \times 100\% \tag{6.4}$$

where η is the charge–discharge efficiency, while W_1 and W_2 represent the gray area (the discharged energy density) and the dark area (the wasted energy density) shown in Figure 6.1b, respectively. The low efficiency results in huge energy dissipation and temperature increment, which will shorten the longevity of capacitors. How to increase the gray area and decrease the dark area is also an important research focus in recent years.

6.3 Energy Storage of Ferroelectric Polymers

As a kind of polymer dielectrics, ferroelectric polymers also enjoy many advantages such as low cost, light weight, ease of processing, flexibility, high breakdown strength, etc. [13]. What is more, the most important character of ferroelectrics is the high permittivity. Compared with the nonpolar polymer dielectrics, ferroelectric polymers have much higher permittivity. For example, PVDF and some of its copolymers have a permittivity of about 10–12 at low frequency. Thus, as the most famous ferroelectric polymer, PVDF will be carefully discussed here.

PVDF can be readily prepared through free radical polymerization with the gaseous monomer 1,1-difluoroethylene. The polar C—F bond from the large electronegativity difference between the fluorine and carbon atoms shows a strong dipole moment of about 6.4×10^{-30} Cm, which is the cause of the large electric polarization of PVDF.

Owing to the small volume of fluorine atom, PVDF can easily crystallize; however, because the inevitable presence of configurational defects (head-to-head ($—CF_2—CF_2—$) or tail-to-tail ($—CH_2—CH_2—$)) and the chain transfer reaction influence the tacticity of molecular chain, the crystalline fraction in PVDF is limited to 50–60%. As is well known, there are four major crystalline phases of PVDF, which are named as α, β, γ, and δ, respectively, which have all been observed experimentally [23–25]. Both α and δ phases exhibit the TG^+TG^- chain conformation. On a macro level, α phase is nonpolar because of the dipole self-cancellation between the two antiparallel chains in a unit cell of lattice, which will surely have negative effects on the ferroelectric properties of PVDF. By contrast, the δ-PVDF crystals are macroscopically polar due to the identical directions of two chains. The all-trans conformation of molecular chains in the γ and β phases shows great potential in ferroelectric field for their huge dipole movements [26]. However, a large portion of dipoles become irreversible even when the external field is removed, which results in a high remnant polarization in turn. Thus, the γ and β phases are bad options for the dielectric application.

To solve this problem, defect modifications of PVDF have been used as a way to improve the microstructure as an effort to stabilize the paraelectric α phase and reduce the remnant polarization [27]. The intermolecular interactions and the physical properties of the polymer matrix can be affected by the molecular defects of PVDF [28]. Thus, a number of PVDF-based copolymers and terpolymers with tailored dielectric properties are prepared to obtain the new polymer system. Using chemical methods or irradiating beams to improve PVDF becomes an effective way that can modify the polymer configuration, which results in promoted dielectric properties [29].

The second monomer such as chlorotrifluor ethylene (CTFE), hexafluoropropylene (HFP), and trifluoroethylene (TrFE) is used to produce P(VDF–CTFE), P(VDF–HFP), and P(VDF–TrFE) through free radical polymerization for the purpose of molecular modification [30, 31]. However, due to the higher ferroelectricity of the all-trans conformation, P(VDF–TrFE) showed much higher remnant polarization than pure PVDF, which resulted in lower energy density $1.13\,J\,cm^{-3}$ at $130\,MV\,m^{-1}$. It has been demonstrated that the Curie temperature

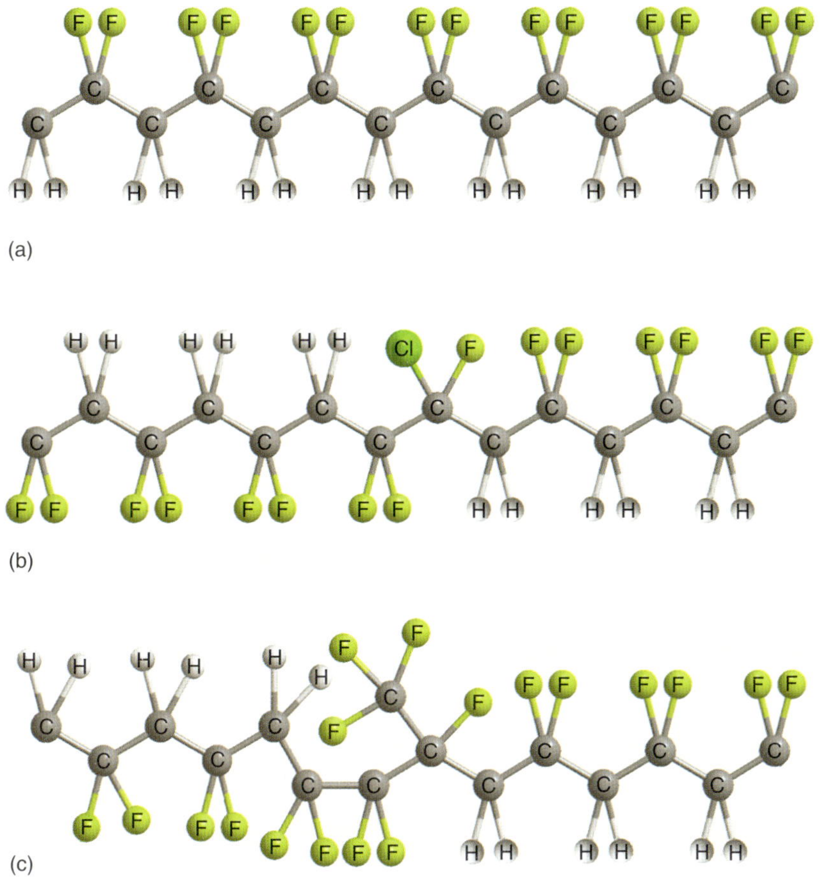

Figure 6.2 The schematic image of (a) all-trans PVDF chain, (b) P(VDF–CTFE) chain, and (c) P(VDF–HFP) chain.

(T_c) of P(VDF–TrFE) copolymers varied from 135 to 60 °C with the increasing molar ratio of 20–50% TrFE [32–35]. As shown in Figure 6.2, for other monomers such as CTFE and HFP, because there were relatively bulky comonomers compared with VDF, the copolymers exhibited larger interchain distances with respect to the PVDF [36–38], and this enlarged interchain distance weakened the interchain interaction, which showed lower ferroelectric-to-paraelectric transition temperature than that of the β-phase PVDF homopolymer [39–41]. The fact that the hydrogen atom is smaller than the CF_3 group or chlorine atom demonstrated that the second monomer showed higher steric hindrance compared with VDF [42]. It has been proved that the resultant random copolymers such as P(VDF–HFP) and P(VDF–CTFE) both exhibited slimmer electric hysteresis loops than pure PVDF [27, 43]. This made for comparatively lower crystallinity than that of PVDF, and TG^+TG^- became the dominating conformation instead of the all-trans conformation [44, 45], which resulted in a reversible phenomenon between polar and nonpolar phases under the alternating electric field [27]. Thus, P(VDF–HFP) had a much higher breakdown strength of about 600 MV m^{-1}, which resulted in higher energy density of about 11–13.5 J cm^{-3} than that of PVDF and P(VDF–TrFE) [46]. Owing to the absence of phase transition and good independence of temperature, P(VDF–CTFE) showed advantages than other PVDF-based copolymers in dielectric fields. The permittivity was up to 13 at 1 kHz and a high energy density of 8 J cm^{-3} with an electric field of 400 MV m^{-1} was observed [47]. Moreover, the energy density of P(VDF–CTFE) (91/9 mol%) uniaxially stretched films was as high as 17 J cm^{-3} at 575 MV m^{-1}, which was from the reduced ionic conduction and improved mechanical properties of thin films [27, 48]. What is more, the energy densities of PVDF copolymers can be further improved up to 25–27 J cm^{-3} by an extrusion-blowing process [49, 50], which was at a level comparable to that of the electrochemical capacitors [14]. However, the charge–discharge efficiency of these mechanically stretched PVDF copolymers was still as low as 50–60% at high electric fields.

To solve this issue with defect modification method, terpolymers of PVDF such as P(VDF–TrFE–CTFE) and P(VDF–TrFE–CFE) were prepared to decrease the energy barrier in phase transition and ferroelectric domain size [51–54]. With the addition of the third monomer into PVDF copolymers to prepare the terpolymers with proper compositions, the polymer chains of PVDF were pinned among bulky comonomers nearby. The third monomer further enlarged the interchange distance, while the physical pinning effect made the polymer chain segments among the neighboring pinning sites rotate freely, which led to excellent reversibility of dipoles under applied low electric field. In addition, because the Curie transition was tailored by introducing the third monomer into the ferroelectric polymer to match the ambient temperature [55], the terpolymers showed improved permittivity of about 15–50, reduced hysteresis loss, and promoted charge–discharge efficiency. For example, the P(VDF–TrFE–CFE) terpolymer film produced by the solution-casting procedure showed a high energy density of about 9 J cm^{-3} at 400 MV m^{-1}, while its charge–discharge efficiency was as high as 70% [56].

Chemical crosslinking of PVDF copolymers also has been proved to be an effective strategy for the highly effective energy storage application [57]. The crosslinking points not only were helpful to turn the aligned crystalline dipoles back to the original state, which led to low ferroelectric loss, but also confined the chain mobility to decrease the conduction loss. As a result, because of the excellent charge–discharge efficiency of about 83% at 400 MV m^{-1}, the energy density was as high as 17 J cm^{-3}.

Grafting other kinds of chains to the PVDF-based copolymers or terpolymers is also a useful way to modify the defects. By irradiating with a beam of electrons, Lee and coworkers have done much work in modifying the PVDF, especially the dielectric properties [29, 58, 59]. They grafted (2-hydroxyethyl methacrylate) (HEMA) onto the PVDF polymer, and the irradiating electron beam tailored the C—F, C—H, and even C—C bonds, which became the initiation sites for the formation of new HEMA chains through termination reaction. An improved permittivity of about 45 at 1 kHz was obtained, and the dielectric loss was as low as 0.0011. Dopamine was also grafted onto PVDF molecule chains by bombardment with electron beams [29]. The dopamine-modified PVDF polymer showed enhanced permittivity of 32 at 1 kHz and reduced dielectric loss of about 0.001, while the energy density reached 2.7 J cm^{-3} at 140 MV m^{-1}. Thus the results indicated that a proper grafting agent further improved the dielectric properties of pristine PVDF polymer. Wang and Zhu successfully grafted the low polarizable PS chains onto the P(VDF–TrFE–CTFE) and P(VDF–CTFE), and an energy density of about 10 J cm^{-3} at 550 MV m^{-1} was gained, while the charge–discharge efficiency reached 82% [60, 61]. This was because the PS side chains on the ferroelectric polymers facilitated the dipole switching, increased the amorphous content of the resulting copolymer, and decoupled the interaction among the ferroelectric domains nearby, which resulted in significantly reduced hysteresis loss. Enhanced dielectric properties of PVDF polymer also were realized through the addition of a proper organic compound. Li et al. added aromatic polythiourea (ArPTU) into the PVDF polymer and found that the PVDF/ArPTU(90/10) film exhibited an improved breakdown strength of about 700 MV m^{-1}, which led to a promoted energy density of 10.8 J cm^{-3}, while the charge–discharge efficiency was up to 83% [62].

Many kinds of PVDF dielectric materials based on defect modification have been found to exhibit improved dielectric properties, energy density, and charge–discharge efficiency. However, these strategies have many disadvantages such as comparatively low permittivity, much higher dielectric loss compared with the ceramic films, and high cost. Another promising way is to introduce different nanofillers as the reinforcement materials into these polymer matrices. Thus, we will discuss dielectric properties of ferroelectric polymer nanocomposites in the following section.

6.4 Energy Storage of Ferroelectric Polymer-Based Nanocomposites

In this part, different kinds of nanofillers including nonconductors, semiconductors, and conductors are incorporated into ferroelectric polymer matrices to improve the dielectric properties of the resulting dielectric nanocomposites. A lot of factors such as crystallinity, chain confirmation, and polymer chain mobility are discussed to demonstrate the mechanisms of the interaction between the nanofillers and the polymer matrices [63, 64]. The interphases formed between the nanofillers and the polymer matrix can greatly affect the dielectric properties of nanocomposites. The permittivities of both individual fillers and polymer matrix can influence the breakdown strength. The differences in electric properties lead to inhomogeneity of the field across the film. When the permittivity of the polymer matrix is much lower than that of the nanofiller, the electric field in the polymer increases to a much higher level, which results in the breakdown strength degradation of the whole nanocomposite material [36, 65].

According to the situation of adding spherical nanoparticles into the polymer matrix, Tanaka and his coworkers proposed a theoretical model for the formation of interphases, which was named as Tanaka's model, as shown in Figure 6.3a [66].

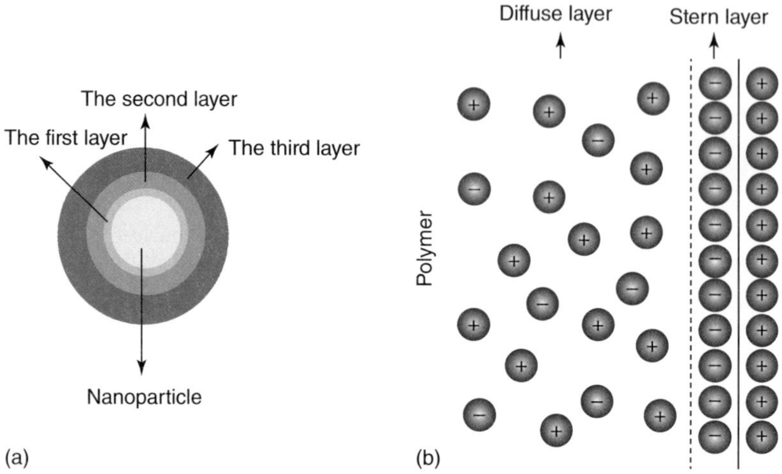

Figure 6.3 The schematic image of (a) Tanaka's model and (b) Lewis's model.

They further divided the interphase into three layers, all of which were bonded layers, The first layer with a thickness of about 1 nm strongly bonded to both inorganic and organic layers, which could be called the transition layer. Polymer chains in the second layer with a thickness of about 2–9 nm were strongly bound to the first layer, and it mainly depended on the interaction between nanofillers and the polymer matrix. The third layer was a loose layer with a thickness of several nanometers, which coupled to the second layer. With respect to mobility, chain conformation, crystallinity, and free volume of the bulk polymer matrix, the polymer in the interphase was different [15, 67]. Tanaka's model provided many possibilities in tuning the dielectric properties of the polymer nanocomposites. The dipole of the polar radicals was in the bound layer. The free volume of the third layer was smaller than that of the bulk polymer matrix, and the orientation of dipoles of the second layer was adversely affected due to its dense structure, both of which showed negative impact on the promotion of permittivity for dielectric nanocomposites. Thus, proper surface modification of nanoparticles was an important parameter for good dielectric performances.

Lewis's model proposed that the surface of nanofillers accumulated charges due to the difference of Fermi levels between the nanoparticles and polymer matrix, and the matrix developed counter charges near the surface of the nanofillers in turn [68]. As is well known, because of Coulomb attraction, the charged nanoparticle surfaces resulted in charge redistribution in the polymer matrix, and this caused formation of an electrical double layer including the Gouy–Chapman diffused layer and the Stern layer, as shown in Figure 6.3b [69]. This electrical double layer had great impact on the dielectric properties of the polymer nanocomposites and became more predominant near the percolation threshold of the fillers [12]. According to the percolation theory, there will be an abrupt transition of properties due to the formation of continuous networks in the polymer nanocomposite at the percolation threshold [70]. The percolation threshold of 1-dimension (1D) and 2-dimension (2D) nanofillers is much lower than the 0-dimension (0D) nanofillers due to the higher aspect ratio. Thus 1D and 2D nanofillers are more effective on the dielectric properties of nanocomposites.

The ferroelectric polymer-based nanocomposites exhibit many advantages compared with the pristine ferroelectric polymer. Different types of nanocomposites using various nanofillers have different dielectric properties. To improve the dielectric performances of ferroelectric polymer-based nanocomposites with full utilization of nanofillers, a number of valuable efforts have been devoted in recent years. In this section, we mainly summarize the recent progress in the ferroelectric polymer-based nanocomposites with 0D, 1D, and 2D nanofillers, respectively. The advantages and mechanisms of different strategies to fabricate ferroelectric polymer-based nanocomposites are discussed carefully here.

6.4.1 Ferroelectric Polymer-Based Nanocomposites Using 0D Nanofillers

0D nanoparticles are the most common nanofillers applied in nanocomposites. Most work reported in dielectric field involves 0D nanoparticles because of their universality and high industrialization. Ceramic nanoparticles such as TiO_2, $BaTiO_3$, and $Ba_{0.6}Sr_{0.4}TiO_3$ have been frequently used as nanofillers in dielectric nanocomposites. It is well known that the neat ceramic nanoparticles show poor dispersion state in a polymer matrix due to their high surface energy, and the corresponding nanocomposites also endure high dielectric loss and low breakdown strength because of the poor compatibility between inorganic nanofillers and organic matrix [71, 72].

6.4.1.1 Surface-Modified 0D Nanofillers

Surface modification of nanofillers adds flavor to the dielectric properties of nanocomposites. Thus, surface modifications are useful strategies to minimize the dielectric loss and promote the breakdown strength. For example, to investigate the dielectric properties of nanocomposites with surface modification nanofillers, Dang and his coworkers used hydrogen peroxide (H_2O_2) to hydroxylate $BaTiO_3$ nanoparticles [15]. The results demonstrated that the hydroxylated $BaTiO_3$ nanoparticles improved the morphology of the composite due to hydrogen bonding between fluorine of PVDF and the hydroxyl group of hydroxylated $BaTiO_3$. Because the movement of polymer chains near the interface and nanoparticles was restricted, the permittivity of nanocomposite with hydroxylated $BaTiO_3$ was much more stable with temperature and frequency than that of neat-$BaTiO_3$ composite. What is more, the dielectric loss was significantly improved. For instance, the nanocomposites with 30 vol.% hydroxylated $BaTiO_3$ had a high permittivity and dielectric loss of about 25 and 0.02 at 20 °C, as compared to neat $BaTiO_3$ nanocomposite, having a permittivity of about 28 and a dielectric loss of 0.03, respectively. In this case, $BaTiO_3$ nanoparticles were also modified using covalently bonded dopamine (PDOPA) via a reflux method by Lee and coworkers [73]. Because the dopamine contained both NH_2 groups and OH, it produced strong interaction with numerous metal oxides, metals, and polymers through hydrogen bonding, which made dopamine a promising surface modification agent. In addition, due to the agglomeration of nanoparticles in polymer matrix, the unmodified $BaTiO_3$/PVDF nanocomposite had pores when the loading of nanofillers was too high, and this caused seriously high dielectric loss and low breakdown strength. The aggregate issue led to the pores in the composite films and reduced the breakdown voltage of the nanocomposite [74]. On the contrary, the dopamine-modified $BaTiO_3$/PVDF nanocomposite imparted good compatibility between the polymer matrix and the nanoparticles, and exhibited much better dispersion state even at higher filler content. As a result, the permittivity and dielectric loss of nanocomposite with 50 wt% dopamine-modified $BaTiO_3$ reached 56.8 and 0.04 at 1 kHz, while the same for the corresponding unmodified $BaTiO_3$/PVDF nanocomposite were 40 and 0.2, respectively.

Wang and coworkers prepared surface-modified $BaTiO_3$ successfully using the coupling agent NXT-105 and introduced it into PVDF polymer matrix to fabricate NXT-105-modified $BaTiO_3$/PVDF nanocomposites with excellent dielectric properties [75]. The permittivity of nanocomposite with 50 vol.% NXT-105 modified $BaTiO_3$ was 53.9 at 1 kHz, while the dielectric loss was as low as 0.036. They also found that the electric displacement of nanocomposite with 20 vol.% NXT-105 modified $BaTiO_3$ was up to 6.28 $\mu C\ cm^{-2}$ at 200 $MV\ m^{-1}$, which was much higher than that of pure PVDF at about 3.5 $\mu C\ cm^{-2}$, and the resultant energy density was 3.54 $J\ cm^{-3}$ at 200 $MV\ m^{-1}$, and this was about 1.5 times higher than that of pure PVDF at the same condition.

Zhang et al. used γ-aminopropyl triethoxysilane (APS) to modify the surfaces of $BaTiO_3$ nanoparticles [76]. Owing to the polar nature of APS and the bridging between the nanofillers and polymer matrix, the permittivity of APS-modified $BaTiO_3$/PVDF nanocomposite with 20 vol.% nanofillers was as high as 27.74 at 1 kHz; however, the dielectric loss was only 0.039. It was demonstrated that proper APS concentration had positive impact on breakdown strength as well as the resistivity, and the optimum value was 1 wt%, while the breakdown strength was up to 245 $MV\ m^{-1}$, which resulted in a high energy density of 7.3 $J\ cm^{-3}$.

Tomer et al. successfully modified kaolinite clay with (3-mercaptopropyl)trime thoxysilane; after that the modified nanofiller was added into P(VDF–HFP) polymer matrix to fabricate the nanocomposites undergoing both unstretching and stretching processes [77]. It has been found that the stretched nanocomposites had more β-phase content than the unstretched nanocomposites; however, α-phase still remained predominant in both of these nanocomposites. As is well known, nanofillers played an important role in the polymer crystallinity as a nucleating agent in many cases, but the promoted crystallinity was not caused by the incorporation of nanofillers. Although the permittivity of nanocomposite with 5 wt% nanofiller loading showed similar dielectric properties, the higher breakdown strength of stretched nanocomposites of about 780 $MV\ m^{-1}$ and lower remnant polarization resulted in a much higher energy density up to 19 $J\ cm^{-3}$ than that of pure polymer (about 9.8 $J\ cm^{-3}$), as shown in Figure 6.4. Thus both the surface modification of the nanofillers and stretching process of nanocomposites had a positive effect on the breakdown strength as well as the energy density.

Kim et al. carried out the surface modification of $BaTiO_3$ using 2,3,4,5,6-pentafluorobenzyl phosphonic acid (PFBPA) [78]. In this work, to select the proper

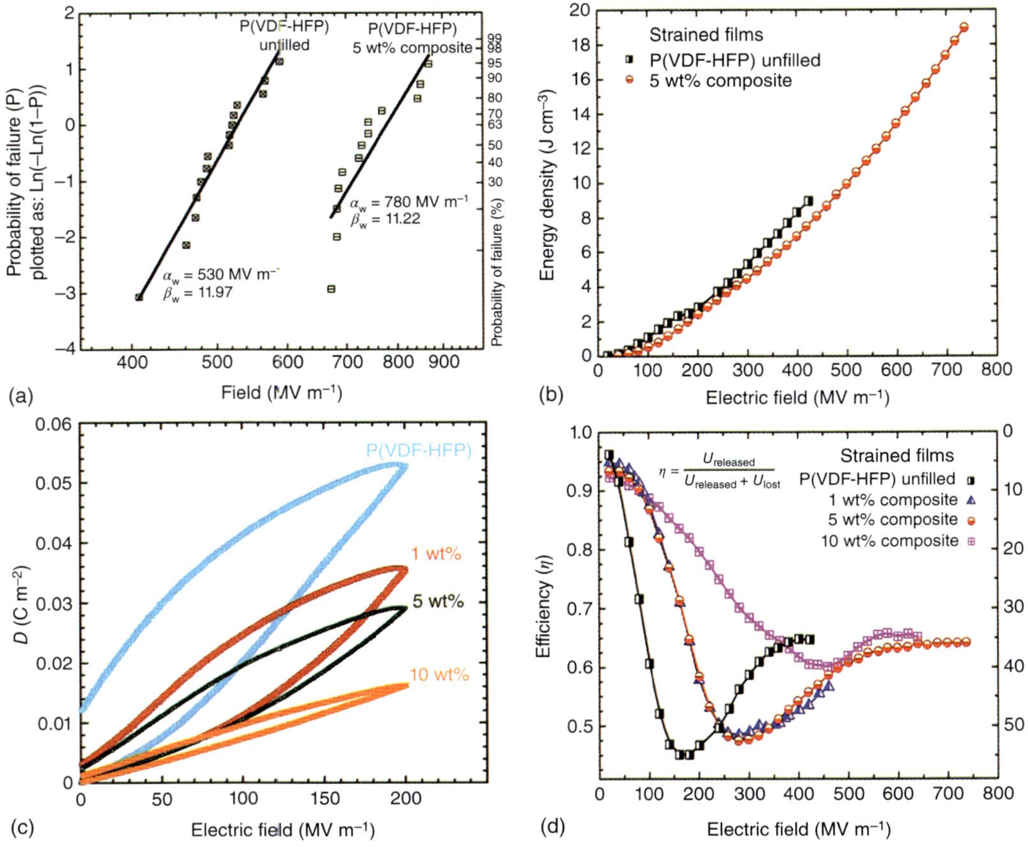

Figure 6.4 (a) Weibull distribution of breakdown strength and (b) energy density of stretched P(VDF–HFP) and nanocomposite with 5 wt% nanofiller; (c) The D–E loops and (d) the charge–discharge efficiency of nanocomposites with different nanofiller loading [77].

ligands for the surface modification procedure, the authors tested four different kinds of ligands for their compatibility with $BaTiO_3$: $PO(OH)_2$(OPA), SO_2ONa(OSA), $Si(OCH_3)_3$ (OTMOS) and CO_2H (OCA). According to the performance of different ligands on the surface of $BaTiO_3$, the phosphonic-acid-based ligand showed excellent compatibility with $BaTiO_3$ and robust surface coverage. Thus the agent PFBPA with the phosphonic acid ligand was selected as the surface modifier for $BaTiO_3$ nanoparticles; in the meantime, the fluorinated aryl group in PFBPA was compatible with the organic solvents and fluoropolymer such as P(VDF–HFP). The permittivity and dielectric loss of nanocomposite with 50 vol.% PFBPA modified $BaTiO_3$ at 1 kHz were about 37 and 0.025, respectively, and the corresponding breakdown strength was as high as $210\,MV\,m^{-1}$, which led to a high energy density of $6.1\,J\,cm^{-3}$. It was assumed that PFBPA not only reduced trap charges and the ionizable hydroxyl groups on the surfaces of $BaTiO_3$ serving as a passivation layer but also promoted uniform dispersion of $BaTiO_3$ nanoparticles. To further study the addition of PFBPA-modified $BaTiO_3$ nanoparticles into P(VDF–HFP) polymer matrix, the nanocomposites with higher nanofiller loading was investigated [74]. Kim et al. found that the permittivity of PFBPA-modified $BaTiO_3$/P(VDF–HFP) varied following the incremental content of PFBPA-modified $BaTiO_3$. The permittivity of nanocomposite with 60 vol.% nanofiller loading at 1 kHz was as high as 31, and then decreased sharply to 19 at 80 vol.%. However, the dielectric loss of nanocomposites with 80 vol.% nanofiller loading was as low as 0.01, which was much smaller even than that of pure P(VDF–HFP) (about 0.04). The phenomenon of dielectric loss continuously decreased with increasing loading, and was attributed to the changes of the semicrystalline polymer structure or the interference in segmental polymer movement caused by the bonding between filler and matrix. To perform a systematic study on the influence of different surfactants, additional dispersants such as BYK-W 9010 (phosphoric acid polyester copolymer with acid groups) or sodium dodecyl sulphate (SDS) were introduced into PFBPA-modified $BaTiO_3$/P(VDF–HFP) or OPA-modified $BaTiO_3$/P(VDF–HFP) [79, 80]. The permittivity of nanocomposites with PFBPA-modified $BaTiO_3$ was higher as compared to that of OPA-modified $BaTiO_3$ at the same content, while both of the P(VDF–HFP) nanocomposites showed similar dielectric loss. For example, the permittivities of nanocomposites with 30 vol.% PFBPA-modified $BaTiO_3$ and OPA-modified $BaTiO_3$ at 1 kHz were about 19 and 15.5, respectively, while the dielectric losses of both were less than 0.1. The PFBPA-modified $BaTiO_3$/P(VDF-HFP) nanocomposites had more uniform dispersion state, which resulted in better dielectric properties. In this case, two different dispersants were used to further improve the interphases between the nanofillers and the polymer matrix. For OPA-modified $BaTiO_3$/P(VDF–HFP) nanocomposites, the addition of SDS provided a better dispersion state than that in PFBPA-modified $BaTiO_3$/P(VDF–HFP) nanocomposites. Therefore, different kinds of surface modifiers and dispersants greatly influenced the interphases as well as the dispersion state.

The dielectric properties of polymer matrix also play an important role in those of the corresponding nanocomposites. The ethylenediamine-modified $BaTiO_3$ nanoparticles were also prepared as fillers and added into both P(VDF–CTFE) and P(VDF–TrFE–CTFE) by Li et al. [81]. Because of the inherently higher permittivity of P(VDF–TrFE–CTFE), its nanocomposites showed much higher permittivity than that of ethylenediamine-modified $BaTiO_3$/P(VDF–CTFE) nanocomposites at the same filler loading. For example, at 20 vol.%, the permittivities of PVDF copolymer and terpolymer-based nanocomposites with 20 vol.% nanofiller loading at 1 kHz were 24 and 50, while their dielectric losses were as low as 0.05 and 0.07, respectively. As a result, the maximum energy density of P(VDF–TrFE–CTFE) nanocomposite reached to 7.0 J cm^{-3}, which was much higher than that of P(VDF–CTFE)-based nanocomposites at about 3.7 J cm^{-3}. Thus, the energy density of ferroelectric polymer-based nanocomposites not only was affected by the surface modifying agents of the fillers but also depended on the synthesis conditions.

6.4.1.2 Core–Shell Structure 0D Nanofillers

As discussed above, the nanoparticle can be surface modified by micromolecules such as hydroxy, surfactants, coupling agents, phosphoric acids, and other organic molecules to improve the dielectric properties of ferroelectric polymer-based nanocomposites. However, because the modifiers themselves usually cannot promote the properties of the nanocomposites, the limitations of the simple modification strategies hinder the realization of full potential for high-k polymer nanocomposites in energy storage application. It has been proved that core–shell strategies are useful and versatile methods to design powerful nanofillers for dielectric nanocomposites. The core–shell strategies are not only helpful to achieve homogeneous dispersion state even at high nanofiller loading, but also can effectively enhance the permittivity of the nanocomposites and maintain high breakdown strength [82].

Wang and her workers used polyvinylpyrrolidone (PVP) as the shell to modify $BaTiO_3$ nanoparticle and fabricated PVP@$BaTiO_3$/PVDF nanocomposites with excellent dielectric properties [83]. The PVP@$BaTiO_3$/PVDF nanocomposite with a loading of 55 vol.% nanofillers exhibited a high permittivity and low dielectric loss of 77 and 0.05 at 1 kHz, respectively. It was demonstrated that the permittivity of nanocomposites showed great stability at the frequency range from 1 kHz to 1 MHz when the loading of PVP@$BaTiO_3$ was below 30 vol.%. However, space charge polarizations and Maxwell–Wagner interfacial polarizations became dominant on the enhanced dielectric properties following further increased filler loading. Thus the permittivity showed obvious decrease in the low-frequency region. They demonstrated that the core–shell structure PVP@$BaTiO_3$ nanoparticles in PVDF matrix was helpful for good compatibility between the nanofillers and the polymer matrix, which resulted in better dispersion state and relatively high breakdown strength. For example, the breakdown strength of nanocomposite with 10 vol.% PVP@$BaTiO_3$ was up to 336 MV m^{-1}, which was 1.38 times higher than that of unmodified $BaTiO_3$/PVDF nanocomposite with the same filler content, and the resultant energy density was as high as 6.8 J cm^{-3}, which was 1.89 times higher than that with untreated $BaTiO_3$.

Huang and coworkers successfully prepared a type of nanocomposites with excellent dielectric properties including core–shell fluoro-polymer@$BaTiO_3$ nanoparticles and PVDF–HFP [84] to investigate the physical interaction between the polymer shell and the polymer matrix. Two types of fluoroalkyl acrylate monomers (trifluoroethyl acrylate, TFEA and 1*H*,1*H*,2*H*,2*H* heptadecafluorodecyl acrylate, HFDA) were grafted onto $BaTiO_3$ through surface-initiated RAFT polymerization, which meant different shell molecular structures. After that, different kinds of core–shell structure fluoro-polymer @$BaTiO_3$ were introduced into P(VDF–HFP) polymer matrix. It has been demonstrated that the shell structure of the fluoro-polymer@$BaTiO_3$ played an important role in the dielectric properties and energy storage capability of the nanocomposites. The low interchain interactions and low chain mobility between PHFDA@$BaTiO_3$ and the P(VDF–HFP) matrix caused by ordered chain structure of PHFDA was found harmful for the enhancement of dielectric properties and breakdown strength, while high interchain interactions as well as high chain mobility between PTFEA@$BaTiO_3$ and the PVDF–HFP matrix induced by the disordered chain structure of PTFEA strongly restricted the dipole moment of P(VDF–HFP) matrix and suppressed the space charge polarization between the nanoparticles and the polymer matrix. As a result, the PTFEA@$BaTiO_3$/P(VDF–HFP) nanocomposites exhibited a lower dielectric loss, higher breakdown strength, and a resultant higher energy density than those of nanocomposites with PHFDA@$BaTiO_3$.

Apart from organic-shell nanoparticles, inorganic-shell nanoparticles also can be used as fillers to prepare high-permittivity ferroelectric polymer-based nanocomposites with high energy density such as TiO_2-shell $BaTiO_3$ nanoparticles (TiO_2@$BaTiO_3$) and SiO_2-shell $BaTiO_3$ nanoparticles (SiO_2@$BaTiO_3$) [10, 85]. The core–shell structure TiO_2@$BaTiO_3$ nanoparticles were prepared via a wet chemical route by Yao and Rahimabady, and added in P(VDF–HFP) to fabricate TiO_2@$BaTiO_3$/P(VDF–HFP) nanocomposites [10]. What is more, the TiO_2@$BaTiO_3$/P(VDF–HFP) nanocomposites exhibited much better dielectric properties than those of the $BaTiO_3$/P(VDF–HFP) nanocomposites at the same filler loading, and both the permittivity and the breakdown strength were improved in the TiO_2@$BaTiO_3$/P(VDF–HFP) nanocomposites. It has been widely accepted from previous studies that the increase of permittivity for the nanocomposites usually resulted in degraded breakdown strength. However, in this work, because the introduction of TiO_2 layer not only promoted the dielectric properties of TiO_2@$BaTiO_3$/P(VDF–HFP) nanocomposites but also decreased the sharp electric field distortion caused by the significant differences of permittivity between $BaTiO_3$ nanoparticles and PVDF matrix, the permittivity and the breakdown strength of TiO_2@$BaTiO_3$/P(VDF–HFP) nanocomposites were both enhanced. The improved dielectric properties were attributed to the Maxwell–Wagner interfacial polarization, which stemmed from the formation of Gouy–Chapman–Stern layers due to the strongly interactional interphases among the multiple dielectric materials and space charge accumulation. As a result, an unexpectedly large permittivity of nanocomposite with 50 vol.% TiO_2@$BaTiO_3$ was as high as 110 at 1 kHz, which was over three times higher than that of the pure $BaTiO_3$/P(VDF–HFP) nanocomposite at the same filler content. In addition, the breakdown field was also promoted up to 340 MV m^{-1} as the loading of TiO_2@$BaTiO_3$ was 30 vol.%; this resulted in an energy density of about 12.2 J cm^{-3}, which is almost three times higher than that of pristine

PVDF (about 4.1 J cm^{-3}), because the TiO_2 middle layer became a buffer as its intermediate dielectric permittivity of about 80 reduced the local electric field. In addition, the dispersion state of nanoparticles in P(VDF–HFP) was greatly improved because of the stronger interaction between the polymer matrix and the TiO_2 shell.

Zhu and coworkers decorated nickel (Ni) shell onto the TiO_2 core by sol–gel method to prepare Ni@TiO_2 nanoparticles, and then added the same into P(VDF–TrFE) polymer matrix (70/30 mol%) to fabricate Ni@TiO_2/P(VDF–TrFE) nanocomposites [86]. It is well known that, as a kind of conducting metal, the electric properties of Ni in polymer matrix followed the percolation theory, and even slight variation in filler concentration near the percolation threshold led to abrupt changes in dielectric properties. The permittivity of Ni@TiO_2/P(VDF–TrFE) nanocomposite with 60 wt% nanofillers at 1 kHz was only 18.56, while the dielectric loss was as low as 0.046–0.064, which was relatively independent of filler content due to the better dispersion state in the P(VDF–TrFE). Because the incorporation of nanofillers hindered the crystallization of the polymer, the crystallinity of the polymer nanocomposite was reduced.

Yu et al. reported that coating a silicon dioxide (SiO_2) layer (7 nm) onto the $BaTiO_3$ nanoparticle's surface is helpful in lowering the dielectric loss [85]. The SiO_2@$BaTiO_3$/PVDF nanocomposite with 10 vol.% filler loading showed a relatively high permittivity and low dielectric loss of about 14.7 and 0.02 at 1 kHz, respectively. Compared to unmodified $BaTiO_3$/PVDF nanocomposite with 10 vol.% $BaTiO_3$, the leakage current of SiO_2@$BaTiO_3$/PVDF nanocomposite was also reduced to 15% due to the lower dielectric loss. The resultant energy density of the nanocomposite reached 6.28 J cm^{-3} at 340 MV m^{-1}.

6.4.1.3 Multilevel Structure Nanocomposites

The multilevel structure shows great potential in the enhancement of dielectric properties, breakdown strength as well as the energy density, because it can combine many unique advantages together to balance the properties discussed above.

Yu et al. introduced a third component (polyacrylate elastomer (AR71)) into the $BaTiO_3$/PVDF system to fabricate the multilevel structure AR71/$BaTiO_3$/PVDF nanocomposites [87]. In this work, the volume fraction of AR71 ranged from 0 to 7 vol.% and the content of $BaTiO_3$ nanoparticles was fixed at 5 vol.%. As we know, AR71 contained some inclusions such as initiators and surfactants due to the free radical polymerization, so the DC conductivity increased with increasing AR71 loading in content because of the residual ions. Lower loading of $BaTiO_3$ nanoparticles barely influenced the permittivity of nanocomposites, which was usually about 12. However, increasing AR71 loading exhibited a negative impact on the dielectric loss because of the increased leakage current. On the base of the polar groups in AR71, this kind of dielectric particles also exhibited interface and space charges in the external electric field. Thus, polarization increased following the incremental elastomer concentration. The nanocomposite with 3 vol.% AR71 had a higher polarization than the $BaTiO_3$/PVDF nanocomposite. As a result, the maximum energy density reached 8.8 J cm^{-3} at 3 vol.% AR71. The addition of AR71 also resulted in expansion and inhomogeneity for local effective electric field in the PVDF polymer matrix. AR71 elastomer shrunk as the electric field decreased, while the space charge polarization and Maxwell–Wagner–Sillars interfacial polarization reduced and led to decreased remnant polarization, which also promoted the energy density.

Aiming to enhance both the dielectric properties and breakdown strength, Wang et al. prepared a sandwich structure using $BaTiO_3$ nanoparticles as the dielectric filler to obtain a greatly improved breakdown strength as well as energy density [88]. They defined the top and bottom layers with higher $BaTiO_3$ loading as the "soft layers" to improve the permittivity, and the middle layer with lower $BaTiO_3$ loading as the "hard layer" to enhance the breakdown strength. The results showed that small amount of $BaTiO_3$ nanoparticles in the middle layer introduced electron traps into the PVDF, which effectively limited the electron paths through the interphases between the nanofillers and the polymer matrix. What is more, the middle layer blocked the growth of electrical trees from both the top and bottom sides, which surely benefited the improvement of breakdown strength. It was discovered that the nanocomposite with a content of 1 vol.% $BaTiO_3$ exhibited the highest breakdown strength over 370 MV m^{-1}; thus the loading of nanofillers in the middle layer was 1 vol.%. The $BaTiO_3$/PVDF nanocomposites with various increasing loading of $BaTiO_3$ from 10 to 50 vol.% were added into the "soft layer" to provide high permittivity. The high permittivity nanoparticles in "soft layers" induced huge electric displacement polarization at the high external electric field, which not only limited the remnant polarization but also provided a high maximum polarization. Finite element simulation also indicated that the growth of electrical tree could be hindered at the interphases between layers. The unique interphases interrupted the conductive paths and promoted a ultrahigh electric breakdown strength of 470 MV m^{-1} when the $BaTiO_3$ loading in "soft layers" was 20 vol.%, which resulted in a highly desirable energy density of about 18.8 J cm^{-3}.

6.4.2 Ferroelectric Polymer-Based Nanocomposites Using 1D Nanofillers

1D and 2D nanofillers with large aspect ratios showed better flexibility than 0D nanofillers due to their much lower loadings. Utilizing various nanowires, including conducting carbon nanotubes (CNTs) and semiconducting ceramic nanowire, is a novel and fast-developing way to promote the dielectric properties and energy density of polymer matrix due to their higher aspect ratios and less aggregation than 0D nanoparticles. Based on the works discussed below, it could be concluded that the lower loading of nanowires (approximately 2.5–7.5 vol.%) usually improves both the dielectric properties and breakdown strength, which results in higher energy density as compared to that of 0D nanoparticles.

6.4.2.1 Surface-Modified 1D Nanofillers

On the base of percolation threshold, the addition of CNTs into polymers can significantly improve their dielectric properties, because the high aspect ratio makes it easy to form the conducting network. However, the easy formation of conductive networks also induces high dielectric loss, low breakdown strength, and high electrical conductivity [89]. The concentration of CNTs has to be carefully controlled. In addition, due to the easily entangled state, it is difficult to disperse CNTs uniformly into the polymer matrix [90]. However, the dispersion state of CNTs can be promoted through a proper processing method, while the dielectric properties will also be greatly improved [91, 92]. Dang and Bai fabricated the CNTs/PVDF nanocomposites with an extruder by melt-mixing and found an interesting phenomenon; they reported the obvious molecular-level interaction at interphases between the CNTs and the polymer matrix [93]. What is more, the sufficient shear force resulted in a certain degree of CNTs alignment and a uniform dispersion state. As a result, the maximum permittivity reached up to 3800, while still maintaining a low conductivity about 6.3×10^{-5} S m^{-1}.

To obtain optimum dielectric properties of CNTs/polymer composites, the CNTs content is often very near the percolation threshold, which is usually below 1 vol.%, and the adjustment "window" of dielectric properties is narrow and hard to control, which is harmful for practical applications [94]. Although ceramic nanowires cannot promote high permittivity as CNTs, the low dielectric loss as well as relatively high breakdown strength are good for the energy storage application.

Nan and Shen successfully prepared PDOPA-modified $Ba_{0.6}Sr_{0.4}TiO_3$ nanowires to improve the dielectric properties of PVDF [95]. Good dispersion and excellent flexibility were observed even in 11.2 vol.% PDOPA-modified $Ba_{0.6}Sr_{0.4}TiO_3$ nanowires, which also exhibited a high permittivity of about 22.5 at 1 kHz. What is more, the frequency stability of permittivity and the decreased dielectric loss at high frequency with increasing filler loading made these nanocomposites more attractive, and it remained at about 0.02 at 1 kHz with different filler concentrations. The strong interaction between the low loading (2.1 vol.%) nanowires and the PVDF matrix resulted in a high breakdown strength of about 406 MV m^{-1}. Both the large polarization caused by high breakdown strength of the nanocomposite and the paraelectric behavior of fillers showed positive influence on the energy density. As a result, the energy density of nanocomposite with 4.4 vol.% PDOPA-modified $Ba_{0.6}Sr_{0.4}TiO_3$ nanowires reached 5.24 J cm^{-3}.

Liu and Zhai used APS to modify $Ba_{0.6}Sr_{0.4}TiO_3$ nanowires and then introduced it into PVDF [96]. The nanocomposites with 7.5 vol.% nanofiller loading showed a high permittivity of 21.78 and a very low dielectric loss of about 0.015 at 1 kHz. The surface modification ensured the uniform dispersion state of nanowires and improved the compatibility between fillers and polymer matrix, which played an important role in the excellent dielectric properties. They also found that the breakdown strength of the nanocomposites containing 7.5 vol.% unmodified $Ba_{0.6}Sr_{0.4}TiO_3$ was 223 MV m^{-1}, which is much lower than that of APS-modified $Ba_{0.6}Sr_{0.4}TiO_3$/PVDF nanocomposites at the same filler content of about 297 MV m^{-1}. The nanocomposite with 2.5 vol.% APS-modified $Ba_{0.6}Sr_{0.4}TiO_3$ showed a higher breakdown strength of 380 MV m^{-1}, which resulted in a maximum energy density of about 6.8 J cm^{-3} in this work.

Tang and Sodano added APS-modified TiO_2 nanowires into the PVDF matrix and found that the promoted γ-phase of PVDF was good for the high breakdown strength and high energy density [97]. As is well known, the addition of fillers usually decreases the breakdown strength; however, low loading of nanofillers could maintain a relatively high breakdown strength, and the 1D nanofillers turned out to be the more effective additives than 0D nanofillers. For instance, the permittivity of nanocomposites with 7.5 vol.% APS-modified TiO_2 nanowires was up to 16, while maintaining a low dielectric loss of about 0.08 at 1 kHz and a high breakdown strength of 450 MV m^{-1}, which resulted in a high energy density of 12.4 J cm^{-3}. Based on the work above, they first modified the $Ba_{0.8}Sr_{0.8}TiO_3$ nanowires using ethylenediamine to improve the dispersion state and compatibility with the PVDF [98]. The γ-phase of PVDF was also achieved by the addition of nanowires. The permittivity of nanocomposites with 7.5 vol.% ethylenediamine-modified $Ba_{0.8}Sr_{0.8}TiO_3$ nanowires increased to 18 at 1 kHz, which was 1.8 times higher than that of pure PVDF. Although the breakdown strength reduced to 450 MV m^{-1} at 7.5 vol.% nanofillers, which was 55 MV m^{-1} lower than that of the nanocomposite with 2.5 vol.% loading, the maximum energy density reached 14.86 J cm^{-3} at 7.5 vol.% nanofillers, which was over 10 times higher than that of BOPP. What is more, a high charge–discharge efficiency of 60% at 450 MV m^{-1} was observed, which was attributed to the quenching procedure of materials and nonferroelectric nature of nanowires.

Zhai and his coworkers fabricated hydroxylated $Ba_{0.6}Sr_{0.4}TiO_3$ nanowires using H_2O_2 [99]. Less air voids due to better compatibility between the fillers and PVDF led to lower dielectric loss than that of unmodified $Ba_{0.6}Sr_{0.4}TiO_3$ nanowires/PVDF; there was no significant improvement in permittivity. For example, the hydroxylated $Ba_{0.6}Sr_{0.4}TiO_3$ nanowires/PVDF nanocomposite with 10 vol.% filler content showed a permittivity of 25 and a low dielectric loss of about 0.01 at 1 kHz. Because of the accumulation of charges at the interphases, the movement and broadening of loss peaks were obtained with incremental filler concentrations and became more and more significant at higher filler content. Moreover, the breakdown strength of the nanocomposite with 2.5 vol.% loading was up to 398 MV m^{-1}, and still maintained 260 MV m^{-1} with 10 vol.% hydroxylated $Ba_{0.6}Sr_{0.4}TiO_3$ nanowires. Because the effective electric field reduced as it was applied in an out-of-plane direction, the large aspect ratio of oriented 1D nanofillers in the plane direction was assumed to be one of the causes for the increased breakdown strength [100]. As a result, the maximum energy density, which was consistent with the highest breakdown strength (398 MV m^{-1}) of nanocomposites, was up to 6.4 J cm^{-3}.

Apart from TiO_2 and $Ba_{0.6}Sr_{0.4}TiO_3$ nanowires, $BaTiO_3$ nanowires are also applied in this field. Sodano et al. studied the influence of different aspect ratios (9.5–45.8) for the permittivity variations of ethylenediamine-modified $BaTiO_3$ nanowires/PVDF nanocomposites [101]. They demonstrated that the permittivity of nanocomposites with 30 vol.% $BaTiO_3$ nanowires at 1 kHz increased from 34 to 45 following the aspect ratio ranging from 9.5 to 45.8. It was thought that the high aspect ratio played an important role in the improvement of permittivity due to the increased dipole moment, easy formation of nanowire network, and good dispersion state. In addition, different aspect ratios barely showed an impact on the dielectric loss and maintained a stable value of about 0.04 at 1 kHz. Therefore, the nanocomposites with higher aspect ratio had a positive influence on the dielectric performances.

Nan and coworkers used PDOPA-modified $BaTiO_3$ nanoparticles and nanowires to further study the properties of 0D and 1D nanofillers in both PVDF and epoxy polymers, respectively [102]. Homogeneous dispersion states of fillers due to the PDOPA surface modification were observed. The permittivity of PDOPA-modified $BaTiO_3$ nanowires/PVDF with 11 wt% nanofiller content reached about 30 at 1 kHz. The breakdown strength did not show much variation with lower nanofiller loading (below 11 vol.%). However, because of the increasing remnant polarization with incremental nanofiller loading of nanocomposites, the energy density decreased. They also introduced PDOPA-modified $BaTiO_3$ nanowires as fillers into P(VDF–TrFE) [103]. The permittivity of the nanocomposites with 10.8 vol.% nanofiller loading at 1 kHz was as high as 30,while a dielectric loss as low as 0.02 was observed. Moreover, the maximum breakdown strength reached 226 MV m^{-1} at 3 vol.% loading, and still remained at a relatively high value of 205 MV m^{-1} at 10.8 vol.%. The covalent bond of PDOPA with the $BaTiO_3$ nanowires reduced mobile charges as well as the electric field distortion [104]. In addition, the nanowires oriented in the perpendicular direction of the electric field, and played a role as scattering centers. Thus, both the dielectric properties and breakdown strength of the nanocomposites were promoted with the incorporation of PDOPA-modified $BaTiO_3$ nanowires.

Zhai and coworkers used APS-modified $BaTiO_3$ nanowires to investigate the improvement of the breakdown strength of PVDF nanocomposites [105]. They found that the breakdown strength of nanocomposite with 2.5 vol.% nanofiller content increased to 330 MV m^{-1} due to the reduced conduction losses, which resulted in a high energy density of about 5.6 J cm^{-3} and also a relatively high charge–discharge efficiency of about 63%. What is more, a better permittivity and dielectric loss of nanocomposites with 7.5 vol.% nanofiller loading at 1 kHz of about 24 and 0.03 were observed, respectively. Thus, proper surface modification of nanowires was good for high energy density and charge–discharge efficiency.

Zhang et al. successfully modified the $BaTiO_3$ nanowires with 1*H*,1*H*,2*H*,2*H*-perfluorooctyltrimethoxysilane [106]. The introduction of fluorosilane onto the surface of $BaTiO_3$ nanowires using silane coupling modified the nanowires with short perfluoroalkyl chain, which was helpful for the uniform dispersion of the surface-modified nanofillers in PVDF polymer matrix. As a result, the permittivity of the nanocomposites with 20 vol.% nanofillers reached 22 at 1 kHz, which was two times more than that of pure PVDF, and the dielectric loss decreased to 0.025. They also discovered that the tendency of permittivity with increasing nanofiller loading was consistent with Lichtenecker logarithmic rule.

Hu and Nan synthesized $Ba_{0.3}Sr_{0.7}TiO_3$ with 5 mol% Bi_2O_3 nanowires as the filler using electrospinning technique to improve the energy density of P(VDF–TrFE) polymer matrix [107]. They demonstrated that the permittivity of the nanocomposites at 27.7 vol.% nanofiller loading achieved a value of 34 at 1 kHz, while the corresponding dielectric loss was only 0.03. They also found a maximum value of breakdown strength of about 228 MV m^{-1} for the nanocomposite with 4.1 vol.% nanofillers, which was 1.4 times higher than that of the pure polymer. The homogeneous dispersion and good compatibility at lower content due to the surface modification method promoted the breakdown strength. The results suggested that the current density increased when the nanofiller loading was over 10 vol.%, which degraded the breakdown strength with increasing content. For example, the leakage current density remained at about 1.3×10^{-6} A cm^{-2} below 10 vol.% loading, and increased up to about 8×10^{-5} A cm^{-2} at 27.7 vol.% nanofiller content.

Ba-OH-modified TiO_2 nanowires were also prepared as fillers and added into both P(VDF–CTFE) and P(VDF–TrFE–CTFE) to fabricate dielectric nanocomposites [108]. Owing to the little difference in permittivity for TiO_2 and P(VDF–TrFE–CTFE), the resulting value at low electric field remained at about 40, while the dielectric loss was 0.05 at 1 kHz even at 30 vol.% nanofiller loading. The permittivity and the dielectric loss of P(VDF–CTFE)-based nanocomposites were 25 and 0.04 at the same situation, respectively. Comparing the energy density, P(VDF–CTFE)-based nanocomposites with 10 vol.% loading had a value of 2.5 J cm^{-3} at 170 MV m^{-1}, while for the P(VDF–TrFE–CTFE)-based nanocomposites with the same content the value was up to 6.9 J cm^{-3} at 200 MV m^{-1}. Because of the higher crystallinity and smaller crystalline domains, the dielectric properties of P(VDF–TrFE–CTFE)-based nanocomposites were much better than those of P(VDF–CTFE)-based nanocomposites.

Nan and Shen incorporated $BaTiO_3$ nanoparticles into TiO_2 nanowires using the electrospinning method, as shown in Figure 6.5a,b, and demonstrated that the hierarchical interphases between $BaTiO_3$ and TiO_2 greatly enhanced the interfacial polarization, which resulted in higher electric displacement in the PVDF-based nanocomposites [18]. It was assumed that the large strain existed

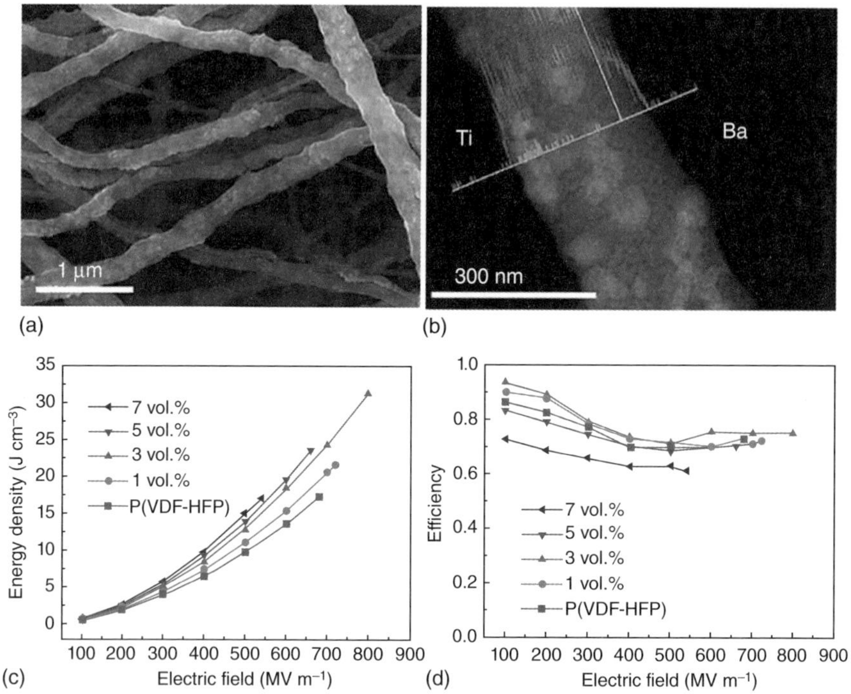

Figure 6.5 (a) SEM image and (b) elements mapping image of $BaTiO_3$@TiO_2 nanowires; (c) energy density and (d) the charge–discharge efficiency of nanocomposites with different nanofiller content [18].

in the interfacial regions because of the large difference in lattice parameter, and the Ba and Ti atoms showed mutual occupation in the interphases between $BaTiO_3$ and TiO_2, which were both helpful for enhancement of polarization and electric displacement. They also found that proper addition of $BaTiO_3$@TiO_2 nanowires, such as 3 vol.%, could significantly increase the breakdown strength to 646 MV m^{-1}, and a high energy of about 21.2 J cm^{-3} was observed. To further promote the energy density, they incorporated the $BaTiO_3$@TiO_2 nanowires into P(VDF–HFP) due to its higher inherent breakdown strength, and demonstrated that the nanocomposites with 3 vol.% nanofiller content exhibited a promoted breakdown strength as high as 798 MV m^{-1}, which led to a maximum energy density of about 31.2 J cm^{-3} in this work and a high charge–discharge efficiency of 78%, as shown in Figure 6.5c,d [109].

6.4.2.2 Core–Shell Structure 1D Nanofillers

Generally speaking, a shell with proper permittivity on the high permittivity core not only creates additional interfacial polarization induced by space charges but also greatly decreases electric field distortion in the nanocomposites.

Zhai and coworkers successfully coated SiO_2 onto the $BaTiO_3$ nanowires through the hydrolysis reaction to fabricate the core–shell structure SiO_2@$BaTiO_3$ nanowires [110]. It was demonstrated that the SiO_2 shell can hinder the movement of charge carriers through the PVDF-based nanocomposites due to the shielding effect of $BaTiO_3$, which weakened the Maxwell–Wagner–Sillars interfacial polarization, reduced the dielectric loss, and improved the energy density. As a result, the nanocomposite with 2.5 vol.% SiO_2@$BaTiO_3$ nanowires showed higher breakdown strength of 330 MV m^{-1} and a maximum energy density of about 6.3 J cm^{-3}, which was 1.2 times larger than that of $BaTiO_3$ nanowires/PVDF nanocomposite at the same condition. To further promote the energy density of PVDF-based nanocomposites, they also prepared Al_2O_3@$BaTiO_3$ nanowires by a coaxial electrospinning technique [111]. The insulating Al_2O_3 shell with moderate permittivity as a buffer greatly improved the breakdown strength as well as the energy density. For example, the breakdown strength of nanocomposites with 5 vol.% Al_2O_3@$BaTiO_3$ nanowires was as high as 400 MV m^{-1}, which resulted in a large energy density of 12.18 J cm^{-3}.

Hu and coworkers coated TiO_2 onto $BaTiO_3$ nanowires through the coaxial electrospinning method [112]. The addition of TiO_2@$BaTiO_3$ nanowires into PVDF polymer matrix significantly enhanced the electric displacement in electric field. Because of the additional polarization induced by the movement of charges in the interphases between TiO_2 and $BaTiO_3$, the permittivity of nanocomposites was obviously improved. What is more, the nanocomposites could maintain relatively high breakdown strength due to the limitation of charge shifting in the buffer zones. Thus, a maximum energy density of 10.94 J cm^{-3} was obtained at 360 MV m^{-1} with 3 vol.% nanofiller content.

6.4.2.3 Multilevel Structure Nanocomposites

The multilevel structure also has been applied in the ferroelectric polymer-based nanocomposites with 1D nanofillers. Nan and coworkers integrated three different layers to prepare sandwich structure nanocomposites using layer-by-layer tape casting procedure [113]. They added 30 vol.% TiO_2 nanoparticles into both the top and bottom layers for the purpose of higher electric displacement, while PDOPA-modified 5 mol% Bi_2O_3-doped $Ba_{0.3}Sr_{0.7}TiO_3$ nanowires were introduced into the middle layer to achieve higher breakdown strength. The permittivity of nanocomposites with 0–5 vol.% nanowires loading in the middle layer at 1 kHz was up to 14.5, while the dielectric loss decreased to 0.03 at 1 vol.% and barely changed at higher loading. They also demonstrated that the middle layer played a positive role on the promoted breakdown strength of nanocomposites. The breakdown strength of the middle layer with 3 vol.% nanowires loading was as high as 385 MV m^{-1}, while the sandwich structure nanocomposites decreased to 300 MV m^{-1}. Owing to the arrangement in parallel to the plane orientation and perpendicular to the external electric field, the nanowires presented in the middle layer hindered the charge from both the top and bottom layers and hence greatly improved the breakdown strength. The resultant energy density of nanocomposites containing 3 vol.% nanowires in the middle layer reached 7.99 J cm^{-3}, which was much higher than that of the two-phase nanocomposites at about 4.5 J cm^{-3} at the same nanowire content. To further study the sandwich structure nanocomposites, they introduced PDOPA-modified $BaTiO_3$ nanoparticles into the top and bottom layers and $BaTiO_3$ nanowires into the middle layer to promote the dielectric properties [114]. Because the nanocomposites with $BaTiO_3$ nanowires exhibited the highest breakdown strength at lower content, the volume fraction of $BaTiO_3$ nanowires was fixed at 2 and 3 vol.% in the middle layer. The loading of $BaTiO_3$ nanoparticles increased from 10 to 30 vol.% to adjust the dielectric properties of the sandwich structure nanocomposites. For example, the maximum permittivity was found to be up to 25 at 1 kHz with 30 vol.% $BaTiO_3$ nanoparticles in the side layers and 3 vol.% $BaTiO_3$ nanowires in the middle layer, while the dielectric loss was as low as 0.04. The nanocomposites with 10 vol.% nanoparticles and 2 vol.% nanowires showed maximum breakdown strength of 453 MV m^{-1} and an energy density of 9.72 J cm^{-3}.

6.4.3 Ferroelectric Polymer-Based Nanocomposites Using 2D Nanofillers

Compared to the 0D nanoparticles and 1D nanowires, 2D nanofillers showed more effective aspect ratios, which means much lower loading for the purpose of high dielectric properties. As a kind of novel nanomaterials, 2D nanofillers, such as graphene nanosheets (GNs) and boron nitride nanosheets (BNNs), have been applied in many fields including the dielectric fields.

The GNs used as nanofillers usually originate from graphene oxide (GO) due to its proper size, ease of production, and low cost. Moreover, because of the existence of numerous oxygenated functional groups on the surfaces, GNs are much facile to react with chemical reagents for the purpose of surface modification [115, 116]. It has been proved that GNs acted as an effective nucleation, which was good for the formation of the β-phase of PVDF [117]. For example, Lu and coworkers introduced GNs into PVDF to fabricate nanocomposites and found that a low content of 0.1 wt% GNs turned the crystal phase into nearly all β-phase [118].

Core–shell structure also has a positive impact on the breakdown strength as well as energy density. Zhang and coworkers used polyaniline (PANI) to modify GNs and then introduced it into PVDF to improve both the dielectric properties and breakdown strength [119]. The PANI@GNs looked like a sandwich structure, while the top and bottom layers were PANI. The PANI shell not only optimized the dispersion state of GNs due to the improvement of compatibility between GNs and PVDF polymer matrix, but also provided buffer zones insulating the direct contact among GNs, which was of great benefit for low dielectric loss and high breakdown strength. What is more, the dielectric properties of PANI@GNs/PVDF nanocomposites became more independent than those of GNs/PVDF nanocomposites. For example, the nanocomposites with 0.5 vol.% PANI@GNs maintained a low dielectric loss of 0.04 ranging from 100 to 10^5 Hz, which also had an enhanced breakdown strength of about 275 MV m^{-1} and a maximum energy density of 3.1 J cm^{-3}.

Jang and coworkers prepared a kind of novel nanofiller named graphene nanodots (GNDs) by excessively oxidating carbon nanowires [120]. After that, they further modified GNDs into NH_2-GNDs to improve the compatibility with PVDF and GNs. GNDs/GNs/PVDF three-phase nanocomposites were fabricated to improve the energy storage capability. The authors demonstrated that the addition of GNs and NH_2-GNDs impeded the movement of the polymer chains, which was helpful for the reduction of the β-relaxation and dipole loss. The synergetic effects from GNDs and the GNs further improved the resulting dielectric properties of the three-phase nanocomposites. The permittivity of nanocomposites with 10 vol.% loading was as high as 60.6 and maintained a relative high breakdown strength of 390 MV m^{-1}, which resulted in a large energy density of 14.1 J cm^{-3}.

As a layered structure similar to graphite, boron nitride (BN) is an insulator with a wide band gap over 6 eV, which shows a high breakdown strength over 800 MV m^{-1} [121, 122]. Moreover, the breakdown strength can be further promoted by reducing the thickness of the BN sheet, which means the BNNs have higher breakdown strength than pristine BN sheet [123]. Wang et al. fabricated three-phase P(VDF–CTFE) nanocomposites using $BaTiO_3$ nanoparticles to improve the dielectric properties and BNNs to enhance the breakdown strength, which resulted in high energy density, great power density, and large charge–discharge efficiency [124]. An improved dispersion state was observed as shown in Figure 6.6a,b. They investigated the breakdown strength of the nanocomposites with different loading of BNNs and discovered that the breakdown strength was greatly improved upon the incorporation of BNNs. For example, the value was up to 649 MV m^{-1} at 12 wt% BNNs, which was 1.7 times higher than the pure P(VDF–CTFE). They also demonstrated that the nanocomposites with 12 wt% BNNs and 15 wt% $BaTiO_3$ nanoparticles displayed an ultrahigh energy density of 21.2 J cm^{-3} (5.9 Wh l^{-1}) and a great power density of 850 MW l^{-1}, which was over eight times as compared to that of BOPP, as shown in Figure 6.6c–e.

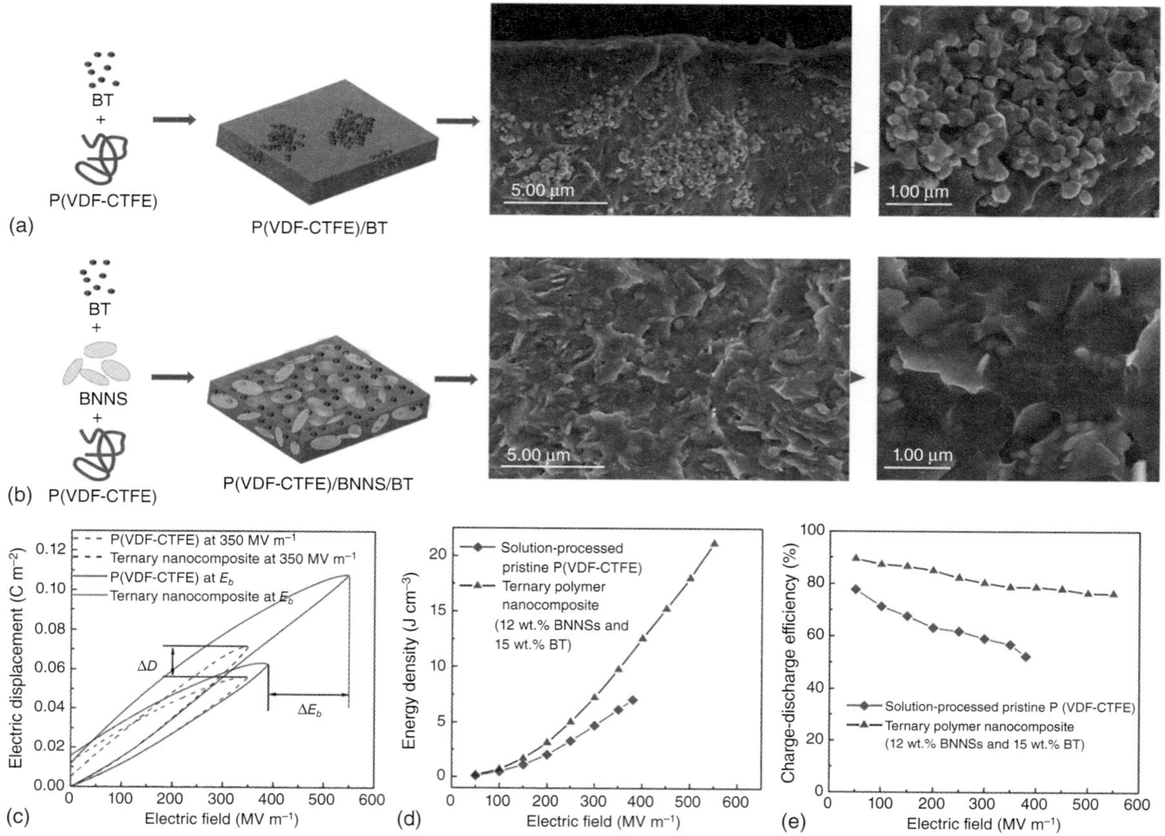

Figure 6.6 Schematic and cross-sectional SEM images of (a) $BaTiO_3$/P(VDF–CTFE) nanocomposite with 15 wt% nanofiller and (b) three-phase nanocomposite with 12 wt% BNNSs and 15 wt% of $BaTiO_3$; (c) D–E loops, (d) energy density, and (e) charge–discharge efficiency of pure P(VDF–CTFE) and three-phase nanocomposite with 12 wt% BNNSs and 15 wt% of $BaTiO_3$ [124].

6.5 Summary

Owing to their large dipoles, ferroelectric polymers usually show high permittivities, which makes them promising candidates in dielectric fields. However, still many issues exist and hinder the development of ferroelectric polymer materials, such as the following:

(1) Numerous dipoles induced by the external electric field turn out to be irreversible even when the field is removed, which results in a high remnant polarization. The huge remnant polarization causes unnecessary energy waste and decreases the charge–discharge efficiency.
(2) The incorporation of nanofillers can significantly promote the permittivity of ferroelectric polymers, but this usually follows high dielectric loss and low breakdown strength, which prevent the improvement of energy density.
(3) Owing to the high surface energy of nanomaterials, the aggregation of fillers in the polymer matrix seriously influences the uniform dispersion state of nanocomposites, which further has a negative impact on the dielectric properties, flexibility, breakdown strength, etc.
(4) It has been demonstrated that there is huge interfacial polarization between the nanofiller and polymer matrix because of the significant differences in permittivity. The interfacial polarization can promote the permittivity in lower frequency; however, it is not desired for energy storage applications because of its high relaxation time.

As discussed above, much work has been devoted to developing novel ferroelectric polymer materials. Defect modifications of PVDF were applied as a way to improve the microstructure, stabilize the reversible α phase, and reduce the remnant polarization. Nanofillers with 0–2 dimensions were introduced into ferroelectric polymers to fabricate high-performance nanocomposites. The surface modification of nanofillers decreased the interfacial polarization and improved the dispersion state, which provided lower dielectric loss, relatively high breakdown strength, and a resultant large energy density. Core–shell structure nanofillers showed more uniform dispersion state, higher breakdown strength as well as better compatibility with polymer matrix due to the existence of buffer layer (the shell). Multilevel structure strategy was also a novel and promising method, because it could combine more advantages to the nanocomposites. Both improved dielectric properties and enhanced breakdown strength can be realized, and a resulting excellent energy density was obtained.

References

1 Cao, Y., Irwin, P.C., and Younsi, K. (2004). The future of nanodielectrics in the electrical power industry. *IEEE Transactions on Dielectrics and Electrical Insulation* 11 (5): 797–807.
2 Jang, J., Song, Y., Yoo, D. et al. (2014). Micro-scale twistable organic field effect transistors and complementary inverters fabricated by orthogonal photolithography on flexible polyimide substrate. *Organic Electronics* 15 (11): 2822–2829.

3 Li, Q., Zhang, G., Liu, F. et al. (2015). Solution-processed ferroelectric terpolymer nanocomposites with high breakdown strength and energy density utilizing boron nitride nanosheets. *Energy & Environmental Science* 8 (3): 922–931.
4 Liu, J., Tian, G., Qi, S. et al. (2014). Enhanced dielectric permittivity of a flexible three-phase polyimide–graphene–$BaTiO_3$ composite material. *Materials Letters* 124: 117–119.
5 Ding, Y., Wu, Q., Zhao, D. et al. (2013). Flexible PI/$BaTiO_3$ dielectric nanocomposite fabricated by combining electrospinning and electrospraying. *European Polymer Journal* 49 (9): 2567–2571.
6 Dang, Z.M., Zhou, T., Yao, S.H. et al. (2009). Advanced calcium copper titanate/polyimide functional hybrid films with high dielectric permittivity. *Advanced Materials* 21 (20): 2077–2082.
7 Dang, Z.M., Lin, Y.Q., Xu, H.P. et al. (2008). Fabrication and dielectric characterization of advanced $BaTiO_3$/polyimide nanocomposite films with high thermal stability. *Advanced Functional Materials* 18 (10): 1509–1517.
8 Huang, X., Iizuka, T., Jiang, P. et al. (2012). Role of interface on the thermal conductivity of highly filled dielectric epoxy/AlN composites. *The Journal of Physical Chemistry C* 116 (25): 13629–13639.
9 Wang, T., Liang, G., Yuan, L., and Gu, A. (2014). Unique hybridized graphene and its high dielectric constant composites with enhanced frequency stability, low dielectric loss and percolation threshold. *Carbon* 77: 920–932.
10 Rahimabady, M., Mirshekarloo, M.S., Yao, K., and Lu, L. (2013). Dielectric behaviors and high energy storage density of nanocomposites with core–shell $BaTiO_3$@TiO_2 in poly (vinylidene fluoride-hexafluoropropylene). *Physical Chemistry Chemical Physics* 15 (38): 16242–16248.
11 Tian, M., Wei, Z., Zan, X. et al. (2014). Thermally expanded graphene nanoplates/polydimethylsiloxane composites with high dielectric constant, low dielectric loss and improved actuated strain. *Composites Science and Technology* 99: 37–44.
12 Thakur, V.K. and Gupta, R.K. (2016). Recent progress on ferroelectric polymer-based nanocomposites for high energy density capacitors: synthesis, dielectric properties, and future aspects. *Chemical Reviews* 116 (7): 4260–4317.
13 Rabuffi, M. and Picci, G. (2002). Status quo and future prospects for metallized polypropylene energy storage capacitors. *IEEE Transactions on Plasma Science* 30 (5): 1939–1942.
14 Burke, A. (2007). R&D considerations for the performance and application of electrochemical capacitors. *Electrochimica Acta* 53 (3): 1083–1091.
15 Zhou, T., Zha, J.W., Cui, R.Y. et al. (2011). Improving dielectric properties of $BaTiO_3$/ferroelectric polymer composites by employing surface hydroxylated $BaTiO_3$ nanoparticles. *ACS Applied Materials & Interfaces* 3 (7): 2184–2188.
16 Ramani, R., Das, V., Singh, A. et al. (2014). Free volume study on the origin of dielectric constant in a fluorine-containing polyimide blend: poly (vinylidene fluoride-co-hexafluoro propylene)/poly (ether imide). *The Journal of Physical Chemistry B* 118 (42): 12282–12296.

17 Kiley, E.M., Yakovlev, V.V., Ishizaki, K., and Vaucher, S. (2012). Applicability study of classical and contemporary models for effective complex permittivity of metal powders. *Journal of Microwave Power and Electromagnetic Energy* 46 (1): 26–38.
18 Zhang, X., Shen, Y., Zhang, Q. et al. (2015). Ultrahigh Energy Density of Polymer Nanocomposites Containing $BaTiO_3$@TiO_2 Nanofibers by Atomic-Scale Interface Engineering. *Advanced Materials* 27 (5): 819–824.
19 MacDougall, F.W., Ennis, J.B., Cooper, R.A. et al. (2003, June). High energy density pulsed power capacitors. In: *Pulsed Power Conference, 2003. Digest of Technical Papers. PPC-2003. 14th IEEE International*, vol. 1, 513–517. IEEE.
20 Wen, H., Xiao, W., Wen, X., and Armstrong, P. (2012). Analysis and evaluation of DC-link capacitors for high-power-density electric vehicle drive systems. *IEEE Transactions on Vehicular Technology* 61 (7): 2950–2964.
21 Sihvola, A. (2000). Mixing rules with complex dielectric coefficients. *Subsurface Sensing Technologies and Applications* 1 (4): 393–415.
22 Li, Q. and Wang, Q. (2016). Ferroelectric polymers and their energy-related applications. *Macromolecular Chemistry and Physics* 217 (11): 1228–1244.
23 Kobayashi, M., Tashiro, K., and Tadokoro, H. (1975). Molecular vibrations of three crystal forms of poly (vinylidene fluoride). *Macromolecules* 8 (2): 158–171.
24 Davis, G.T., McKinney, J.E., Broadhurst, M.G., and Roth, S. (1978). Electric-field-induced phase changes in poly (vinylidene fluoride). *Journal of Applied Physics* 49 (10): 4998–5002.
25 Davies, G.R. and Singh, H. (1979). Evidence for a new crystal phase in conventionally poled samples of poly (vinylidene fluoride) in crystal form II. *Polymer* 20 (6): 772–774.
26 Lovinger, A.J. (1983). Ferroelectric polymers. *Science* 220 (4602): 1115–1121.
27 Chu, B., Zhou, X., Ren, K. et al. (2006). A dielectric polymer with high electric energy density and fast discharge speed. *Science* 313 (5785): 334–336.
28 Lovinger, A.J., Davis, D.D., Cais, R.E., and Kometani, J.M. (1987). The role of molecular defects on the structure and phase transitions of poly (vinylidene fluoride). *Polymer* 28 (4): 617–626.
29 Thakur, V.K., Lin, M.F., Tan, E.J., and Lee, P.S. (2012). Green aqueous modification of fluoropolymers for energy storage applications. *Journal of Materials Chemistry* 22 (13): 5951–5959.
30 Furukawa, T., Date, M., Fukada, E. et al. (1980). Ferroelectric behavior in the copolymer of vinylidenefluoride and trifluoroethylene. *Japanese Journal of Applied Physics* 19 (2): L109.
31 Yamada, T., Ueda, T., and Kitayama, T. (1981). Ferroelectric-to-paraelectric phase transition of vinylidene fluoride-trifluoroethylene copolymer. *Journal of Applied Physics* 52 (2): 948–952.
32 Kim, K.J., Kim, G.B., Vanlencia, C.L., and Rabolt, J.F. (1994). Curie transition, ferroelectric crystal structure, and ferroelectricity of a VDF/TrFE (75/25) copolymer 1. The effect of the consecutive annealing in the ferroelectric state on curie transition and ferroelectric crystal structure. *Journal of Polymer Science Part B: Polymer Physics* 32 (15): 2435–2444.

33 Yagi, T., Tatemoto, M., and Sako, J.I. (1980). Transition behavior and dielectric properties in trifluoroethylene and vinylidene fluoride copolymers. *Polymer Journal* 12 (4): 209–223.
34 Tashiro, K., Takano, K., Kobayashi, M. et al. (1981). Structural study on ferroelectric phase transition of vinylidene fluoride-trifluoroethylene random copolymers. *Polymer* 22 (10): 1312–1314.
35 Tashiro, K., Takano, K., Kobayashi, M. et al. (1984). Structural study on ferroelectric phase transition of vinylidene fluoride-trifluoroethylene copolymers (III) dependence of transitional behavior on VDF molar content. *Ferroelectrics* 57 (1): 297–326.
36 Zhu, L. (2014). Exploring strategies for high dielectric constant and low loss polymer dielectrics. *The journal of physical chemistry letters* 5 (21): 3677–3687.
37 Zhu, L. and Wang, Q. (2012). Novel ferroelectric polymers for high energy density and low loss dielectrics. *Macromolecules* 45 (7): 2937–2954.
38 Zhang, Q.M., Bharti, V., and Zhao, X. (1998). Giant electrostriction and relaxor ferroelectric behavior in electron-irradiated poly (vinylidene fluoride-trifluoroethylene) copolymer. *Science* 280 (5372): 2101–2104.
39 Nakamura, K.I. and Wada, Y. (1971). Piezoelectricity, pyroelectricity, and the electrostriction constant of poly (vinylidene fluoride). *Journal of Polymer Science Part A-2: Polymer Physics* 9 (1): 161–173.
40 Glass, A.M., McFee, J.H., and Bergman, J.G. Jr., (1971). Pyroelectric properties of polyvinylidene fluoride and its use for infrared detection. *Journal of Applied Physics* 42 (13): 5219–5222.
41 Lovinger, A.J., Davis, D.D., Cais, R.E., and Kometani, J.M. (1986). On the Curie temperature of poly (vinylidene fluoride). *Macromolecules* 19 (5): 1491–1494.
42 Furukawa, T. (1997). Structure and functional properties of ferroelectric polymers. *Advances in Colloid and Interface Science* 71: 183–208.
43 Guan, F., Wang, J., Pan, J. et al. (2010). Effects of polymorphism and crystallite size on dipole reorientation in poly (vinylidene fluoride) and its random copolymers. *Macromolecules* 43 (16): 6739–6748.
44 Su, H., Strachan, A., and Goddard, W.A. III, (2004). Density functional theory and molecular dynamics studies of the energetics and kinetics of electroactive polymers: PVDF and P (VDF–TrFE). *Physical Review B* 70 (6): 064101.
45 Sousa, R.E., Nunes-Pereira, J., Ferreira, J.C.C. et al. (2014). Microstructural variations of poly (vinylidene fluoride co-hexafluoropropylene) and their influence on the thermal, dielectric and piezoelectric properties. *Polymer Testing* 40: 245–255.
46 Guan, F., Pan, J., Wang, J. et al. (2009). Crystal orientation effect on electric energy storage in poly (vinylidene fluoride-co-hexafluoropropylene) copolymers. *Macromolecules* 43 (1): 384–392.
47 Zhang, Z., Meng, Q., and Chung, T.M. (2009). Energy storage study of ferroelectric poly (vinylidene fluoride-trifluoroethylene-chlorotrifluoroethylene) terpolymers. *Polymer* 50 (2): 707–715.

48 Gadinski, M.R., Han, K., Li, Q. et al. (2014). High energy density and breakdown strength from β and γ phases in Poly (vinylidene fluoride-co-bromotrifluoroethylene) copolymers. *ACS Applied Materials & Interfaces* 6 (21): 18981–18988.
49 Zhou, X., Zhao, X., Suo, Z. et al. (2009). Electrical breakdown and ultrahigh electrical energy density in poly (vinylidene fluoride-hexafluoropropylene) copolymer. *Applied Physics Letters* 94 (16): 162901.
50 Zhou, X., Chu, B., Neese, B. et al. (2007). Electrical energy density and discharge characteristics of a poly (vinylidene fluoride-chlorotrifluoroethylene) copolymer. *IEEE Transactions on Dielectrics and Electrical Insulation* 14 (5).
51 Yang, L., Li, X., Allahyarov, E. et al. (2013). Novel polymer ferroelectric behavior via crystal isomorphism and the nanoconfinement effect. *Polymer* 54 (7): 1709–1728.
52 Xia, F., Cheng, Z.Y., Xu, H.S. et al. (2002). High electromechanical responses in a poly (vinylidene fluoride–trifluoroethylene–chlorofluoroethylene) terpolymer. *Advanced Materials* 14 (21): 1574–1577.
53 Bobnar, V., Vodopivec, B., Levstik, A. et al. (2003). Dielectric properties of relaxor-like vinylidene fluoride– trifluoroethylene-based electroactive polymers. *Macromolecules* 36 (12): 4436–4442.
54 Chung, T.C. and Petchsuk, A. (2002). Synthesis and properties of ferroelectric fluoroterpolymers with Curie transition at ambient temperature. *Macromolecules* 35 (20): 7678–7684.
55 Lu, Y., Claude, J., Neese, B. et al. (2006). A modular approach to ferroelectric polymers with chemically tunable curie temperatures and dielectric constants. *Journal of the American Chemical Society* 128 (25): 8120–8121.
56 Chu, B., Zhou, X., Neese, B. et al. (2006). Relaxor ferroelectric poly (vinylidene fluoride-trifluoroethylene-chlorofluoroethylene) terpolymer for high energy density storage capacitors. *IEEE Transactions on Dielectrics and Electrical Insulation* 13 (5): 1162–1169.
57 Khanchaitit, P., Han, K., Gadinski, M.R. et al. (2013). Ferroelectric polymer networks with high energy density and improved discharged efficiency for dielectric energy storage. *Nature Communications* 4.
58 Thakur, V.K., Tan, E.J., Lin, M.F., and Lee, P.S. (2011). Poly (vinylidene fluoride)-graft-poly (2-hydroxyethyl methacrylate): a novel material for high energy density capacitors. *Journal of Materials Chemistry* 21 (11): 3751–3759.
59 Thakur, V.K., Tan, E.J., Lin, M.F., and Lee, P.S. (2011). Polystyrene grafted polyvinylidenefluoride copolymers with high capacitive performance. *Polymer Chemistry* 2 (9): 2000–2009.
60 Guan, F., Wang, J., Yang, L. et al. (2011). Confinement-induced high-field antiferroelectric-like behavior in a poly (vinylidene fluoride-co-trifluoroethylene-co-chlorotrifluoroethylene)-graft-polystyrene graft copolymer. *Macromolecules* 44 (7): 2190–2199.
61 Guan, F., Yang, L., Wang, J. et al. (2011). Confined ferroelectric properties in poly (vinylidene fluoride-co-chlorotrifluoroethylene)-graft-polystyrene graft copolymers for electric energy storage applications. *Advanced Functional Materials* 21 (16): 3176–3188.

62 Li, W., Jiang, L., Zhang, X. et al. (2014). High-energy-density dielectric films based on polyvinylidene fluoride and aromatic polythiourea for capacitors. *Journal of Materials Chemistry A* 2 (38): 15803–15807.

63 Li, Y., Krentz, T.M., Wang, L. et al. (2014). Ligand engineering of polymer nanocomposites: from the simple to the complex. *ACS Applied Materials & Interfaces* 6 (9): 6005–6021.

64 Veena, M.G., Renukappa, N.M., Suresha, B., and Shivakumar, K.N. (2011). Tribological and electrical properties of silica-filled epoxy nanocomposites. *Polymer Composites* 32 (12): 2038–2050.

65 Wang, J., Guan, F., Cui, L. et al. (2014). Achieving high electric energy storage in a polymer nanocomposite at low filling ratios using a highly polarizable phthalocyanine interphase. *Journal of Polymer Science Part B: Polymer Physics* 52 (24): 1669–1680.

66 Tanaka, T., Kozako, M., Fuse, N., and Ohki, Y. (2005). Proposal of a multi-core model for polymer nanocomposite dielectrics. *IEEE Transactions on Dielectrics and Electrical Insulation* 12 (4): 669–681.

67 Lewis, T.J. (2004). Interfaces are the dominant feature of dielectrics at the nanometric level. *IEEE Transactions on Dielectrics and Electrical Insulation* 11 (5): 739–753.

68 Kao, K.C. (2004). Charge carrier injection from electrical contacts. In: *Dielectric Phenomena in Solids*, 1e, 327–380. San Diego: Academic Press.

69 Lewis, T.J. (1994). Nanometric dielectrics. *IEEE Transactions on Dielectrics and Electrical Insulation* 1 (5): 812–825.

70 Wilkinson, D., Langer, J.S., and Sen, P.N. (1983). Enhancement of the dielectric constant near a percolation threshold. *Physical Review B* 28 (2): 1081.

71 Mao, Y.P., Mao, S.Y., Ye, Z.G. et al. (2010). Size-dependences of the dielectric and ferroelectric properties of $BaTiO_3$/polyvinylidene fluoride nanocomposites. *Journal of Applied Physics* 108 (1): 014102.

72 Xu, H. and Gao, L. (2003). Tetragonal nanocrystalline barium titanate powder: preparation, characterization, and dielectric properties. *Journal of the American Ceramic Society* 86 (1): 203–205.

73 KumaráThakur, V., JináTan, E., and SeeáLee, P. (2011). Surface functionalization of $BaTiO_3$ nanoparticles and improved electrical properties of $BaTiO_3$/polyvinylidene fluoride composite. *RSC Advances* 1 (4): 576–578.

74 Kim, P., Doss, N.M., Tillotson, J.P. et al. (2009). High energy density nanocomposites based on surface-modified $BaTiO_3$ and a ferroelectric polymer. *ACS Nano* 3 (9): 2581–2592.

75 Yu, K., Wang, H., Zhou, Y. et al. (2013). Enhanced dielectric properties of $BaTiO_3$/poly (vinylidene fluoride) nanocomposites for energy storage applications. *Journal of Applied Physics* 113 (3): 034105.

76 Zhang, Y., Li, L., Wang, B. et al. (2014). Influence of coupling agent on microcosmic morphology and dielectric properties of energy-storage nanocomposite. *Journal of Materials Science: Materials in Electronics* 25 (2): 805–810.

77 Tomer, V., Manias, E., and Randall, C.A. (2011). High field properties and energy storage in nanocomposite dielectrics of poly (vinylidene fluoride-hexafluoropropylene). *Journal of Applied Physics* 110 (4): 044107.

78 Kim, P., Jones, S.C., Hotchkiss, P.J. et al. (2007). Phosphonic acid-modified barium titanate polymer nanocomposites with high permittivity and dielectric strength. *Advanced Materials* 19 (7): 1001–1005.

79 Ehrhardt, C., Fettkenhauer, C., Glenneberg, J. et al. (2013). $BaTiO_3$–P (VDF-HFP) nanocomposite dielectrics—Influence of surface modification and dispersion additives. *Materials Science and Engineering B* 178 (13): 881–888.

80 Baik, S., Kim, M.G., Shin, N., and Kim, I. (2007). Synthesis of tetragonal barium titanate nanoparticles via alkoxide–hydroxide sol-precipitation: effect of water addition. *Journal of the American Ceramic Society* 90 (1): 311–314.

81 Li, J., Claude, J., Norena-Franco, L.E. et al. (2008). Electrical energy storage in ferroelectric polymer nanocomposites containing surface-functionalized $BaTiO_3$ nanoparticles. *Chemistry of Materials* 20 (20): 6304–6306.

82 Huang, X. and Jiang, P. (2015). Core–shell structured high-k polymer nanocomposites for energy storage and dielectric applications. *Advanced Materials* 27 (3): 546–554.

83 Yu, K., Niu, Y., Zhou, Y. et al. (2013). Nanocomposites of Surface-Modified $BaTiO_3$ Nanoparticles Filled Ferroelectric Polymer with Enhanced Energy Density. *Journal of the American Ceramic Society* 96 (8): 2519–2524.

84 Yang, K., Huang, X., Huang, Y. et al. (2013). Fluoro-polymer@ $BaTiO_3$ hybrid nanoparticles prepared via RAFT polymerization: toward ferroelectric polymer nanocomposites with high dielectric constant and low dielectric loss for energy storage application. *Chemistry of Materials* 25 (11): 2327–2338.

85 Yu, K., Niu, Y., Bai, Y. et al. (2013). Poly (vinylidene fluoride) polymer based nanocomposites with significantly reduced energy loss by filling with core-shell structured $BaTiO_3/SiO_2$ nanoparticles. *Applied Physics Letters* 102 (10): 102903.

86 Kuang, X., Lin, S., and Zhu, H. (2013). Synthesis and characterization of poly (vinylidene-trifluoroethylene)/Ni-TiO_2. *Physica Status Solidi (a)* 210 (3): 570–573.

87 Yu, K., Bai, Y., Zhou, Y. et al. (2014). Poly (vinylidene fluoride) polymer based nanocomposites with enhanced energy density by filling with polyacrylate elastomers and $BaTiO_3$ nanoparticles. *Applied Physics Letters* 104 (8): 082904.

88 Wang, Y., Cui, J., Yuan, Q. et al. (2015). Significantly enhanced breakdown strength and energy density in sandwich-structured barium titanate/poly (vinylidene fluoride) nanocomposites. *Advanced Materials* 27 (42): 6658–6663.

89 Simoes, R., Silva, J., Vaia, R. et al. (2008). Low percolation transitions in carbon nanotube networks dispersed in a polymer matrix: dielectric properties, simulations and experiments. *Nanotechnology* 20 (3): 035703.

90 Dalmas, F., Dendievel, R., Chazeau, L. et al. (2006). Carbon nanotube-filled polymer composites. Numerical simulation of electrical conductivity in three-dimensional entangled fibrous networks. *Acta Materialia* 54 (11): 2923–2931.

91 Pötschke, P., Dudkin, S.M., and Alig, I. (2003). Dielectric spectroscopy on melt processed polycarbonate—multiwalled carbon nanotube composites. *Polymer* 44 (17): 5023–5030.

92 Pötschke, P., Abdel-Goad, M., Alig, I. et al. (2004). Rheological and dielectrical characterization of melt mixed polycarbonate-multiwalled carbon nanotube composites. *Polymer* 45 (26): 8863–8870.

93 Yuan, J.K., Yao, S.H., Dang, Z.M. et al. (2011). Giant dielectric permittivity nanocomposites: realizing true potential of pristine carbon nanotubes in polyvinylidene fluoride matrix through an enhanced interfacial interaction. *The Journal of Physical Chemistry C* 115 (13): 5515–5521.

94 Dang, Z.M., Zheng, M.S., and Zha, J.W. (2016). 1D/2D carbon nanomaterial-polymer dielectric composites with high permittivity for power energy storage applications. *Small* 12 (13): 1688–1701.

95 Song, Y., Shen, Y., Hu, P. et al. (2012). Significant enhancement in energy density of polymer composites induced by dopamine-modified $Ba_{0.6}Sr_{0.4}TiO_3$ nanofibers. *Applied Physics Letters* 101 (15): 152904.

96 Liu, S. and Zhai, J. (2014). A small loading of surface-modified $Ba_{0.6}Sr_{0.4}TiO_3$ nanofiber-filled nanocomposites with enhanced dielectric constant and energy density. *RSC Advances* 4 (77): 40973–40979.

97 Tang, H. and Sodano, H.A. (2013). High energy density nanocomposite capacitors using non-ferroelectric nanowires. *Applied Physics Letters* 102 (6): 063901.

98 Tang, H. and Sodano, H.A. (2013). Ultra high energy density nanocomposite capacitors with fast discharge using $Ba_{0.2}Sr_{0.8}TiO_3$ nanowires. *Nano Letters* 13 (4): 1373–1379.

99 Shaohui, L., Jiwei, Z., Jinwen, W. et al. (2014). Enhanced energy storage density in poly (vinylidene fluoride) nanocomposites by a small loading of suface-hydroxylated $Ba_{0.6}Sr_{0.4}TiO_3$ nanofibers. *ACS Applied Materials & Interfaces* 6 (3): 1533–1540.

100 Dang, Z.M., Wang, H.Y., Zhang, Y.H., and Qi, J.Q. (2005). Morphology and dielectric property of homogenous $BaTiO_3$/PVDF nanocomposites prepared via the natural adsorption action of nanosized $BaTiO_3$. *Macromolecular Rapid Communications* 26 (14): 1185–1189.

101 Tang, H., Zhou, Z., and Sodano, H.A. (2014). Relationship between $BaTiO_3$ nanowire aspect ratio and the dielectric permittivity of nanocomposites. *ACS Applied Materials & Interfaces* 6 (8): 5450–5455.

102 Song, Y., Shen, Y., Liu, H. et al. (2012). Improving the dielectric constants and breakdown strength of polymer composites: effects of the shape of the $BaTiO_3$ nanoinclusions, surface modification and polymer matrix. *Journal of Materials Chemistry* 22 (32): 16491–16498.

103 Song, Y., Shen, Y., Liu, H. et al. (2012). Enhanced dielectric and ferroelectric properties induced by dopamine-modified $BaTiO_3$ nanofibers in flexible poly (vinylidene fluoride-trifluoroethylene) nanocomposites. *Journal of Materials Chemistry* 22 (16): 8063–8068.

104 Wang, Y.U. (2010). Phase field model of dielectric and magnetic composites. *Applied Physics Letters* 96 (23): 232901.

105 Liu, S., Xue, S., Zhang, W., and Zhai, J. (2014). Enhanced dielectric and energy storage density induced by surface-modified $BaTiO_3$ nanofibers in poly (vinylidene fluoride) nanocomposites. *Ceramics International* 40 (10): 15633–15640.

106 Zhang, X., Ma, Y., Zhao, C., and Yang, W. (2014). High dielectric constant and low dielectric loss hybrid nanocomposites fabricated with ferroelectric polymer matrix and $BaTiO_3$ nanofibers modified with perfluoroalkylsilane. *Applied Surface Science* 305: 531–538.

107 Hu, P., Song, Y., Liu, H. et al. (2013). Largely enhanced energy density in flexible P (VDF–TrFE) nanocomposites by surface-modified electrospun $BaSrTiO_3$ fibers. *Journal of Materials Chemistry A* 1 (5): 1688–1693.

108 Li, J., Seok, S.I., Chu, B. et al. (2009). Nanocomposites of ferroelectric polymers with TiO_2 nanoparticles exhibiting significantly enhanced electrical energy density. *Advanced Materials* 21 (2): 217–221.

109 Zhang, X., Shen, Y., Xu, B. et al. (2016). Giant energy density and improved discharge efficiency of solution-processed polymer nanocomposites for dielectric energy storage. *Advanced Materials* .

110 Liu, S., Xue, S., Shen, B., and Zhai, J. (2015). Reduced energy loss in poly (vinylidene fluoride) nanocomposites by filling with a small loading of core-shell structured $BaTiO_3/SiO_2$ nanofibers. *Applied Physics Letters* 107 (3): 032907.

111 Pan, Z., Yao, L., Zhai, J. et al. (2016). Excellent energy density of polymer nanocomposites containing $BaTiO_3$@ Al_2O_3 nanofibers induced by moderate interfacial area. *Journal of Materials Chemistry A* 4 (34): 13259–13264.

112 Lin, X., Hu, P., Jia, Z., and Gao, S. (2016). Enhanced electric displacement induces large energy density in polymer nanocomposites containing core–shell structured $BaTiO_3$@ TiO_2 nanofibers. *Journal of Materials Chemistry A* 4 (6): 2314–2320.

113 Hu, P., Wang, J., Shen, Y. et al. (2013). Highly enhanced energy density induced by hetero-interface in sandwich-structured polymer nanocomposites. *Journal of Materials Chemistry A* 1 (39): 12321–12326.

114 Hu, P., Shen, Y., Guan, Y. et al. (2014). Topological-structure modulated polymer nanocomposites exhibiting highly enhanced dielectric strength and energy density. *Advanced Functional Materials* 24 (21): 3172–3178.

115 Dreyer, D.R., Park, S., Bielawski, C.W. et al. (2010). The chemistry of graphene oxide[J]. *Chemical Society Reviews* 39 (1): 228–240.

116 Zhu, Y., Murali, S., Cai, W. et al. (2010). Graphene and graphene oxide: synthesis, properties, and applications. *Advanced Materials* 22 (35): 3906–3924.

117 Zhang, Y., Jiang, S., Fan, M. et al. (2013). Piezoelectric formation mechanisms and phase transformation of poly (vinylidene fluoride)/graphite nanosheets nanocomposites. *Journal of Materials Science: Materials in Electronics* 24 (3): 927–932.

118 Huang, L., Lu, C., Wang, F., and Wang, L. (2014). Preparation of PVDF/graphene ferroelectric composite films by in situ reduction with hydrobromic acids and their properties. *RSC Advances* 4 (85): 45220–45229.

119 Shang, J., Zhang, Y., Yu, L. et al. (2013). Fabrication and enhanced dielectric properties of graphene–polyvinylidene fluoride functional hybrid films with a polyaniline interlayer. *Journal of Materials Chemistry A* 1 (3): 884–890.
120 Cho, S., Lee, J.S., and Jang, J. (2015). Poly (vinylidene fluoride)/NH2-treated graphene nanodot/reduced graphene oxide nanocomposites with enhanced dielectric performance for ultrahigh energy density capacitor. *ACS Applied Materials & Interfaces* 7 (18): 9668–9681.
121 Dean, C.R., Young, A.F., Meric, I. et al. (2010). Boron nitride substrates for high-quality graphene electronics. *Nature Nanotechnology* 5 (10): 722–726.
122 Young, A.F., Dean, C.R., Meric, I. et al. (2012). Electronic compressibility of layer-polarized bilayer graphene. *Physical Review B* 85 (23): 235458.
123 Padilha, J.E., Pontes, R.B., and Fazzio, A. (2012). Bilayer graphene on h-BN substrate: investigating the breakdown voltage and tuning the bandgap by electric field. *Journal of Physics: Condensed Matter* 24 (7): 075301.
124 Li, Q., Han, K., Gadinski, M.R. et al. (2014). High energy and power density capacitors from solution-processed ternary ferroelectric polymer nanocomposites. *Advanced Materials* 26 (36): 6244–6249.

7

Pyroelectric Energy Harvesting: Materials and Applications

Chris R. Bowen[1], *Mengying Xie*[1], *Yan Zhang*[1], *Vitaly Yu. Topolov*[2], *and Chaoying Wan*[3]

[1] *University of Bath, Materials and Structures Centre, Department of Mechanical Engineering, Claverton Down, Bath, BA2 7AY, UK*
[2] *Southern Federal University, Department of Physics, 5 Zorge Street, Rostov-on-Don, 344090, Russia*
[3] *University of Warwick, International Institute for Nanocomposites Manufacturing (IINM), WMG, CV4 7AL, UK*

7.1 Introduction to Pyroelectric Energy Harvesting

In energy harvesting systems, heat has always been an attractive and available source of energy owing to its ubiquity in various industrial applications as well as in the environment. The two most popular approaches to harvest energy from heat rely on thermoelectric and pyroelectric effects. Thermoelectric devices rely on converting temperature gradients into useful electric energy and have been widely investigated in recent years to convert waste thermal energy into an electric form. Pyroelectric energy harvesting, on the other hand, refers to the ability to convert temperature fluctuations into useful energy. The merits of both approaches have been presented by Sebald et al. [1], and the potential to use the high conversion efficiency of pyroelectric energy harvesting was highlighted.

While the piezoelectric effect is the generation of bound (polarization) charge with applied stress, the pyroelectric effect is the generation of charge with a temperature change. Pyroelectrics are polar dielectrics and exhibit spontaneous polarization in the absence of an applied electric field. Pyroelectric effect is observed in all of ferroelectrics, and they are often regarded as "soft pyroelectrics." In ferroelectric polymers polarization is a consequence of the alignment of polarized covalent bonds. Polarization leads to the presence of bound charge on each surface of the material, as in Figure 7.1a, and the origin of pyroelectric behavior stems from the change in the polarization level with material temperature. If a pyroelectric sample is heated ($dT/dt > 0$), as in Figure 7.1b, there is a decrease in its polarization as dipoles within the material lose their orientation due to thermal vibrations. This fall in the polarization P_s leads to a decrease in the number of free charges bound to the material surface; see Figure 7.1c. If the material is under an open circuit condition, the free charges remain at the electrode surface and an electric potential is generated across the material. If the material is under short-circuit conditions, an electric

Ferroelectric Materials for Energy Applications, First Edition.
Edited by Haitao Huang and James F. Scott.

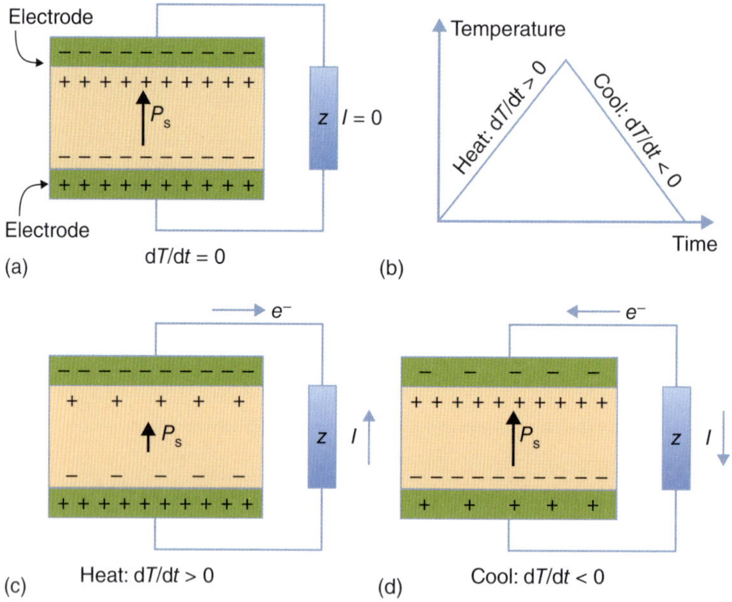

Figure 7.1 The pyroelectric effect: (a) poled material and corresponding bound surface charge, (b) applied thermal cycle; (c) heating leads to a decrease in polarization and free surface charge for current generation; (d) cooling leads to increase of polarization leading to current reversal. P_s is the spontaneous polarization, and I is the currency caused by a change in bound charge. *Source:* Wan and Bowen 2017 [2]. Reproduced with permission of Royal Society of Chemistry.

current I flows between the two polar surfaces of the material. If the pyroelectric is then cooled ($\mathrm{d}T/\mathrm{d}t < 0$), as in Figure 7.1b, the dipoles regain their orientation, leading to an increase in the level of spontaneous polarization, thus reversing the electric current flow under short-circuit conditions as free charges are now attracted to the polar surfaces; see Figure 7.1d.

Equation (7.1) defines the relationship between the pyroelectric charge (Q) developed due to a change in temperature (ΔT) for a material of surface area (A) with a *pyroelectric coefficient* (p) as follows:

$$Q = pA\,\Delta T \tag{7.1}$$

Since current is $\mathrm{d}Q/\mathrm{d}t$, Eq. (7.2) provides the short-circuit pyroelectric current (i_p) as a function of the rate of temperature change ($\mathrm{d}T/\mathrm{d}t$) with electrodes orientated normal to the polar direction.

$$i_\mathrm{p} = \frac{\mathrm{d}Q}{\mathrm{d}t} = pA\frac{\mathrm{d}T}{\mathrm{d}t} \tag{7.2}$$

The pyroelectric coefficient of an unclamped material sample, under a constant stress and electric field, is defined by Eq. (7.3):

$$p^{\sigma,E} = \left(\frac{\mathrm{d}P_\mathrm{s}}{\mathrm{d}T}\right)_{\sigma,\mathrm{E}} \tag{7.3}$$

In Eq. (7.3) subscripts σ and E correspond to conditions of constant stress and electric field, respectively.

7.2 Nanostructured and Microscale Materials and Devices

For pyroelectric harvesters the operating frequency is often small (typically much less than 1 Hz). In fluid-based systems the frequency of operation is limited by heat transfer between the pyroelectric and the working fluid that is oscillating between hot and cold sources. The output power of a pyroelectric harvester can be enhanced by subjecting it to an increased operating frequency [1]. Frequency can be increased with nanoscale radiative heat transfer. Energy transfer by thermal radiation between two semi-infinite solids is almost instantaneous and can be enhanced by several orders of magnitude from the conventional Stefan–Boltzmann law if their separation is a distance smaller than a characteristic wavelength. As an example, Fang et al. predicted that a device using 60/40 porous poly(vinylidene fluoride–trifluoroethylene) achieved a 0.2% efficiency and a 0.84 mW cm^{-2} electrical power output for the cold (273 K) and hot sources (388 K) [3]; porous materials will be discussed in more detail later in the chapter. A pyroelectric plate made from lead magnesium niobate–lead titanate (0.9PMN–PT)-based composite thin films have a higher efficiency (1.3%) and a larger power output (6.5 mW cm^{-2}) for a temperature oscillation amplitude of 10 K at a temperature of 343 K at a relatively high frequency of 5 Hz. A ZnO–poly(3-hexylthioohene) heterojunction solar cell was used for harvesting solar energy while a PVDF-based pyroelectric and piezoelectric nano-generator was built on its bottom surface for harvesting thermal and mechanical energy [4]. Hereby, a pyroelectric coefficient of about 44 μC (m^2 K)$^{-1}$ was measured. Using a lithium ion battery to store the harvested energy, the device could drive light emitting diode (LED) devices. Pyroelectric "nano-generators" based on ZnO nanowire arrays [5] have higher pyroelectric voltage and current coefficients compared to bulk and film materials. The energy conversion characteristic coefficient of this nano-generator is about 0.05–0.08 V m^2 W^{-1}. An active temperature change sensor [6] consisted of a lead zirconate titanate (PZT) microwire on a thin glass substrate with electrical contacts at its ends and packaged in polydimethylsiloxane (PDMS). The ferroelectric wire was poled at room temperature, and the output current and voltage increased linearly with the rate of temperature change; to demonstrate its potential application the harvested energy was used to power a liquid crystal display (LCD). A pyroelectric generator based on a PZT thin film, which was 175 μm thick, exhibited a pyroelectric coefficient of approximately 800 maximum power density of 0.22 mW cm^{-3} based on the open-circuit voltage and short-circuit current density [7]. The power of the pyroelectric generator was used to charge a lithium ion battery, and a single output pulse could charge an LCD. A lead-free $KNbO_3$ nanowire/PDMS polymer system [8] with silver and indium tin oxide (ITO) electrodes has also been fabricated as a flexible "nano-generator" where the output could be tuned by the electric field due to changes in orientations of ferroelectric domains. The nanowires were grown by a hydrothermal method, and the diameter of the nanowire was approximately 150 nm. The bulk pyroelectric coefficient of $KNbO_3$ was 50 μC $(m^2\ K)^{-1}$, while the effective coefficient of the nanowire–polymer mixture was 8 μC $(m^2\ K)^{-1}$ due to the presence of the nonpolar PDMS; the proposed advantage of the PDMS was to provide flexibility.

At the microscale, Hsiao et al. [9–12] have reported details on the etching of pyroelectric surfaces and on etching the electrode structure to improve an energy harvesting performance. This is based on the observation that partially covered top electrodes provided a higher current and voltage responsivity than a fully covered electrode since it allows the pyroelectric layer to be in a closer contact with the heat source [13]. The authors of work [13] showed that a meshed top electrode and trenched pyroelectric improved the responsivity of the PZT ceramic. One issue is that using a thicker pyroelectric element leads to a larger total heat capacity that reduces the temperature change [10] for a specific energy input. The purpose of trenching the PZT was to enhance the rate of the temperature change due to a lateral temperature gradient as a result of the trenched architecture. A vortex-like electrode with a deep structure was also produced [11] by sandblasting, which improved the harvested power by 11% compared to a fully covered electrode. Zabek et al. [13] reported a flexible, freestanding micro-patterned aluminum electrode with holes on polyvinylidene fluoride (PVDF film), as shown in Figure 7.2. This meshed electrode on pyroelectric

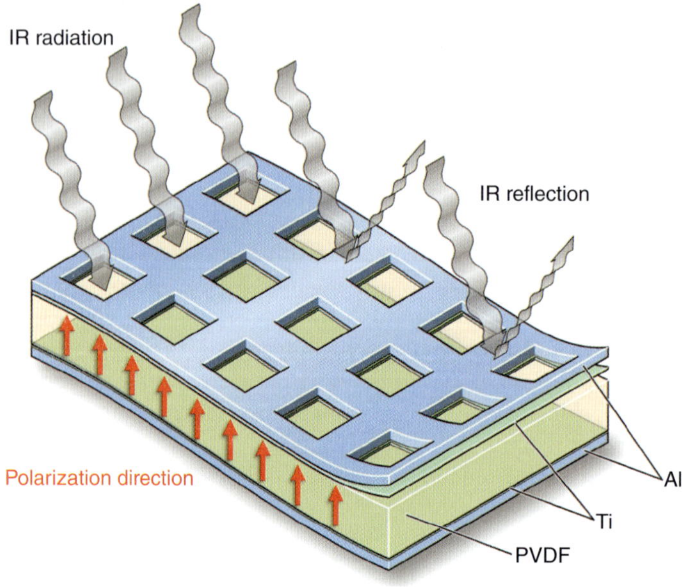

Figure 7.2 Pyroelectric generator with patterned top electrode exposed to infrared (IR) radiation, which is reflected at the aluminum electrode but absorbed at the exposed PVDF area. *Source:* Wei et al. 2006 [12]. Reproduced with permission of John Wiley and Sons.

materials enables a significant increase in pyroelectric voltage (380%), current (420%), and pyroelectric energy harvesting transformation (1080%). A specific generator performance of 66.9 μJ cm^{-3} per cycle is presented at oscillation temperatures of 2.8 °C.

The size of the pyroelectric element has also been used to tailor the phase transition temperatures in ferroelectric nanowires, enabling a "giant" pyroelectric response [14]. It was shown using phenomenological Landau–Ginzburg–Devonshire theory that it is possible to tune the pyroelectric coefficient p by changing the nanowire radius and the nature of the surrounding media, e.g. template material, gas, or gel, since the ferroelectric–ambient interface determines the surface energy properties. While the predicted efficiency of these nanoscale materials was low at room temperature, it was noted that as the temperature decreased, the efficiency tends to the Carnot cycle efficiency, making it suitable for low temperatures, e.g. space applications and systems exploited in the Arctic.

Nanowires of GaN and ZnO have been examined using first-principles-based density functional theory (DFT) calculations; the size dependency of the piezoelectric coefficients was considered and a "giant piezoelectric size effect" was identified [15]. It would be of interest to examine the effect of scaling on pyroelectric coefficients and performance of nano-structured materials. The reader is referred to an excellent review by Lingam et al. [16] for a further discussion on nano/microscale pyroelectric harvesters.

7.3 Hybrid Pyroelectric Generators

The efficiency of an energy harvesting system can be improved by employing various energy sources. Hybrid energy harvesters are designed by coupling two or more conversion mechanisms (pyroelectric, piezoelectric, thermoelectric, photothermal, etc.) and ensuring that each harvesting method does not inhibit the others.

7.3.1 Hybrid Piezoelectric and Pyroelectric System

Since the pyroelectric effect originates from spontaneous polarization within the material when subjected to a temperature change, all pyroelectric materials are also piezoelectric; therefore, it is not surprising that researchers have attempted to combine both pyroelectric and piezoelectric harvesting. Table 7.1 compares the relevant equations for a pyroelectric subjected to a temperature change (ΔT) and piezoelectric subjected to a stress ($\Delta\sigma$) with similarities in the relationships between current, voltage, and stored energy between temperature change and applied stress under both open- and closed-circuit conditions. Owing to their similarities, there is interest for potential hybrid piezoelectric–pyroelectric harvesting systems [4, 17–21] whereby a combination of temperature change and stress is applied. In such systems care must be taken to ensure that the changes in polarization are constructive and enhance the power generation of the harvesting device [22]. As discussed by Sebald et al., the frequencies of temperature and

Table 7.1 Comparison of relevant equations for pyroelectric $p = \mathrm{d}P_s/\mathrm{d}T$ (C m^{-2} K^{-1}) and piezoelectric systems $d_{ij} = \mathrm{d}P_s/\mathrm{d}\sigma$ (C N^{-1}) and ε_{33}^T is dielectric permittivity at constant stress. *A* is area and h is thickness.

	Pyroelectric	**Piezoelectric**
Charge (Q)	$Q = p\,A\,\Delta T$	$Q = d_{ij}\,A\,\Delta\sigma$
Short-circuit current ($i = p\,A\,\Delta T/\Delta t$)	$i = p A \dfrac{\Delta T}{\Delta t}$	$i = d_{ij} A \dfrac{\Delta\sigma}{\Delta t}$
Open-circuit voltage ($V = Q/C$)	$V = \dfrac{p}{\varepsilon_{33}^T} h\ \Delta T$	$V = \dfrac{d_{ij}}{\varepsilon_{33}^T} h\,\Delta\sigma$
Stored energy ($\frac{1}{2}CV^2$)	$E = \dfrac{1}{2}\dfrac{p^2}{\varepsilon_{33}^T} A\,h\,(\Delta T)^2$	$E = \dfrac{1}{2}\dfrac{d_{ij}^{\ 2}}{\varepsilon_{33}^T} A\,h\,(\Delta\sigma)^2$

vibration may be different and therefore there is a need to optimize the electronics for such a hybrid system [23].

Lee et al. [17] fabricated a stretchable, hybrid piezoelectric–pyroelectric nano-generator based on a micro-patterned piezoelectric polyvinylidene fluoride (PVDF)–polytrifluoroethylene (PTrFE) polymer, micro-patterned polydimethylsiloxane–carbon nanotube (PDMS–CNTs) composite, and graphene nanosheets. The PDMS–CNT was used to make the device flexible and also serve as a robust electrode on the base of the device. Graphene was used as a top flexible electrode to allow a fast temperature gradient on the device due to its high thermal conductivity. The potential of the material to harvest both mechanical loads (σ) and temperature changes (ΔP) was examined. The total change in polarization is expressed as follows:

$$\Delta P = d{\cdot}\sigma + p{\cdot}\Delta T \tag{7.4}$$

In Eq. (7.4), ΔP is the change in polarization, d is the piezoelectric coefficient, σ is stress, and p is the pyroelectric coefficient.

Similarly, Ko et al. [20] fabricated a flexible PZT film-based hybrid nano-generator with high piezoelectric (140 pC N^{-1}) and pyroelectric (50 nC cm^{-2} K^{-1}) coefficients at room temperature. Wang et al. [24] demonstrated that a hybrid nano-generator, which has the structure of PVDF nanowires–PDMS composite/ITO/PVDF/ITO, individually/simultaneously scavenges mechanical and thermal energies using piezoelectric, triboelectric, and pyroelectric effects. Erturun et al. [22] examined combined harvesting using a heating lamp directed at a vibrating beam. Both effects were initially investigated independently and subsequently coupled. In some cases, the combination of beam vibration with thermal cycling had a negative effect on scavenged energy and this indicates the potential complexities in such an approach, especially due to differences in frequency of temperature and mechanical oscillations. However, the use of "piezoelectric–pyroelectric harvesters" potentially offers an interesting method of enhancing power.

7.3.2 Hybrid Pyroelectric and Solar Systems

The heat generated by light could also provide viable static energy sources from the living environment, through several energy conversion methods including thermoelectric and pyroelectric methods. A solar cell hybridized with a pyroelectric harvester shows an enhanced photo-conversion efficiency due to the additional pyroelectric conversion of solar heat. Obviously, the photo-conversion efficiency of the solar cell could be improved further when the efficiency of solar heat conversion become higher by employing multiple thermal energy harvesters. Yang et al. [4] demonstrated a flexible hybrid energy cell for simultaneously harvesting thermal, mechanical, and solar energies to power electronics. Park et al. [25] fabricated a hybrid photothermal–pyroelectric–thermoelectric energy cell; see Figure 7.3. The poly(3,4-ethylenedioxythiophene) (PEDOT)

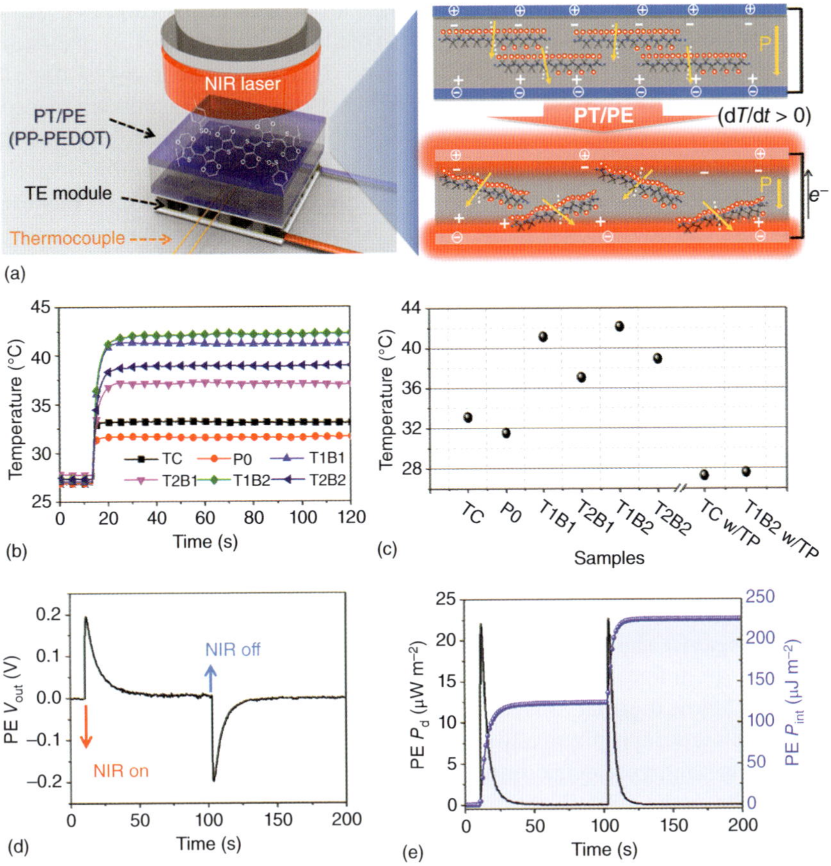

Figure 7.3 (a) Illustration of photothermally driven pyroelectric and thermoelectric device, (b) temperature of the interface for different electrode samples under NIR exposure, (c) temperature measured for all samples, including samples covered with thermal paste (w/TP), (d) output voltage of the photothermally activated pyroelectric film device under NIR on/off, and (e) output power density and integrated power (blue line) of pyroelectric energy harvester. *Source:* Park et al. 2015 [25]. Reprinted with permission of American Chemical Society.

films worked as heat source upon light exposure as well as electrodes for the output energy from the pyroelectric layer in an energy harvester hybridized with a thermoelectric layer. This energy cell, as in Figure 7.3, showed more than six times higher thermoelectric output with the additional pyroelectric output compared to a bare thermoelectric system under near-infrared (NIR) irradiation. The photo-conversion efficiency of the hybrid device was increased more than 20% under sunlight irradiation (AM 1.5G) utilizing the transmitted light through the photovoltaic cell as a heat source whose energy was converted into pyroelectric and thermoelectric output.

7.4 Pyroelectric Oscillator Systems

Oak Ridge National Laboratory [26] designed an MEMS-based cantilever harvesting system. This system is based on a thermally cycled pyroelectric capacitor that acts as a bimorph cantilever. The bimorph operates between two surfaces, one heated by waste heat and the other a cold heat sink. Proof masses are placed at the cantilever tip to ensure a good thermal contact to the hot and cold surfaces. When the cantilever is heated, it deforms due to a thermal expansion mismatch between the bimorph layers that leads to it contacting the cold surface, making the structure cool and deform in the reverse direction and then making a contact to the hot surface. This cyclic deformation leads to the cantilever alternately contacting the hot and cold surfaces at the resonant frequency of the cantilever to generate a pyroelectric current. The use of an MEMS approach means that large arrays of devices could be used to increase power and this interesting approach allows the device to potentially operate at high frequencies, up to 20 Hz or higher [27, 28].

Another approach to increase the operating frequency uses liquid-based switchable thermal interfaces to convert a spatial temperature gradient into temporal temperature oscillations [29]. A plate with a pyroelectric material oscillates up and down between a high-temperature source and a cold heat sink and repeatedly makes thermal contact with the thermal sinks to undergo temperature oscillations. In the thermally conducting state, the pyroelectric is pressed against the hot or cold surface using a linear actuator, and liquid droplets at the interface deform to make them merge into a continuous thin liquid layer of low thermal resistance. In the non-thermally conducting state, the pyroelectric material is physically separated from the hot and cold surfaces, and the liquid on the pyroelectric interface exists as discrete droplets. By creating a hydrophilic pattern on the surfaces the rupture distance was reduced, thus reducing the distance required for increasing the operating frequency. A device was demonstrated at frequencies of the order of 1 Hz with a power density of 110 mW cm^{-3}.

A self-sustaining polymeric pyroelectric nano-generator driven by water vapor was designed and fabricated; see Figure 7.4 [30]. When water vapor that is at a typical temperature range of 70–90 °C condenses on the surface of a polymeric nano-generator, the temperature of the pyroelectric nano-generator increases. When low-speed air flow passes over the generator, the temperature of the

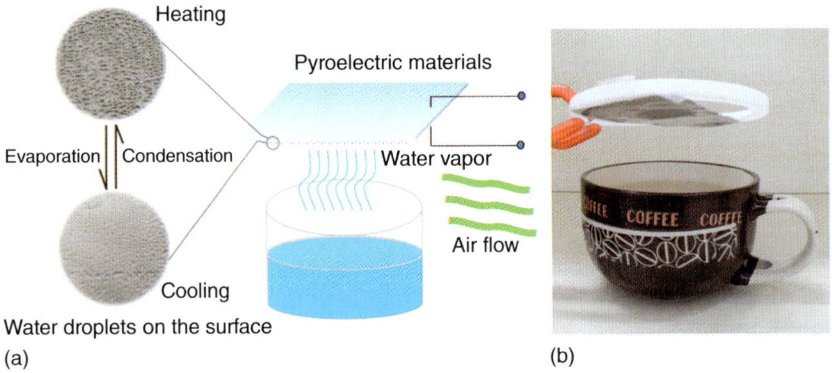

Figure 7.4 (a) Schematic illustration of a water-vapor-driven nano-generator. (b) A photograph of an assembled pyroelectric nano-generator held above a cup of hot coffee. *Source:* Gao et al. 2016 [30]. Reprinted with permission of Elsevier.

generator decreases, primarily due to the rapid evaporation of the small water droplets on the nano-generator. These temperature fluctuations are then used to generate power via the pyroelectric effect. Owing to the high latent heat of water vapor, a fast temperature oscillation up to $23\,\mathrm{K\,s^{-1}}$ was achieved by a process of water condensation and evaporation on the surface of the nano-generator. A nano-generator based on commercial polyvinylidene difluoride polymer was able to output an open-circuit voltage of 145 V and a short-circuit current of $0.12\,\mu\mathrm{A\,cm^{-2}}$. The peak power density was $1.47\,\mathrm{mW\,cm^{-3}}$ by volume and $4.12\,\mu\mathrm{W\,cm^{-2}}$ by area.

Research has also been focused on using other energy sources, such as wind, to induce thermal fluctuations in pyroelectric materials. As an example, Goudarzi et al. [19] examined a cantilever beam with PZT and lead magnesium niobate–lead titanate (PMN–0.25PT) elements that were subjected to sinusoidal mechanical vibration with heat loads of the same frequency. A wind-assisted design was reported by Krishnan et al. [31] that concentrated solar radiation using a Fresnel lens onto a fixed pyroelectric material and used a chopper disc that was driven by a wind turbine to achieve thermal modulation. The generated energy and power densities produced by PZT-5H ceramic samples were reported to be $6.9\,\mathrm{mJ\,cm^{-3}}$ per cycle and $421\,\mu\mathrm{W\,cm^{-3}}$, respectively. Zhang et al. [32] employed wind fluctuation to induce temperature variation in a PZT disk heated by solar radiation. A hybrid device that encompasses the effects of pyroelectric, piezoelectric, and shape memory alloys to enhance the thermal energy harvesting was presented by Zakharov et al. [33]. Shape memory alloys were chosen owing to their ability to generate relatively large strains in a narrow temperature range, thereby making the device especially useful for environments with small temperature variations. Xie et al. [34] experimentally investigated a wind-driven pyroelectric energy harvester in which a low-speed steady wind stream cycled a pyroelectric material between stationary hot and cold zones at frequencies ranging from 2.9 to $5.7\,\mathrm{cycles\,min^{-1}}$. Unlike conventional wind turbines, no mechanical power is drawn from the transmission and hence the system can operate at low wind speeds ($<2\,\mathrm{m\,s^{-1}}$).

7.5 Pyroelectric Coupling with Electrochemical Systems

Recently, it has been reported that the piezoelectric effect can be coupled with electrochemical processes [35–38] and allows the engineering of charge-carrier conduction characteristics at the heterojunction between a strained piezoelectric material and a chemical solution. Hong et al. presented a "piezo-electro-chemical" effect where mechanical energy could be converted to hydrogen and oxygen [36]. In an ultrasonic water bath, the mechanical vibration of barium titanate, and zinc oxide microfibers was reported to lead to the generation of strain-induced electric charges, and the electric potential across the strained fibers was said to be sufficient to trigger the redox reaction and produce hydrogen and oxygen from water.

Starr et al. [37] studied a "piezo-catalytic" effect for a PMN–PT single-crystal cantilever coated with gold electrodes, and this cantilever was subjected to vibration in water. Their results show that the rate of hydrogen production is dependent on the frequency of the vibration and amplitude of the strain that the cantilever subjected to, with the hydrogen production increasing with increased piezoelectric potential, increased holding time, and a decrease of the electrolyte concentration. In addition to the piezoelectric effects, "pyro-electro-catalysis" which combines the pyroelectric effect and an electrochemical oxidation–reduction reaction, has been studied and used for the disinfection of bacteria and decomposition of various toxic hazardous organic compounds in aqueous environments [39–41]. Gutmann et al. [39] and Benke et al. [41] focused on disinfection with nano- and micro-crystalline lithium niobate, lithium tantalite powder, and barium titanate/palladium nanoparticles. When the powder was subjected to a temperature change, reactive oxygen species were created at the surface of the pyroelectric powders and reacted due to the uncompensated screening charge carriers. Later, Wu et al. [40] used bismuth ferrite nanoparticles to degrade a Rhodamine B dye solution using pyroelectric effects and it achieved a 99% degradation efficiency for a temperature change from 27 to 38 °C. Furthermore, water-splitting using the pyroelectric effect has recently been theoretically studied by Kakekhani and Ismail-Beigi [42] using DFT. In their model, the surface phase of lead titanate switches between the ferroelectric and paraelectric phases when subjected to cyclical temperature changes below and above the Curie temperature of the material respectively. In the lower temperature ferroelectric state, H_2O molecules are dissociated on the negatively poled surface to produce bound atomic hydrogen. When the material is switched to the higher temperature paraelectric phase, the hydrogen atoms recombine to form weakly bound H_2, thereby creating a clean surface that is ready for the next thermal cycle.

7.6 Porous Pyroelectric Materials

A variety of figures of merit (FOMs) have been derived for pyroelectric materials selection based on consideration of the thermal and electrical circuits employed [43, 44]. The most common FOMs are based on the generation of maximum current or voltage for a given power input for applications such as thermal imaging sensors.

For high-voltage responsivity (F_V), the relevant figure of merit to maximize pyroelectric voltage for a given temperature change is represented by Eq. (7.5):

$$F_V = \frac{p}{c_E \, \varepsilon_{33}^{\sigma}} = \frac{p}{\rho \, c_p \, \varepsilon_{33}^{\sigma}} \tag{7.5}$$

To maximize the performance of infrared detection devices based on current responsivity (F_i) in order to maximize the pyroelectric current generated, the FOM is given by Eq. (7.6):

$$F_i = \frac{p}{c_E} = \frac{p}{\rho \, c_p} \tag{7.6}$$

The two FOMs from Eqs. (7.5) and (7.6) are often used to select pyroelectric materials for heat and infrared detection, but these are not to be confused with energy harvesting applications where generated power is a key criterion as well as the overall efficiency of the conversion of thermal energy into electrical energy.

In order to assess the properties of materials for energy harvesting a variety of performance *FOMs* containing combinations of physical properties have been developed to describe the ability of materials to generate energy for practical applications. For pyroelectric energy harvesting applications, the following two FOMs have been widely used for materials selection and materials design [43, 44]:

$$F_E' = \frac{p^2}{\varepsilon_{33}^{\sigma} \, (c_E)^2} \tag{7.7}$$

$$F_E = \frac{p^2}{\varepsilon_{33}^{\sigma}} \tag{7.8}$$

Efforts to improve the pyroelectric performance often concentrate on dense materials that undergo chemical modification (doping, substitution) [45, 46] and the employment of single crystals [47, 48]. Owing to the complexity of developing new formulations and the high cost, low Curie temperature and poor mechanical properties of single crystals, the applications of pyroelectric harvesting materials are limited. Moreover, one widely used way to increase the FOM is to increase the pyroelectric coefficient, which is limited because of the inherent property of the material itself. Furthermore, increasing the pyroelectric coefficient will also increase the permittivity at the same time, which is non-beneficial to the FOM.

The relationship between the polarization (P) of the unclamped pyroelectric material and the pyroelectric coefficient p (defined as $p = \mathrm{d}P/\mathrm{d}T$) for a temperature fluctuation ($\mathrm{d}T$) is similar to the that of the piezoelectric coefficient d (defined as $d = \mathrm{d}P/\mathrm{d}\sigma$), which corresponds to the change in the polarization at an applied stress ($\mathrm{d}\sigma$); this is especially the longitudinal piezoelectric charge coefficient (d_{33}), which is measured parallel to the poling direction and links the action and response on the same poling direction. The porous structure can offer not only lower acoustic impedance, but also higher hydrostatic coefficients, higher piezoelectric sensitivity, lower density, and stiffness than monolithic poled ferroelectric ceramics in the aspect of piezoelectric applications. At the same time, the porosity has the potential to decrease the heat capacity, which leads to an improvement of the thermal response; however, porosity can also decrease the electrical resistivity and pyroelectric coefficient [49].

This complex relationship between pyroelectric, dielectric, and thermal properties means that there is a potential to tune the porosity to achieve the optimum response for a given application. Consequently, by utilizing a porous structure whose permittivity decreases while the pyroelectric coefficient is not reduced significantly would have beneficial consequences for the pyroelectric energy harvesting applications.

From the microstructural point of view, the porous structures can be classified as two different types in terms of isotropic and anisotropic characteristics. Different manufacturing methods can directly control the microstructural features and porous structure, which ultimately determine many physical properties and related parameters of porous materials.

7.6.1 Manufacture of Isotropic Porous Pyroelectric Materials

The porous materials with an isotropic structure whose properties remain the same at different directions can be readily fabricated by traditional methods, such as replica technique (including positive and negative templates) and direct pore forming method.

7.6.1.1 Lost Wax Replication of a Coral Skeleton (Positive Template)

In the coral replication technique, a cubic coral structure is impregnated with wax under vacuum, followed by dissolving of the coral using the acid. After filling the wax with the slip, such as a PZT-slip, the green body is dried and sintered to remove the wax and densify the structure of the sample. The resulting structure formed by this method typically has large pores, similar to that of the coral, which are then filled with epoxy resin, resulting in 3–3 connectivity. This is complete connectivity of both the PZT ceramic and epoxy components where each component is distributed continuously along the three coordinate axes. The large pore size leads to increase in the flexibility of the ceramic, and an increase in pyroelectric voltage responses is noted. A 0–3 structure can be obtained when the PZT coral structure is cracked since this leads to isolation of the ceramic regions from one another; this can be achieved by exerting a high strain on the material [50]. Moreover, since coral is not abundant and the porosity structure, including the pore size, pore distribution, and their interconnection, is not easy to be controlled, this method is not widely used.

7.6.1.2 Polymeric Sponge (Positive Template)

In this method, reticulated polymeric sponge that acts as a template is infiltrated by a ceramic slip with a specific viscosity. This is followed by drying and sintering to an appropriate temperature. During the sintering process, the burnout of the polymer sponge and binder contributes to the formation of micro-size pores and cracks, which are mainly at the edge of the struts. Among the possible polymeric sponge materials that can be employed, polyurethane is used the most. In addition, various levels of elongation of the pores can be achieved by stretching the sponge at high temperature (c. 200 °C) and the elongated pores offer a route to tailor the fraction of material in the polarization direction, influencing the piezoelectric properties and permittivity [51, 52]. This approach is helpful for evaluating the effect of the pore anisotropy on piezoelectric properties. The pore size of the spongedetermines the pore size of the final product after firing shrinkage. These conditions exposed the porous ceramic to low strength and fracture toughness, reducing the structural integrity and thus limiting their application. Moreover, the porous material produced by this method exhibits open-cell structures, a uniform degree of porosity, no combustion residue, and a good ability to recover its original shape, while sponges of small pore size and closed porosity cannot be obtained.

7.6.1.3 Burned Out Plastic Spheres (BURPS) (Negative Template)

This method utilizes a pore-forming agent (PFA) that is mixed with the ceramic powder. The pore-former is then allowed to burn out while sintering so as to leave pores that are a negative image of the morphology of the PFA in the final porous material. This method was originally developed by Shrout et al. [53] and offers an advantage in providing the ability to control porosity and pore size with ease and enabling fabrication at a large scale. There are a number of potential pore-formers than can be explored; these include inorganic additives (carbon, salt, etc.) [54, 55] and natural organic matter [56–58] (wheat flour, corn starch, cotton, poppy seed, etc.). Since the porosity volume fraction can be changed to a significant degree, there can also be different fractions of closed and open porosity in the final microstructure, ranging from a 3–3 connectivity (fully interconnected pores) to a 3–0 connectivity type (isolated pores) [59, 60]. Although uniform and isolated pores can be obtained using this approach, the formation of connected pores can lead to a reduction of the flexural strength of the sintered body. Moreover, the use of volatile additives is suitable for lower levels of porosity (<60%) as a high volatile content often results in deformation and destruction of the green body during the burnout cycle.

7.6.1.4 Direct Pore Forming

In the direct foaming method, bubbles can be created in the suspension, causing it to foam, either by mechanical frothing [61], or by the injection of gases into ceramic suspensions [62], or by adding a suitable reactant to the ceramic suspension to develop a gas phase [63]. The subsequent drying and firing stages lead to the consolidation of the porous structure with the complete burnout of the foaming agent. As a result, the porosity is proportional to the amount of gas incorporated into the suspension during the foaming process. Generally, direct foaming process is considered to be one of the most effective approaches to obtain closed pores in porous ceramics. This is a relatively low-cost method of forming porous ceramics. However, due to the thermodynamic instability of gas bubbles, they readily coalesce, resulting in large pores in the final ceramics with wide pore size distribution [64, 65].

7.6.1.5 Gel Casting

Gel casting was initially developed to form dense ceramics with superior shape retention from the green body to the sintered form [66] but was adapted to synthesize porous materials with relatively high mechanical strength [67]. In gel casting a ceramic powder is mixed with an organic monomer binder and a dispersant in a solvent to form a slip. A commonly used pore-former in gel casting is tertiary-butyl alcohol (TBA) [68–70]. An initiator is then added to initiate polymerization of the monomer, leading to in situ formation of gel spheres that burn out during the sintering stage, leaving spherical pores. Research into gel casting piezoelectric composites has been a reasonably recent phenomenon, and a comparison between gel casting and BURPS processes has shown gel casting to be advantageous, mainly due to the more uniform and spherical pores that can be obtained [68]. More recently, gel casting has been used in combination with the direct foaming method [71] and freeze casting [72]. Combining direct foaming method with gel casting, the stability of foams can be improved, leading to a more even pore size distribution [73].

The processing methods described above are all used to generate 3–3 composite structures, which tend toward 3–0 as porosity levels are decreased, with pores randomly distributed through the bulk (see Table 7.2).

7.6.2 Manufacture of Anisotropic Porous Pyroelectric Materials

Owing to the reduced mechanical properties of porous ceramic materials with an isotropic structure, their application has been largely restrained. There is an almost linear relationship between porosity and compliance in 3–3 composite-like ceramics [78]. One possible way of improving the mechanical properties of porous ceramics is to utilize structures with a ceramic phase that is aligned in the direction in which the material is stressed.

7.6.2.1 Freeze Casting

Owing to many levels of hierarchical structures from macroscopic to microscopic length scales of the microstructure, directional freezing process has attracted much interest in the fabrication of porous ceramic with lamellar oriented architectures recently. Freeze casting is a method for forming porous ceramics that has come to attention in the last 15 years [79] and has the capability to form complex pore structures [80]. Taking advantage of the principle of unidirectional solidification of a liquid vehicle such as water, freeze casting, also called "ice templating" [81, 82], is also combined with other methods, such as gel casting [70, 83] and tape casting [84]. These methods were developed as simple and novel methods with low production cost for the preparation of oriented porous structures. Ceramic powders are mixed with a dispersant and a solvent and cooled from one side to stimulate directional freezing of the solvent, which grows into dendrites that form at the cooled surface. These dendrites form the template for the porous structure. During solidification, ceramic particulates are forced into the channels within the crystallizing solvent, leaving behind high density "pillars" also aligned in the direction of freezing [85]. The freezing vehicle is then removed at low pressure so as not to degrade the unsintered ceramic.

The resulting structure can be thought of as having 3–1 connectivity, with a fully interconnected ceramic phase containing isolated pore channels, or 2–2, with a "nacre-like" structure. For a given porosity, freeze cast ceramics exhibit

Table 7.2 Processing parameters and selected properties of isotropic porous piezoelectric and pyroelectric ceramics.

Method	Connectivity	Porosity (%)	Composition	d_{33} (pC/N)	$\frac{\varepsilon^T_{33}}{\varepsilon_0}$	PyroFOM, F'_E (pm^3 J^{-1})	References
Lost wax replication of coral skeleton	3–3		PZT/polymer	220	≈100	—	Skinner et al. [50]
Polymeric sponge	3–0 3–3	9–92	PZT	210	316	—	Kara et al. [74]
Burnable plastic spheres	3–0 3–3	5–75	PZT, $BiScC_3$–0.64$PbTiO_3$ (BS–0.64PT), $BaTiO_3$ (BT)	d_h from 8 to 58 for PZT, 485 to 423 for BS–0.64PT, 128 to 37 for BT	145 to 90 for PZT, 1733 to 1194 for BS–0.64PT, 1600 to 180 for BT	1.8–2 (1.2 for dense samples)	Kumar et al. [75], Lang et al. [76], Roscow et al. [77]
Gel casting	3–3	36–56	PZT	504 to 415	1566 to 742	—	Yang et al.[70]
Direct pore forming–Gel casting	3–3	33–69	PZT	515 to 330	600 to 250	—	Liu et al.[73]

superior mechanical properties to 3–3 structured materials when the force is applied in the direction of alignment [86]. This would be favorable for vibration energy harvesting because of an increased electromechanical coupling coefficient, as well as making it easier to further process and handle the materials.

Freeze casting allows good control over porosity, simply by adjusting the solid loading in the suspension [87–90]. Altering the freezing solvent can also change the shape of the pore channels: water tends to form flat ellipsoidal pores [82], whereas camphene can yield pores with a near-circular cross section [91] and TBA produces hexagonal channels [92]. Freezing vehicles that allow room temperature processing, such as near-eutectic naphthalene–camphor, have also been investigated [93]. The width of the pore channels [94] and ceramic walls [79] can be controlled by altering the degree of undercooling and the size of the ceramic particles in the suspension [95]. Freeze casting is a promising processing route for piezoelectric composites for harvesting as it enables control over the microstructure through relatively easy changes in the freezing agent, solid loading, and freezing temperature that may have benefits for the electromechanical properties. It may also allow high levels of porosity to be achieved (>80%) [79], while maintaining structural integrity.

Porous PZT with a 2–2 structure formed by freeze casting [96] has been found to exhibit a higher pyroelectric coefficient and a lower permittivity than that found in 3–3/3–0 PZT structures made using a BURPS process by Lang and Ringgaard [76]. The aligned structure may yield improved thermal properties for pyroelectric harvesting as well as superior compressive strengths (250–300%) [96] that have been found for these structures compared with 3–3 ceramic–air composites. Most of the work on freeze cast porous ceramics has been on non-ferroelectric material systems, although in the last few years freeze casting has been shown to work with barium titanate [97, 98], PZT [69, 99], and PZT–lead zinc niobate (PZN) [100]. Notable results include reported d_{33} values of 81.8–93% of the dense value despite the porosity being over 65% [94], with d_{33} values at around 75% of dense values reported elsewhere for 82% porous samples [101]. This is typically much better than would be expected with 3–3 samples manufactured by means of the BURPS method, and these samples often exhibit d_{33} values that are smaller than 60% of d_{33} of the dense sample for similar levels of porosity [102]. However, the permittivity of porous materials usually correlates somewhat with the piezoelectric strain coefficient, and is therefore often found to be higher in freeze cast materials.

7.6.2.2 3D Rapid Prototyping

Another relatively recent development in the production of porous ceramics is the use of 3D printing technology, although it may be referred to under a number of different names, such as fused deposition technology (FDM) or solid freeform fabrication (SFF) [103, 104]. A ceramic slurry is mixed with a thermoplastic binder to make the filament, which is printed to form the desired structure. The green structure is then sintered to remove the thermoplastic and densify the ceramic phase. This can be used to form complex geometries and connectivities and, although early attempts led to relatively coarse (>250 μm) structures, advances in the technology mean microscale separations and pore sizes are possible [105]. In addition, the use of 3D printing in combination with electrospinning technology can be used to form structures with nanoscale piezoelectric filaments [106]. Excellent control over the structure can be achieved using these methods so that good quality 3–1, 1–3, or 2–2 piezoelectric composite samples could be formed with properties tailored for energy harvesting.

Details are summarized in Table 7.3.

7.7 Figures of Merit and Applications Concerned with Radiations

Of additional interest are results presented in [109] on perovskite-type ferroelectric ceramics that are suitable for pyroelectric applications whereby a source of radiation is available. Borisenok et al. [109] studied FOMs and materials for transducing heat energy from an impulsive ionizing radiation into an electrical form. This particular application is based on an effective absorption of energy of radiation from an external source (either γ-quants or neutrons) in the volume of the pyroelectric transducer, which leads to the pyroelectric effect due to temperature changes. It is assumed that the duration of the radiation impulse is much less than the thermal constant of the transducer. The following FOMs are used to evaluate effectivity of the transduction process:

$$F_{i,q} = p\chi / c_V \tag{7.9}$$

and

$$F_V = p\chi / (\varepsilon_{33}^T c_V) \tag{7.10}$$

In Eqs. (7.9) and (7.10), p is the pyroelectric coefficient, χ is the density of energy absorbed by the material, c_V is the thermal capacity of the unit volume of the material, and ε_{33}^T is the dielectric permittivity at constant stress. The FOM $F_{i,q}$ is used when a current or charge is regarded as a main output characteristic and F_V is applied when the main output characteristic is voltage in the non-closed circuit [109]. Table 7.4 shows data on the FOMs from Eqs. (7.9) and (7.10) for a range of materials. The smaller values of the FOMs are observed in a case of neutron radiation stimulating the pyroelectric effect in the ceramic sample. This is a result of the smaller density of absorbed energy for a unit portion of the neutron radiation in comparison to the γ radiation.

It should also be noted that the largest $F_{i,q}$ value is achieved in the PCR-7M ceramic composition with the large piezoelectric coefficient $d_{33} = 760$ pC N^{-1} [111]. The F_V value of the same material is small because of the large value of ε_{33}^T. The PCR-7M composition is also characterized by a large pyroelectric coefficient p compared to the other listed materials; however, it is of interest to note that no strong correlation between d_{33} and p is observed. For instance, for the PCR-1 composition $d_{33} = 220$ pC N^{-1} [111] is about 3.5 times smaller than d_{33} of PCR-7M; however, a difference between the p values of these materials (see the second column in Table 7.4) is approximately 30%. This data may be of interest for harvesting radiation-induced temperature fluctuations.

Table 7.3 Processing parameters and selected properties of anisotropic porous piezo/pyroelectric ceramics.

Method		Connectivity	Porosity (%)	Composition	d_{33} (pC N^{-1})	Relative permittivity	Pyro FOM, F'_E (pm^3 J^{-1})	References
Freeze casting	Water-based	2–2	25–45	PZT	—	—	3.1–6.8 (1.7 for dense samples)	Zhang et al. [96]
	Camphene-based	3–1	50–82, 65	PZT–PZN, BT	475 to 380, 238 to 303	853 to 284, 890 to 1564	—	Lee et al. [101], Xu and Wang [97]
	TBA-based	3–1	25–65, 28–69	PZT	>595, 690–608	—, 3500 to 1200	—	Xu and Wang [107], Guo et al. [108]
3D rapid prototype		2–2, 3–3, 1–3,3–2	—	PZT, PLZT ($Pb_{0.93}La_{0.07}(Zr_{0.65}Ti_{0.35})_{0.9825}O_3$)	344–397, 347 to 335 (481 for dense)	—, 325 to 256 (1986 for dense samples)	—	Safari and Akdogan [103], Li et al. [105]

Table 7.4 Pyroelectric coefficients p and FOMs $F_{i,q}$ and F_V of some ferroelectric ceramics [109] suitable for pyroelectric devices that transduce radiation energy into electric energy.

Ceramics	Pyroelectric coefficient p (10^4 C m^{-2} K^{-1})	FOM $F_{i,q}$ (10^{-22} C)	FOM F_V (10^{-24} C)
PCR-1[a)]	4.30	8.7 (1.5)[b)]	1.2 (0.2)[b)]
PCR-3	4.67	8.8 (2.5)	2.6 (0.7)
PCR-5	3.60	7.0 (1.9)	1.8 (0.5)
PCR-6	4.48	8.7 (2.3)	0.4 (0.1)
PCR-7M	5.66	11.6 (2.0)	0.2 (0.04)
PCR-8	3.88	7.1 (1.6)	0.5 (0.1)
ZTS-19[c)]	2.87	2.8 (1.3)	0.2 (0.1)

a) PCR is the abbreviation for the group "piezoelectric ceramics from Rostov-on-Don" (Russia). Full sets of electromechanical constants of various PCRs were published in work [110, 111].
b) The first value in the third and fourth columns is related to a transducer under the γ-quant radiation. The second value in the third and fourth columns is related to a transducer under the neutron radiation.
c) ZTS means "zirkonat-titanat svintsa," i.e. ceramic based on $Pb(Zr, Ti)O_3$ (Russia). The full set of electromechanical constants of poled ZTS-19 ceramics were published in work [111].

7.8 Conclusions

Compared to other forms of energy harvesting and thermal harvesting such as thermoelectric generators, the use of pyroelectric harvesting to generate electrical energy from temperature fluctuations and thermal fields is studied to a lesser degree. The inability to induce high-frequency temperature fluctuations currently limits the amount of power that can be harvested; this is in contrast to mechanical oscillations where mechanical vibrations over 10^2 Hz are relatively easy to implement. The creation of highly effective pyroelectric harvesting materials and systems at the nanoscale may offer opportunities for operation at higher frequencies. This can be coupled with the development of new materials with improved pyroelectric coefficients especially for harvesting applications. In the present chapter, we considered the system of FOMs that need be taken into account in specific pyroelectric applications and lead to better performance of novel pyroelectric materials.

Composite material systems or design of porous materials with high FOMs to tune the pyroelectric response and mechanical and thermal properties are also potential future avenues of research. Attempts to increase the operational frequency are often undertaken, such as the generation of mechanical oscillations from a temperature gradient. Recent development involving the coupling of pyroelectric effects with electrochemical systems or heat transfer due to water vapor is also a topic that is attracting attention. It is believed that some issues in the pyroelectric area will be solved at the development of modern harvesting technologies and devices for specific applications.

Since the pyroelectric effect originates from spontaneous polarization within the polar dielectric material, all pyroelectric materials are also piezoelectric; therefore, hybrid pyroelectric and piezoelectric harvesting systems are of interest. While the power levels are low compared to other harvesting methods, in locations where there are low levels of mechanical vibrations or low levels of light, it is an intriguing option to generate useable power.

Acknowledgments

Dr. Y. Zhang would like to acknowledge the European Commission's Marie Skłodowska-Curie Actions (MSCA), through the Marie Skłodowska-Curie Individual Fellowships (IF-EF) (H2020-MSCA-IF-2015-EF-703950-HEAPPs) under Horizon 2020. Prof. C. R. Bowen would like to acknowledge funding from the European Research Council under the European Union's Seventh Framework Programme (FP/2007–2013)/ERC Grant Agreement no. 320963 on Novel Energy Materials, Engineering Science and Integrated Systems (NEMESIS). Results on project No. 11.1627.2017/4.6 PCh have been represented within the scientific activity at the Southern Federal University, and Prof. Dr. V. Yu. Topolov acknowledges financial support.

References

1 Sebald, G., Guyomar, D., and Agbossou, A. (2009). On thermoelectric and pyroelectric energy harvesting. *Smart Materials and Structures* 18: 125006.
2 Wan, C. and Bowen, C.R. (2017). Multiscale-structuring of polyvinylidene fluoride for energy harvesting: the impact of molecular-, micro- and macro-structure. *Journal of Materials Chemistry* doi: 10.1039/C6TA09590A.
3 Fang, J., Frederich, H., and Pilon, L. (2010). Harvesting nanoscale thermal radiation using pyroelectric materials. *Journal of Heat Transfer* 132: 092701.
4 Yang, Y., Zhang, H., Zhu, G. et al. (2012). Flexible hybrid energy cell for simultaneously harvesting thermal, mechanical, and solar energies. *ACS Nano* 7: 785–790.
5 Yang, Y., Guo, W., Pradel, K.C. et al. (2012). Pyroelectric nanogenerators for harvesting thermoelectric energy. *Nano Letters* 12: 2833–2838.
6 Yang, Y., Zhou, Y., Wu, J.M., and Wang, Z.L. (2012). Single micro/nanowire pyroelectric nanogenerators as self-powered temperature sensors. *ACS Nano* 6: 8456–8461.
7 Yang, Y., Wang, S., Zhang, Y., and Wang, Z.L. (2012). Pyroelectric nanogenerators for driving wireless sensors. *Nano Letters* 12: 6408–6413.
8 Yang, Y., Jung, J.H., Yun, B.K. et al. (2012). Flexible pyroelectric nanogenerators using a composite structure of lead-free $KNbO_3$ nanowires. *Advanced Materials* 24: 5357–5362.

9 Hsiao, C.-C., Ciou, J.-C., Siao, A.-S., and Lee, C.-Y. (2011). Temperature field analysis for PZT pyroelectric cells for thermal energy harvesting. *Sensors* 11: 10458–10473.
10 Hsiao, C.-C., Siao, A.-S., and Ciou, J.-C. (2012). Improvement of pyroelectric cells for thermal energy harvesting. *Sensors* 12: 534–548.
11 Hsiao, C.-C. and Siao, A.-S. (2013). Improving pyroelectric energy harvesting using a sandblast etching technique. *Sensors* 13: 12113–12131.
12 Wei, C., Lin, Y., Hu, Y. et al. (2006). Partial-electroded ZnO pyroelectric sensors for responsivity improvement. *Sensors and Actuators A: Physical* 128: 18–24.
13 Zabek, D., Taylor, J., Boulbar, E.L., and Bowen, C.R. (2015). Micropatterning of flexible and free standing polyvinylidene difluoride (PVDF) films for enhanced pyroelectric energy transformation. *Advanced Energy Materials* 5: 1401891.
14 Morozovska, A., Eliseev, E., Svechnikov, G., and Kalinin, S.V. (2010). Pyroelectric response of ferroelectric nanowires: size effect and electric energy harvesting. *Journal of Applied Physics* 108: 042009.
15 Agrawal, R. and Espinosa, H.D. (2011). Giant piezoelectric size effects in zinc oxide and gallium nitride nanowires. A first principles investigation. *Nano Letters* 11: 786–790.
16 Lingam, D., Parikh, A.R., Huang, J. et al. (2013). Nano/microscale pyroelectric energy harvesting: challenges and opportunities. *International Journal of Smart and Nano Materials* 4: 229–245.
17 Lee, J.H., Lee, K.Y., Gupta, M.K. et al. (2014). Highly stretchable piezoelectric–pyroelectric hybrid nanogenerator. *Advanced Materials* 26: 765–769.
18 Zi, Y., Lin, L., Wang, J. et al. (2015). Triboelectric–pyroelectric–piezoelectric hybrid cell for high-efficiency energy-harvesting and self-powered sensing. *Advanced Materials* 27: 2340–2347.
19 Goudarzi, M., Niazi, K., and Besharati, M.K. (2013). Hybrid energy harvesting from vibration and temperature gradient by PZT and PMN–0.25 PT ceramics. *Materials Physics and Mechanics* 16: 55–65.
20 Ko, Y.J., Kim, D.Y., Won, S.S. et al. (2016). Flexible $Pb(Zr_{0.52}Ti_{0.48})O_3$ films for a hybrid piezoelectric–pyroelectric nanogenerator under harsh environments. *ACS Applied Materials & Interfaces* 8: 6504–6511.
21 Chen, X., Shao, J., Li, X., and Tian, H. (2016). A flexible piezoelectric–pyroelectric hybrid nanogenerator based on P(VDF-TrFE) nanowire array. *IEEE Transactions on Nanotechnology* 15: 295–302.
22 Erturun, U., Waxman, R., Green, C. et al. (2010). Energy scavenging combining piezoelectric and pyroelectric effects. ASME 2010 Conference on Smart Materials, Adaptive Structures and Intelligent Systems, 53–59.
23 Sebald, G., Lefeuvre, E., and Guyomar, D. (2008). Pyroelectric energy conversion: optimization principles. *IEEE Transactions on Ultrasonics, Ferroelectrics, and Frequency Control* 55: 538–551.
24 Wang, S., Wang, Z.L., and Yang, Y. (2016). A one-structure-based hybridized nanogenerator for scavenging mechanical and thermal energies by

triboelectric–piezoelectric–pyroelectric effects. *Advanced Materials* 28: 2881–2887.

25 Park, T., Na, J., Kim, B. et al. (2015). Photothermally activated pyroelectric polymer films for harvesting of solar heat with a hybrid energy cell structure. *ACS Nano* 9: 11830–11839.

26 Mostafa, S.L., Nickolay, V., Bannuru, T. et al. (2011). A finite element model of self-resonating bimorph microcantilever for fast temperature cycling in a pyroelectric energy harvester. Presented at the 2011 MRS Spring Meeting, San Francisco, CA, USA.

27 Hunter, S.R., Lavrik, N.V., Bannuru, T. et al. (2011). Development of MEMS based pyroelectric thermal energy harvesters. SPIE Defense, Security, and Sensing, 80350V–80350V-12.

28 Hunter, S.R., Lavrik, N.V., Mostafa, S. et al. (2012). Review of pyroelectric thermal energy harvesting and new MEMs-based resonant energy conversion techniques. SPIE Defense, Security, and Sensing, 83770D–83770D-14.

29 Cha, G. and Ju, Y.S. (2013). Pyroelectric energy harvesting using liquid-based switchable thermal interfaces. *Sensors and Actuators A: Physical* 189: 100–107.

30 Gao, F., Li, W., Wang, X. et al. (2016). A self-sustaining pyroelectric nanogenerator driven by water vapour. *Nano Energy* 22: 19–26.

31 Krishnan, S.H., Ezhilarasi, D., Uma, G., and Umapathy, M. (2014). Pyroelectric-based solar and wind energy harvesting system. *IEEE Transactions on Sustainable Energy* 5: 73–81.

32 Zhang, Q., Agbossou, A., Feng, Z., and Cosnier, M. (2011). Solar micro-energy harvesting with pyroelectric effect and wind flow. *Sensors and Actuators A: Physical* 168: 335–342.

33 Zakharov, D., Gusarov, B., Gusarova, E. et al. (2013). Combined pyroelectric, piezoelectric and shape memory effects for thermal energy harvesting. *Journal of Physics: Conference Series* 476: 012021.

34 Xie, M., Zabek, D., Bowen, C. et al. (2016). Wind-driven pyroelectric energy harvesting device. *Smart Materials and Structures* 25: 125023.

35 Starr, M.B. and Wang, X. (2015). Coupling of piezoelectric effect with electrochemical processes. *Nano Energy* 14: 296–311.

36 Hong, K.-S., Xu, H., Konishi, H., and Li, X. (2010). Direct water splitting through vibrating piezoelectric microfibers in water. *The Journal of Physical Chemistry Letters* 1: 997–1002.

37 Starr, M.B., Shi, J., and Wang, X. (2012). Piezopotential-driven redox reactions at the surface of piezoelectric materials. *Angewandte Chemie International Edition* 51: 5962–5966.

38 Hong, K.-S., Xu, H., Konishi, H., and Li, X. (2012). Piezoelectrochemical effect: a new mechanism for azo dye decolorization in aqueous solution through vibrating piezoelectric microfibers. *The Journal of Physical Chemistry C* 116: 13045–13051.

39 Gutmann, E., Benke, A., Gerth, K. et al. (2012). Pyroelectrocatalytic disinfection using the pyroelectric effect of nano- and microcrystalline $LiNbO_3$ and $LiTaO_3$ particles. *The Journal of Physical Chemistry C* 116: 5383–5393.

40 Wu, J., Mao, W., Wu, Z. et al. (2016). Strong pyro-catalysis of pyroelectric $BiFeO_3$ nanoparticles under a room-temperature cold–hot alternation. *Nanoscale* 8: 7343–7350.
41 Benke, A., Mehner, E., Rosenkranz, M. et al. (2015). Pyroelectrically driven OH generation by barium titanate and palladium nanoparticles. *The Journal of Physical Chemistry C* 119: 18278–18286.
42 Kakekhani, A. and Ismail-Beigi, S. (2016). Ferroelectric oxide surface chemistry: water splitting via pyroelectricity. *Journal of Materials Chemistry A* 4: 5235–5246.
43 Mangalam, R.V.K., Agar, J.C., Damodaran, A.R. et al. (2013). Improved pyroelectric figures of merit in compositionally graded $PbZr_{1-x}Ti_xO_3$ thin films. *ACS Applied Materials and Interfaces* 5: 13235–13241.
44 Bowen, C.R., Taylor, J., LeBoulbar, E. et al. (2015). A modified figure of merit for pyroelectric energy harvesting. *Materials Letters* 138: 243–246.
45 Wei, H. and Chen, Y. (2014). Ferroelectric and pyroelectric properties of Mn-doped lead zirconate titanate ceramics. *Ceramics International* 41: 6158–6163.
46 Bahri, F., Khemakhem, H., Simon, A. et al. (2003). Dielectric and pyroelectric studies on the $Ba_{1-3a}Bi_{2a}TiO_3$ classical and relaxor ferroelectric ceramics. *Solid State Sciences* 5: 1235–1238.
47 Sun, R., Wang, J., Wang, F. et al. (2014). Pyroelectric properties of Mn-doped $94.6Na_{0.5}Bi_{0.5}TiO_3$–$5.4BaTiO_3$ lead-free single crystals. *Journal of Applied Physics* 115: 074101–074104.
48 Malyshkina, O.V., Lisitsin, V.S., Dec, J., and Łukasiewicz, T. (2014). Pyroelectric and dielectric properties of calcium barium niobate single crystals. *Physics of the Solid State* 56: 1824–1827.
49 Navid, A., Lynch, C.S., and Pilon, L. (2010). Purified and porous poly(vinylidene fluoride-trifluoroethylene) thin films for pyroelectric infrared sensing and energy harvesting. *Smart Materials and Structures* 19: 055006.
50 Skinner, D.P., Newnham, R.E., and Cross, L.E. (1978). Flexible composite transducers. *Materials Research Bulletin* 13: 599–607.
51 In-Hwan Jo, K.-H.S., Soon, Y.-M., Koh, Y.-H. et al. (2009). Highly porous hydroxyapatite scaffolds with elongated pores using streched polymeric sponges as novel template. *Materials Letters* 63: 1702–1704.
52 Creedon, M.J. and Schulze, W.A. (1994). Axially distorted 3–3 piezoelectric composites for hydrophone applications. *Ferroelectrics* 153: 333–339.
53 Shrout, T.R., Schulze, W.A., and Biggers, J.V. (1979). Simplified fabrication of PZT/polymer composites. *Materials Research Bulletin* 14: 1553–1559.
54 Maca, K., Dobsak, P., and Boccaccini, A.R. (2001). Fabrication of graded porous ceramics using alumina–carbon powder mixtures. *Ceramics International* 27: 577–584.
55 Jalila, J., Moncef, B., and Hatem, E. (2016). Salt removal from soil using the argil porous ceramic. *Desalination* 379: 53–67.
56 Gregorová, E., Pabst, W., Živcová, Z. et al. (2010). Porous alumina ceramics prepared with wheat flour. *Journal of the European Ceramic Society* 30: 2871–2880.

57 Pabst, W., Gregorová, E., Sedlářová, I., and Černý, M. (2011). Preparation and characterization of porous alumina–zirconia composite ceramics. *Journal of the European Ceramic Society* 31: 2721–2731.

58 Chen, L., Ji, T., Mu, L., and Zhu, J. (2017). Cotton fabric derived hierarchically porous carbon and nitrogen doping for sustainable capacitor electrode. *Carbon* 111: 839–848.

59 Gregorová, E. and Pabst, W. (2007). Porous ceramics prepared using poppy seed as a pore-forming agent. *Ceramics International* 33: 1385–1388.

60 Kara, H., Perry, A., Stevens, R., and Bowen, C.R. (2001). Interpenetrating PZT/polymer composites for hydrophones: models and experiments. *Ferroelectrics* 265: 317–332.

61 Sarkar, N., Lee, K.S., Park, J.G. et al. (2016). Mechanical and thermal properties of highly porous Al_2TiO_5–mullite ceramics. *Ceramics International* 42: 3548–3555.

62 Katayose, A., Yokose, R., and Makuta, T. (2016). Fabrication of closed-cell porous metals by using ultrasonically generated microbubbles. *Materials Letters* 185: 211–213.

63 Liu, P.S. and Chen, G.F. (2014). Chapter Five – Fabricating Porous Ceramics. In: *Porous Materials*, 221–302. Boston, MA: Butterworth-Heinemann.

64 H-Ya, S., Islam, S.a., and Lee, B.-T. (2008). A novel method to fabricate unidirectional porous hydroxyapatite body using ethanol bubbles in a viscous slurry. *Journal of the American Ceramic Society* 91: 3125–3127.

65 Barg, S., Soltmann, C., Andrade, M. et al. (2008). Cellular ceramics by direct foaming of emulsified ceramic powder suspensions. *Journal of the American Ceramic Society* 91 (9): 2823.

66 Young, A.C., Omatete, O.O., Janney, M.A., and Menchhofer, P.A. (1991). Gelcasting of alumina. *Journal of the American Ceramic Society* 74: 612–618.

67 Chen, R., Wang, C.A., Huang, Y. et al. (2007). Ceramics with special porous structures fabricated by freeze-gel casting: using *tert*-butyl alcohol as a template. *Journal of the American Ceramic Society* 90: 3478–3484.

68 Yang, A., Wang, C.A., Guo, R. et al. (2010). Effects of sintering behavior on microstructure and piezoelectric properties of porous PZT ceramics. *Ceramics International* 36: 549–554.

69 Yang, A., Wang, C.A., Guo, R. et al. (2010). Porous PZT ceramics with high hydrostatic figure of merit and low acoustic impedance by TBA-based gel-casting process. *Journal of the American Ceramic Society* 93: 1427–1431.

70 Yang, A.K., Wang, C.A., Guo, R., and Huang, Y. (2011). Effects of porosity on dielectric and piezoelectric properties of porous lead zirconate titanate ceramics. *Applied Physics Letters* 98: 152904–152903.

71 Liu, W., Xu, J., Wang, Y. et al. (2013). Processing and properties of porous PZT ceramics from particle-stabilized foams via gel casting. *Journal of the American Ceramic Society* 96: 1827–1831.

72 Zhang, D., Zhang, Y., Xie, R., and Zhou, K. (2012). Freeze gelcasting of aqueous alumina suspensions for porous ceramics. *Ceramics International* 38: 6063–6066.

73 Liu, W., Du, L., Wang, Y. et al. (2013). Effects of foam composition on the microstructure and piezoelectric properties of macroporous PZT ceramics from ultrastable particle-stabilized foams. *Ceramics International* 39: 8781–8787.

74 Kara, H., Ramesh, R., Stevens, R., and Bowen, C.R. (2003). Porous PZT ceramics for receiving transducers. *IEEE Transactions on Ultrasonics, Ferroelectrics, and Frequency Control* 50: 289–296.

75 Kumar, B.P., Kumar, H.H., and Kharat, D.K. (2006). Effect of porosity on dielectric properties and microstructure of porous PZT ceramics. *Materials Science and Engineering: B* 127: 130–133.

76 Lang, S.B. and Ringgaard, E. (2009). Measurements of the thermal, dielectric, piezoelectric, pyroelectric and elastic properties of porous PZT samples. Electrical Insulation and Dielectric Phenomena, 2009 CEIDP '09 IEEE Conference on 2009, 739–742.

77 Roscow, J.I., Taylor, J., and Bowe, C.R. (2016). Manufacture and characterization of porous ferroelectrics for piezoelectric energy harvesting applications. *Ferroelectrics* 498: 40–46.

78 Zeng, T., Dong, X., Chen, S., and Yang, H. (2007). Processing and piezoelectric properties of porous PZT ceramics. *Ceramics International* 33: 395–399.

79 Deville, S., Saiz, E., Nalla, R.K., and Tomsia, A.P. (2006). Freezing as a path to build complex composites. *Science* 311: 515–518.

80 Fukasawa, T., Deng, Z.Y., Ando, M. et al. (2002). Synthesis of porous silicon nitride with unidirectionally aligned channels using freeze-drying process. *Journal of the American Ceramic Society* 85: 2151–2155.

81 Munch, E., Launey, M.E., Alsem, D.H. et al. (2008). Bio-inspired hybrid materials. *Science* 322: 1516–1520.

82 Zhang, Y., Zhou, K.C., Zeng, J., and Zhang, D. (2013). Control of pore structures and sizes in freeze cast ceramics. *Advances in Applied Ceramics* 112: 405–411.

83 Kim, J.H.L., Yang, J.H., Yoon, T.Y. et al. (2011). TBA-based freeze/gel casting of porous hydroxyapatite scaffolds. *Ceramics International* 37: 2317–2322.

84 Chen, Y., Bunch, J., Li, T. et al. (2012). Novel functionally graded acicular electrode for solid oxide cells fabricated by the freeze-tape-casting process. *Journal of Power Sources* 213: 93–99.

85 Araki, K. and Halloran, J.W. (2005). Porous ceramic bodies with interconnected pore channels by a novel freeze casting technique. *Journal of the American Ceramic Society* 88: 1108–1114.

86 Deville, S. (2008). Freeze-casting of porous ceramics: a review of current achievements and issues. *Advanced Engineering Materials* 10: 155–169.

87 Yan Zhang, K.Z., Bao, Y., and Zhang, D. (2013). Effects of rheological properties on ice-templated porous hydroxyapatite ceramics. *Materials Science and Engineering: C* 33: 340–346.

88 Deville, S., Maire, E., Lasalle, A. et al. (2010). Influence of particle size on ice nucleation and growth during the ice-templating process. *Journal of the American Ceramic Society* 93: 2507–2510.

89 Deville, S. and Stevenson, A.J. (2015). Mapping ceramics research and its evolution. *Journal of the American Ceramic Society* 98: 2324–2332.

90 Li, D. and Li, M. (2012). Preparation of porous alumina ceramic with ultra-high porosity and long straight pores by freeze casting. *Journal of Porous Materials* 19: 345–349.

91 Moon, Y.-W., Shin, K.-H., Koh, Y.-H. et al. (2011). Production of highly aligned porous alumina ceramics by extruding frozen alumina/camphene body. *Journal of the European Ceramic Society* 31: 1945–1950.

92 Yang, T.Y., Ji, H.B., Yoon, S.Y. et al. (2010). Porous mullite composite with controlled pore structure processed using a freeze casting of TBA-based coal fly ash slurries. *Resources, Conservation and Recycling* 54: 816–820.

93 Araki, K. and Halloran, J.W. (2004). Room-temperature freeze-casting for ceramics with nonaqueous sublimable vehicles in the naphthalene–camphor eutectic system. *Journal of the American Ceramic Society* 87: 2014–2019.

94 Guo, R., Wang, C.-A., and Yang, A. (2011). Effects of pore size and orientation on dielectric and piezoelectric properties of 1–3 type porous PZT ceramics. *Journal of the European Ceramic Society* 31: 605–609.

95 Zhou Kechao, Z.Y., Dou, Z., Xiaoyong, Z. et al. (2011). Porous hydroxyapatite ceramics fabricated by an ice-templating method. *Scripta Materialia* 64: 426–429.

96 Zhang, Y., Bao, Y., Zhang, D., and Bowen, C.R. (2015). Porous PZT ceramics with aligned pore channels for energy harvesting applications. *Journal of the American Ceramic Society* 98: 2980–2983.

97 Xu, T. and Wang, C.-A. (2016). Effect of two-step sintering on micro-honeycomb $BaTiO_3$ ceramics prepared by freeze-casting process. *Journal of the European Ceramic Society* 36: 2647–2652.

98 Gorzkowski, E.P. and Pan, M.J. (2009). Barium titanate–polymer composites produced via directional freezing. *IEEE Transactions on Ultrasonics, Ferroelectrics, and Frequency Control* 56: 1613–1616.

99 Zhou, Q.F., Zhu, B.P., Hu, C.H. et al. (2009). Novel piezoelectric ceramic-polymer aligned composites via the freeze casting method for high frequency transducer applications. Proceedings – IEEE Ultrasonics Symposium, Art No 5441756.

100 Lee, S.H., Jun, S.H., Kim, H.E., and Koh, Y.H. (2008). Piezoelectric properties of PZT-based ceramic with highly aligned pores. *Journal of the American Ceramic Society* 91: 1912–1915.

101 Lee, S.-H., Jun, S.-H., Kim, H.-E., and Koh, Y.-H. (2007). Fabrication of porous PZT–PZN piezoelectric ceramics with high hydrostatic figure of merits using camphene-based freeze casting. *Journal of the American Ceramic Society* 90: 2807–2813.

102 Bowen, C.R., Perry, A., Lewis, A.C.F., and Kara, H. (2004). Processing and properties of porous piezoelectric materials with high hydrostatic figures of merit. *Journal of the European Ceramic Society* 24: 541–545.

103 Safari, A. and Akdogan, E.K. (2006). Rapid prototyping of novel piezoelectric composites. *Ferroelectrics* 331: 153–179.

104 Allahverdi, M., Hall, A., Brennan, R. et al. (2002). An overview of rapidly prototyped piezoelectric actuators and grain-oriented ceramics. *Journal of Electroceramics* 8: 129–137.

105 Li, Y.-Y., Li, L.-T., and Li, B. (2015). Direct ink writing of 3–3 piezoelectric composite. *Journal of Alloys and Compounds* 620: 125–128.

106 Chang, C., Tran, V.H., Wang, J. et al. (2010). Direct-write piezoelectric polymeric nanogenerator with high energy conversion efficiency. *Nano Letters* 10: 726–731.

107 Xu, T. and Wang, C.-A. (2016). Control of pore size and wall thickness of 3-1 type porous PZT ceramics during freeze-casting process. *Materials & Design* 91: 242–247.

108 Guo, R., Wang, C.-A., and Yang, A. (2011). Piezoelectric properties of the 1–3 type porous lead zirconate titanate ceramics. *Journal of the American Ceramic Society* 94: 1794–1799.

109 Borisenok, V.A., Koshelev, A.S., and Novitsky, E.Z. (1996). Pyroelectric materials for transducers of energy of impulsive ionising radiation into electric energy. *Izvestiya RAN. Seriya Fizicheskaya* 60: 195–198. (in Russian).

110 Dantsiger, A.Ya., Razumovskaya, O.N., Reznitchenko, L.A. et al. (1994). *Highly Effective Piezoceramic Materials (Handbook).* Rostov-on-Don: Kniga (in Russian).

111 Bowen, C.R., Topolov, V.Yu, and Kim, H.A. (2016). *Modern Piezoelectric Energy-Harvesting Materials*. Switzerland: Springer International Publishing.

8

Ferroelectrics in Electrocaloric Cooling

Biaolin Peng[1] *and Qi Zhang*[2]

[1] *Guangxi University, School of Physical Science and Technology, Nanning, 530004, PR China*
[2] *Cranfield University, Department of Manufacturing and Materials, Bedfordshire, MK43 0AL, UK*

8.1 Fundamentals of Electrocaloric Effects

8.1.1 Maxwell Relations and Coupled Electrocaloric Effects

A crystal is depicted as a solid in which atoms, molecules, or ions are periodically arranged in the three crystallographic orientations i, j, and k of an (x, y, z) orthogonal coordinate system. Therefore, in an individual thermal system, heat capacity C of a crystal can be defined by entropy variation dS driven by a small temperature change dT as follows:

$$\mathrm{d}S = \left(\frac{C}{T}\right)\mathrm{d}T \tag{8.1}$$

Similarly, in an individual dielectric system, dielectric permittivity ε_{ij} with a second rank tensor is defined by electric displacement variation dD_i driven by a small electric field change dE_j as follows:

$$\mathrm{d}D_i = \varepsilon_{ij}\,\mathrm{d}E_j \tag{8.2}$$

Also, in a mechanical system, elasticity s_{ijkl} with a fourth rank tensor is defined by strain variation dx_{ij} driven by a small stress variation dX_{kl} as follows:

$$\mathrm{d}x_{\mathrm{ij}} = s_{ijkl}\,\mathrm{d}X_{kl} \tag{8.3}$$

The properties of a crystal are described in equilibrium with its surroundings, i.e. the state of the crystal and its surroundings do not vary with time. In individual thermal, dielectric, and mechanical systems, Eqs. (8.1)–(8.3) can characterize the physical properties of a crystal very well. However, in fact, the case in coupled thermal, dielectric, and mechanical systems will become more complicated.

Figure 8.1 shows the physical properties of crystals. A cross-coupled correlation between thermal, electrical, and mechanical properties can be seen clearly. It can be found that electric displacement may be promoted through mechanical stress (piezoelectric effect) or temperature variation (pyroelectric effect). These phenomena are the so-called electromechanical and electrothermal effects, respectively, and their cross-coupled correlation is represented in Figure 8.1

Ferroelectric Materials for Energy Applications, First Edition.
Edited by Haitao Huang and James F. Scott.

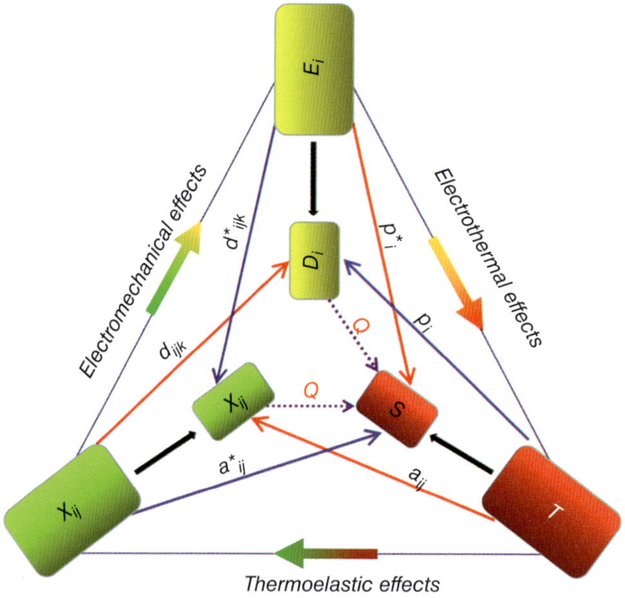

Figure 8.1 The correlation diagram between elastic, electrical, and thermal properties in a crystal. The a_{ij}, d_{ijk}, and p_i are the thermal expansion, piezoelectric, and pyroelectric coefficients, respectively. The a^*_{ij}, d^*_{ijk}, and p^*_i are the corresponding converse coefficients.

by lines and arrows. These coupled effects are classified as electromechanical, thermoelastic, and electrothermal properties.

Considering the coupled interaction between thermal, electrical, and elastic variables of a material and the first-order approximation of the coupled effects, the following state equations are valid when strain, electric displacement, and entropy are small.

$$\mathrm{d}x_{ij} = \left(\frac{\partial x_{ij}}{\partial X_{kl}}\right)_{E,T} \mathrm{d}X_{kl} + \left(\frac{\partial x_{ij}}{\partial E_k}\right)_{X,T} \mathrm{d}E_k + \left(\frac{\partial x_{ij}}{\partial T}\right)_{X,E} \mathrm{d}T \tag{8.4}$$

$$\mathrm{d}D_i = \left(\frac{\partial D_i}{\partial X_{jk}}\right)_{E,T} \mathrm{d}X_j + \left(\frac{\partial D_i}{\partial E_j}\right)_{X,T} \mathrm{d}E_j + \left(\frac{\partial D_i}{\partial T}\right)_{X,E} \mathrm{d}T \tag{8.5}$$

$$\mathrm{d}S = \left(\frac{\partial S}{\partial X_{ij}}\right)_{E,T} \mathrm{d}X_{ij} + \left(\frac{\partial S}{\partial E_i}\right)_{X,T} \mathrm{d}E_i + \left(\frac{\partial S}{\partial T}\right)_{X,E} \mathrm{d}T \tag{8.6}$$

The partial derivatives $\left(\left(\frac{\partial x_{ij}}{\partial X_{kl}}\right)_{E,T}, \left(\frac{\partial x_{ij}}{\partial E_k}\right)_{X,T}, \text{and } \left(\frac{\partial x_{ij}}{\partial T}\right)_{X,E}\right)$ in Eq. (8.4) are elastic compliances, piezoelectric, and thermal expansion, respectively. The partial derivatives $\left(\left(\frac{\partial D_i}{\partial X_{jk}}\right)_{E,T}, \left(\frac{\partial D_i}{\partial E_j}\right)_{X,T}, \text{and } \left(\frac{\partial D_i}{\partial T}\right)_{X,E}\right)$ in Eq. (8.5) are direct piezoelectricity, dielectric permittivity, and pyroelectricity, respectively. The partial derivatives $\left(\left(\frac{\partial S}{\partial X_{ij}}\right)_{E,T}, \left(\frac{\partial S}{\partial E_i}\right)_{X,T}, \text{and } \left(\frac{\partial S}{\partial T}\right)_{X,E}\right)$ in Eq. (8.6) are piezoelectric effect, electrocaloric effect, and heat capacity, respectively.

Based on the first and second law of thermodynamics, a change in the integral energy ($\mathrm{d}U$) of a material is given by

$$\mathrm{d}U = X_{ij}\,\mathrm{d}x_{ij} + E_i\,\mathrm{d}D_i + T\,\mathrm{d}S \tag{8.7}$$

Considering the potential Gibbs free energy (G),

$$G = U - X_{ij}x_{ij} - E_iD_i - TS \tag{8.8}$$

Accordingly, $\mathrm{d}G$ is given by

$$\mathrm{d}G = \mathrm{d}U - (X_{ij}\,\mathrm{d}x_{ij} + x_{ij}\,\mathrm{d}X_{ij} + D_i\,\mathrm{d}E_i + E_i\,\mathrm{d}D_i + S\,\mathrm{d}T + T\,\mathrm{d}S) \tag{8.9}$$

Considering Eqs (8.1)–(8.9), $\mathrm{d}G$ can be rewritten as

$$\mathrm{d}G = -x_{ij}\,\mathrm{d}X_{ij} - D_i\,\mathrm{d}E_i - S\,\mathrm{d}T \tag{8.10}$$

From the above equation, one obtains the relations

$$\left(\frac{\partial G}{\partial X_{ij}}\right)_{E,T} = -x_{ij} \tag{8.11}$$

$$\left(\frac{\partial G}{\partial E_i}\right)_{X,T} = -D_i \tag{8.12}$$

$$\left(\frac{\partial G}{\partial T}\right)_{E,T} = -S \tag{8.13}$$

By using Eqs. (8.4)–(8.6) and Eqs. (8.11) and (8.13), Maxwell relations are obtained as follows:

$$-\left(\frac{\partial^2 G}{\partial X_{ij}\,\partial E_k}\right)_T = \left(\frac{\partial x_{ij}}{\partial E_k}\right)_{X,T} = \left(\frac{\partial D_k}{\partial X_{ij}}\right)_{E,T} = d_{ijk} \tag{8.14}$$

$$-\left(\frac{\partial^2 G}{\partial X_{ij}\,\partial T}\right)_E = \left(\frac{\partial x_{ij}}{\partial T}\right)_{X,E} = \left(\frac{\partial S}{\partial X_{ij}}\right)_{E,T} = \alpha_{ij} \tag{8.15}$$

$$-\left(\frac{\partial^2 G}{\partial E_i\,\partial T}\right)_X = \left(\frac{\partial D_i}{\partial T}\right)_{X,E} = \left(\frac{\partial S}{\partial E_i}\right)_{X,T} = p_i \tag{8.16}$$

From Eqs. (8.14)–(8.16), it can be found that the converse piezoelectric effect is equal to the direct effect, i.e. d_{ijk}. Similarly, the thermal expansion coefficient is the same as that for piezocaloric effect, i.e. α_{ij}, and pyroelectric coefficient is equal to that for electrocaloric effect, i.e. p_i.

$$\mathrm{d}x_{ij} = s_{ijkl}\,\mathrm{d}X_{kl} + d_{kij}\,\mathrm{d}E_k + \alpha_{ij}\,\mathrm{d}T \tag{8.17}$$

$$\mathrm{d}D_i = d_{ijk}\,\mathrm{d}X_{jk} + \varepsilon_{ij}\,\mathrm{d}E_j + p_i\,\mathrm{d}T \tag{8.18}$$

$$\mathrm{d}S = \alpha_{ij}\,\mathrm{d}X_{ij} + p_i\,\mathrm{d}E_i + \frac{C}{T}\,\mathrm{d}T \tag{8.19}$$

After integrating, the linear state equations (8.6)–(8.8) can be rewritten as

$$x_{ij} = s_{ijkl}X_{ij} + d_{kij}E_k + \alpha_{ij}\Delta T \tag{8.20}$$

$$D_i = d_{ijk}X_{jk} + \varepsilon_{ij}E_j + p_i\Delta T \tag{8.21}$$

$$\Delta S = \alpha_{ij}X_{ij} + p_i E_i + \frac{C}{T}\Delta T \tag{8.22}$$

Under isothermal ($\Delta T = 0$) and stress-free ($X_{kl} = 0, X_{jk} = 0, X_{ij} = 0$) conditions, the mechanical strain, electric displacement, and entropy variations driven by the change in E can be given by the following equation:

$$x_{ij} = -\int_{E_1}^{E_2} d_{kij}\, \mathrm{d}E \tag{8.23}$$

$$D = -\int_{E_1}^{E_2} \varepsilon_{ij}\, \mathrm{d}E \tag{8.24}$$

$$\Delta S = -\int_{E_1}^{E_2} p\, \mathrm{d}E \tag{8.25}$$

where E_1 and E_2 are the starting and final applied electric fields, respectively.

Considering Eq. (8.22) and the adiabatic ($\Delta S = 0$) and the stress-free ($X_{ij} = 0$) conditions, the change in temperature driven by the change in electric field in a material is then given by

$$\Delta T = -\frac{T}{C}\int_{E_1}^{E_2} p\, \mathrm{d}E \tag{8.26}$$

The bigger the p and the smaller C for a crystal dwelling in a thermal equilibrium system T, the bigger the ΔT driven by the electric field ($\Delta E = E_2 - E_1$).

So far, Eqs. (8.25) and (8.26) have been used to assess electrocaloric effect in most dielectric materials indirectly, especially for thin films. However, in some ferroelectric relaxors, the results of direct measurements of electrocaloric heat disagreed with those of indirect measurements. One possible explanation is the fact that Eqs. (8.25) and (8.26) are derived from state equations that consider a system in equilibrium with its surroundings. For ferroelectric relaxors, their dielectric responses are strongly dependent on the thermal histories; therefore Eqs. (1.25) and (1.26) cannot describe electrocaloric effect accurately. Another possible reason is the fact that all equations are derived from the first-order approximation of the coupled effects (see Eqs. (8.4)–(8.6)). In a nonpolar dielectric and ferroelectric material in its paraelectric phase, the first-order approximation is valid. However, for a relaxor in its ferroelectric phase, apart from the first-order approximation, an extension to higher order differentials is required and nonlinear compliances such as electrostriction and electro-optics effect are essential. Additional contribution to electrocaloric effect from higher order approximations should be included. Nevertheless, the evaluation of electrocaloric effect using Eqs. (8.25) and (8.26) does not take into account the contribution of nonlinear effects from higher order approximations; as a result, the disagreement of the results between the indirect and direct measurements is inevitable.

It should be noted that Eqs. (8.25) and (8.26) are obtained under the precondition of the stress-free ($X_{ij} = 0$) or ($x_{ij} = 0$). For thin films, especially for those deposited on substrates, the pyroelectricity P_i^x measured in a clamped condition can be quite different from that measured under stress-free conditions because of the piezoelectric contribution caused by the thermal expansion.

The pyroelectric coefficient caused by the thermal expansion is termed as the secondary pyroelectric coefficient. Considering the secondary pyroelectric coefficient, the P_i^X of the thin film can be written as [1]

$$P_i^X = P_i^x + d_{ijk}^X s_{jklm}^{X,E} \alpha_{lm}^E \tag{8.27}$$

Hence, the first term in Eq. (8.27) represents the primary pyroelectric effect and the second term is the secondary effect.

In some materials, especially for polymetric materials, the secondary effect can be comparable to or even larger than the primary effect. In the assessment of their electrocaloric effect, the secondary effect should be included in the considerations.

8.1.2 Electrocaloric Effect Derived from the Landau–Devonshire Phenomenological Theory

In most cases, for ferroelectric polar materials, the macroscopic phenomena near their phase transitions can be described well by using the Landau–Devonshire (L–D) phenomenological theory. Here, the electrocaloric effect of ferroelectrics is estimated by the L–D phenomenological theory. From the L–D theory, the Gibbs free energy of a ferroelectric material can be expressed as an expansion of the polarization P, which is equivalent to D for most ferroelectrics [1–5]:

$$G = \frac{1}{2}\alpha P^2 + \frac{1}{4}\xi P^4 + \frac{1}{6}\zeta P^6 - EP \tag{8.28}$$

where $\alpha = \beta(T - T_0)$ and β, ξ, and ζ are temperature-independent phenomenological coefficients, which can be deduced from the experimental data on the dielectric permittivity and the polarization versus temperature [6]. From $\left(\frac{\partial G}{\partial T}\right)_{E,T} = -S$ (see Eq. (8.13)) and at temperatures above FE–PE transition (at paraelectric phase) where a single P value exists under an applied field E, one obtains

$$\Delta S = -\frac{1}{2}\beta P^2 \tag{8.29}$$

and the adiabatic temperature change ($\Delta T = -T\Delta S/C_E$) can be obtained:

$$\Delta T = \frac{1}{2C_E}\beta T P^2 \tag{8.30}$$

Equation (8.29) indicates that the entropy of the material will be reduced when an external action (temperature, electric field, or stress) is applied to increase its polarization. ΔS and ΔT are proportional to β and P^2. Usually, a large electrocaloric effect can be generated near the FE–PE phase transition temperature due to the large β and P.

Generally, β for ferroelectric ceramics (for $PbZr_xTi_{1-x}O_3$, $\beta \sim 1.88 \times 10^5$ K^{-1} and $P \sim 39$ μC cm^{-2}) is about 2 orders of magnitude smaller than that for the ferroelectric polymer of P(VDF–TrFE) ($\beta \sim 3.5 \times 10^7$ K^{-1} and $P \sim 8$ μC cm^{-2}) [7–9]. Even though P is higher for ceramics, βP^2 (2.29×10^8 (μC cm^{-2})^{2}K^{-1}) is still about one order of magnitude smaller than that (2.24×10^9 (μC cm^{-2})^2 K^{-1}) of P(VDF–TrFE).

It should be noted that the heat (Q) of FE–PE phase transition can also be used to assess the electrocaloric effect ($\Delta S = Q/T$) of a ferroelectric material at its paraelectric phase. The bigger the Q is, the bigger the ΔS.

8.1.3 Physical Upper Bounds on the Electrocaloric Effect Derived from the Statistical Thermodynamics Theory

Electrocaloric effect is characterized by a temperature change as observed in insulating polar materials under the adiabatic application and/or withdrawal of an electric field. Figure 8.2 illustrates the adiabatic process schematically, as shown by the path AC at constant entropy S leading from the initial point $S_1 = S(E_1, T_1)$ to the final point $S_2 = S(E_2, T_2)$, where $S_2 = S_1$ and $T_2 = T_1 + \Delta T$. Generally, the change of entropy consists of two contributions, one from the dipolar subsystem and another from the remaining degrees of freedom, such as lattice vibrations and electrons. In the case of relaxor ferroelectrics, polar nanoregions (PNRs) are the effective dipolar entities. Obviously, the entropy change along the path AC can be replaced by the sum of AB and BC since entropy is a function of state. $\Delta S(AB)$ and $\Delta S(BC)$ are given by, respectively,

$$\Delta S(AB) = S(E_2, T_1) - S(E_1, T_1) \tag{8.31}$$

$$\Delta S(BC) = \int_{T_1}^{T_2} \frac{C_{E2}}{T} \mathrm{d}T \cong \frac{C_{E2}}{T} \Delta T \tag{8.32}$$

where C_E is the volume specific heat of the entire system at constant field; usually, C_E is a weak function of T in the interval between $(T, T + \Delta T)$, where $\Delta T \ll T$. Considering that the path AC is an adiabatic process ($\Delta S(AC) = 0$) and $\Delta S(AC) = \Delta S(AB) + \Delta S(BC)$, and choosing $E_1 = 0$ and $E_2 = E$, one obtains

$$\Delta T = \frac{T}{C_E}[S(0, T) - S(E, T)] \tag{8.33}$$

Assuming that the change of entropy from lattice vibrations and electrons is electric field independent, the difference $S(0, T) - S(E, T)$ in Eq. (8.33) is

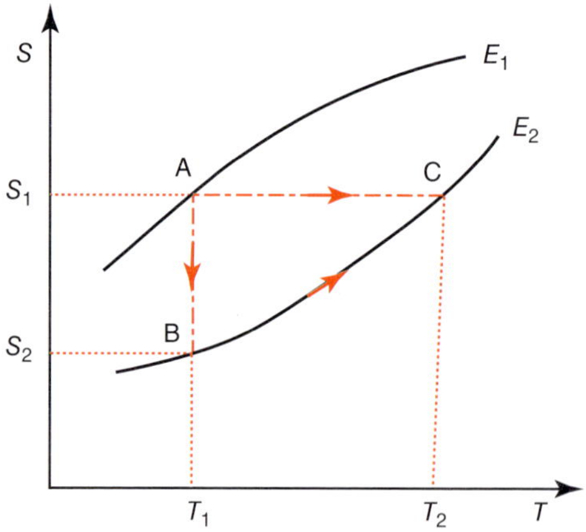

Figure 8.2 Schematic entropy–temperature diagram in an electrocaloric adiabatic process. *Source:* Pirc et al. 2011 [10]. Reproduced with the permission of AIP.

determined solely by the dipolar entropy density S_{dipolar}. Based on the statistical thermodynamics theory, S_{dipolar} can be given by the Gibbs–Shannon entropy of mixing [11]:

$$S_{\text{dip}} = -\frac{k}{V}\sum_{i=1}^{\Omega} N_i \ln \frac{N_i}{N} \tag{8.34}$$

where N_i is the number of dipoles along symmetry direction i with the condition that $N = \sum_{i=1} N_i$ is the sum of dipolar entities in a system, Ω is the number of the discrete equilibrium orientations for each dipolar entity, and V is the volume of the system.

Assuming that the maximum of the entropy $S_{\text{dip}}(0, T)$ is close to $E = 0$ and the minimum of the entropy $S_{\text{dip}}(E, T)$ occurs in a field E, which is sufficiently large to induce polarization saturation, the upper bound on the value of ΔT (see Eq. (8.33)) in the saturation limit is given by

$$\Delta T_{\text{sat}} = \frac{kT \ln \Omega}{\upsilon C_E} \tag{8.35}$$

The value of υ can be estimated from the Curie constant Θ, which is defined by the asymptotic behavior of the linear dielectric susceptibility, $\chi_1 \sim \Theta/(T - T_0)$, where T_0 is the Curie–Weiss temperature in ferroelectrics. At saturation, the polarization $P(E, T)$ reaches the value $P_{\text{sat}} = Np/V = p/\upsilon$, where p is the average dipole moment of a dipolar entity. Utilizing $P_{\text{sat}} = p/\upsilon$ and the relation $\Theta = p^2/(3k\upsilon\varepsilon_0)$ derived within a molecular field-type approach for a system of interacting dipole in three dimensions, one can obtain $\upsilon = 3\varepsilon_0 k\Theta/(P_{\text{sat}}^2)$. Substituting $\upsilon = 3\varepsilon_0 k\Theta/(P_{\text{sat}}^2)$ into Eq. (8.35), the upper bound on the value of ΔT under complete saturation can be rewritten as

$$\Delta T_{\text{sat}} = \frac{T \ln \Omega}{3\varepsilon_0 \Theta C_E} P_{\text{sat}}^2 \tag{8.36}$$

where the value of P_{sat} can be estimated from the polarization–electric field hysteresis (P–E) loops, and the Curie constant Θ from the dielectric constant relation $\chi_1 \sim \Theta/(T - T_0)$.

Considering the smallest natural value $\upsilon = \upsilon_0$ for ferroelectric dipolar solids in its paraelectric phase, an ultimate upper bound of ΔT in Eq. (8.35) can be obtained:

$$\Delta T_{\text{max}} = \frac{RT \ln \Omega}{C_{\text{molar}}} \tag{8.37}$$

where N_{A} is the Avogadro number, $R = N_{\text{A}}k$ is the gas constant, and $C_{\text{molar}} = N_{\text{A}}\upsilon_0 C_E$ is the molar specific heat. Equation (8.37) can be regarded as a universal law for the theoretical maximum value of the electrocaloric effect in electrically polarizable bulk materials with orientational degeneracy Ω. Considering the perovskite ferroelectric relaxor with molecular unit ABO_3 at their paraelectric state and the Dulong–Petit law limit ($C_{\text{molar}} = 3n_m R = 15R$), the ΔT_{max} around room temperature is about $T \ln \Omega/15$. Especially, for $\Omega = 8$ and $T = 300$ K, ΔT_{max} is about 41.58 K when a large electric field is permitted. Generally, the estimated values of ΔT_{sat} for perovskites are much smaller than that of ΔT_{max}. The main reason is that the actual size of the volume υ is larger than the molecular unit volume υ_0.

8.1.4 ECE Measurement Methods

Usually, ECE measurement methods can be divided into two groups. One is the indirect method where ΔT and ΔS are obtained using the Maxwell relation; the other is the direct method, where ΔT and ΔS are acquired from specially designed calorimeters.

Although the Maxwell relation has been widely used in deducing ECE from ferroelectric materials, there is concern on its validity when applying it to relaxor ferroelectrics since these are non-ergodic material systems (polar glasses). The Maxwell relation is derived based on the assumption that the thermodynamic system is ergodic. Therefore, one needs to be cautious when using the Maxwell relation to deduce the ECE in material systems that are not thermodynamically reversible such as the relaxor ferroelectric polymers [12].

The direct method (measuring entropy change (extensive) or temperature change (intensive)) can thus be categorized into two subgroups. Since there is no standard commercial equipment for measuring the ECE, various calorimeters have been developed by several research groups to measure the electrocaloric effect under different electric fields over a wide temperature range. The commercial DSC system can be modified so that electric fields can be applied to the sample, and the heat absorbed or ejected by the sample is simultaneously obtained.

A calorimeter using a heat flux sensor was invented by Zhang's group at Penn State, as shown in Figure 8.3 [12, 13], where the heat generated by the ECE of the sample is calibrated using the heat generated by a standard reference heater, from which ΔS is obtained. By replacing the heat flux sensor with a thermal couple or infrared sensor (or high resolution infrared camera), the adiabatic temperature change can be acquired using the same calorimeter, as shown in Figure 8.4 [12–19]. Some factors that may influence the accuracy of the direct ECE measurement include the sensitivity, signal-to-noise ratio and response time of the thermal sensor, the stability of temperature controller, and the ability of the system to provide good thermal contact during the measurement.

8.1.5 Positive and Negative Electrocaloric Effects

Recent theoretical and experimental studies in various FE/AFE bulk and film systems have revealed that there exist two electrocaloric effects, namely, the positive ($\Delta T > 0$) and the negative ($\Delta T < 0$) electrocaloric effects. The theoretical background of the positive electrocaloric effects can be readily understood from the basic knowledge on thermodynamic degrees of freedom. In the case of a positive electrocaloric effect, the sign of $(\partial P/\partial T)_E$ is positive, and as such, positive ΔT (or ΔS) values are achieved. As shown in Figure 8.5a–d, the positive ΔT of the prototypical ferroelectrics $BaTiO_3$ and $PbTiO_3$, relaxor $0.9Pb(Mg_{1/3}Nb_{2/3})O_3$–$0.1PbTiO_3$, and relaxor-ferroelectric solid solution $0.65Pb(Mg_{1/3}Nb_{2/3})O_3$–$0.35PbTiO_3$ single crystals increase with the applied electric field.

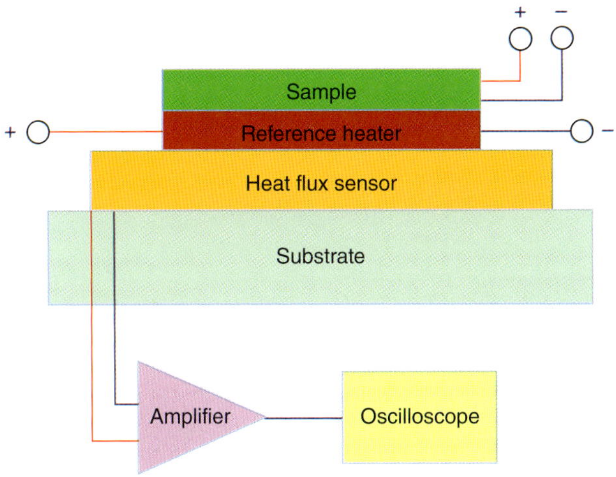

Figure 8.3 Schematic configuration of direct ECE measurement setup with heat flux sensor [12, 13].

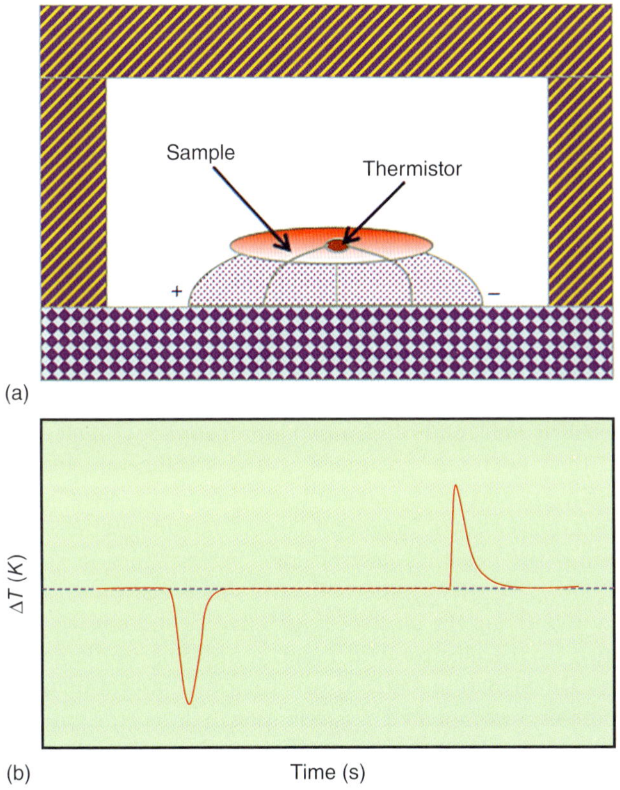

Figure 8.4 (a) Schematic configuration of direct ECE measurement setup with thermistor. (b) Signal of ΔT_{sat} versus time. *Source:* Lu et al. 2010 [13]. Reproduced with the permission of AIP.

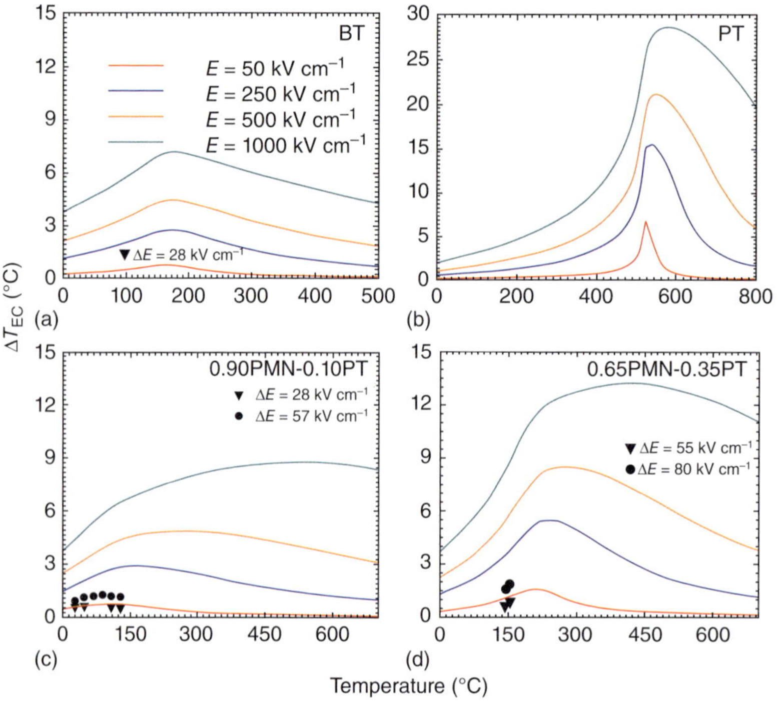

Figure 8.5 Positive theoretical and experimentally measured adiabatic electrocaloric temperature changes of (a) $BaTiO_3$, (b) $PbTiO_3$, (c) $0.9Pb(Mg_{1/3}Nb_{2/3})O_3$–$0.1PbTiO_3$, and (d) $0.65Pb(Mg_{1/3}Nb_{2/3})O_3$–$0.35PbTiO_3$ single crystals. *Source:* Khassaf et al. 2016 [20].

By contrast, the sign of $(\partial P/\partial T)_E$ is negative in the case of negative electrocaloric effect. Accordingly, negative ΔT (or ΔS) values are achieved. As shown in Figure 8.6, a maximum of negative $= -5.76\,°C$ at 59 °C under an applied field change of $\Delta E = 338\,kV\,cm^{-1}$ is obtained in a $(Pb_{0.97}La_{0.02})(Zr_{0.95}Ti_{0.05})O_3$ antiferroelectric thin film fabricated on a Pt(111)/Ti/SiO_2/Si substrate by using the sol–gel method. The absolute values of $\Delta T(|\Delta T|)$ increase first with the increase of the applied electric field, and reach a maximum at a modest electric field ($338\,kV\,cm^{-1}$) and then decrease. Enlargement and stabilization of the AFE–FE switching field E_{AFE-FE} and FE–AFE switching field E_{FE-AFE} may be an effective way to push the modest electric field ($338\,kV\,cm^{-1}$) to a higher electric field, and finally to improve the negative electrocaloric performance (a larger $|\Delta T|$) of the $(Pb_{0.97}La_{0.02})(Zr_{0.95}Ti_{0.05})O_3$ antiferroelectric thin films.

So far, the mechanism underlying the electrocaloric effect is still elusive, but entropy change stimulated by the electric-field-induced ordering of dipoles is the widely accepted physical explanation. Usually, an isothermal application of an electric field orders the dipoles, accompanied by a reduction in the corresponding entropy, which causes the increase in the temperature. The subsequent adiabatic depolarization process randomizes the dipoles again, causing an increase in electrical entropy, which finally leads to the cooling of the materials.

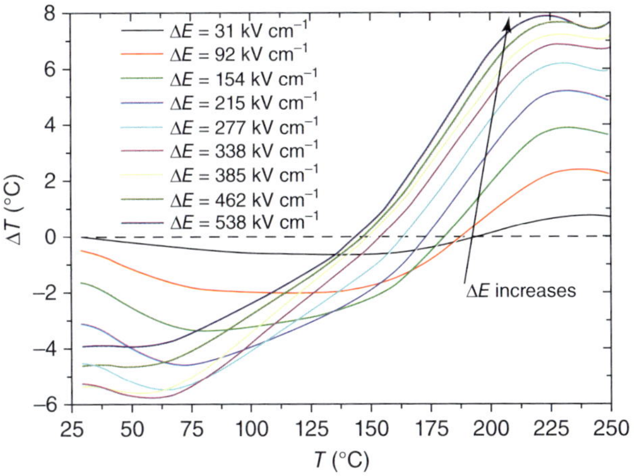

Figure 8.6 Negative electrocaloric temperature changes in antiferroelectric $(Pb_{0.97}La_{0.02})(Zr_{0.95}Ti_{0.05})O_3$ thin films. *Source:* Li et al. 2013 [1]. Reproduced with the permission of Royal Society of Chemistry.

Antiferroelectrics are defined by a crystal structure with two or more sublattices of spontaneous polarization of opposite directions, demonstrating a zero macroscopic polarization as shown in Figure 8.7a, and a field-induced AFE to FE phase transition at high-enough electric field. When a modest electric field E is applied, as shown in Figure 8.7b, a canting between the P_+ and P_- will increase the entropy of the dipoles, thereby yielding a negative electrocaloric effect. Simply put, the dipoles are non-collinear in a local region, but are collinear only on average at a finite temperature. Local polarization P_{local} moves around because of temperature fluctuations. When an electric field is applied along the z axis there is no significant change in the magnitude of polarization (the length of the arrows), but instead the fluctuation circle for the P_+ sublattice becomes narrower (increasing the P_+ component in the z direction) while that of the P_- sublattice becomes wider (thus decreasing P_- component in the z direction). In addition to the change of the polarization in the z direction, the decrease (or increase) of the polarization in x and y directions also increases the entropy of the system since the total quadratic of the polarization $(\overline{P^2})$ in x (or y) direction increases. In other words, assuming P_+ and P_- increase and decrease by the same amount, $\overline{P}$ remains unchanged but $\overline{P^2}$ increases.

Moreover, the negative electrocaloric effect has also been studied in other relaxor ferroelectric ceramics, thin films, and single crystals [22–29]. For different systems, the underlying mechanism may not be the same. Therefore, more theoretical insights using the first principles calculation method are highly desired to understand the origin of the negative electrocaloric in AFEs and relaxors, for instance, the subtle coupling between AFE, FE, and oxygen tilting instabilities [30–33].

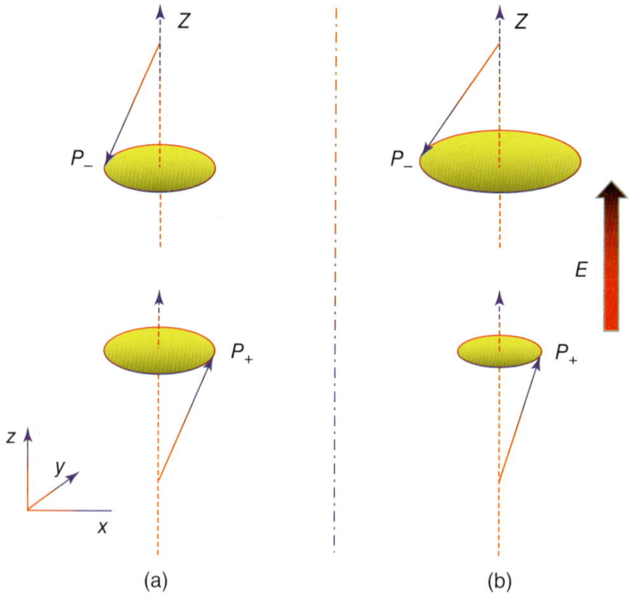

Figure 8.7 A mechanism schematic of negative electrocaloric effect in antiferroelectrics. (a) $\Delta E = 0$, (b) $\Delta E \neq 0$. *Source:* Geng et al. 2015 [21]. Reproduced with the permission of John Wiley & Sons.

8.2 Electrocaloric Devices

8.2.1 Electrocaloric Refrigerator Prototype

Just as the change of inherent entropy in a body of gas can be utilized to do mechanical work when the gas is heated and cooled, the change of entropy in a ferroelectric material close to a phase transition can also be used to extract electrical energy or, alternatively, to provide refrigeration. By means of some cycles of electrical displacement versus field across ferroelectric materials such as capacitors (C), electrical energy from heat can be converted directly. These cycles that are analogous to the Carnot cycle include the resistive cycle, two-diode cycle, Stirling cycle, and Ericsson cycle. Among these cycles, the Stirling cycle is regarded as the most effective one that can be used to extract heat by being run in reverse. Figure 8.8 shows the principal thermodynamic process in a Stirling cycle. Starting at point *1*, the switch is closed with the electric field E_1 across C (ferroelectric material). The switch is opened and C connected to a heat source, raising the temperature to T_2 at constant displacement. The permittivity of the ferroelectric material is reduced as T increases, so the electric field increases to E_{max} (point *2*). The switch is closed, and C discharges isothermally (point *3*) delivering energy to the load. The switch is opened again, C is connected to the heat sink so the temperature reduces to T_1, again at constant displacement, and the field reduces to E_{min} as the permittivity of the ferroelectric material increases (point *4*). Finally, the switch is closed, and C recharges through the load, delivering more energy to it and returning to point *1*.

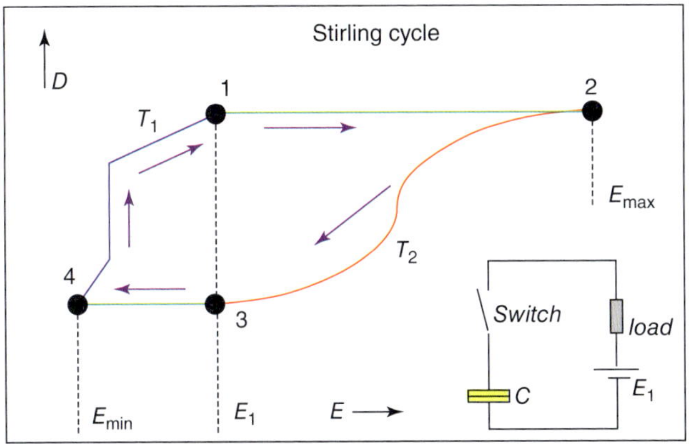

Figure 8.8 Diagram of the Stirling cycle in the pyroelectric energy recovery and harvesting. The inset: a simple circuit schematic. The capacitor (*C*) is the ferroelectric element, and the direct current bias field is provided by the battery. Each *C* cycle starts at point 1 and moves around to point 4, and is cycled between two temperatures T_1 and T_2 ($T_1 < T_2$). *Source:* Alpay et al. 2014 [34]. Reproduced with the permission of Cambridge University Press.

The ideal Carnot cycle efficiency (η_{Carnot}) is independent of heat engine design and is given by

$$\eta_{\text{Carnot}} = 1 - \frac{T_1}{T_2} \tag{8.38}$$

where T_1 and T_2 are the heat sink and heat source, respectively.

According to the principle diagram of the Stirling cycle in Figure 8.8, a schematic (Figure 8.9) of a solid-state refrigeration system operating in a regenerative manner is proposed by Alpay et al. In this system, a fluid is employed as the regenerator to minimize irreversible heat flows between two heat exchangers held at T_1 and T_2 ($T_1 < T_2$). The fluid derived by a pump flows past the ferroelectric module in an insulated region, so that all the heat exchange is between the fluid and the EC element.

In addition to liquid materials (the fluid), solid-state materials (generally metals) can also be employed as a regenerator. The ideal regenerating material should have high thermal conductivity and low thermal capacity. Sophisticated carbon nanotube composites should be a promising candidate to increase the thermal conductivity of the regenerator.

Parasitic losses are a serious issue for electrocaloric refrigerator devices. They diminish the coefficients of performance (COP $= Q/W$, where $Q = T_2 \int_{E_1}^{E_2} p\,\mathrm{d}E$ is the work performed and $W = -\eta_{\text{Carnot}} Q$ is the work supplied) of actual physical systems, and care must be taken to minimize their impacts on the overall coefficient of performance. Typical parasitic losses mainly come from the contributions of the devices and the ferroelectric materials. The former includes electrical contact resistance, charge pinning at trap states from the interface between the electrode and ferroelectric, and electrode loss. The latter includes grain boundary conduction, dielectric loss, and domain switching.

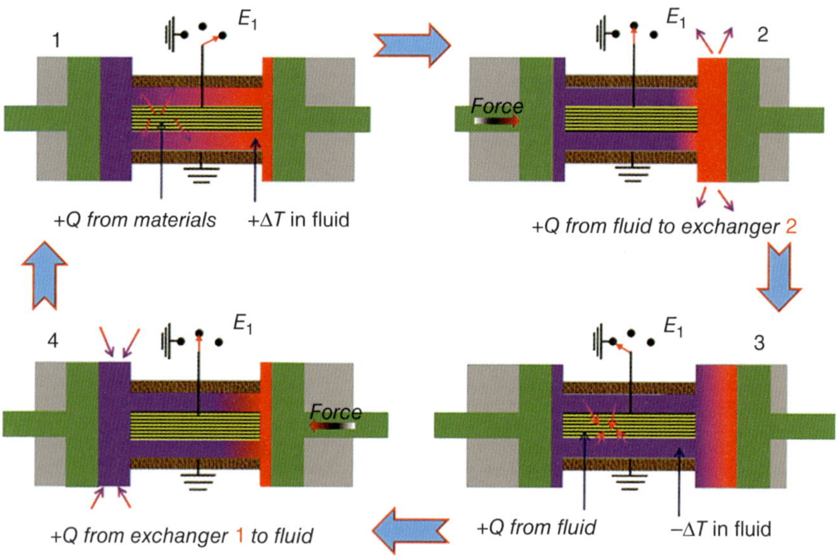

Figure 8.9 (1) An electric field E_1 is applied to the EC element, and the heat from the EC materials is released into the fluid and a $+\Delta T$ is caused in the fluid, heating the fluid. (2) The electric field E_1 is disconnected, and the fluid is pumped past the EC element so that the excess heat is lost at the heat exchanger *2*. (3) The EC element is shorted so that the field returns to zero and a $-\Delta T$ is caused in the fluid, cooling the fluid. (4) The EC element is again put to open circuit, and the fluid is pumped back into the heat exchanger *1*, so that the fluid then absorbs heat from the heat exchanger *1*. *Source:* Alpay et al. 2014 [34]. Reproduced with the permission of Cambridge University Press.

It should be noted that a new conceptual design for solid-state refrigerators without periodical application and withdrawal of the electric field has also been proposed, as shown in Figure 8.10. By using 180 domain structures in ferroelectric thin films, the cooling process in solid-state refrigerators can be completed in one step with a sustained applied electric field [35]. The heat in hot reservoirs is absorbed by an adjacent ferroelectric thin film and is then expelled to the cold reservoirs under the applied electric field.

8.2.2 MLCC and MLPC EC Refrigerator Modules

So far, the reported works on electrocaloric effect of ferroelectric/antiferroelectric materials mainly focused on bulk ceramics and thin films. Because of the low breakdown strength (BDS), the electrocaloric effects of bulk ceramics are usually very small. For example, it was reported that the maximum electrocaloric effect obtained in the AFE $(Pb_{0.87}Ba_{0.1}La_{0.02})$ $(Zr_{0.65}Sn_{0.3}Ti_{0.05})O_3$ bulk ceramics is only 2.5 K. Improved electrocaloric effects could be realized in ferroelectric thin films due to the large BDS, which usually could be improved to 1–2 orders of magnitude of that in their counterpart bulk ceramics. For example, Mischenko et al. reported a giant electrocaloric effect of $\Delta T = 12$ K in 350-nm-thick $PbZr_{0.95}Ti_{0.05}O_3$ AFE films near the Curie temperature. Very recently, Peng et al. reported a much larger electrocaloric effect of $\Delta T = 45.3$ K

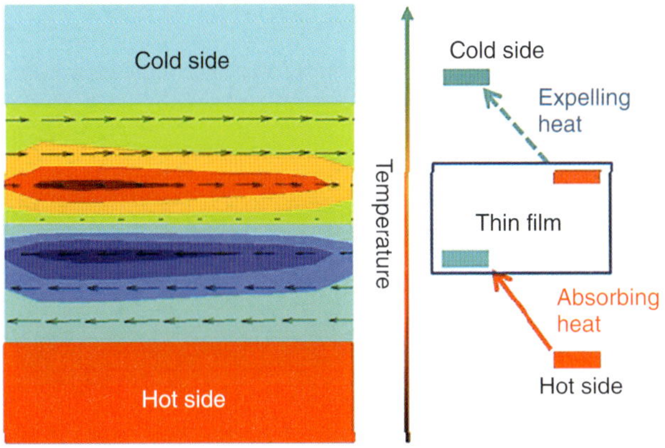

Figure 8.10 Conceptual design schematic of ferroelectric thin film with 180 domain structures as solid-state refrigerators. *Source:* Li et al. 2013 [35].

at room temperature in 320-nm-thick $Pb_{0.8}Ba_{0.2}ZrO_3$ films. These results raised hopes for the realization of solid-state cooling devices based on the electrocaloric effect. However, due to their small thickness (<1 μm), the overall heat-sinking capacity of thin films is still not so high as to satisfy the requirement of practical applications. For security reasons, the operational voltage cannot be too high, which should be less than 250 V. Therefore, multilayer ceramic capacitors (MLCC) or multilayer polymer capacitors (MLPC) with an appropriate thickness (~1–10 μm) and high BDS might overcome the shortcomings of bulk ceramics and thin films and are suitable for practical applications in solid-state cooling devices.

Based on the electrocaloric effect in a $BaTiO_3$ MLCC and its performance, some researcher predicted that an MLCC array of about 0.6 m^2 total area could provide up to 20 kW of cooling power. It should be noted that the penetration of heat along the metal electrodes in the interleaved multilayer structures (see Figure 8.11) is key to determining how fast the system can be cycled, and that the metal thermal diffusivity, $\alpha = K/c'$ (where K is the thermal conductivity), is a key parameter. Simulation results by Crossley et al. indicated that the EC performance of the MLCC structures with Ag electrodes ($\alpha = 173 \times 10^{-6}\ m^2\ s^{-1}$) would be roughly five times faster than those with Ni electrodes ($\alpha = 24.6 \times 10^{-6}\ m^2\ s^{-1}$), giving a better power handling capability.

8.3 Electrocaloric Materials

8.3.1 EC in Ferroelectric Ceramics

8.3.1.1 In Bulk Ceramics and Single Crystals

The first study on ECE was mainly focused on Rochelle salt by Kobeko and Kurtschatov in 1930. The ECE of these materials was measured in the low temperature range of 4–15 K where the specific heat is low for them and a large adiabatic temperature change was expected to be obtained. However, no

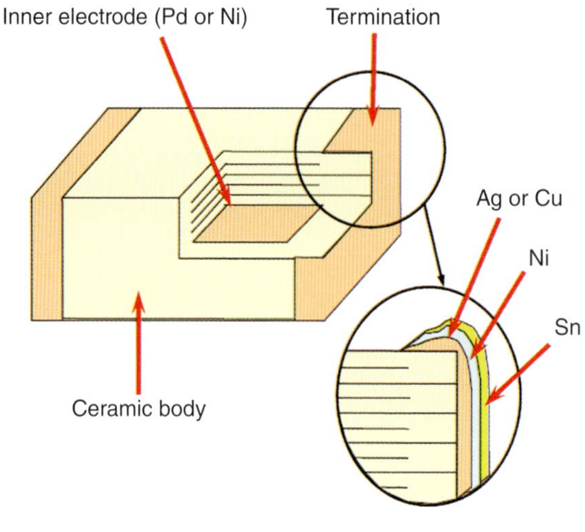

Figure 8.11 Front and cross-sectional views of a multilayer ceramic capacitors.

numerical values were reported. In 1963, the ECE of Rochelle salt was measured by Wiseman and Kuebler again. By means of the Maxwell relation, a ΔT of about 0.0036 K was obtained at 22.2 °C and 1.4 kV cm^{-1} and also a ΔS of 0.001 56 J kg^{-1} K^{-1}. The largest ECE effect ($\Delta T = 0.3$ K) at low temperature was found in $SrTiO_3$ at 10 K.

After that, the ECEs were studied in many perovskite ceramics and single crystals, which include normal ferroelectrics (FE), antiferroelectrics (AFE), and relaxor ferroelectrics/antiferroelectrics. The properties of these ceramics can be customized by element doping, providing good methods to understand the key factors that may influence the ECE.

In normal ferroelectrics, a first-order FE–PE transition usually can cause a large ECE in a narrow temperature range. For example, the largest electrocaloric effect ($\Delta T = 2.5$ K, $\Delta S = 2.88$ J kg^{-1} K^{-1}) in bulk ceramics was observed in $Pb_{0.98}Nb_{0.02}(Zr_{0.75}Sn_{0.2}Ti_{0.05})_{0.98}O_3$ ceramics at 161 °C and 30 kV cm^{-1}. However, in relaxor ferroelectrics or antiferroelectrics, a second-order FE–PE or AFE–PE transition should achieve an ECE in a wider temperature range. For example, a ΔT of 2.5 K in the temperature range from 16 to 67 °C under an electric field of 90 kV cm^{-1} was maintained in the prototypical relaxor ferroelectric $Pb(Mg_{1/3}Nb_{2/3})TiO_3$ (PMN), which exhibits a diffused dielectric phase transition around −13 °C.

In addition to the element doping, the ECE in a broad temperature range can be tailored by the regulation of the ordering–disordering degree of the B ions in some solid solutions by thermal treatment. For instance, a high degree of long-range order yields a high ECE maximum ($\Delta T = 1.7$ °C) as well as a high phase transition temperature [36], while a decreased order leads to a lower ECE in a broader temperature range and diffused phase transition.

The ECEs in some lead-free ceramic and single crystal systems such as $BaTiO_3$, $Ba(Ti_{1-x}Zr_x)O_3$, $BaHf_xTi_{1-x}O_3$, $(Ba_{1-x}Sr_x)TiO_3$, $(Na_{1/2}Bi_{1/2})TiO_3$, $Ba_{0.85}Ca_{0.15}Ti_{(1-x)}Hf_xO_3$, and $Sr_xBa_{(1-x)}Nb_2O_6$ are also investigated [4, 26, 37–50].

Although ECE is small, lead-free materials are attractive because of their environment friendliness.

All these results indicate that ECE in bulk ceramics is not large (below 3 °C) [4, 26, 37–53], which makes them unimpressive for practical applications. The main reason is that the breakdown electric field in bulk ceramics is too low and the fabrication of bulk ceramics into small thickness (below 10 μm) is difficult. From Eqs. (8.25) and (8.26), it can be seen that a larger ΔT and ΔS can be obtained when a higher electric field is applied. By fabricating thick films by using the tape casting method and using an interdigitated multilayer geometry such as the MLCC, the thickness of each layer can be reduced to ~1 μm, which allows for the application of high electric field, and a large ECE can be obtained [19, 54].

For films fabricated using the sol–gel, PLD, MOCVD, and ALD methods [55, 56], the thickness of films can be smaller, indicating that the film can be applied under a higher electric field, and consequently larger ΔT and ΔS. Therefore, bulk ceramics and single crystals with large EC strength ($|\Delta T|/|\Delta E|$ and $|\Delta S|/|\Delta E|$) should yield a large ΔT and ΔS in the geometry of their thin or thick films.

For example, a large $|\Delta T|/|\Delta E|$ of 44.97 mK cm kV^{-1} was obtained in the $Pb[(Ni_{1/3}Nb_{2/3})_{0.6}Ti_{0.4}]O_3$ (PNNT) single crystal with composition close to the morphotropic phase boundary (MPB) [57, 58], as shown in Figure 8.12. The EC strength of PNNT single crystal is much higher and comparable to that (5 mK cm kV^{-1}) of the antiferroelectric $PbZr_{0.95}Ti_{0.05}O_3$ ceramics [59], that (6 mK cm kV^{-1}) of the ferroelectric polymer P(VDF–TrFE), and even that (25 mK cm kV^{-1}) of the ferroelectric single crystal $0.9Pb(Mg_{1/3}Nb_{2/3})O_3$ $-0.1PbTiO_3$ [6, 60–62], etc., indicating that the PNNT single crystal is a promising EC material for application in cooling system. Predictably, an intrinsic EC effect with bigger ΔT in the PNNT thin or thick film could be obtained due to a significant improvement in its dielectric BDS.

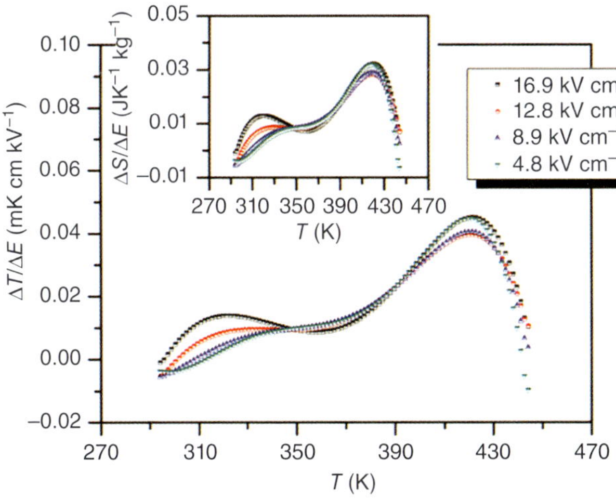

Figure 8.12 $|\Delta T|/|\Delta E|$ and $|\Delta S|/|\Delta E|$ (inset) of the $Pb[(Ni_{1/3}Nb_{2/3})_{0.6}Ti_{0.4}]O_3$ single crystal. *Source:* Peng et al. 2015 [57]. Reproduced with the permission of Elsevier.

8.3.1.2 In Thin Films

As mentioned above, thin ceramic films allow for the application of high electric fields and consequently a large ECE. Here thin films refer to those with a thickness of less than a micron.

In 2006, Mischenko et al. first reported the ECE in $PbZr_{0.95}Ti_{0.05}O_3$ antiferroelectric ceramic thin films with a thickness of 350 nm. An adiabatic ΔT of 12 K (as deduced from Eq. (8.26)) near the Curie temperature of 222 °C was obtained under the electric field of 776 kV cm^{-1}, as shown in Figure 8.13. The isothermal ΔS is estimated to be 8 J kg^{-1} K^{-1} according to Eq. (8.25). A ΔT of about 11 K was reported in a 700-nm $PbZrO_3$ thin film under the electric field of 510 kV cm^{-1} near the phase transition temperature of 235 °C. Similarly, a ΔT of about 11 K was also observed in the prototypical $PbZr_{0.52}Ti_{0.48}O_3$ ferroelectric thin film under the electric field of 577 kV cm^{-1} near the phase transition temperature of 387 °C. Both the high electric field applied to these thin films and the operation temperature near the ferroelectric–paraelectric (F–P) transition contribute to the large ΔT. However, the phase transitions in these thin films are first-order transitions and the high ECE can only exist in a narrow temperature range. Moreover, the phase transition temperature is too high for applications near room temperature.

To obtain a large ECE at a wide range of temperatures near room temperature, many thin films with a low phase transition temperature were studied. A large ΔT of ~31 K under the electric field of 747 kV cm^{-1} in $Pb(Mg_{1/3}Nb_{2/3})_{0.65}Ti_{0.35}O_3$ thin films was reported at 140 °C because of the contribution of the MPB. A maximum ΔT of ~5 K under the electric field of 895 kV cm^{-1} in $0.9PbMg_{1/3}Nb_{2/3}O_3$–$0.1PbTiO_3$ thin film was observed at 75 °C, where a pseudocubic relaxor ferroelectric transforms to a cubic paraelectric phase. A large ΔT of ~40 K under the electric field of 1200 kV cm^{-1} in $Pb_{0.88}La_{0.08}Zr_{0.65}Ti_{0.35}O_3$ thin films was reported at 45 °C. A ΔT of ~9 K under the electric field of 723 kV cm^{-1} in $0.93PbMg_{1/3}Nb_{2/3}O_3$–$0.07PbTiO_3$

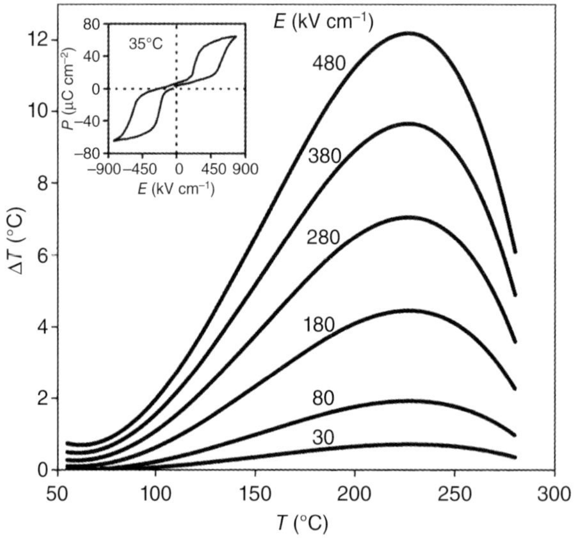

Figure 8.13 ΔT of the $Pb(Zr_{0.95}Ti_{0.05})O_3$ thin films. *Source:* Mischenko et al. 2006 [59].

thin films was observed at the depolarizing temperature 18 °C instead of the phase transition temperature of 35 °C, where the dielectric permittivity peak appeared, indicating a dipolar glass–relaxor phase transition. A ΔT of ~13.4 K with a ΔS of 16.7 J K^{-1} kg^{-1} under the ultrahigh electric field of 3260 kV cm^{-1} in 9.2 nm $Hf_{0.2}Zr_{0.8}O_3$ ultrathin films fabricated by atomic layer deposition (ALD) was reported at 15 °C. Interestingly, a giant ECE (ΔT = 45.3 K and ΔS = 46.9 J K^{-1} kg^{-1} at 598 kV cm^{-1}) in relaxor $Pb_{0.8}Ba_{0.2}ZrO_3$ thin films was reported at room temperature (17 °C) rather than at its Curie temperature (135 °C), as shown in Figure 8.14a,b. Field-induced nanoscale antiferroelectric to ferroelectric phase transitions may play a key role in the dramatic EC effect.

Other thin films such as $PbSc_{0.5}Ta_{0.5}O_3$ relaxor with a wide temperature range of ECE were also investigated [36]. A ΔT of 6.2 K with ΔS of 6.3 J K^{-1} kg^{-1} under the electric field of 774 kV cm^{-1} was obtained at 68 °C [36]. The ECE in

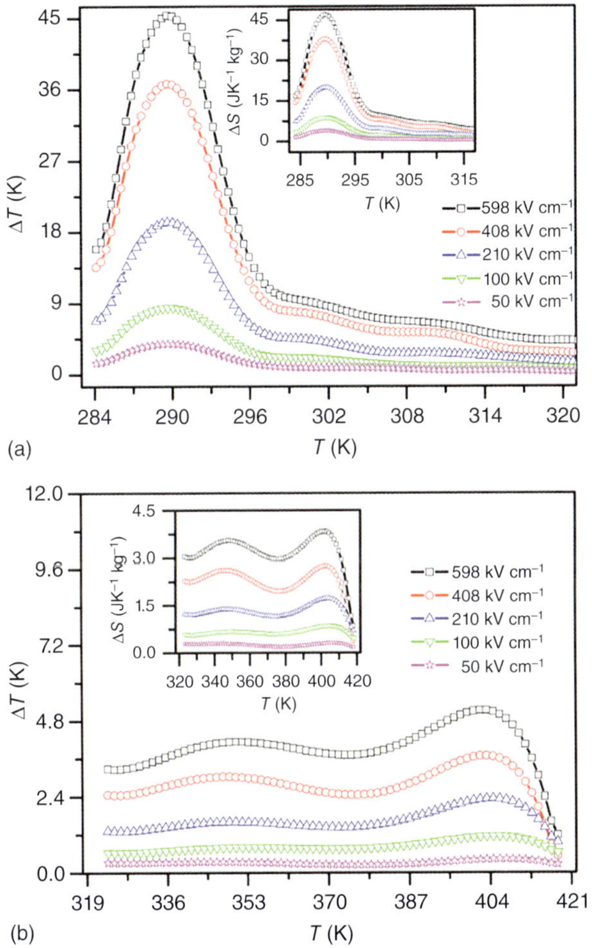

Figure 8.14 ΔT and ΔS (inset) of the PBZ thin films. (a) Around room temperature, and (b) around T_{dp} and T_c. *Source:* Peng et al. 2013 [63]. Reproduced with the permission of John Wiley & Sons.

a 200-nm perovskite-type bismuth layered oxide lead-free $SrBiTa_2O_9$ thin films was investigated. A ΔT of 4.93 °C under the electric field of 600 kV cm^{-1} was obtained at 288 °C [64].

It should be noted that the ECE in thin films with a thickness of more than a micron (the so-called thick films) has also been investigated. For instance, a large ΔT of 28.1 °C with ΔS of 31.6 J K^{-1} kg^{-1} under the electric field of 900 kV cm^{-1} was obtained at 21 °C in a 1.5-μm antiferroelectric thick film with (111) orientation. The ECE in $BaTiO_3$ multilayer thick film (~1.4 μm) was also investigated by direct calorimetric measurement [54, 65]. The maximum ECE occurs above Tc and shifts to higher temperature with increasing applied field. A giant ECE of $\Delta T \sim 7.1$ K and $\Delta S \sim 10.1$ J K^{-1} kg^{-1} at 80 °C was obtained under an ultrahigh field of 800 kV cm^{-1} [54].

Recently, negative ECE also was reported in some antiferroelectric thin films [21, 24, 66, 67]. For example, a ΔT of ~−5.76 K under a modest electric field of 338 kV cm^{-1} in $(Pb_{0.97}La_{0.02})(Zr_{0.95}Ti_{0.05})O_3$ antiferroelectric thin films fabricated on a Pt(111)/Ti/SiO_2/Si substrate by sol–gel method was observed at 59 °C. A ΔT of ~−6.62 K with a ΔS of ~−5.42 K under an electric field of 709 kV cm^{-1} in 4 mol% Eu-doped $PbZrO_3$ antiferroelectric thin films was also observed at 130 °C. The negative ECE in these thin films could be microscopically induced within a domain whose polarization is non-collinear with the applied electric field.

Additionally, a ΔT of ~−10.8 K with a ΔS of −10.9 J K^{-1} kg^{-1} under the ultrahigh electric field of 3260 kV cm^{-1} in 9.2 nm $Hf_{0.5}Zr_{0.5}O_3$ ultrathin ferroelectric films fabricated by ALD was reported at 175 °C, as shown in Figure 8.15. The negative ECE could be ascribed to the non-polarization of t-phase and the polycrystalline nature of the $Hf_{0.5}Zr_{0.5}O_3$ thin film. The nonpolar t-phase modified the U–P curve where one local minimum at $P = 0$ was induced in addition to two stable minima at $P = \pm P_r$. The potential barrier between the two stable minima decreased rapidly when the temperature increased, accompanied by a decrease in E_c. As a result, the polarization under a given electric field can increase with increasing temperature and consequently a negative ECE.

It is suggested that the combination of positive and negative ECEs in these thin films can complete the cooling process in one step with a sustained applied electric field in the cooling process [24, 67]. Li et al. suggested that the 180° domain structure of $PbTiO_3$ can be utilized as a combined nano-laminate structure to include positive and negative EC materials [35]. As a result, it could provide a new pathway to design solid-state refrigerators without any moving parts and without the need for the periodic application and withdrawal of an electric field.

8.3.2 EC in Ferroelectric Polymer Materials

The breakdown electric field of dielectric polymers can be several orders of magnitude higher than in ceramics, suggesting that a large ECE can be achieved in polymer thin films. Compared to ceramics, the PVDF-based polymers can be more easily fabricated into thin films and large size devices, indicating that they are the most likely ECE materials to be developed into commercial applications. Very large ECE has been achieved in PVDF-based polymers, especially for ΔS, which is required for high-performance refrigerants. The following sections will summarize the main achievements in this field.

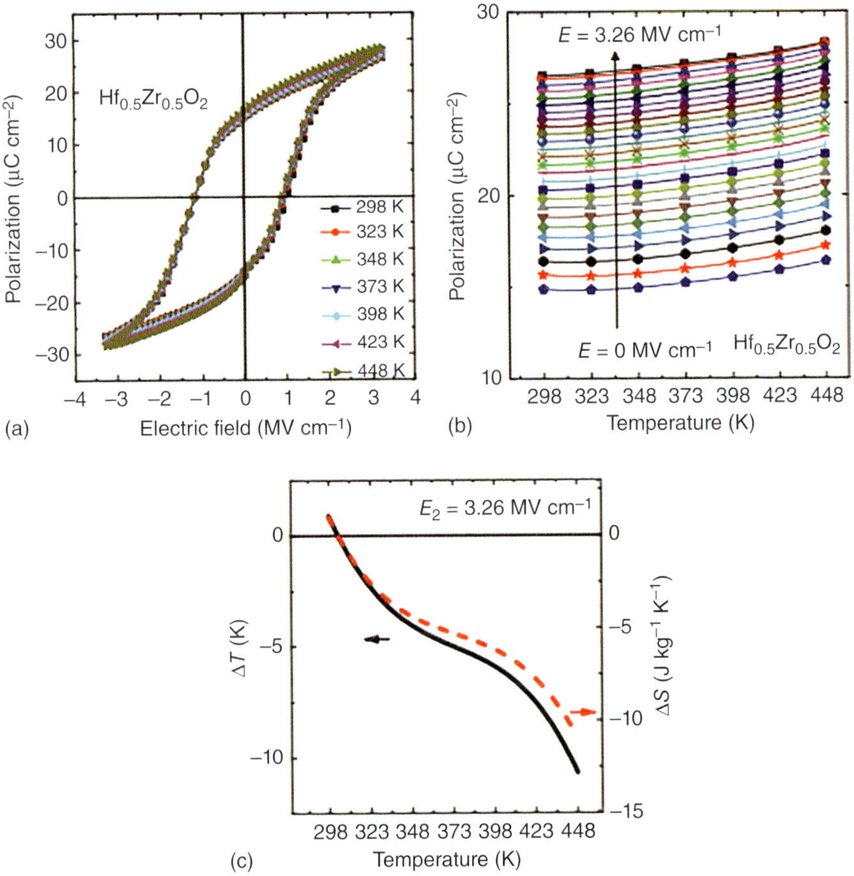

Figure 8.15 (a) Temperature-dependent polarization–electric field hysteresis, (b) polarization–temperature plot, and (c) ΔT and ΔS of the $Hf_{0.5}Zr_{0.5}O_2$ ferroelectric thin films. *Source:* Park et al. 2016 [24]. Reproduced with the permission of John Wiley & Sons.

8.3.2.1 In Normal Ferroelectric Polymers

Similar to the ferroelectric ceramics, a large ECE associated with the order–disorder transition of the dipole states would be obtained near the FE–PE phase transition temperature of ferroelectric polymers. The normal P(VDF–TrFE) 55/45 mol% copolymer ferroelectric thin films with a thickness range from 0.4 to 2 μm was studied as an electrocaloric material due to the continuity of FE–PE phase transition, which can avoid the hysteresis effect associated with a first-order phase transition. As shown in Figure 8.16a,b, ΔS deduced from the Maxwell relation increases from 20 to 60 J K^{-1} kg^{-1} and the corresponding ΔT increases from 6 to 12 K as the applied electric field increases from 1340 to 2090 kV cm^{-1}. ΔS (60 J K^{-1} kg^{-1}) in the ferroelectric polymers is much higher than that in ferroelectric/antiferroelectric ceramic thin films; for example, ΔS is only 8 J K^{-1} kg^{-1} in the $PbZr_{0.95}Ti_{0.05}O_3$ antiferroelectric thin films. An underlying reason for the small ΔS is that the small ionic displacements in these ferroelectric/antiferroelectric ceramics may not yield large entropy

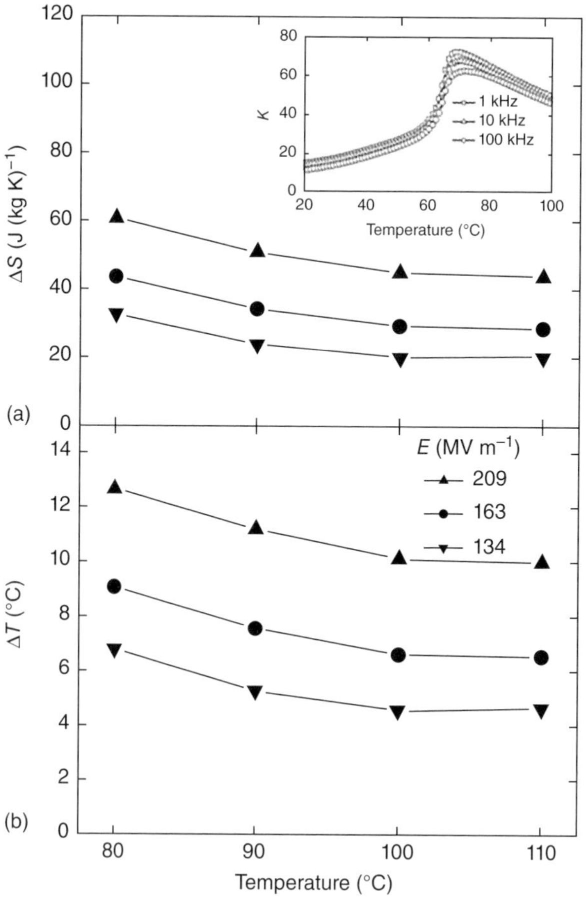

Figure 8.16 (a) ΔT and (b) ΔS of the P(VDF-TrFE) 55/45 mol% ferroelectric copolymer thin films. Inset in (a): the dielectric permittivity as a function of temperature. *Source:* Neese et al. 2008 [6].

changes. By contrast, the ordering and disordering of dipoles in ferroelectric polymers can induce a large entropy change. The ECE in normal P(VDF–TrFE) 55/45 mol% copolymer was also measured by using the direct method as well as by comparison with the ECE results deduced from the Maxwell relation. The results indicated that similar rules between the two methods were obtained, namely, the ECE reaches a maximum at the FE–PE transition, where a $\Delta T \sim 12$ K can be induced under a 1200 kV cm^{-1}. As the temperature above the FE–PE transition increases, ΔT drops sharply, indicating a characteristic feature for normal ferroelectric materials. The comparison result reveals that the ECE deduced from the Maxwell relation is consistent with that directly measured, suggesting that the Maxwell relation is valid for deducing ECE for normal ferroelectric materials at temperatures above their FE–PE transition.

It should be noted that ΔT and ΔS based on the L–D phenomenological theory (see Eqs. (8.29) and (8.30)) are higher than the experimental values by about 15%

at 80 °C and 25% at 110 °C. The most likely reason is that the experimental films contain an amorphous phase (>25% wt%), whereas the phenomenological theory only applies to the crystalline phase.

8.3.2.2 In Relaxor Ferroelectric Terpolymers

Based on the statistical thermodynamics theory, in a dipolar disordered state at $E = 0$, such as the relaxor at temperatures above the broad dielectric permittivity peak, the entropy $S_p(0, T)$ is proportional to $P^2 \ln(\Omega)$ (see Eq. (8.36)). By utilizing defect modification, local states can be introduced and the number of available polar states (Ω) could be enhanced. As a result, it can be expected that relaxor ferroelectrics have the potential to achieve a large ECE around room temperature.

Recently, experimental research results have shown that by employing high-energy electron irradiation, the normal ferroelectric P(VDF-TrFE-CFE) 59.2/33.6/7.2 mol% terpolymer can be converted into a relaxor ferroelectric polymer, which possesses high dielectric permittivity (~55 at 1 kHz), large reversible polarization change, and high electrostriction near room temperature, suggesting a high ECE in these polymers near room temperature. The large ECE response of ΔT (~7 K) and ΔS (~35 J K^{-1} kg^{-1}) measured directly under an electric field of 1000 kV cm^{-1} in the P(VDF-TrFE-CFE)59.2/33.6/7.2 mol% non-stretched relaxor ferroelectric terpolymer is nearly temperature independent from 0 to 45 °C as shown in Figure 8.17a,b, which is in sharp contrast to that in normal ferroelectrics where the ECE peaks at the FE–PE transition and exhibits strong temperature dependence. These results revealed that a large ECE can be tailored by the modification of defects, such as a temperature-independent ECE over a broad range, which is attractive for practical cooling device applications. As has been shown in Figure 8.17c,d, the temperature dependence of the ECE of the relaxor ferroelectric terpolymer thin film depends critically on the film fabrication conditions. ΔT or ΔS in the stretched thin film is more than 15% smaller than that in the non-stretched thin film, and displays a pronounced temperature dependence on the ECE. Such a difference is likely to be caused by the changes in possible polar states and polar correlation length due to film stretching.

The ECEs acquired in the relaxor ferroelectric polymer by direct and indirect methods are very different, as shown in Figure 8.18. The indirectly measured ECE deduced from the Maxwell relation is much smaller than that directly measured. Moreover, directly measured ECE displays a much stronger temperature dependence on the ECE at $E < 70$ kV cm^{-1}. These results indicate that the Maxwell relation is not suitable to deduce ECE for the relaxor ferroelectric polymers even at temperatures above the broad dielectric permittivity maximum. One possible explanation is the fact that Eqs. (8.25) and (8.26) are derived from state equations that consider an ergodic system in equilibrium with its surroundings. For ferroelectric relaxor polymers (non-ergodic systems), their dielectric responses are strongly dependent on the thermal histories even at temperatures above the dielectric permittivity maximum; therefore, Eqs. (8.25) and (8.26) cannot describe electrocaloric effect in ferroelectric relaxor polymers accurately.

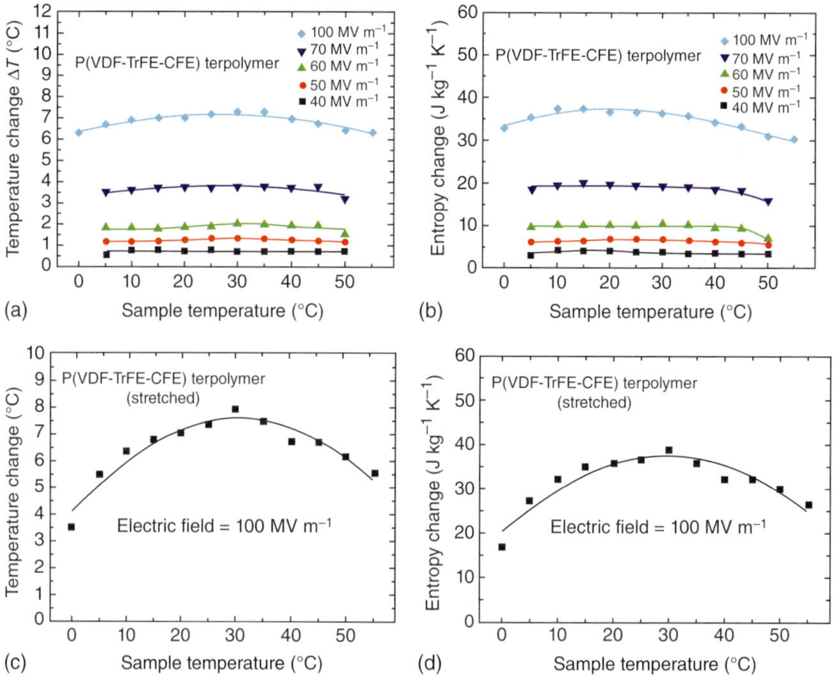

Figure 8.17 Directly measured ΔT (a) and ΔS (b) in the non-stretched P(VDF-TrFE-CFE) 59.2/33.6/7.2 mol% relaxor ferroelectric terpolymer thin films; ΔT (c) and ΔS (d) in the stretched thin films.

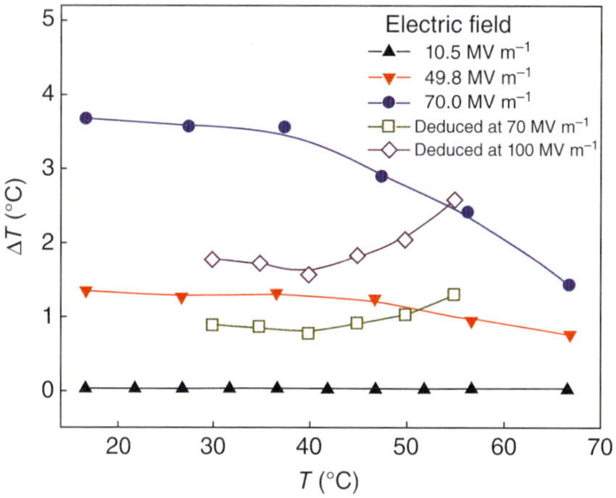

Figure 8.18 Indirectly and directly measured ΔT in P(VDF-TrFE-CFE) 59.2/33.6/7.2 mol% relaxor ferroelectric terpolymer thin films.

8.3.3 EC in Other Materials

8.3.3.1 In Composites

P(VDF–TRFE) with Coexistence of Ferroelectric and Relaxor Electrocaloric effect refers to the change in temperature and/or entropy of a dielectric material due to the electric-field-induced change in dipolar states and is characterized by the isothermal entropy change $S(E = 0, T) - S(E, T)$ under adiabatic application or removal of an electric field. Generally for the positive electrocaloric effect, when E is high, $S(E, T)$ is much smaller than $S(0, T)$. Hence, a large electrocaloric effect requires dielectric materials possessing a large $S(0, T)$. Some recent research works indicated that polar dielectrics with a large number of equivalent polar orientations and short polar correlation can cause a large $S(0, T)$ and consequently a large electrocaloric effect. Based on this point, Casar et al. speculate that a large $S(0, T)$ could be higher in mixed or partially ordered systems. By changing the irradiation doses from 0 to 60 Mrad in PVDF–TrFE-based polymers (Figure 8.19), they demonstrated that a large electrocaloric response of $\Delta T \sim 13$ K around the Curie temperature is achieved in a sample with moderate irradiation dose (20 Mrad), where relaxor and normal ferroelectric states coexist. In contrast, the electrocaloric response in systems without irradiation (0 Mrad) and with excess of irradiation (40, 60 Mrad) is weak, $\Delta T < 6$ K. This phenomenon can be well understood according to Eq. (8.36), $\Delta T_{sat} = \frac{T \ln \Omega}{3\varepsilon_0 \Theta C_E} P_{sat}^2$, which is deduced from the statistical thermodynamics theory. On the one hand, while the highest polarization can be induced by the external electric field around the Curie temperature in ferroelectric systems (irradiation dose with 0 Mrad), the electron irradiation, which gradually transforms the ferroelectric chain orientation into a relaxor system, increases the number of possible polar states (P) due to the fact that there are more preferred orientations in relaxors than in ferroelectrics. On the other hand, it has in fact been shown that uniaxial stretching of a relaxor terpolymer, which causes preferred polymer chain orientation, reduces the value of Ω. Therefore, there is a competing effect between the contributions of P and Ω for the increase of the ΔT in systems with coexisting relaxor and ferroelectric states.

Such a promotion of electrocaloric response in mixed systems, particularly prominent in the temperature range around the Curie temperature, makes a

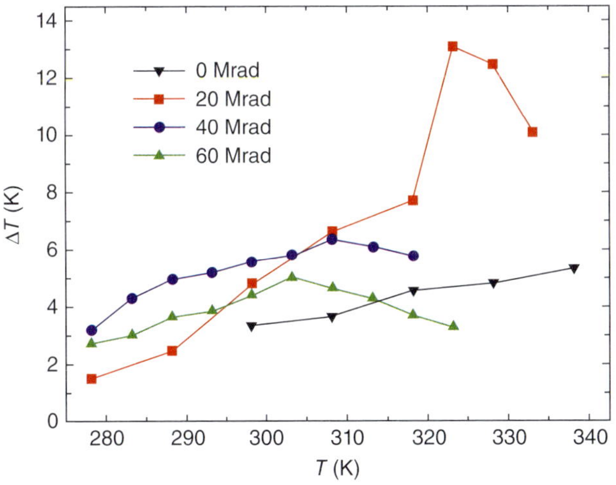

Figure 8.19 The electrocaloric effect for the ΔT of the ferroelectric and irradiated P(VDF-TRFE) copolymers. ΔT is measured using a specially designed calorimeter.

suggestion for the development of a new generation of electrocaloric materials that might provide an effective means to realize solid-state cooling devices for a broad range of applications.

Composite of Ferroelectric Ceramics and Polymers Many research works have indicated that large ΔT can be achieved in thin films of ferroelectric in the vicinity of the ferroelectric–paraelectric phase transition, because their large spontaneous electrical polarization (P) can be oriented under the application and/or withdrawal of an external electric field, making them the leading materials for ECE refrigeration. For instance, a ΔT of 12 °C and a ΔS of 60 kJ m^{-3} K^{-1} have been observed in $Pb(Zr_{0.95}Ti_{0.05})O_3$ thin films near T_c of 222 °C [59]. Similarly, for instance, a ΔT of 16.6 °C and a ΔS of 62 kJ m^{-3} K^{-1} were obtained in a ferroelectric P(VDF-TrFE) copolymer thin film near T_c of 102 °C. Compared with the ferroelectric ceramics, ferroelectric polymers in a sense are more appropriate for EC cooling applications because of their unparalleled processability, scalability, weight, and cost advantages. However, large electric fields (E) that may be several times or 10 times of that in thin films of ferroelectric ceramics are required to induce ECE in ferroelectric polymer thin films, as indicated by much lower EC strengths represented by the values of $|\Delta T|/|\Delta E|$ and $|\Delta S|/|\Delta E|$ in comparison to those of ferroelectric ceramics thin films and single crystals. In addition, low thermal conductivity in ferroelectric polymers (~0.2 W m^{-1} K^{-1}) would limit their performance in the corresponding cooling devices in spite of large EC responses.

Recently, Zhang et al. demonstrated that by utilizing two fillers with complementary functions, i.e. boron nitride nanosheets (BNNSs) employed to improve E_b and thermal conductivity while $Ba_{0.67}Sr_{0.33}TiO_3$ (BST67) utilized to enhance EC strength, the ferroelectric polymer nanocomposites P(VDF–TrFE–CFE)/BNNSs/BST67 can be effectively operated under both modest and high electric fields to yield significant ECE at room temperature, as shown in Figure 8.20. For example, ΔT, ΔS, and EC strength of the P(VDF–TrFE–CFE)/BNNSs/BST67 nanocomposite with 9 vol.% BNNSs and 6 vol.% BST67 increase by threefold when compared to those of P(VDF–TrFE–CFE) at 50 MV m^{-1}. Moreover, ultrahigh ΔT of 50.5 °C and ΔS of 426 kJ m^{-3} K^{-1} can be obtained at 250 MV m^{-1}. These values are more than three times higher than those of P(VDF–TrFE–CFE) measured at 150 MV m^{-1}. The excellent ECE performance of ferroelectric polymer nanocomposites opens up new perspectives for development of light, compact, and environmentally friendly solid-state EC cooling devices.

It should be noted that nanowire (NW) array, i.e. the BST NW array, which is developed from these ferroelectric polymer nanocomposites, might be suitable for wearable cooling technologies, due to the highly bendable, stretchable, and excellent ECE under low electric field and in rigorous stretching–bending conditions, as shown in Figure 8.21. In order to maintain comfortable conditions for moderate exercise under a safe applied voltage of 36 V and an operating frequency of 1 Hz, it is estimated that only 200 g of the BST NW array is needed to produce a cooling power of 300 W, yielding a sizable ECE of a ΔT of 3 °C and a

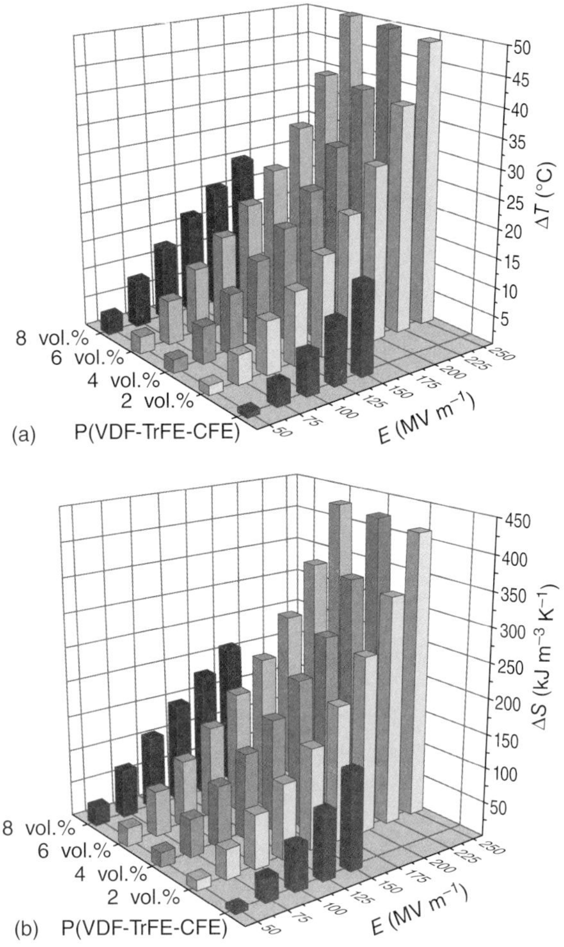

Figure 8.20 ΔT and ΔS of P(VDF-TrFE-CFE)/BNNSs/BST67 ternary polymer nanocomposites at 9 vol.% BNNSs and different BST contents versus P(VDF-TrFE-CFE). *Source:* Zhang et al. 2015 [68]. Reproduced with the permission of John Wiley & Sons.

ΔS of 4.9 J kg^{-1} K^{-1}. Comparatively, 6.4 kg of ice is required to supply 300 W of cooling for two hours.

8.3.3.2 In Liquid Crystals

Theoretically, a dielectric fluid possessing a high density of dipoles could achieve a large electrocaloric effect. Constrained by the relative low operating voltage (usually less than 2 or 3 V), the electrocaloric effect should be very weak in many dielectric liquids such as water, which possesses high dielectric constants, making it difficult to reach a large electrocaloric effect for practical applications. Liquid crystals that have been used widely in optic displays are dipolar liquids in which the molecular orientation and consequently dipolar states can be easily controlled by external electric signals with a long operation life under high voltage. Hence, liquid crystals may be ideal dielectric liquids to realize large ECE.

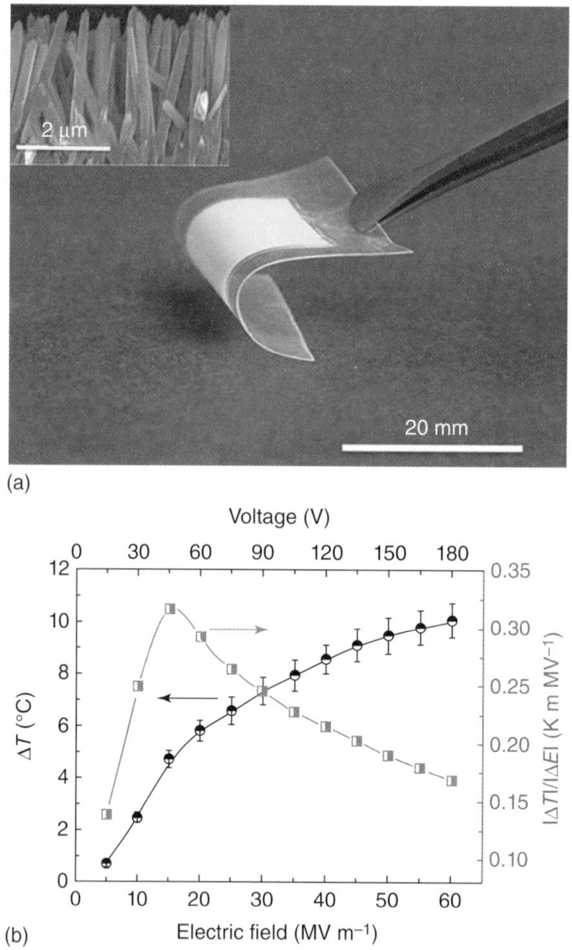

Figure 8.21 (a) The BST nanowire array on scotch tape and bended with the tweezers. Inset: cross-sectional view SEM image. (b) ΔT and $|\Delta T|/|\Delta E|$ of the BST nanowire array as a function of electric field. *Source:* Zhang et al. 2016 [69]. Reproduced with the permission of John Wiley & Sons.

Similar to the paraelectric state in ferroelectrics, rod-like, anisotropic liquid crystal molecules, the mesogens, will form a randomly oriented state, the isotropic phase (Figure 8.22a), at high temperatures. With decrease of temperature, liquid crystal molecules can transform into a nematic phase or a smectic phase (Figure 8.22b,c). In the nematic phase, the mesogens are aligned, on average, along a preferred direction as defined by a director $\boldsymbol{n}$. In a smectic phase, the mesogens, besides being aligned along a preferred direction, form a layered structure while in each layer, the molecular centers are randomly arranged (2D liquids). Applying electrical fields may induce a transition from an isotropic (I) phase to a nematic (N) or smectic (S) phase when near the N–I or S–I transition, and thus may lead to a large ECE.

Considering that the large dielectric anisotropy and the large electric-field-induced polarization change near the nematic–isotropic (N–I) phase transition in 5CB (Figure 8.22d), Qian et al. find that a large isothermal entropy change of $23.6\,\mathrm{J\,kg^{-1}\,K^{-1}}$ based on the Maxwell relation (8.16) and a large adiabatic temperature change of $3.35\,\mathrm{K} < \Delta T < 5.26\,\mathrm{K}$ under $90\,\mathrm{MV\,m^{-1}}$ electric field can be

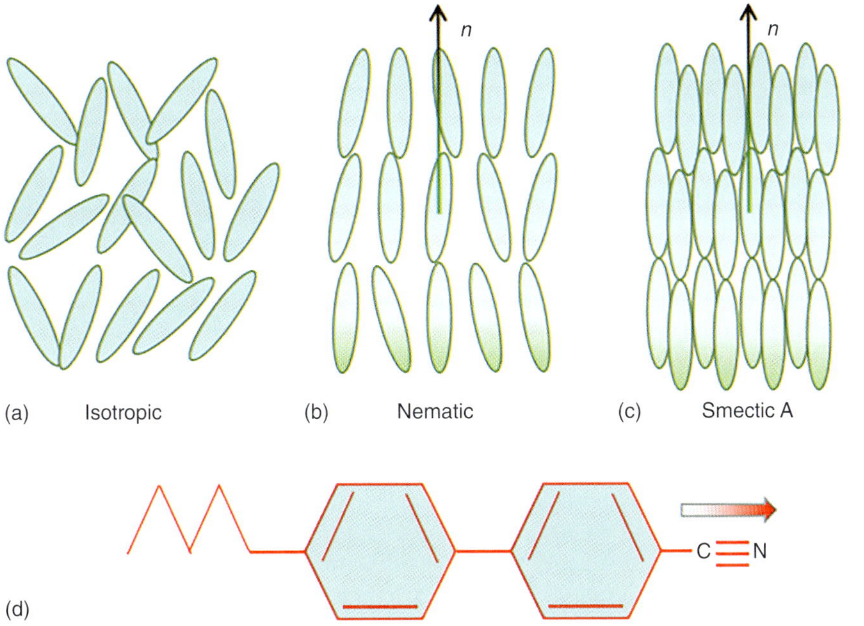

Figure 8.22 Schematic of various mesophases for rod-like liquid crystal molecules for positive ECE: (a) isotropic, (b) nematic, and (c) smectic A phases. (d) 5CB Molecular structure. *Source:* Qian et al. 2013 [70]. Reproduced with the permission of John Wiley & Sons.

achieved near 39 °C in homogeneously aligned cells (HA-Cells) [70], as shown in Figure 8.23. The ECE in the 5CB can also be estimated from the phenomenological Landau–de Gennes formulation (LG) and the agreement between the experimental data and LG phenomenological theory estimation is quite good. These results indicate that giant ECE can be obtained from electrical-field-induced orientational order–disorder transition in liquid crystals near the N–I and S–I transitions. Therefore, liquid crystals with a large operating voltage could be promising materials for refrigerators based on their large electrocaloric effect.

8.3.3.3 In Fast Ion Conductors

By means of molecular dynamics simulations, Cazorla et al. find that the onset of superionicity in archetypal fast ion conductor CaF_2 can be reduced by about 100 K when relatively small electric fields of $50\,kV\,cm^{-1}$ are applied [71]. A large increase of entropy can be realized during the normal to superionic phase transition. The resulting change of temperature $|\Delta T/T|$ is about 1% under $E < 100\,kV\,cm^{-1}$ and 1350 K conditions. This effect has the potential to change our paradigm in the search of new electrocaloric materials, which are promising candidates for solid-state cooling applications. In addition to the CaF_2-based fast ion conductors, the electrode materials such as the CGO and YSZ that are used in solid oxide fuel cell may be another promising candidate due to their low superionic phase transition temperature [72, 73].

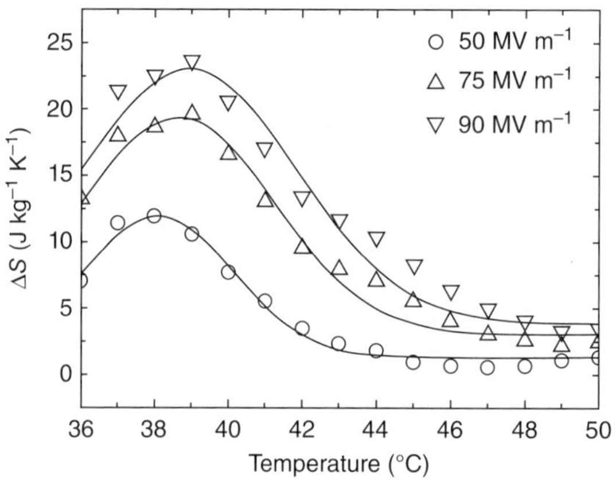

Figure 8.23 The electrocaloric effect for ΔS of 5CB liquid crystals in homogeneously aligned cells. *Source:* Qian et al. 2013 [70]. Reproduced with the permission of John Wiley & Sons.

Unlike the positive electrocaloric effect (Figure 8.24a) in prototypical ferroelectrics, the adiabatic switch of an electric field causes a negative electrocaloric effect (Figure 8.24b) in the fast ion conductors because the entropy of the disordered superionic state is larger than that of the normal ($\Delta S > 0$ hence $\Delta T < 0$). Conversely, the material gets cooler ($\Delta S < 0$ hence, $\Delta T > 0$) when the electric field is adiabatically removed.

8.4 Summary and Outlook

The Maxwell relation is valid only for the study of ECE in ergodic systems, which are thermodynamically reversible near first-order phase transitions. One should be cautious when using the indirect method to deduce ECE in non-ergodic systems such as ferroelectric relaxor polymers and when interpreting the experimental results. Based on the L–D phenomenological theory, a large electrocaloric effect can be generated near the FE–PE phase transition temperature with big Q, β, and P. By increasing the number of polar states in the disordered phase, a larger ECE can be induced in the relaxor ferroelectrics than that in the normal ferroelectrics. Working with dielectrics near MPB may also lead to higher ECE due to the large entropy $S_p(0, T)$. A large ECE can be obtained over a broad temperature range in relaxor ferroelectrics, however, only over a narrow temperature range about the ferroelectric transition in normal ferroelectrics. Large ECE in a broad range of temperatures near room temperature has been observed in several material systems, indicating that it is highly feasible to exploit them for solid-state cooling applications. An MLCC or MLPP prepared by using electrocaloric thin film materials with both large ECE and high thermal conductivity may be a promising candidate for practical cooling device applications. The combination of positive and negative ECEs could be helpful for the enhancement of the refrigeration efficiency. The combination of ECE and other caloric effect such as magnetocaloric effect could be also a new method to realize more effective refrigeration effect. New ECE materials possessing better properties will be expected.

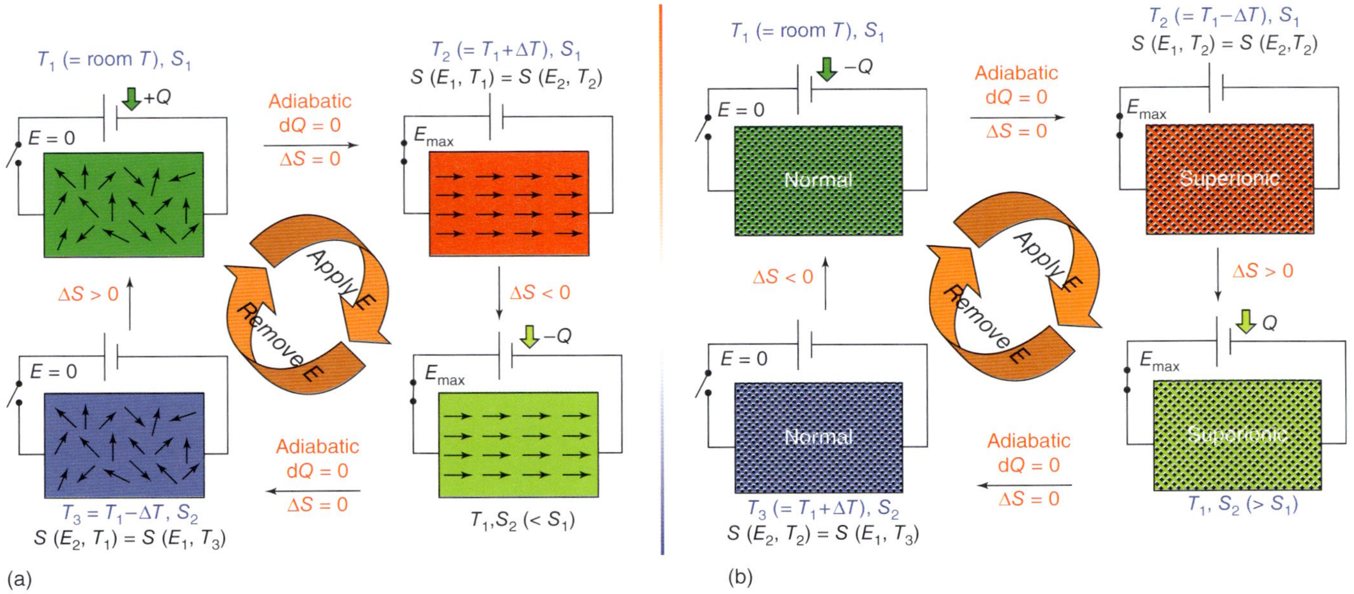

Figure 8.24 Positive and negative electrocaloric cooling cycles in (a) ferroelectrics and (b) fast ion conductors, respectively.

References

1 Li, X., Lu, S.-G., Chen, X.-Z. et al. (2013). *Journal of Materials Chemistry C* 1 (1): 23–37.
2 Akcay, G., Alpay, S.P., Rossetti, G.A., and Scott, J.F. (2008). *Journal of Applied Physics* 103 (2): 024104.
3 Zhang, D.W., Yao, Y.G., Fang, M.X. et al. (2016). *Acta Materialia* 103: 746–753.
4 Li, J.N., Zhang, D.W., Qin, S.Q. et al. (2016). *Acta Materialia* 115: 58–67.
5 Woldman, A.Y. and Landis, C.M. (2016). *Smart Materials and Structures* 25 (3): 035033.
6 Neese, B., Chu, B., Lu, S.-G. et al. (2008). *Science* 321 (5890): 821–823.
7 Amin, A., Cross, L.E., and Newnham, R.E. (1981). *Ferroelectrics* 37 (1–4): 647–650.
8 Amin, A., Newnham, R.E., Cross, L.E., and Cox, D.E. (1981). *Journal of Solid State Chemistry* 37 (2): 248–255.
9 Furukawa, T. (1984). *Ferroelectrics* 57 (1–4): 63–72.
10 Pirc, R., Kutnjak, Z., Blinc, R., and Zhang, Q.M. (2011). *Applied Physics Letters* 98 (2): 021909.
11 Furukawa, T., Tajitsu, Y., Zhang, X., and Johnson, G.E. (1992). *Ferroelectrics* 135 (1–4): 401–417.
12 Kar-Narayan, S. and Mathur, N.D. (2010). *Journal of Physics D-Applied Physics* 43 (3): 032002.
13 Lu, S.G., Rozic, B., Zhang, Q.M. et al. (2010). *Applied Physics Letters* 97 (20): 202901.
14 Lu, S.G., Rozic, B., Zhang, Q.M. et al. (2010). *Applied Physics Letters* 97 (16): 162904.
15 Li, X., Qian, X.-S., Lu, S.G. et al. (2011). *Applied Physics Letters* 99 (5): 052907.
16 Lu, S.G., Rozic, B., Zhang, Q.M. et al. (2011). *Applied Physics Letters* 98 (12): 122906.
17 Lu, S.G., Rozic, B., Zhang, Q.M. et al. (2012). *Applied Physics a-Materials Science & Processing* 107 (3): 559–566.
18 Li, J.T., Bai, Y., Qin, S.Q. et al. (2016). *Applied Physics Letters* 109 (16): 162902.
19 Lu, B., Wen, X., Tang, Z. et al. (2016). *Science China-Technological Sciences* 59 (7): 1054–1058.
20 Khassaf, H., Mantese, J.V., Bassiri-Gharb, N. et al. (2016). *Journal of Materials Chemistry C* 4 (21): 4763–4769.
21 Geng, W., Liu, Y., Meng, X. et al. (2015). *Advanced Materials* 27 (20): 3165–3169.
22 Wu, H.-H., Zhu, J., and Zhang, T.-Y. (2015). *Physical Chemistry Chemical Physics* 17 (37): 23897–23908.
23 Wu, H.-H., Zhu, J., and Zhang, T.-Y. (2015). *Nano Energy* 16: 419–427.
24 Park, M.H., Kim, H.J., Kim, Y.J. et al. (2016). *Advanced Materials* 28 (36): 7956–7961.

25 Zhuo, F., Li, Q., Gao, J. et al. (2016). *Applied Physics Letters* 108 (8): 082904.
26 Bai, Y., Zheng, G.P., and Shi, S.Q. (2011). *Materials Research Bulletin* 46 (11): 1866–1869.
27 Zheng, G.P., Uddin, S., Zheng, X.C., and Yang, J.H. (2016). *Journal of Alloys and Compounds* 663: 249–255.
28 Vats, G., Kumar, A., Ortega, N. et al. (2016). *Energy & Environmental Science* 9 (4): 1335–1345.
29 Qian, X., Yang, T., Zhang, T. et al. (2016). *Applied Physics Letters* 108 (14): 142902.
30 Ponomareva, I. and Lisenkov, S. (2012). *Physical Review Letters* 108 (16).
31 Herchig, R., Chang, C.M., Mani, B.K., and Ponomareva, I. (2015). *Scientific Reports* 5.
32 Prosandeev, S., Ponomareva, I., and Bellaiche, L. (2008). *Physical Review B* 78 (5).
33 Lisenkov, S. and Ponomareva, I. (2012). *Physical Review B* 86 (10).
34 Alpay, S.P., Mantese, J., Trolier-McKinstry, S. et al. (2014). *MRS Bulletin* 39 (12): 1099–1109.
35 Li, B., Wang, J.B., Zhong, X.L. et al. (2013). *EPL* 102 (4): 47004.
36 Correia, T.M., Kar-Narayan, S., Young, J.S. et al. (2011). *Journal of Physics D-Applied Physics* 44 (16): 165407.
37 Wang, X., Wu, J., Dkhil, B. et al. (2017). *Rsc Advances* 7 (10): 5813–5820.
38 Qian, X.S., Ye, H.J., Zhang, Y.T. et al. (2014). *Advanced Functional Materials* 24 (9): 1300–1305.
39 Bai, Y., Ding, K., Zheng, G.P. et al. (2012). *Physica Status Solidi A-Applications and Materials Science* 209 (5): 941–944.
40 Bhaumik, I., Ganesamoorthy, S., Bhatt, R. et al. (2014). *EPL* 107 (4): 47001.
41 Le Goupil, F., Axelsson, A.-K., Dunne, L.J. et al. (2014). *Advanced Energy Materials* 4 (9): 1301688.
42 Han, F., Bai, Y., Qiao, L.-J., and Guo, D. (2016). *Journal of Materials Chemistry C* 4 (9): 1842–1849.
43 Kaddoussi, H., Gagou, Y., Lahmar, A. et al. (2015). *Solid State Communications* 201: 64–67.
44 Asbani, B., Gagou, Y., Dellis, J.L. et al. (2016). *Solid State Communications* 237: 49–54.
45 Le Goupil, F., McKinnon, R., Koval, V. et al. (2016). *Scientific Reports* 6.
46 Bai, Y., Han, X., Zheng, X.-C., and Qiao, L. (2013). *Scientific Reports* 3.
47 Zhou, Y., Lin, Q., Liub, W., and Wang, D. (2016). *Rsc Advances* 6 (17): 14084–14089.
48 Singh, G., Bhaumik, I., Ganesamoorthy, S. et al. (2013). *Applied Physics Letters* 102 (8): 082902.
49 Le Goupil, F., Axelsson, A.K., Valant, M. et al. (2014). *Applied Physics Letters* 104 (22): 222911.
50 Liu, Y., Wei, J., Janolin, P.-E. et al. (2014). *Applied Physics Letters* 104 (16): 162904.
51 Kriaa, I., Abdelmoula, N., Maalej, A., and Khemakhem, H. (2015). *Journal of Electronic Materials* 44 (12): 4852–4856.
52 Rozic, B., Malic, B., Ursic, H. et al. (2011). *Ferroelectrics* 421: 103–107.

53 Ye, H.-J., Qian, X.-S., Lu, J. et al. (2015). *Ieee Transactions on Dielectrics and Electrical Insulation* 22 (3): 1501–1505.
54 Bai, Y., Zheng, G.P., Ding, K. et al. (2011). *Journal of Applied Physics* 110 (9).
55 Zhang, L., Liu, M., Ren, W. et al. (2017). *Rsc Advances* 7 (14): 8388–8393.
56 Zhao, Y., Hao, X., and Zhang, Q. (2015). *Journal of Materials Chemistry C* 3 (8): 1694–1699.
57 Peng, B., Zhang, Q., Ke, S. et al. (2015). *Ceramics International* 41 (8): 9344–9349.
58 Peng, B., Zhang, Q., Li, X. et al. (2015). *Journal of Alloys and Compounds* 646: 597–602.
59 Mischenko, A.S., Zhang, Q., Scott, J.F. et al. (2006). *Science* 311 (5765): 1270–1271.
60 Luo, L., Chen, H., Zhu, Y. et al. (2011). *Journal of Alloys and Compounds* 509 (32): 8149–8152.
61 Qiu, J.-H., Wang, X.-Q., Yuan, N.-Y., and Ding, J.-N. (2015). *Communications in Theoretical Physics* 64 (1): 113–118.
62 Luo, L., Dietze, M., Solterbeck, C.-H. et al. (2012). *Applied Physics Letters* 101 (6): 062907.
63 Peng, B., Fan, H., and Zhang, Q. (2013). *Advanced Functional Materials* 23 (23): 2987–2992.
64 Chen, H., Ren, T.L., Wu, X.M. et al. (2009). *Applied Physics Letters* 94 (18): 182902.
65 Bai, Y., Zheng, G.P., and Shi, S.Q. (2010). *Applied Physics Letters* 96 (19): 192902.
66 Pirc, R., Rozic, B., Koruza, J. et al. (2014). *EPL* 107 (1): 17002.
67 Ye, M., Li, T., Sun, Q. et al. (2016). *Journal of Materials Chemistry C* 4 (16): 3375–3378.
68 Zhang, G., Li, Q., Gu, H. et al. (2015). *Advanced Materials* 27 (8): 1450.
69 Zhang, G., Zhang, X., Huang, H. et al. (2016). *Advanced Materials* 28 (24): 4811–4816.
70 Qian, X.-S., Lu, S.-G., Li, X. et al. (2013). *Advanced Functional Materials* 23 (22): 2894–2898.
71 Cazorla, C. (2015). *Results in Physics* 5: 262–263.
72 Solovyev, A.A., Shipilova, A.V., Ionov, I.V. et al. (2016). *Journal of Electronic Materials* 45 (8): 3921–3928.
73 Solov'ev, A.A., Shipilova, A.V., Koval'chuk, A.N. et al. (2016). *Russ. J. Electrochem.* 52 (7): 662–668.

9

Ferroelectrics in Photocatalysis

Liang Fang[1], *Lu You*[2], *and Jun-Ming Liu*[3]

[1] *Soochow University, School of Physical Science and Technology and Jiangsu Key Laboratory of Thin Films, 333 Ganjiang Road, Suzhou 215006, PR China*
[2] *Nanyang Technological University, School of Materials Science and Engineering, 50 Nanyang Avenue, Singapore 639798, Singapore*
[3] *Nanjing University, Laboratory of Solid State Microstructures and Innovation Center of Advanced Microstructures, 22 Hankou Road, Nanjing 210093, PR China*

9.1 Introduction

Since the seminal work of Fujishima and Honda in 1972 [1], semiconductor-based photocatalytic technology has been extensively investigated in order to address the environmental issues and energy crisis due to the decomposition of toxic pollutants [2], including photo-generation of electricity and photoelectrochemical production of renewable hydrogen from water [3, 4]. Indeed, substantial progress has been made, and the relevant research activities constitute a major portion of the discipline of semiconducting materials and applications. From the point of view of semiconductor photochemistry, the role of photocatalysis is to initiate or accelerate specific reduction and oxidation reactions using photoactive semiconductors as the catalysts. In these systems, a flux of photons is absorbed by a semiconductor, generating electron–hole pairs in the semiconductor. These pairs are effectively separated, allowing the two types of charge carriers, electrons and holes, to diffuse toward catalytically active sites at the semiconductor/reaction species interface where the chemical transformations are driven.

Although various semiconducting oxide materials (TiO_2, $BiVO_4$, WO_3, and Fe_2O_3) have been explored for a long time in terms of many photocatalytic applications [5–8], significant challenges remain for realistic industrialization. The central topic has been the improvement of degradation or energy conversion efficiency, which is closely related to the energy band mismatch of the semiconductor materials, the effective separation and transfer of the photoexcited electrons and holes, and the surface energy or chemisorption properties at the photocatalyst surfaces. In general, the separation and transfer of charge carriers lies at the core of the whole photocatalysis process [9]. Traditional strategies employ the built-in electric fields associated with the hetero-interfaces, such

Ferroelectric Materials for Energy Applications, First Edition.
Edited by Haitao Huang and James F. Scott.

as photocatalyst/reaction species interfaces [10], photocatalyst/metal junctions [11], and p–n junctions in heterojunction photocatalyst systems [12].

Aside from the built-in fields in heterojunctions, a permanent and controllable internal electric field if any is highly appreciated. This electric field can be established from ferroelectric (semiconducting) materials because of their spontaneous ferroelectric polarization [13]. The local field can be very large, thus effectively separating the photoexcited carriers with opposite polarities [14]. Moreover, switchable polarization provides an additional degree of freedom for controlling the local band alignment at ferroelectric/reaction species interface, and may promote movement of different carriers to spatially specific locations [15–18]. This may lead to some interesting interactions at the surface such as spatially selective adsorption and surface redox reactions. These unique properties of ferroelectric semiconductors make them promising candidates for efficient photocatalysis.

In this chapter, we start with a brief introduction of the fundamental physics of photocatalysis, especially the mechanisms of photocatalytic degradation and photoelectrochemical water-splitting. Then, we focus on the effect of an internal electric field/polarization on charge carrier separation in ferroelectric semiconductors, followed by a discussion of the effect of ferroelectric polarization on surface photochemistry, addressing the processes of adsorption and spatially selective oxidation/reduction reactions. Subsequently, the most recent advances in the research on ferroelectric materials as photocatalytic candidates are highlighted, especially pertinent to photocatalytic degradation and photoelectrochemical water-splitting.

9.2 Fundamental Principles of Semiconductor Photocatalysis

The fundamentals of photophysics and photochemistry underlying the photocatalysis process have been discussed extensively in textbooks and literature, and will only be briefly summarized here [19–22]. Without losing the generality, the semiconductor photocatalytic degradation of water pollutants comprises three steps (see Figure 9.1). First is photoexcitation: when photons with energy ($h\nu$) above the band gap of semiconductor irradiate (illumination), electrons will be excited from the valence band (VB) to the conduction band (CB), generating equal numbers of free electrons and vacant sites (holes). Second is the separation and transport of charged carriers: the excited electrons and holes are separated and migrate to the surfaces, while the electron–hole annihilation becomes inevitable and should be suppressed as much as possible. Third is the collection of charged carriers: the photo-generated electrons could reduce the dye or react with electron acceptors such as O_2 adsorbed on the semiconductor surface or dissolved in water, resulting in superoxide radical anion $^{\cdot}O_2^-$. The photo-generated holes can oxidize the organic molecules, or react with OH^- or H_2O into $^{\cdot}OH$ radicals. The as-generated $^{\cdot}OH$ radical, a very strong oxidizing agent (standard redox potential +2.8 V), can oxidize most of azo dyes into mineral end-products.

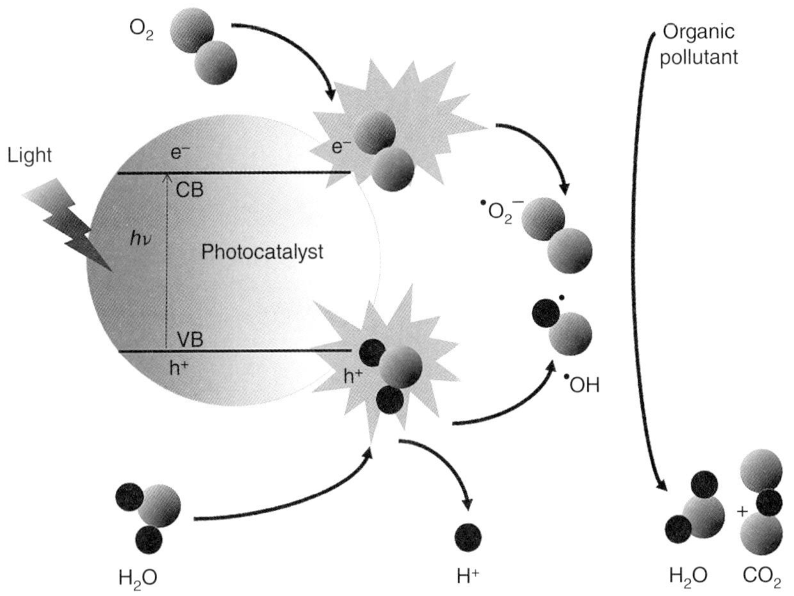

Figure 9.1 A schematic of the photoinduced generation mechanism for electron–hole pairs in semiconductors with presence of a water pollutant.

This set of chain oxidative–reductive reactions (Eqs. (9.1)–(9.8)) have been widely postulated as the following:

$$\text{Semiconductor} + h\nu \rightarrow \text{Semiconductor}(e^- + h^+) \tag{9.1}$$

$$\text{Semiconductor}(h^+) + H_2O \rightarrow \text{Semiconductor} + H^+ + {}^{\cdot}OH \tag{9.2}$$

$$\text{Semiconductor}(h^+) + OH^- \rightarrow \text{Semiconductor} + {}^{\cdot}OH \tag{9.3}$$

$$\text{Semiconductor}(e^-) + O_2 \rightarrow \text{Semiconductor} + {}^{\cdot}O_2{}^- \tag{9.4}$$

$${}^{\cdot}O_2{}^- + H^+ \rightarrow {}^{\cdot}HO_2 \tag{9.5}$$

$$\text{Dye} + {}^{\cdot}OH \rightarrow \text{degradation products} \tag{9.6}$$

$$\text{Dye} + h^+ \rightarrow \text{oxidation products} \tag{9.7}$$

$$\text{Dye} + e^- \rightarrow \text{reduction products} \tag{9.8}$$

Solar water-splitting is another important application of photocatalysis, and the relevant approaches can be divided into two major categories:

1. *Photocatalytic generation of hydrogen*: By this term, we usually mean the generation of hydrogen by heterogeneous photocatalysis using powder photocatalysts with a co-catalyst [23]. Such a reaction process is similar to photocatalytic degradation of the water pollutant, as shown in Figure 9.2a.

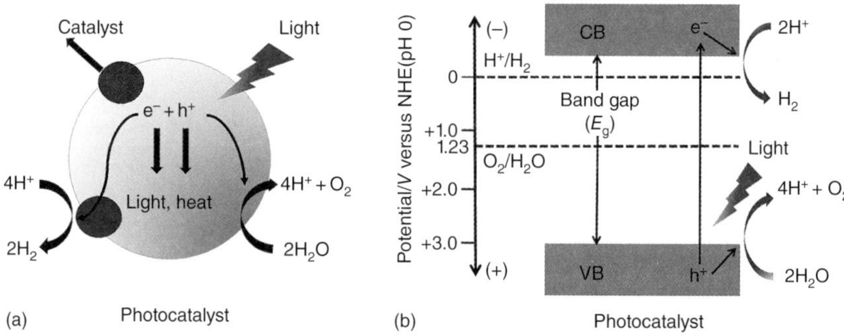

Figure 9.2 (a) A schematic illustration of the fundamental mechanism for a semiconducting photocatalytic water-splitting process. (b) The energy diagrams for photocatalytic water-splitting.

The photocatalyst is excited by absorbing photons, creating electron–hole pairs. The photo-generated electrons and holes cause redox reactions similar to electrolysis. Water molecules are reduced by the electrons to form H_2 and are oxidized by the holes to form O_2, constituting the water-splitting. The important characteristic of a semiconductor photocatalyst for water-splitting is that the energy band of the semiconductor should straddle the redox potentials of water. Specifically, the bottom level of the conduction band must be more negative than the redox potential of H^+/H_2 (0 V versus normal hydrogen electrode (NHE)), while the top level of the valence band must be more positive than the redox potential of O_2/H_2O (1.23 V), as shown in Figure 9.2b. Therefore, the theoretically predicted minimum band gap for water-splitting is 1.23 eV, but not all semiconductors that satisfy the above conditions can decompose water into H_2 and O_2. In addition, the powder photocatalyst system may be simple; the gas products H_2 and O_2 are mixed during the splitting reaction, thus requiring a separation process before the as-generated H_2 can be effectively utilized.

2. *Photoelectrochemical generation of hydrogen*: In this case, hydrogen is produced in a photoelectrochemical cell [24, 25], which consists of three components: (i) the operating electrode (photoelectrode) based on a semiconductor photocatalyst; (ii) the counter electrode, which is typically Pt or graphite; (iii) the electrolyte, which is sandwiched by the two electrodes in order to enhance the electrical conductivity and maintain the required pH value. Figure 9.3 illustrates the fundamental processes in a photoelectrochemical cell as an example of such a two-electrode system. In the n-type semiconductor photoelectrode (photoanode), photo-excited electrons and holes are generated in the conduction band and valence band, respectively. When an anodic potential is applied through an external circuit, these electrons are driven via the real Ohmic contact toward an electrical connection onto the surface of the counter electrode, and then the hydrogen evolution reaction occurs, thereby reducing H^+ into H_2. Meanwhile, the holes drive the oxygen evolution reaction at the surface of the semiconductor operating electrode, generating O_2. Conversely, a p-type semiconductor usually acts as

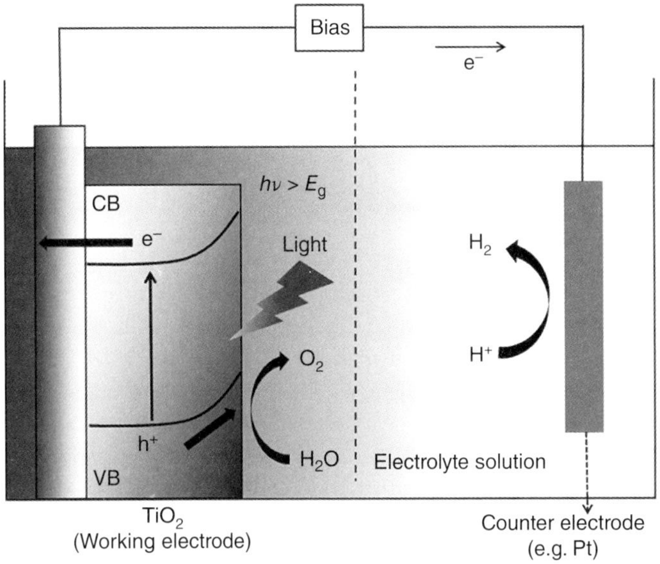

Figure 9.3 A schematic illustration of the fundamental mechanism for a semiconducting photoelectrochemical water-splitting process.

the photocathode for hydrogen evolution and the counter electrode acts as a photoanode for oxygen evolution [26]. Since the oxidation site is spatially separated from the reduction site, O_2 and H_2 can be easily separated in a photoelectrochemical cell.

9.3 Advances in Understanding Ferroelectric Photocatalytic Mechanisms

Ferroelectric materials are characterized by a spontaneous electric polarization that can be switched by electric field. This spontaneous polarization is permanent and originates from spontaneous spatial inversion symmetry breaking modes such as off-center ionic displacement (displacive type)/dipole ordering (order–disorder type) in the non-centrosymmetric crystals. From crystal symmetry consideration, out of the total 32 crystalline classes, 21 classes lack the center of symmetry (non-centrosymmetric), within which 20 are piezoelectric. Of these, 10 classes are polar with spontaneous polarization (pyroelectric). A material is ferroelectric only if the polarization is switchable. By its physical origin, ferroelectrics can be further categorized into "proper" and "improper" types. For the former case, structural-instability-driven polar state is the main order parameter in the phase transitions, whereas the polarization in the latter case is the consequence of other order parameters involved in the phase transitions [27–30].

Spontaneous polarization generates bound charges on the ferroelectric surfaces and interfaces. The sign and magnitude of the bound charges depend

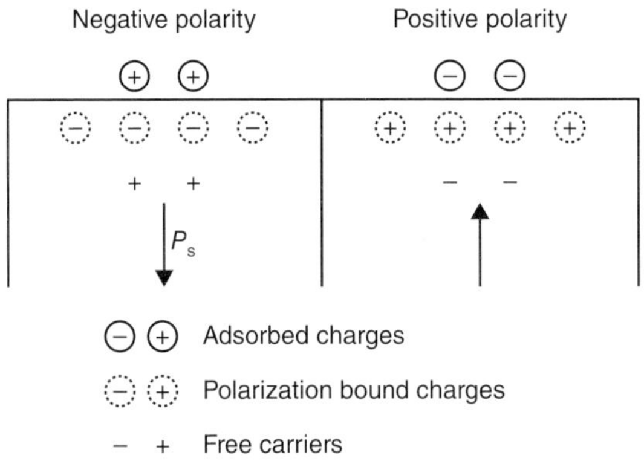

Figure 9.4 A schematic diagram of external and internal screening mechanisms by adsorbing charges and free carriers/defects in ferroelectric materials. *Source:* Yang et al. 2005 [31]. Reproduced with permission of IOP.

on the orientation of polarization that couples to the crystallographic axis of ferroelectric materials for most cases. These bound charges are not energetically stable and will be compensated mainly in two ways: internal screening and/or external screening [31]. As shown in Figure 9.4, internal screening is normally realized by adsorption of free charge carriers from the ambient or electrode and charged defects with opposite sign within the materials, whereas external screening is achieved by the adsorption of oppositely charged ions and molecules from the environment.

In ferroelectrics, internal screening by free charge carriers and defects in the bulk will make the charge carriers near the surface to redistribute, causing band bending and a space charge region. The direction of the band bending depends on the polarity of surface-bound charges. Figure 9.5 shows schematically the band bending near the ferroelectric surface/interface upon different spontaneous polarizations [32]. For example, on a surface with negative polarity (C^- domain), electrons are further depleted from the surface, giving rise to a spatial charge layer with upward band bending. This charge layer is thus called the depletion layer in which the motion of photo-generated electrons toward the surface is blocked but the movement of photo-generated holes toward the

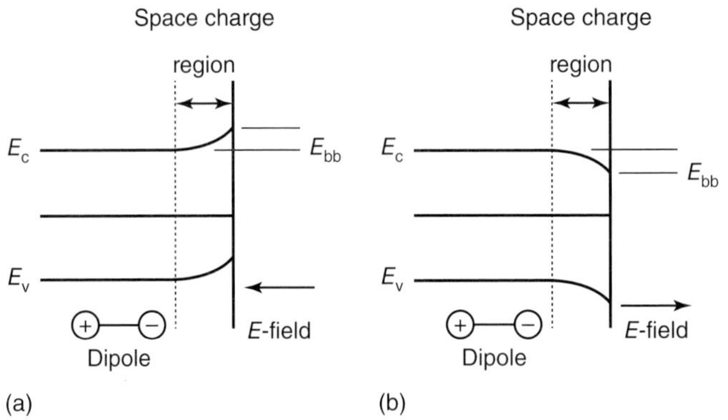

Figure 9.5 A schematic diagram of band bending in a ferroelectric material: a surface with negative polarity (a) and a surface with positive polarity (b). *Source:* Jones and Dunn 2007 [32]. Reproduced with permission of IOP.

surface is driven. In contrast, on the surface with positive polarity (C^+ domain), electrons are accumulated for screening and consequently the downward band bending is formed, accompanied by a spatial charge layer called the accumulation layer. The deformed movements of electrons and holes represent the unique photochemical properties of ferroelectric materials. Additionally, the width of the space charge region (L_d is the depletion layer, L_a is the accumulation layer) can be written as a function of surface potential V_s and Debye length L_D [33]:

$$L_d = \sqrt[2]{\frac{2eV_s}{k_B T}} L_D \tag{9.9}$$

$$L_a = \sqrt{2}\left(1 - \exp\left[\frac{eV_s}{2k_B T}\right]\right) L_D \tag{9.10}$$

where k_B is the Boltzmann constant and T is the absolute temperature. The surface potential V_s depends on not only the polarization but also the position of conduction band edge as well as the adsorption layer on the ferroelectric surface. The Debye length L_D can be written as [33]

$$L_D = \left(\frac{\varepsilon_0 \varepsilon_r k_B T}{e^2 N_D}\right)^{1/2} \tag{9.11}$$

where ε_0 is the permittivity of free space, ε_r is the dielectric constant of the material, and N_D is the donor density. According to Eq. (9.11), we can see that a higher dielectric constant and smaller charge carrier density allow a larger Debye length and a wider space charge region, thus benefiting the separation of photo-generated carriers.

If the space charge region due to the internal screening in ferroelectric materials does not fully compensate the spontaneous polarization, the depolarization field will penetrate more deeply into the environment. Therefore, external screening mechanism by adsorption of charged molecules or ions from the atmosphere will occur, leading to the so-called Stern layer on the surface [34]. The Stern layer consists of the inner Helmholtz layer and outer Helmholtz layer, which will bias the interactions between the charge carriers underneath the surface and the cations/anions in the external environment.

The above discussion allows us to conclude that all these ingredients, including the spontaneous polarization, associated surface charges, band bending, and external/internal screening effects, may take effect simultaneously, contributing to some particular surface properties of the ferroelectric materials. Along this line, an appropriate control of the spontaneous polarization would possibly direct the photo-generated charge carriers to the sites for oxidation and reduction. In the next several sections, we will have a review on recent literature about the effect of ferroelectric polarization on redox reactions, and photocatalytic and photoelectrochemical activities.

9.4 Photochemistry of Ferroelectric Materials

The photochemical properties of ferroelectric materials were originally studied by Nasby and Quinn [35], who showed the pH dependence of the flat-band potential of $BaTiO_3$, which has excellent photostability and a large photocurrent. This study raised a number of open questions at that time, such as how the ferroelectric nature of $BaTiO_3$ modulates the surface photochemistry. The surface photochemistry of ferroelectrics has remained mostly unexplored for two decades until around 2000 when Giocondi and Rohrer published a series of papers on the variations of surface photochemistry for $BaTiO_3$ [36, 37].

From then on, many experiments have utilized the so-called marker reactions that leave insoluble products at the reaction site in order to determine the correlation between polarization and reactivity. For example, for a $BaTiO_3$ single crystal under the illumination of UV light in the presence of aqueous $AgNO_3$ and $Pb(C_2H_3O_2)$ [36, 37], the oxidized Pb and reduced Ag accumulate on the surfaces of domains with opposite polarization, as shown in Figure 9.6. If the domain polarization has its positive end directed toward the surface Ag is reduced, and Pb is oxidized if the polarization is reversed. The image in Figure 9.6b shows that silver is reduced on the positive domains and the image in Figure 9.6c shows that lead is oxidized on the negative domains, indicating that it is the ferroelectric polarization that determines the location at which the redox reaction occurs and separates the two half reactions at different domains. The reactions occurring on different domains are described by

$$Ag^+ + e^- = Ag \tag{9.12}$$

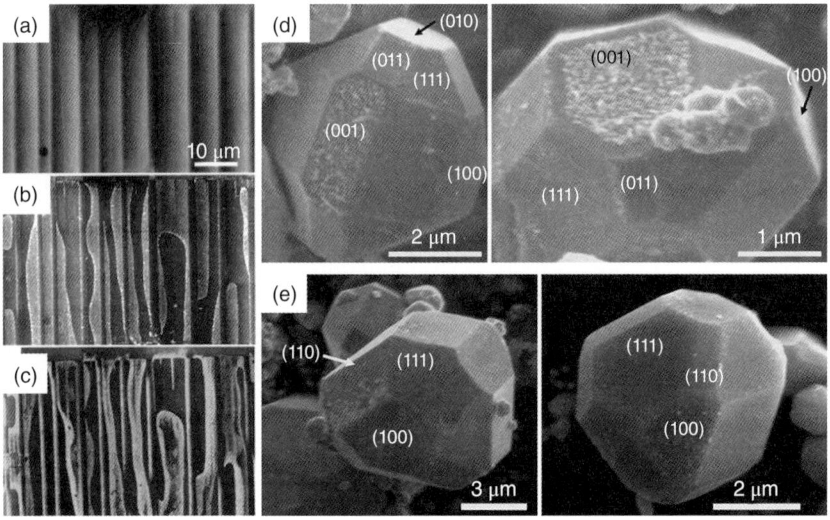

Figure 9.6 Topographic atomic force microscopy (AFM) images of the {001} surfaces of a $BaTiO_3$ single crystal: (a) Clean surfaces before the reactions. (b) The same surface area as in (a) after illumination in an aqueous $AgNO_3$ solution. The white contrast corresponds to the deposited Ag particles. (c) The same area as in (b) after a cleaning of those Ag particles and then illumination in an aqueous lead acetate solution. The white contrast corresponds to Pb-containing deposits [36]. (d) The SEM images of faceted $BaTiO_3$ crystals after reaction in aqueous $AgNO_3$, where the speckled white contrast corresponds to Ag-containing deposits. (e) The SEM images of faceted $BaTiO_3$ crystals after reaction in aqueous $Pb(C_2H_3O_2)_2$. The white contrast corresponds to oxidized Pb-containing deposits. *Source:* Giocondi and Rohrer 2008 [38]. Reproduced with permission of Springer Nature.

$$Pb^{2+} + 2H_2O + 2h^+ = PbO_2 + 4H^+ \tag{9.13}$$

The same process can also occur in $BaTiO_3$ microcrystals or ceramics [38, 39]. Evidence for the silver reduction was most prominently found on the {001} surface, while that for lead oxidation was found on the {100}, {110}, and {111} surfaces. These images show that the patterns of reduced silver on the surface were consistent with the orientations of the underlying ferroelectric domains.

Rohrer and coworkers also studied the photochemical reduction of Ag by $Ba_{1-x}Sr_xTiO_3$ solid solution with $x = 0–1.0$ [33]. Although the reduction of Ag on $BaTiO_3$ is spatially selective and correlated to the locations of positive ferroelectric domains, the reduction of Ag on $SrTiO_3$ is spatially homogeneous due to the non-ferroelectric nature of $SrTiO_3$ [40]. In order to quantify the reactivity, atomic force microscopy (AFM) was used to measure the relative heights of the silver deposits. A selection of AFM images recorded after the reactions on $Ba_{1-x}Sr_xTiO_3$ samples is shown in Figure 9.7. The white contrast corresponds to the deposited silver. For the sample with $x = 0.2$ (Figure 9.7a), the silver is concentrated in stripe-like pattern, similar to some of the features on pure $BaTiO_3$. With increasing amount of strontium, the selectivity for silver reduction in active versus inert domains is weakened. For the sample with $x = 0.26$ (Figure 9.7b), which showed the highest reactivity, silver deposition seems to be spatially homogeneous on the surface. It is difficult to distinguish the domain pattern although the Ag stripe-like pattern can still be more or less identified along the vertical direction. For the sample with $x = 0.27$, the spatially preferred reactivity at some locations rather than others can be observed, as shown in Figure 9.7c. For the sample with $x = 0.4$, the reactivity is spatially uniform at all locations, as shown in Figure 9.7d.

The transition from spatially selective reactivity to spatially uniform reactivity is at least partially associated with reduced polarization as the strontium concentration increases. As implied in Eqs. (9.9)–(9.11), the reduced polarization in both domains will lower the surface potential, leading to the shrinking of the space charge region (width). For larger x, the polarization decreases precipitously and disappears, leading to loss of reactivity. This explanation agrees with the overall trend, although the local maximum in the reactivity at $x = 0.26$, corresponding to the cubic-to-tetragonal phase transformation, remains an issue to be understood. The maximum is most likely explained by an anomalously high dielectric constant at this phase-boundary composition that enlarges the space charge region, which was also observed in $Ba_xSr_{1-x}TiO_3$ epitaxial films [41].

The polarization-dependent photochemical properties of ferroelectric materials have also been revealed in $Pb(Zr,Ti)O_3$ (PZT) [32, 34, 42–48]. Various metal salts on the surface of PZT were found to be photochemically reactive, and the location of reaction products depends on the domain structure. Chen et al. reported that the preferential orientation of polarization in the as-grown PZT epitaxial films on $SrTiO_3$ (STO) substrate could be manipulated by choosing proper bottom electrodes [47]. As shown in Figure 9.8, the PZT epitaxial films deposited on the $SrRuO_3$ (SRO)-buffered STO substrates preferred upward polarization (C^+), while the epitaxial films, if deposited on the $(La,Sr)CoO_3$ (LSCO) buffered STO substrates, are polarized downward (C^-). The experiments

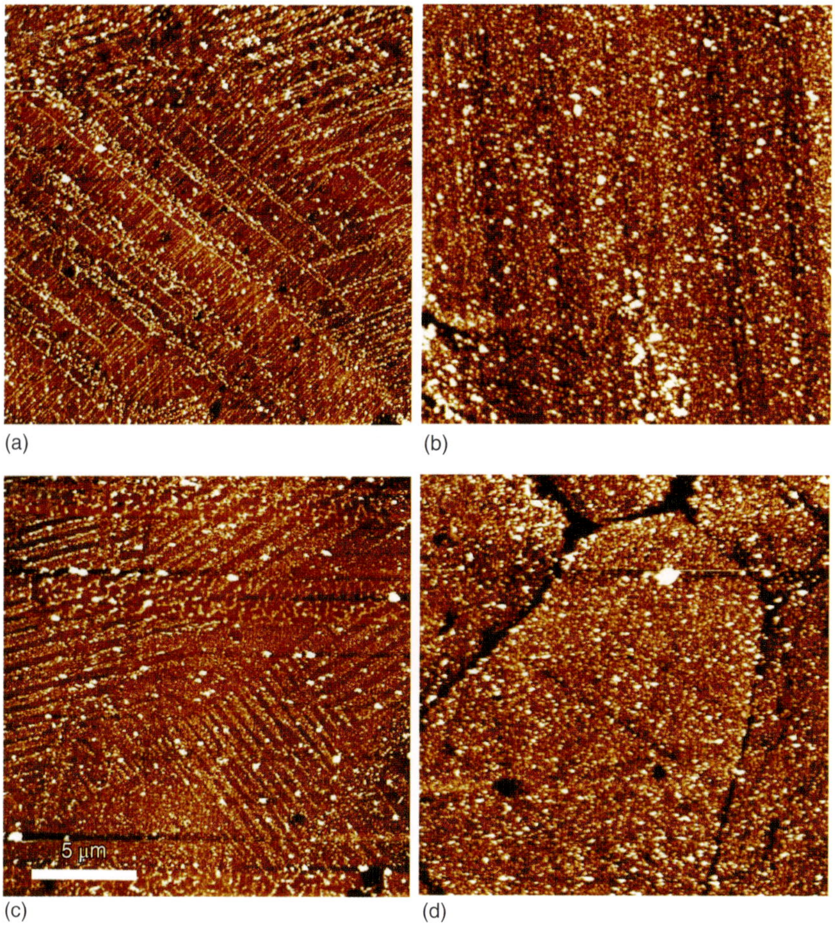

Figure 9.7 AFM images of four $Ba_{1-x}Sr_xTiO_3$ samples with different compositions after reaction and then being covered with silver particles: (a) $x = 0.2$, the black-to-white contrast is 95 nm in height difference; (b) $x = 0.26$, the black-to-white contrast is 60 nm in height difference; (c) $x = 0.27$, the black-to-white contrast is 60 nm in height difference; and (d) $x = 0.4$, the black-to-white contrast is 110 nm in height difference. All the images have a 20-µm field of view and a scale mark is shown in (c). *Source:* Bhardwaj et al. 2010 [33]. Reproduced with permission of ACS.

on UV irradiation of the PZT films in aqueous $AgNO_3$ did reveal more active sites for Ag reduction on the C^+-surfaces than on the C^--surfaces, evidenced by a large difference in the density of deposited silver nanoparticles. Such large difference could be obviously attributed to the band bending at the bottom interface, which depends on the difference in work function between PZT and the bottom electrodes (SRO or LSCO). Another example is the PZT polycrystalline films deposited on indium tin oxide (ITO) coated quartz substrates [48], where both the barrier at the film/electrode interface and the depolarization field have influence on the photoreduction properties of Ag^+. The particle size and density

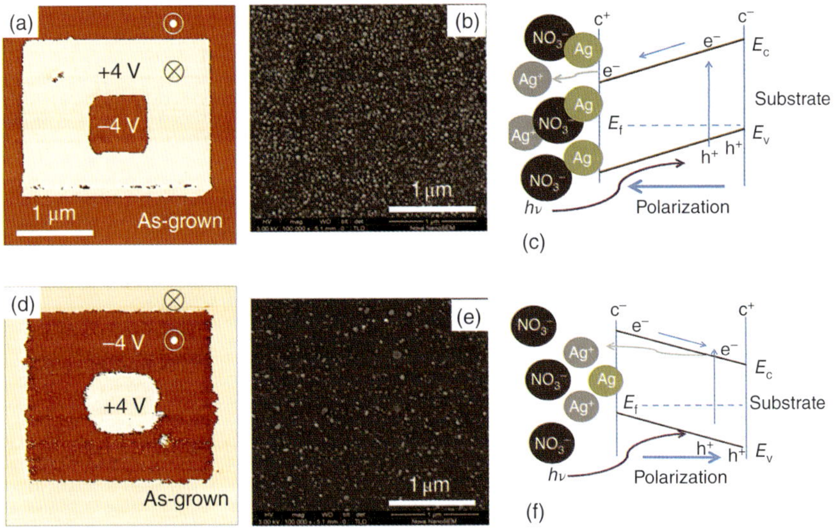

Figure 9.8 Piezoelectric force microscopy (PFM) phase images (a and d) of the PZT/SRO/STO and PZT/LSCO/STO heterostructures respectively. The PFM images indicate that the virgin state in PZT/SRO/STO heterostructure is positively polarized and the virgin state in PZT/LSCO/STO heterostructures is negatively polarized. Scanning electron microscopy (SEM) images (b and e) taken from the surfaces of the PZT/SRO/STO and PZT/LSCO/STO heterostructures. The SEM images obtained after UV irradiation show that the particle density on the surface of PZT/SRO/STO heterostructures is visibly higher. (c) The band structure for a positively polarized PZT thin film, showing the accumulation of electrons underneath the surface, which allows the reduction of Ag ions to occur more readily. (f) The band structure is sloped in the opposite direction for a negatively polarized PZT, causing the electrons to travel inside the film. *Source:* Chen et al. 2013 [47]. Reproduced with permission of AIP.

depend on the concentration of Zr and polarization direction of PZT, similar to the observations on $Ba_{1-x}Sr_xTiO_3$ [33, 41].

Similarly, ferroelectric $LiNbO_3$ was reported to be effective in driving the spatially selective reduction [49–51]. Nemanich et al. investigated the wavelength dependence of the as-deposited Ag nanostructures via photoinduced deposition using $LiNbO_3$ single crystals as templates. Unlike with $BaTiO_3$ and PZT, the Ag particles are reduced mainly on the domain boundaries rather than inside the domains. Figure 9.9 shows the AFM images of the Ag photoinduced deposition on $LiNbO_3$ upon different illumination times. It is evident that the Ag particles most favor the boundaries, then the surfaces of positive domains, and least favor the negative domain surfaces. In other words, more electrons are available at the boundaries than on the domain surfaces. It was proposed that $LiNbO_3$ has weak internal screening and the bound charges are mainly screened externally by the surface absorption of charged molecules. Therefore, the electric field is enhanced at the domain boundaries, driving the electrons toward the domain boundaries, and thus favors deposition of Ag nanoparticles on the boundaries.

The ferroelectrics mentioned above have relatively wide band gaps, and they absorb only a small fraction of the solar spectrum in the UV range. Given this fact, the overall efficiency suffers from the poor match of the energy levels with

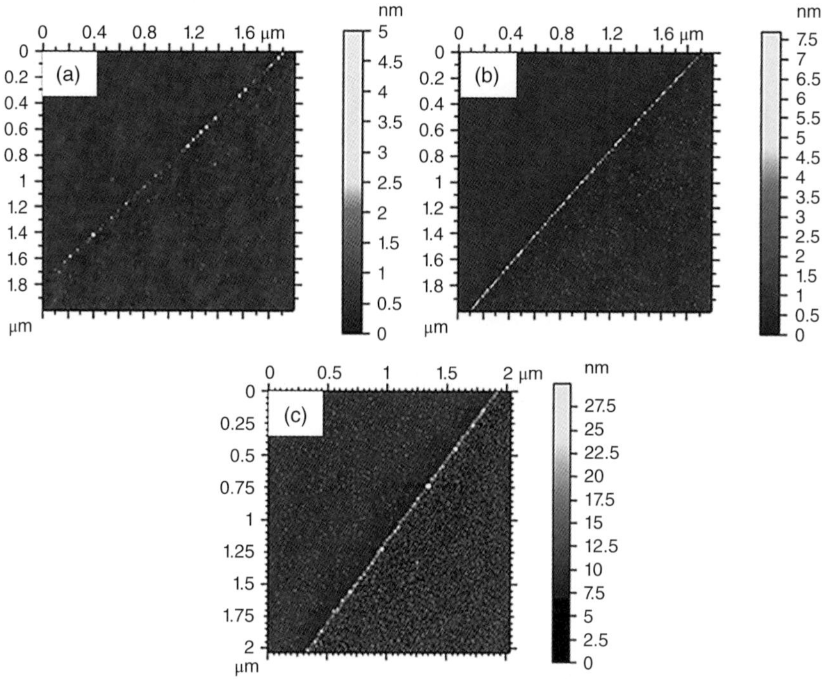

Figure 9.9 AFM images after Ag photoinduced deposition on $LiNbO_3$ surface using 314 nm illumination for 60 s (a), 100 s (b), and 240 s (c). *Source:* Sun and Nemanich 2011 [50]. Reproduced with permission of AIP.

the solar spectrum even if losses due to recombination and back reaction of photo-generated carriers are reduced. An exception is $BiFeO_3$, which is a narrow band gap ferroelectric with a broad spread from 2.0 to 3.0 eV [52, 53]. Schultz et al. reported the visible light reduction of silver from an aqueous solution on $BiFeO_3$ ceramic surfaces, and investigated the influences of crystalline and domain orientations [54]. Visible light excitation does allow $BiFeO_3$ to photochemically reduce silver cations from solution, and the deposited Ag structure shows some correspondence with the underlying ferroelectric domain structure. The observations revealed the preferred silver deposition on domains of positive polarization directed toward the surface, and a one-to-one correspondence between the amount of reduced silver and a polarization component normal to the surface were found, confirming the earlier discussed model proposed based on $BaTiO_3$ [36]. However, contrary to the observations on PZT thin films [38], the observed reduction behavior here is relatively insensitive to the crystal orientation, suggesting that the ferroelectric domain structure is more important than grain orientation in determining the surface reactivity. This work also explains how the band bending direction inside the $BiFeO_3$ near the interface with solution changes as a function of polarization, and how this facilitates the transfer of electrons to solution when the polarization is positive and suppresses it when the polarization is negative, as shown in Figure 9.10.

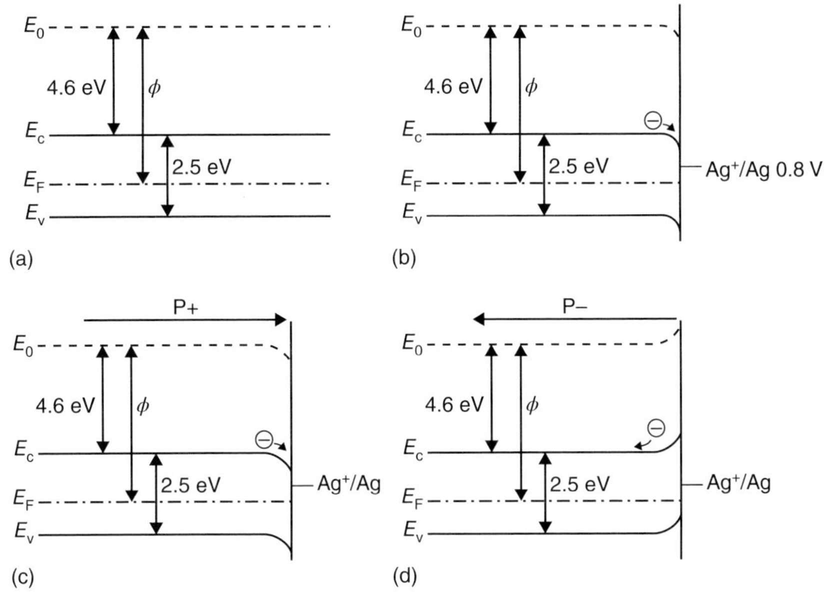

Figure 9.10 Schematic band diagrams for $BiFeO_3$ where ϕ is the work function, E_0, E_c, E_F, and E_v are the energy levels of a free electron, the conduction band edge, the Fermi level, and the valence band edge, respectively. (a) Bands of bulk $BiFeO_3$. (b) Bands of $BiFeO_3$ in contact with solution, with the standard redox potential for Ag^+/Ag versus normal hydrogen electrode. (c) Bands in contact with solution and a positive out-of-plane polarization. (d) Bands in contact with solution and a negative out-of-plane polarization. *Source:* Schult et al. 2011 [54]. Reproduced with permission of ACS.

Ferroelectric polarization can also influence the reactivity of oxide heterostructures as photochemical materials, specifically for cases of non-ferroelectric films on ferroelectric substrates where photoexcitation occurs on the surface of non-ferroelectric films. The physics is based on the fact that the effective electric field imposed by the ferroelectric substrate cannot be screened by carriers around the interface or from non-ferroelectric films. In this case, the charge carriers on the surface of non-ferroelectric films will be modulated. A representative case is TiO_2 thin films on ferroelectric $BaTiO_3$ substrates. Burbure et al. conducted a series of experiments to examine the photochemical behaviors of this heterostructure [55, 56]. TiO_2 films of 15–100 nm thickness were grown by pulsed laser deposition. As shown in Figure 9.11, the photo-generated charge carriers separated by ferroelectric polarization will participate in redox reactions (silver reduction and lead oxide oxidation) on different domains of the ferroelectric surface, and this influence can also be extended to the surface of TiO_2. In the (001)-oriented $BaTiO_3$ substrate with 15-nm anatase TiO_2 film, the reaction products show spatial patterns similar to the ferroelectric domains, noting that the ferroelectric surface is not exposed to the solution. Subsequently, photochemical reactivity as a function of not only the ferroelectric domain orientation of the $BaTiO_3$ substrate but also the thickness/orientation of the TiO_2 thin film was investigated. Certainly, one understands that the influence of

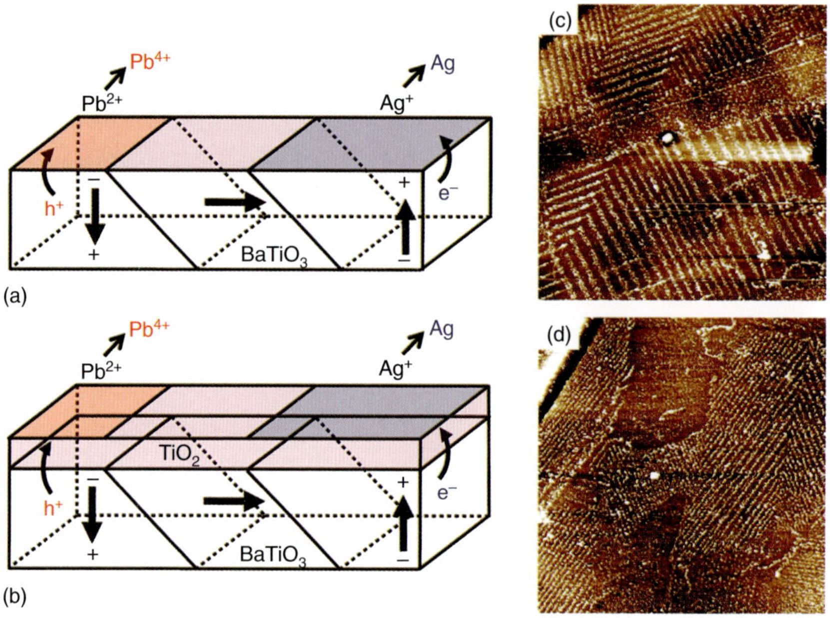

Figure 9.11 Schematic illustrations of domains in the substrate, indicating the promotion of the same half reactions on both the bare substrate (a) and the film surface (b). (c) The AFM images of 15-nm-thick (001)-oriented anatase TiO_2 film supported by (100)-oriented $BaTiO_3$ after a photochemical reaction in aqueous silver nitrate solution. (d) The AFM images of 15-nm-thick (100)-oriented rutile TiO_2 surface supported by (111)-oriented $BaTiO_3$ after a photochemical reaction in aqueous silver nitrate solution [55, 56].

substrate would eventually disappear when the TiO_2 film becomes very thick. It was shown that this thickness is larger than 100 nm. Zhang et al. also studied the oxide heterostructures consisting of thin TiO_2 films (10 nm) on $BiFeO_3$ substrates [57]. Similarly, the reduced silver structure on the TiO_2 surface does mimic the domain structure of $BiFeO_3$ substrate, under illumination with visible light. The observations indicated that electrons generated in the low band gap of $BiFeO_3$ are modulated by the dipolar fields in the ferroelectric domains, and they are transported through the thin TiO_2 film to assist the reduction of silver on the TiO_2 surface.

Furthermore, progress has been made in top–bottom approaches to ferroelectric photochemistry beyond the discussions above. Researchers utilize lithography to pattern the ferroelectric semiconductors into well-aligned nanostructures, so that the photochemical process can be enhanced and maximized [42, 58–61]. In these approaches, ferroelectric polarization is manipulated to control the local electronic structure and chemical activity, using electron beams or probe tips to pattern the domain structure. By utilizing the photoreduction process from aqueous solution, metal nanoparticles are deposited on predefined locations. Then the sample is exposed to a solution of functionalized molecules. The molecules are terminated on one end with a group that will react preferentially with the metal nanoparticles rather than the ferroelectric substrate. Here, a wide range of reactions are possible since a variety of metals can be deposited.

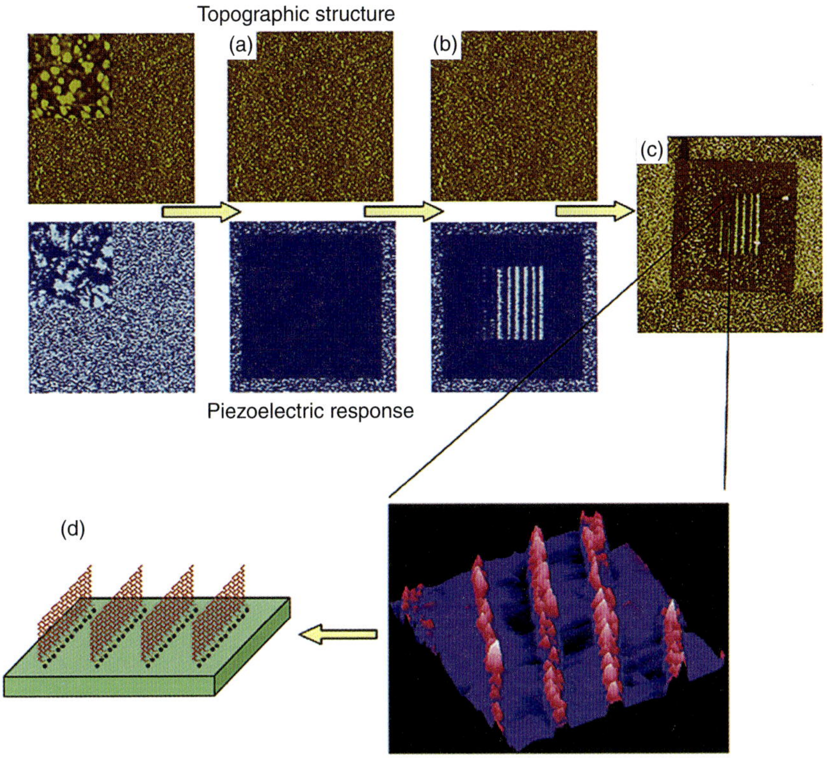

Figure 9.12 A clean background on PZT in the negative orientation is "written" by poling a large square area with +10 V (a). The nanostructure pattern inside this square area is produced with a local negative field −10 V (b). Subsequent Ag photo-deposition results in a pattern of nanoparticles (c). Particle size and spacing can be controlled by deposition time and conditions. Wire-like structures of aligned nanoparticles are fabricated on the smallest diameter domains (d). Reaction of this structure with dodecanethiol results in attachment of the long chained molecules to the nanoparticles. *Source:* Kalinin et al. 2002 [42]. Reproduced with permission of ACS.

Along this line, pioneering work was done by Kalinin et al. who investigated the domain-controlled deposition of various metals such as Au, Ag, and Pd via photoreduction on PZT and $BaTiO_3$ surfaces [42, 58]. Using the AFM tip poling/writing, they fabricated a set of ferroelectric domains that were 100 nm to a few microns in dimension. Consequently, a set of deposited metal nanostructures were obtained. As illustrated in Figure 9.12, a clean square background is first created from the random as-grown state by poling the polarization downward, followed by upward poling of an array of stripe-domains at the center of the square box. Afterwards, it was found that the Ag nanoparticles can be preferentially deposited on the upward-oriented stripe domains. Finally, this structure was exposed to a solution of dodecanethiol in isooctane, and a well-controlled substrate–particle–molecule structure was obtained.

To end this section, we would like to mention that so far all the experiments on photochemical growth of Ag and other metals on the surfaces of ferroelectrics including $BaTiO_3$, PZT, and $LiNbO_3$ were performed using illuminating

Figure 9.13 (a–c) AFM images of periodically poled $LiNbO_3$ single crystal surfaces after Ag photo-deposition at three excitation wavelength bands (3.4, 2.6, 2.3 eV) using the evanescent illumination geometry. (d–f) The zoomed-in images of (a–c). *Source:* Seal et al. 2014 [62]. Reproduced with permission of John Wiley & Sons.

sources of energy larger than the band gap of ferroelectrics. This is because the photo-deposition of metal nanoparticles depends on the generation and transport of illumination-excited free electrons, which can only occur upon above-band-gap excitation. On the other hand, ferroelectrics such as $BaTiO_3$, $LiNbO_3$, and $KTiOPO_4$ are widely utilized in nonlinear optical applications. A question deserving attention is whether the nonlinear optical effect affects the photochemical properties of ferroelectrics. This question was addressed by Seal et al. who reported anomalous photo-deposition of Ag on $LiNbO_3$ surface using the sub-band-gap excitation [62]. They observed that illumination of photons with energy below the band gap was also effective in driving the photoreduction of Ag nanoparticles on periodically poled $LiNbO_3$ single crystals, as shown in Figure 9.13. This anomalous photoreduction at the ferroelectric/reaction species interface could be explained by the generation of photons with higher energy via the mechanism of second harmonic generation, while details remain to be explored.

9.5 Photocatalytic Degradation Using Ferroelectric Materials

The unique polarization-dependent surface properties (e.g. bound charges) of ferroelectric materials and their semiconductor properties have been discussed in previous sections. These properties are certainly useful for photocatalysis applications, which are attractive. A number of investigations along this line have been reported. As a typical ferroelectric, $BaTiO_3$ was once employed in organic dye photocatalytic degradation. However, large band gap (~3.2 eV) is always a significant drawback that impedes the wide use of $BaTiO_3$, and this band gap can be overcome only by UV light excitation. Moreover, high recombination rate of photo-generated carriers and their low mobility during the photocatalytic process are also issues to be considered for using $BaTiO_3$ as a photocatalyst. In order to overcome these limitations, a few studies have been reported, including photocatalytic enhancement by chemical doping, nanostructuring, band gap engineering, and noble metals decorating, among many others. We highlight some of these works, probably with the authors' personal bias.

Cui et al. once studied the photocatalytic degradation of $BaTiO_3$ using Rhodamine B (RhB) as degradation target under UV–visible light illumination [39]. A simple thermal treatment was carried out to synthesize non-ferroelectric cubic and ferroelectric tetragonal $BaTiO_3$ samples respectively. The results revealed a three times enhancement of the decolorization rate in the ferroelectric sample with respect to the non-ferroelectric one, as shown in Figure 9.14. This enhancement is attributed to the fact that the ferroelectricity ensures a tightly bound layer of dye molecules, and also makes the photoexcited charge carriers to separate via the internal space charge layer. In addition, it was shown that nanostructured Ag photochemically deposited on the surface of $BaTiO_3$ assists the photochemical reaction by additional enhancement in the reaction rate. This enhancement can be attributed to the facile electron transport from $BaTiO_3$ to Ag nanoparticles on the $BaTiO_3$ surface, suppressing carrier recombination. However, the enhancement depends on the photo-deposition time of Ag nanoparticles. The photocatalytic performance will deteriorate if the photo-deposition time is long or if the amount and the size of the Ag nanoparticles become too large. The overloaded

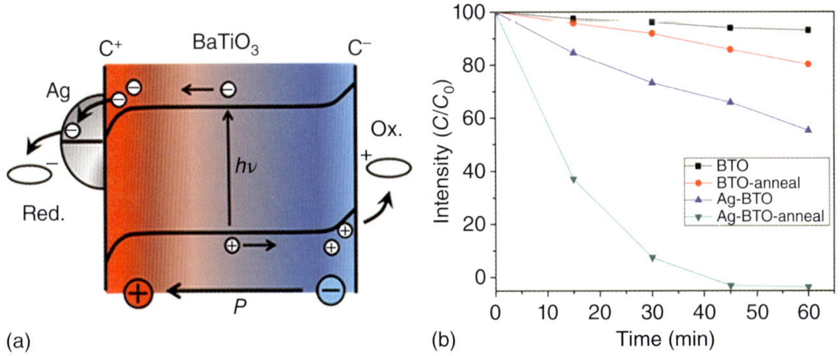

Figure 9.14 A schematic drawing of a ferroelectric material showing the effect of free carrier reorganization on band structure and photoexcited carriers (a). Photo-decolorization profiles of RhB with different catalysts under solar simulator (b). The catalysts consisting of higher tetragonal $BaTiO_3$ phase after annealing show higher activity, especially in case of modification by Ag nanoparticles. *Source:* Cui et al. 2013 [39]. Reproduced with permission of ACS.

Ag nanoparticles can trap the holes, thus enhancing the recombination and/or blocking the surface-active sites for the redox reaction process.

However, it is not easy to decouple multi-sources for the improved adsorption and reactivity on the surface of annealed $BaTiO_3$ (BTO) samples. Crystallinity improvement and ferroelectric polarization may be the sources, and the claim from Cui et al. should be clarified. Yang and coworkers proposed another approach to investigate the effect of ferroelectric polarization on photocatalytic dye degradation [63]. They prepared homogeneous $BaTiO_3$ nanoparticles of difference sizes down to 7.5 nm. The smallest size $BaTiO_3$ nanoparticles can be chosen as the benchmark for checking the ferroelectric model system to investigate the effect of photocatalysis at different temperatures. Using piezoelectric force microscopy (PFM) and Raman microscopy, they found that such small nanoparticles have very diffuse ferroelectric transitions, and thus the polarization must be small in magnitude (close to zero at 80 °C), in contrast to a sharp phase transition in bulk $BaTiO_3$. The decolorization rate of RhB under UV light illumination with the 7.5 nm $BaTiO_3$ nanoparticles as photocatalysts decays by ~12% upon temperature variation from 30 to 80 °C, as shown in Figure 9.15. When large-size $BaTiO_3$ particles (>500 nm) were used as photocatalysts and its photo-decolorization was investigated as a control sample, control results showed that the photo-decolorization rate does not change much at different temperatures. Because of the strong temperature dependence of polarization for the 7.7 nm $BaTiO_3$ nanoparticles, it is suggested that ferroelectric polarization does promote the separation of photo-generated carriers and benefits the efficiency of photocatalysis.

Besides $BaTiO_3$, $LiNbO_3$ was also chosen for photocatalytic investigation. Stock and Dunn measured the photocatalytic decolorization of acid black 1 and RhB with $LiNbO_3$ powder as the photocatalyst, noting that the samples were doped with iron or magnesium to form the n-type or p-type semiconductors, respectively [51]. It was revealed that the surface depolarization field does modify the surface chemistry by changing the Stern and inner Helmholtz plane due to the interaction with the catalyst surface charges and the polar nature of solvated

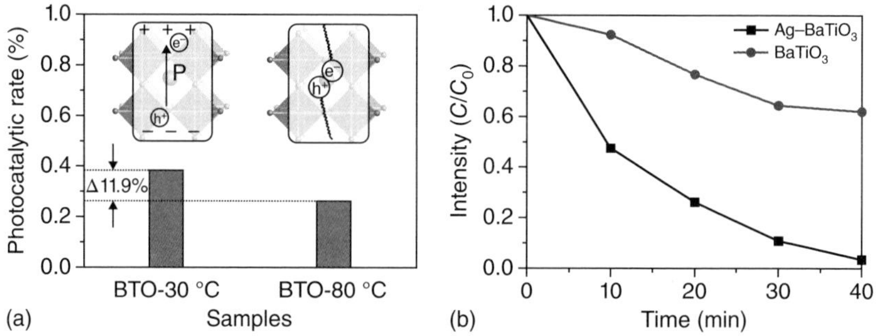

Figure 9.15 (a) Photocatalytic activity of $BaTiO_3$ nanoparticles at 30 and 80 °C. The inset schematic shows how the internal spontaneous polarization of ferroelectric materials affects photocatalytic reaction. (b) Photocatalytic performance of degradation RhB under UV irradiation. *Source:* Su et al. 2015 [63]. Reproduced with permission of John Wiley & Sons.

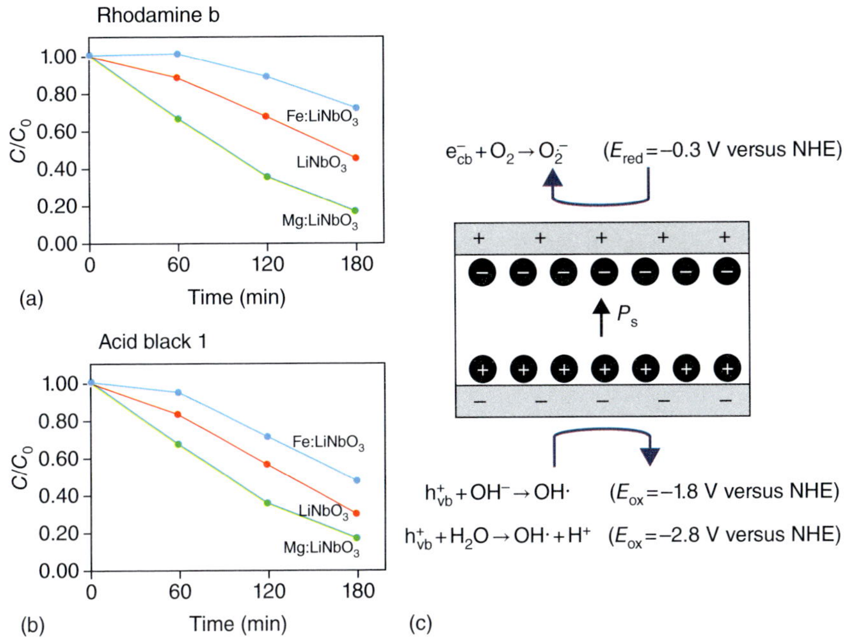

Figure 9.16 Decolorization curves of (a) RhB and (b) acid black 1 under simulated solar light using $LiNbO_3$, Fe-doped $LiNbO_3$, and Mg-doped $LiNbO_3$ powder as the catalyst. (c) Proposed mechanism for formation of reactive species on the surface of $LiNbO_3$. Electrons reduce oxygen to form the superoxide anion radical at the C^+ face, and hole carriers oxidize water or hydroxyl ions to form the hydroxyl radical at the C^- face. *Source:* Stock and Dunn 2012 [51]. Reproduced with permission of ACS.

species. In spite of the large band gap of $LiNbO_3$, the spatial separation of redox reactions in $LiNbO_3$ powder benefits the reaction kinetics, enabling a high rate of decolorization, as shown in Figure 9.16. In addition, the decolorization rate of the dye solutions was found to be the highest and lowest for the p-type and n-type $LiNbO_3$. This dependence is essentially associated with the difference of majority carriers (holes and electrons), making the ratio of reactive species different.

The performance of ferroelectric photocatalysts can be enhanced by various technical approaches, making the photocatalytic materials science rich and fruitful. For example, the photocatalytic dye degradation upon using ferroelectric/non-ferroelectric heterostructures (composites) as catalyst was explored [64–66]. Li et al. prepared a $PbTiO_3/TiO_2$ composite photocatalyst with nanostructured heterojunctions by a simple sol–gel method, as shown in Figure 9.17a [66]. Under the assistance of visible light ($\lambda > 420$ nm), the heterostructured powder can degrade MB at a rate 4.8 times that in the cases using $PbTiO_3$, TiO_2, or their mechanical mixtures as photocatalysts, as shown in Figure 9.17b. A schematic of the energy level diagrams for both negative and positive polarizations is shown in Figure 9.17c,d, respectively. The negative surface polarization (the direction of the polarization points away from the interface) is screened by the free holes in $PbTiO_3$, leading to upward band bending. This negative surface polarization is also screened externally by the

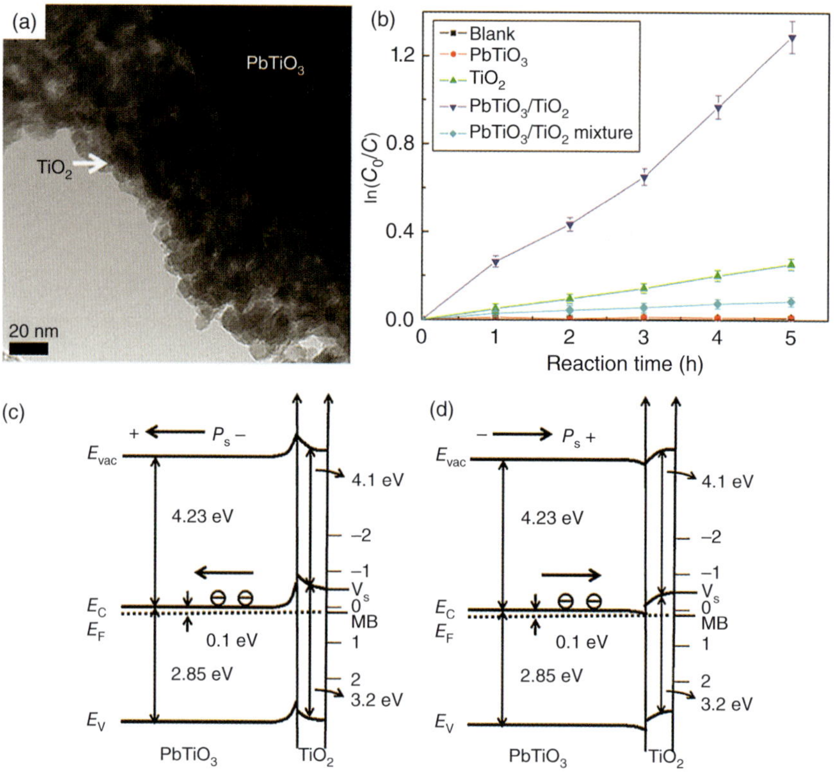

Figure 9.17 (a) Transmission electron microscopy (TEM) image of heterostructured particles composed of $PbTiO_3$ and TiO_2. (b) Photocatalytic degradation with different samples during irradiation by visible light ($\lambda > 420$ nm). Blank refers to data from a control experiment without the addition of a catalyst. The schematic energy level diagrams of $PbTiO_3$–TiO_2 with (c) negative polarization and (d) positive polarization normal to the heterostructural interface. E_{vac}, E_C, E_F, E_V, and E_s are the energies of vacuum level, conduction band, Fermi level, valance band, and surface potential, respectively. *Source:* Li et al. 2012 [66]. Reproduced with permission of Royal Society of Chemistry.

holes in TiO_2 shells, leading to upward band bending too. The opposite process will occur on the positive surface polarization, where electrons will be drawn toward the interface with downward band bending at $PbTiO_3$–TiO_2 interfaces. The enhanced photodegradation reactivity is thus attributed to the absorption of visible light by $PbTiO_3$ cores, and the separation of photo-generated carriers by internal fields at the interface, and reactions at the surface of the nanostructured TiO_2 shells.

The photocatalytic effect of $BaTiO_3$-graphene nanocomposites was also tried recently by Wang et al., and here the nanocomposites were synthesized using the facile one-pot hydrothermal route [67]. As shown in Figure 9.18, upon visible light irradiation ($\lambda > 420$ nm), no electron–hole pairs are generated because of the wide band gap of $BaTiO_3$. However, the as-prepared $BaTiO_3$-graphene nanocomposites do show the visible light photocatalytic degradation of methylene blue (MB), and the highest photocatalytic activity was found for the sample

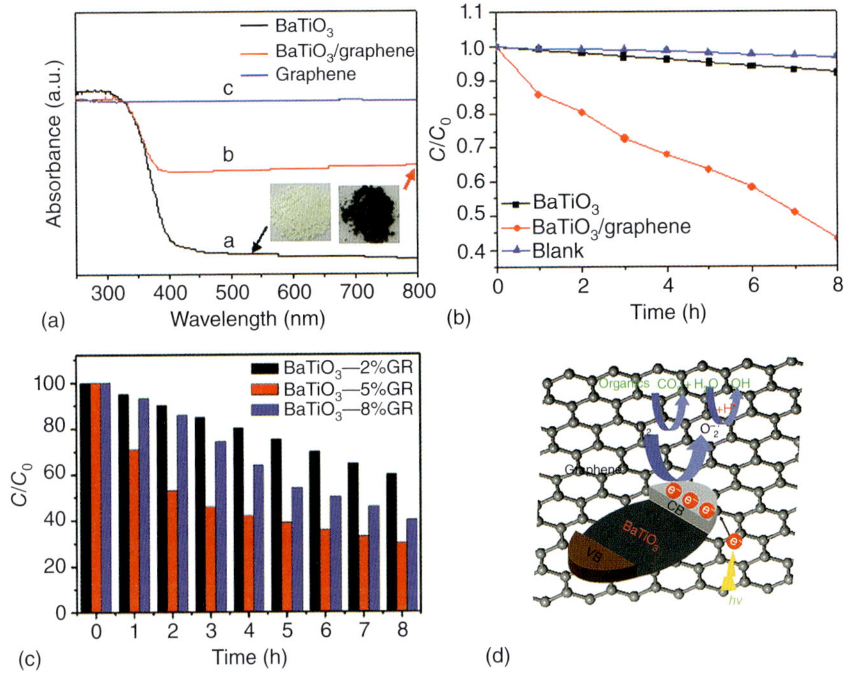

Figure 9.18 (a) UV–Visible spectra of pure $BaTiO_3$ nanoparticles and $BaTiO_3$/5 wt% graphene nanocomposites and graphene (insets are images). (b) Visible light photocatalytic degradation of MB by pure $BaTiO_3$, $BaTiO_3$/5 wt% graphene nanocomposites, and no photocatalyst. (c) Photocatalytic degradation of MB under the irradiation of visible light over the $BaTiO_3$–graphene composites with different contents of graphene. (d) A schematic illustration showing the reaction mechanism for the photocatalytic degradation of organic pollutants over the $BaTiO_3$–graphene composites. *Source:* Wang et al. 2015 [67]. Reproduced with permission of Royal Society of Chemistry.

with a nominal content of 5 wt% graphene. It was demonstrated that the graphene phase in this composite sample acts as an organic dye-like macromolecular "photosensitizer." A possible mechanism for the visible light photocatalysis can be explained as follows: visible light irradiation excites graphene from the ground state to the excited state, similar to the photoexcitation of organic dyes adsorbed on semiconductors. The graphene in the excited state injects electrons into the conduction band of $BaTiO_3$, and the electrons can be trapped by molecular oxygen in solution to form $^{\bullet}O_2^-$ and other oxidative species. Nevertheless, the details on this mechanism should be clarified.

Among ferroelectric materials, the most studied system addressing photocatalysis is $BiFeO_3$. On one hand, $BiFeO_3$ has a band gap much smaller than most available ferroelectric materials, and it is environment friendly and cost competitive on the other hand. Gao et al. synthesized $BiFeO_3$ nanoparticles with size ranging from 80 to 120 nm by the simple sol–gel method and demonstrated their good photocatalytic properties, as shown in Figure 9.19 [68]. Although both $BiFeO_3$ bulk and nanoparticles exhibit photocatalytic activity to decompose methyl orange under UV–Vis light irradiation, degradation with

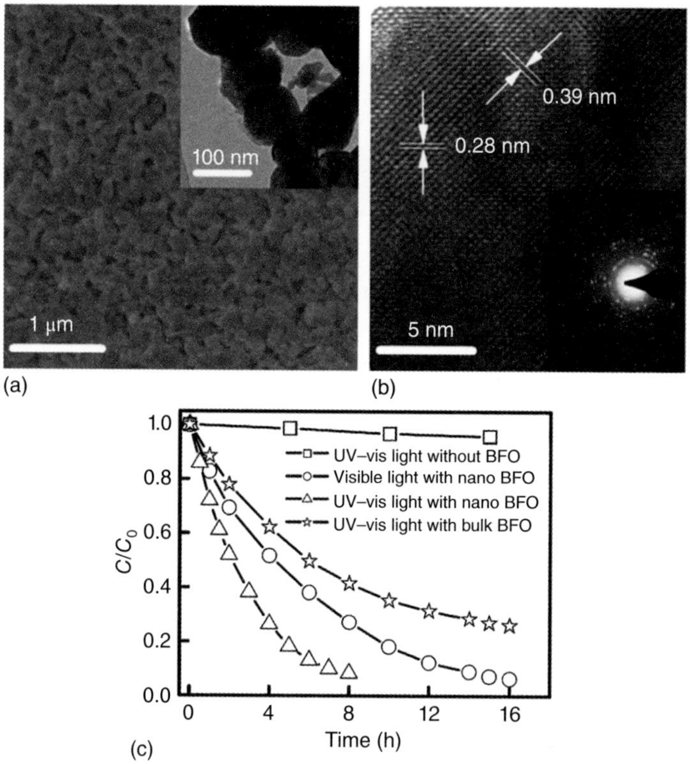

Figure 9.19 (a) SEM image of $BiFeO_3$ nanoparticles with the TEM image as the inset. (b) High resolution transmission electron microscopy (HRTEM) image of a nanoparticle with the selected area electron diffraction (SAED) pattern as the inset. (c) Photocatalysis of $BiFeO_3$ nanoparticle samples on degradation of methyl orange under UV–Vis light irradiation and visible light irradiation. *Source:* Gao et al. 2006 [68]. Reproduced with permission of John Wiley & Sons.

$BiFeO_3$ nanoparticles is significantly more efficient than that of bulk $BiFeO_3$ due to the higher surface area of nano-sized $BiFeO_3$. Most importantly, $BiFeO_3$ nanoparticles show significant degradation ability under visible light irradiation, enabling practical use.

He et al. then studied the photocatalytic reaction mechanism of $BiFeO_3$ nanoparticles for pollutant degradation using a number of diagnostic experiments [69]. As shown in Figure 9.20, when an efficient $^{\cdot}OH$ radical quencher (*tert*-butyl alcohol, TBA) was added into the $BiFeO_3$ photocatalytic degradation system, the degradation rate of RhB was markedly reduced, suggesting that the $^{\cdot}OH$ radicals are the primary oxidant species in the photocatalytic process under visible light irradiation. In order to distinguish the origin of the $^{\cdot}OH$ radicals, different additives were used in their experiments. For instance, if the holes play a crucial role in the pollutant degradation, addition of a hole scavenger (methanol) would significantly inhibit the pollutant degradation. However, the degradation of RhB is only slightly retarded in the presence of methanol, which indicates that only small amounts of $^{\cdot}OH$ radicals originate from the holes. On

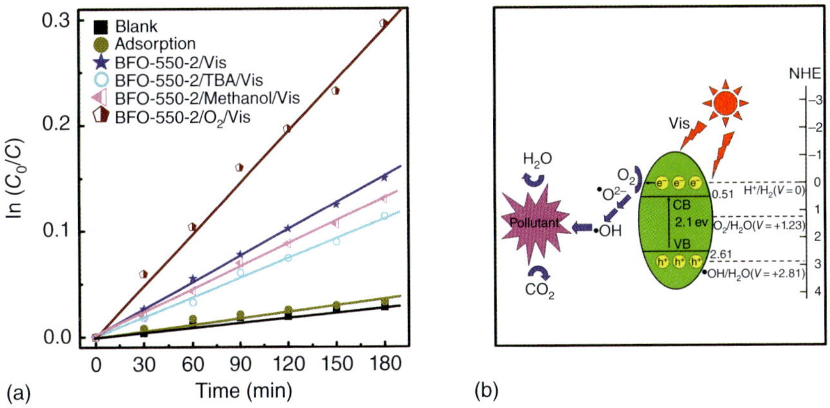

Figure 9.20 (a) The kinetics of RhB degradation using $BiFeO_3$ nanoparticles with various additives under visible light irradiation. (b) A schematic illustration of the pollutants degradation mechanism using $BiFeO_3$ nanoparticles under visible light irradiation. *Source:* He et al. 2013 [69]. Reproduced with permission of Elsevier.

the other hand, the degradation of RhB in the O_2 saturated aqueous solution is significantly enhanced, which implies that more O_2 molecules scavenge more electrons, suppressing the charge recombination.

It should be noted that $BiFeO_3$ can absorb effectively the visible light, but the photocatalytic activity of $BiFeO_3$ is still low, hindering its commercialization in photocatalytic oxidation technology. Many efforts have been made to improve the photocatalytic activity of $BiFeO_3$. It is known that the photocatalytic performance is strongly dependent on the morphology of synthesized particles. Many techniques have been developed to fabricate various $BiFeO_3$ nanostructures [70–74]. For example, Li et al. fabricated uniform $BiFeO_3$ microcrystals with various morphologies (microspheres and micro/submirocubes) by a controlled hydrothermal method [71]. They demonstrated the photocatalytic activity of these regular-shaped structures in degradation of Congo red under visible light irradiation ($\lambda > 400$ nm). Moreover, Fei et al. reported that pills and rods with highly exposed $\{111\}_c$ show enhanced visible light response in comparison with $\{100\}_c$ dominant cubes [73].

Another approach to improve the photocatalytic properties of $BiFeO_3$ is chemical doping, tentatively to modulate the band gap and carrier density of $BiFeO_3$ [75–77]. Guo et al. prepared the Gd-doped $BiFeO_3$ ($Bi_{1-x}Gd_xFeO_3$) nanoparticles by a sol–gel method followed by a rapid thermal annealing process [75]. Their studies indicated that all the samples show the rhombohedral structure up to the moderate doping level $x = 0.1$, while an impurity phase of $GdFeO_3$ appeared at $x = 0.15$, illustrating that the soluble limitation is ~10% for Gd3+ doping. The pure BFO nanoparticles exhibit visible light photocatalytic ability against RhB, while much enhanced photocatalytic activity was obtained by adding an appropriate amount of HCl, which could be ascribed to the extended adsorption of RhB on the catalyst surface. Furthermore, increasing Gd doping allows the photocatalytic activity to increase initially to the maximal level and then decrease, as shown in Figure 9.21. The maximal photocatalytic activity

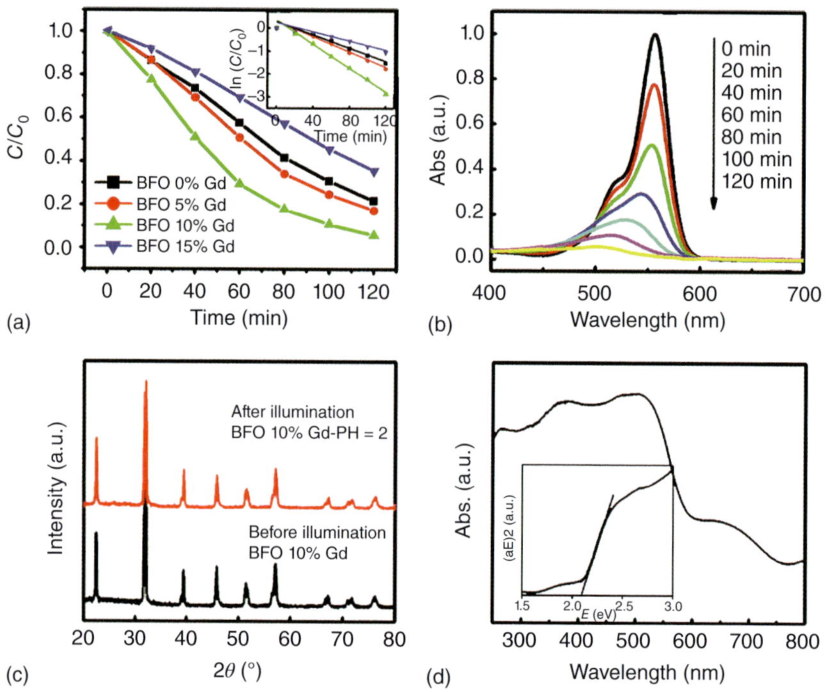

Figure 9.21 (a) Photocatalytic degradation efficiencies of RhB using BGFO*x* nanoparticle samples. (b) The changes of temporal UV–Vis spectral of RhB aqueous solution over BGFO ($x = 0.1$) nanoparticle sample. (c) XRD patterns of the BGFO ($x = 0.1$) nanoparticles before and after the degradation experiment. (d) UV–Vis diffuse reflectance spectrum of the BGFO ($x = 0.1$) nanoparticles; the inset shows the calculation of band gap. *Source:* Guo et al. 2010 [75]. Reproduced with permission of ACS.

appears at $x = 0.1$, which could be explained by an anomalously high dielectric constant at the rhombohedral–orthorhombic phase boundary, which enlarges the space charge region on the interface of the particles and solution.

Finally, we address the issue of effectively separating the photo-generated electrons and holes, which is also certainly essential to improve the photocatalytic activity of semiconductors. A heterojunction consisting of two kinds of materials is generally believed to be of benefit to the photocatalytic activity, due to the reduced recombination rate of photo-generated electrons and holes. Along this line, various heterojunctions of $BiFeO_3$ with metal oxides including TiO_2, Fe_2O_3, CuO, and $BiVO_4$ have been checked [78–82]. Guo et al. synthesized $BiFeO_3$ nanoparticles with γ-Fe_2O_3 parasitic phase by a sol–gel method with different gel-drying temperatures [79]. The parasitic phase of γ-Fe_2O_3 was found to nucleate at higher gel-drying temperatures and remained stable during the annealing process, responsible for the enhanced magnetization. This system could be used as a magnetically separable photocatalyst. Indeed, the $BiFeO_3$ nanoparticles with γ-Fe_2O_3 parasitic phase show higher photocatalytic activity than pure $BiFeO_3$ nanoparticles, which was explained by the existence of $BiFeO_3$/γ-Fe_2O_3 heterojunctions, as shown in Figure 9.22. According to the band energy level

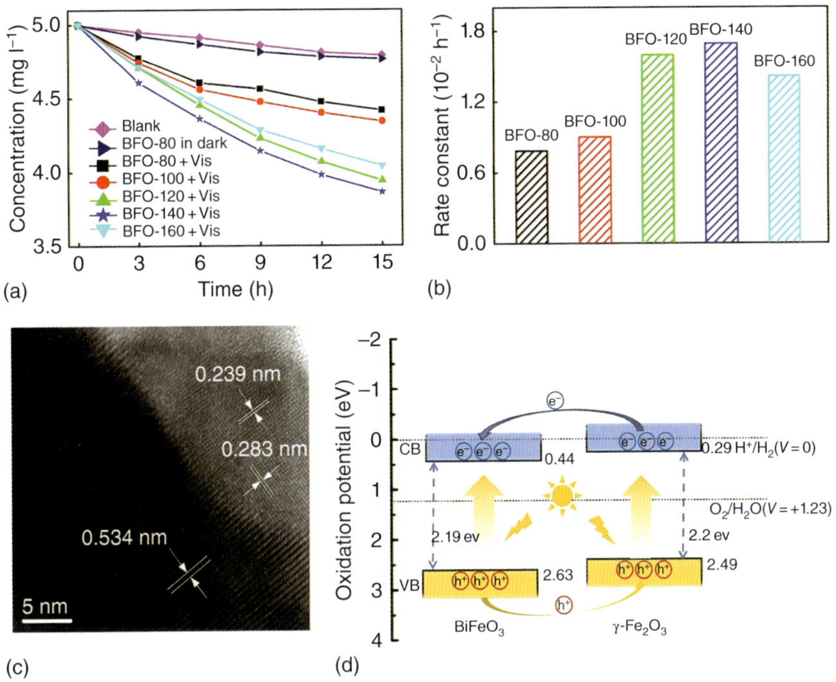

Figure 9.22 Photocatalytic degradation efficiencies of RhB using different samples. (b) The relationship between the rate constants and different samples. (c) HRTEM image of the composite heterojunction structure. (d) A schematic illustration for the calculated energy level diagram of $BiFeO_3/\gamma\text{-}Fe_2O_3$. *Source:* Guo et al. 2011 [79]. Reproduced with permission of Royal Society of Chemistry.

diagram of $BiFeO_3/\gamma\text{-}Fe_2O_3$, the excited electrons on the conductance band of $BiFeO_3$ transfer to that of $\gamma\text{-}Fe_2O_3$, and meanwhile, holes on the valence band of $\gamma\text{-}Fe_2O_3$ can be transferred to that of $BiFeO_3$ under the potential of the band energy difference. Therefore, the recombination of electron–hole pairs can be reduced and the photocatalytic reaction can be enhanced. Besides the metal oxides, graphene oxide, carbon nanotube, and metal organic framework (MOF) have also been combined with $BiFeO_3$ to form heterojunctions [83–85], and the enhanced photocatalytic activity for organic dye degradation under UV or visible light was revealed.

9.6 Photocatalytic Water-splitting Using Ferroelectric Materials

The very early work on photocatalytic water-splitting or photoelectrochemical cells using ferroelectric materials was done by Inoue et al. [86–88]. They fabricated TiO_2 thin films on four types of substrates: an electro-poled single crystal of ferroelectric $LiNbO_3$ with (i) C^+ domain, (ii) C^- domain, (iii) a similarly electro-poled single crystal of non-ferroelectric $LiTaO_3$, and (iv) Al_2O_3 single crystal [86]. The ferroelectric substrates were poled for different directions prior to the TiO_2 thin film deposition. After depositing a thin Pt film on the TiO_2 film surface, photocatalytic water-splitting was demonstrated under irradiation from a Xe lamp. The measured hydrogen production rates of different samples are in the order of $LiNbO_3$ C^--substrate > $LiNbO_3$ C^+-substrate > $LiTaO_3$ substrate > Al_2O_3 substrate. This order suggests the role of ferroelectric polarization in the charge carrier separation and photocatalytic activity enhancement, which can be explained based on a band bending mechanism driven by the ferroelectric polarization.

The pioneering work by Inoue and his coworkers triggered research activities in developing ferroelectric materials for photocatalytic water-splitting. These efforts were driven by academic curiosity rather than any realistic application due to the limited light absorption ($E_g = 3–4$ eV) and poor bulk conductivities of TiO_2-based materials [89]. This expectation seemed to hinder the overall performance. Thanks to a relatively narrow band gap and robust polarization, multiferroic $BiFeO_3$ has emerged as a very promising ferroelectric photocatalytic material and has been subjected to water-splitting research too recently. We highlight several examples along this line. Luo and Maggard investigated the photocatalytic water-splitting activity of $SrTiO_3$-coated $BiFeO_3$ [90]. Although $SrTiO_3$ was reported to be able to produce oxygen in the UV light, it exhibits nearly no detectable visible light activity due to the large band gap. In contrast, $SrTiO_3$-coated $BiFeO_3$ exhibits visible light photocatalytic activity. Based on the optical absorption measurement, it was concluded that the role of $BiFeO_3$ in this system is to sensitize $SrTiO_3$ shell with regard to visible light. Gao et al. then studied the photocatalytic performances of $BiFeO_3$ nanowires [91]. They found that $BiFeO_3$ nanowires failed to produce H_2 either from pure water or from $Pt/CH_3OH/H_2O$ solution under visible light irradiation and UV irradiation, consistent with the results on $BiFeO_3$ nanoparticles. However, $BiFeO_3$ nanowires have the photoinduced oxidization ability to produce O_2 in $AgNO_3/H_2O$ system with considerable efficiency, suggesting that $BiFeO_3$ may be a good candidate for photoelectrode as well as photocatalytic decomposition. Further enhancement of this photoactivity was realized by Li et al. using modified hybrid $BiFeO_3$ nanowires [92], which contained Au nanoparticles adsorbed on their surfaces, as shown in Figure 9.23. In this structure, the Au nanoparticles play the role of electron-trap centers for photo-generated electrons injected from $BiFeO_3$ nanowires. Additionally, Au nanoparticles also act as catalytic sites for gas generation from water oxidation under visible light illumination. It was found that the $Au/BiFeO_3$ hybrid nanowires show significantly better photocatalytic performance over pure $BiFeO_3$ nanowires, where the amount of oxygen generated from $Au/BiFeO_3$ nanowires is about three times that produced using pure $BiFeO_3$ nanowires during 10 hours of measurement.

Li et al. studied the effect of ferroelectric polarization on photocatalytic water-splitting activity using heterostructures comprising of ferroelectric $BaTiO_3$ core and TiO_2 nanoshell [93]. This was similar to their earlier work on dye degradation using the same heterostructured catalysts [66]. The $BaTiO_3/TiO_2$ core–shell photocatalysts have the rates of photocatalytic hydrogen production from water/methanol solutions much greater than those solely using TiO_2

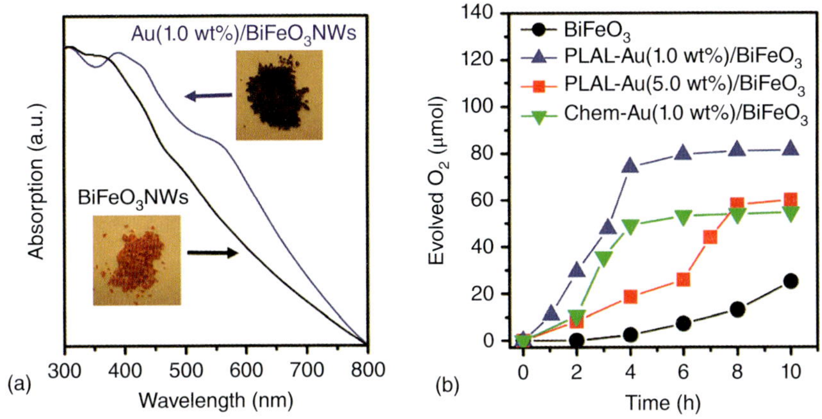

Figure 9.23 (a) Normalized UV–Vis absorption spectra of $BiFeO_3$ nanowires and the $Au/BiFeO_3$ nanocomposite dispersed in ethanol. (b) Oxygen evolved upon visible light ($\lambda > 380$ nm) illumination of the $FeCl_3$ suspension containing different photocatalysts. *Source:* Li et al. 2013 [92]. Reproduced with permission of Royal Society of Chemistry.

or $BaTiO_3$. They suggested that the internal electric field in the ferroelectric core could enhance the charge carrier separation in the TiO_2 nanoshell, and the movement of photoexcited electrons and holes is domain dependent and depends on the band bending at the interface of $BaTiO_3$ and TiO_2. Park et al. demonstrated the effect of electro-poling on powder ferroelectric catalysts and discussed the effect on photocatalytic performance [94]. They used $K_xNa_{1-x}NbO_3$ powder as photocatalyst. To study the effect of polarization, the powder ferroelectric photocatalyst was poled using a corona-poling method as depicted in Figure 9.24a. This system primarily consists of two electrodes: one is designed with a sharp-cornered shape (corona needle), and the other is designed with a smooth, large-diameter rounded shape (Cu disk). When the electric field is applied onto the needle electrode, the electric field intensity around the sharp point becomes much larger than elsewhere and strong enough to ionize a neutral fluid (air). Accordingly, a plasma region filled with positive ions and free electrons is formed. If the needle electrode is more negative than the disk electrode, the ions are attracted to the needle and the electrons are repelled to the disk (negative corona), inducing a negative charge layer on the upper surface of the material. The H_2 evolution rate of the polarized ferroelectric powder is much higher than that of the non-polarized $K_{0.5}Na_{0.5}NbO_3$ powder as seen in Figure 9.24b. Moreover, Figure 9.24c shows that the H_2 evolution activity of the polarized $K_{0.5}Na_{0.5}NbO_3$ powder is still greater than 80% after reuse in three cycles for a total of 15 hours, confirming the good stability of the polarized $K_{0.5}Na_{0.5}NbO_3$ powder. These results demonstrated that the poling process allows a permanent internal electric field inside the powder due to the remnant polarization. Such internal field promotes the separation of electrons and holes.

Subsequently, Yang et al. reported a largely enhanced photoelectrochemical performance of photo-anodes by means of ferroelectric polarization endowed band engineering based on the $TiO_2/BaTiO_3$ core/shell nanowires. The synthesis

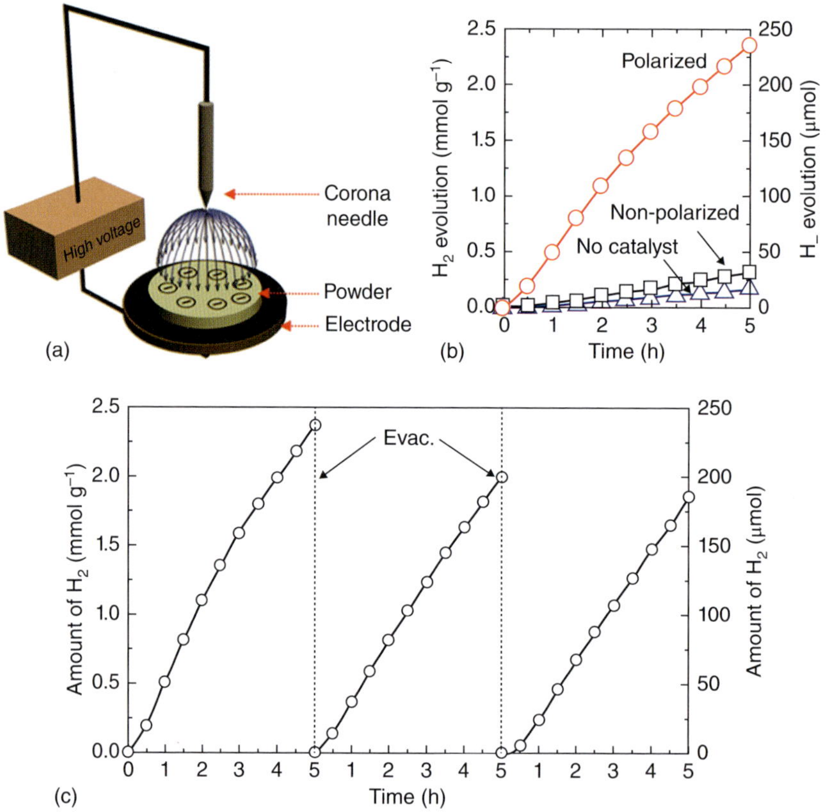

Figure 9.24 (a) Illustration of the corona-poling system. (b) Amount of hydrogen evolution for polarized and non-polarized $Na_{0.5}K_{0.5}NbO_3$ powder under UV-light irradiation. (c) Temporal profile of H_2 evolution for polarized $Na_{0.5}K_{0.5}NbO_3$ powder. *Source:* Park et al. 2014 [94]. Reproduced with permission of Royal Society of Chemistry.

is simple and only needs the one-step hydrothermal process, and then a uniform, epitaxial, and spontaneously poled $BaTiO_3$ layer can be created on single crystalline TiO_2 nanowires. Compared to pristine TiO_2 nanowires, the 5-nm $BaTiO_3$-coated TiO_2 nanowires achieved a 67% enhancement of the photocurrent density. By numerically calculating the potential distribution across the TiO_2/$BaTiO_3$/electrolyte heterojunctions and systematically investigating the light absorption, the enhanced photoelectrochemical performance was proved to be a result of the increased charge separation efficiency induced by the ferroelectric polarization of $BaTiO_3$ shell. In addition, ferroelectric polarization could be switched by external electric field poling, yielding photoelectrochemical performance gain or loss based on the direction of polarization, as shown in Figure 9.25.

However, it should be pointed out that control of ferroelectric polarization in powder and nanostructure samples could be a challenge, and a clarification of how ferroelectric polarization affects the photocatalytic water-splitting needs an additional approach. Along this line, photoelectrochemical cells based on

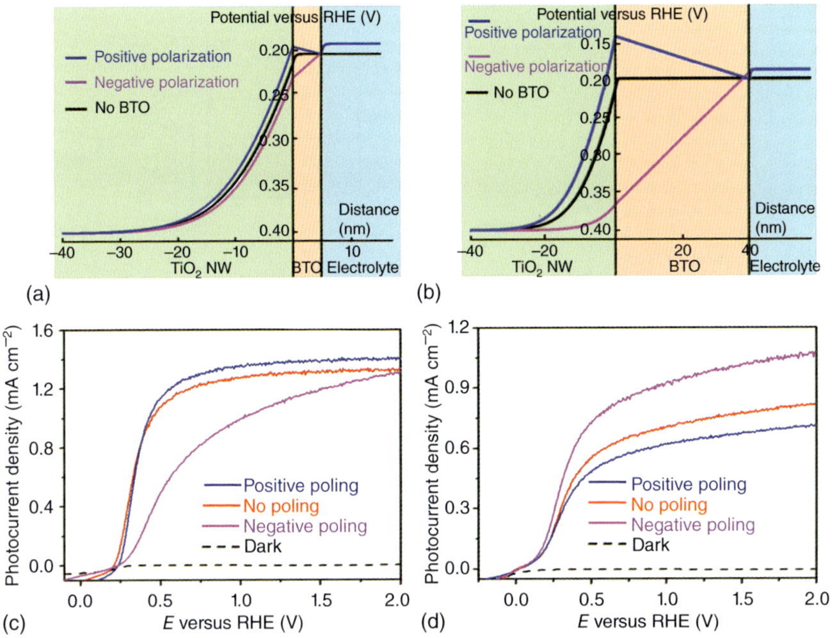

Figure 9.25 Calculated potential distributions and electric poling test of the ferroelectric enhancement. (a) Potential distribution of 150 °C TiO_2/5 nm $BaTiO_3$/NaOH heterojunction, where the spontaneous polarization of $BaTiO_3$ shell induces an upward band bending of TiO_2 core and facilitates the charge separation and transportation inside TiO_2. (b) Potential distribution of the 210 °C TiO_2/40 nm $BaTiO_3$/NaOH heterojunctions, showing a strong electric field inside $BaTiO_3$, which facilitates the hole transport toward solution. (c) *J–V* curves of the as-prepared (red), positively poled (blue), and negatively poled (magenta) 150 °C TiO_2/$BaTiO_3$ NWs. (d) *J–V* curves of the as-prepared (red), positively poled (blue), and negatively poled (magenta) 210 °C TiO_2/$BaTiO_3$ NWs. *Source:* Yang et al. 2015 [95]. Reproduced with permission of ACS.

ferroelectric polycrystalline thin films were fabricated [96–101]. Chen et al. reported the significant photocurrent performance of $BiFeO_3$ polycrystalline films under visible light irradiation [96], allowing $BiFeO_3$ to be one of the hot ferroelectric photoelectrode materials. Improving the film crystallinity could reduce the dark current and enhance the photocurrent density. In addition, the polycrystalline $BiFeO_3$ photoelectrode shows an efficiency comparable with traditional excellent photoelectrode materials such as WO_3 and $BiVO_4$. Subsequently, Cao et al. investigated the strong polarization-dependent photoelectrochemical responses in $BiFeO_3$ polycrystalline films [97]. It is seen from Figure 9.26 that the wavelength-dependent external quantum yield spectra, measured without any external bias, undoubtedly demonstrate the direct correlation of the solar-energy conversion efficiency with the polarization states of the films. The +8 V poling results in the highest external quantum efficiency (EQE), owing to the lowest charge recombination rate in comparison with the cases of fresh electrode or −8 V poled electrode. The external quantum yield of the as-grown sample is higher than the same sample with −8 V poling treatment, implying that the ferroelectric domains in the as-grown polycrystalline $BiFeO_3$

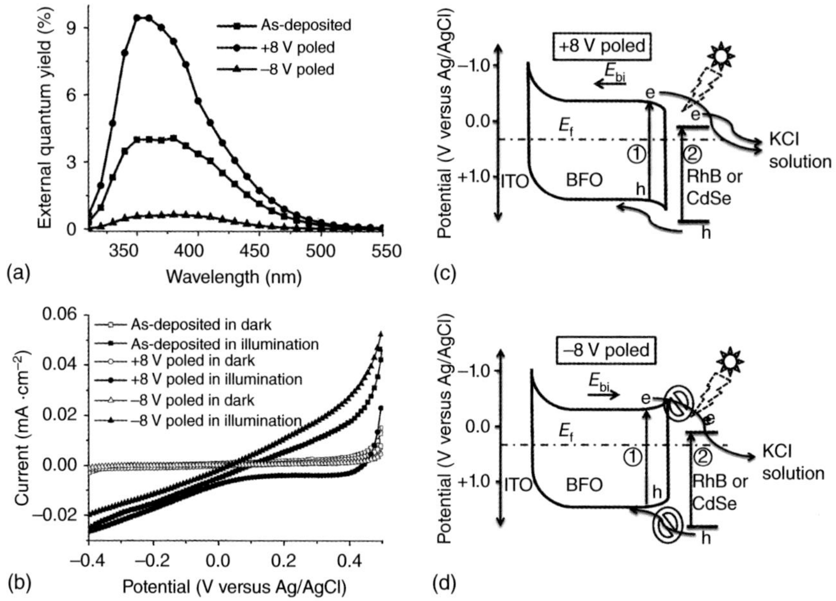

Figure 9.26 (a) External quantum yield spectra measured for $BiFeO_3$ photoelectrodes before poling and after +8 and −8 V poling. (b) Photocurrent–potential characteristics of the photoelectrodes with different polarization states. Schematic representations of the mechanisms in photoexcited charge transfer from $BiFeO_3$ films to the electrolyte ① and from excited surface modifiers to the $BiFeO_3$ films ② after the $BiFeO_3$ films are (c) positively and (d) negatively poled. *Source:* Cao et al. 2014 [97]. Reproduced with permission of John Wiley & Sons.

films are randomly distributed and not optimized. Furthermore, at 0 V versus Ag/AgCl (reference electrode), the photocurrent could be enhanced from 0 to 10 mA cm^{-2} and the open-circuit potential changes from 33 to 440 mV, upon the variation of the poling bias from −8 to +8 V. The possible mechanism underlying these behaviors is related to the different band bending structures arising from the ferroelectric polarization. Besides polycrystalline thin films, the epitaxial thin films also show the ferroelectric controlled photoelectrochemical behaviors [102, 103]. Ji et al. observed the significant anodic photocurrent in epitaxial $BiFeO_3$ films [102]. It seems that the electro-poling control in aqueous solution was not properly addressed, and the response of photoelectrochemical effect to the variation of ferroelectric polarization was not remarkable.

As a general comment, investigations on ferroelectric films as photoelectrodes in photoelectrochemical cells remain yet in the early stage and the photoelectrode performances are far from sufficient in terms of industry application potentials. The core issues include the low light absorption ability and rapid recombination of photo-generated electrons and holes (carrier density, mobility, charge separation). To improve the light harvesting ability, the development of new ferroelectric oxide perovskites with narrowed band gap and enhanced visible light absorption is crucial. A conspicuous step in band gap engineering has been recently marked by the synthesis of Bi_2FeCrO_6 double perovskite. Recent experimental study successfully demonstrated that tailoring the Fe/Cr cationic order allows the band gap to be remarkably tuned, mainly by modifying the B—O bonding [104]. It was found that improving the Fe/Cr cationic order parameter extends the absorption spectrum of Bi_2FeCrO_6 toward the longer wavelength region. Enlarging the domain size can also reduce the band gap.

Furthermore, Li et al. deposited Bi_2FeCrO_6 thin films on various $CaRuO_3$-coated (100)-oriented substrates including $(LaAlO_3)_{0.3}(Sr_2AlTaO_6)_{0.7}$, $LaAlO_3$, and $SrTiO_3$ [105]. They observed no dependence of the band gap of the thin film on the compressive strain. Photoelectrochemical measurements showed that the highest photocurrent in the p-type Bi_2FeCrO_6 photo-cathode is up to ~−1.0 mA cm^{-2}, suggesting that such double-perovskite materials with relatively narrow band gap find promising opportunities as photocathodes for H_2 generation. Most importantly, an effective tuning of photoelectrochemical water-splitting by modulating the internal electric field was identified, as shown in Figure 9.27. A twofold enhancement of photocurrent density was obtained after negatively poling the Bi_2FeCrO_6 thin film, as a result of modulation of the band structure by suitable control of the internal electric field gradient originating from the ferroelectric polarization in the Bi_2FeCrO_6 films.

An efficient way to improve the charge migration is to combine ferroelectric materials with noble metal (e.g. Ag, Au, Pt, etc.) nanostructures [98, 100, 101, 106, 107]. Since metals and semiconductors have different work functions, the Schottky barrier can be formed in the interface between the metal and

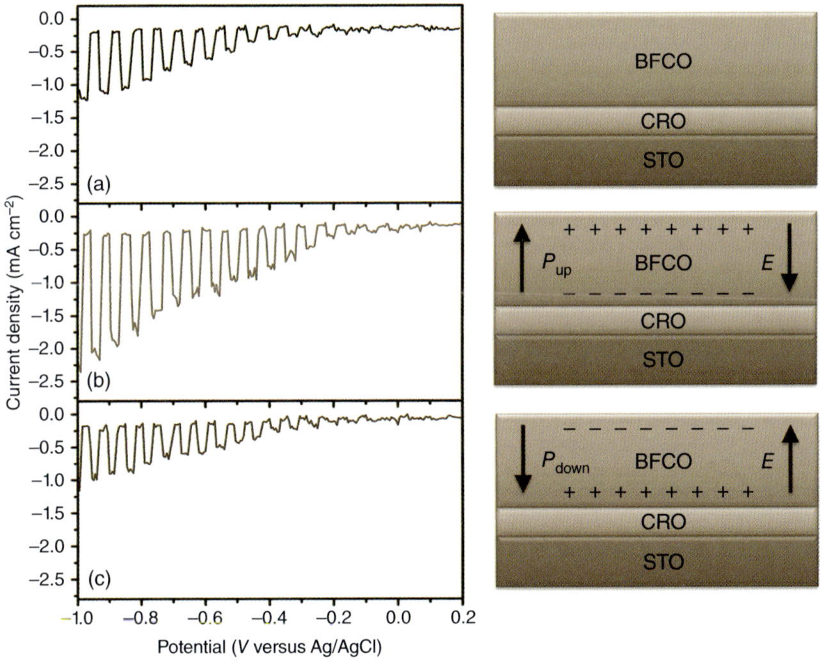

Figure 9.27 Variations of the photocurrent density with applied voltage (V versus Ag/AgCl) under chopped simulated sunlight illumination on $Bi_2FeCrO_6/CaRuO_3/SrTiO_3$ sample: (a) before, (b) after negative poling, and (c) after positive poling. Schematic illustrations are shown on the right of each figure. *Source:* Li et al. 2015 [105]. Reproduced with permission of John Wiley & Sons.

ferroelectric materials [108, 109]. Therefore, metal can act as a sink for photoinduced electrons, and decrease the recombination of photo-generated electrons and holes. Furthermore, some noble metal nanostructures induce the extraordinary localized surface plasmon resonance effect, which can enhance the electric field near the metal/materials interface, increase the photon absorption of semiconductor, and control the charge carrier generation [110]. Wang et al. reported on manipulation of the charge transfer and transport in nano-Au/Pb($Zr_{0.20}Ti_{0.80}$)O_3 hybrids by placing a nano-Au array in different positions within ITO/Pb($Zr_{0.20}Ti_{0.80}$)O_3 and by poling the Pb($Zr_{0.20}Ti_{0.80}$)O_3 films with different potentials [106]. Among the photoelectrochemical electrodes, the structure of ITO/nano-Au/Pb($Zr_{0.20}Ti_{0.80}$)O_3 provides the best performance among the three structures. The transport studies accompanied by femtosecond transient absorbance study revealed that the hot-charge transfer from Au nanoparticles to Pb($Zr_{0.20}Ti_{0.80}$)O_3 films is the main reason for the enhanced photoelectrochemical performance. In addition, using the ITO/Pb($Zr_{0.20}Ti_{0.80}$)O_3/nano-Au/Pb($Zr_{0.20}Ti_{0.80}$)O_3 electrodes, they indicated that when the pre-poling bias was switched from +10 to −10 V, the short-circuit photocurrent could be tuned by nearly an order of magnitude, as shown in Figure 9.28. The simultaneous manipulation of the charge transfer and interface-related photoelectrochemical phenomena presents a route to optimally

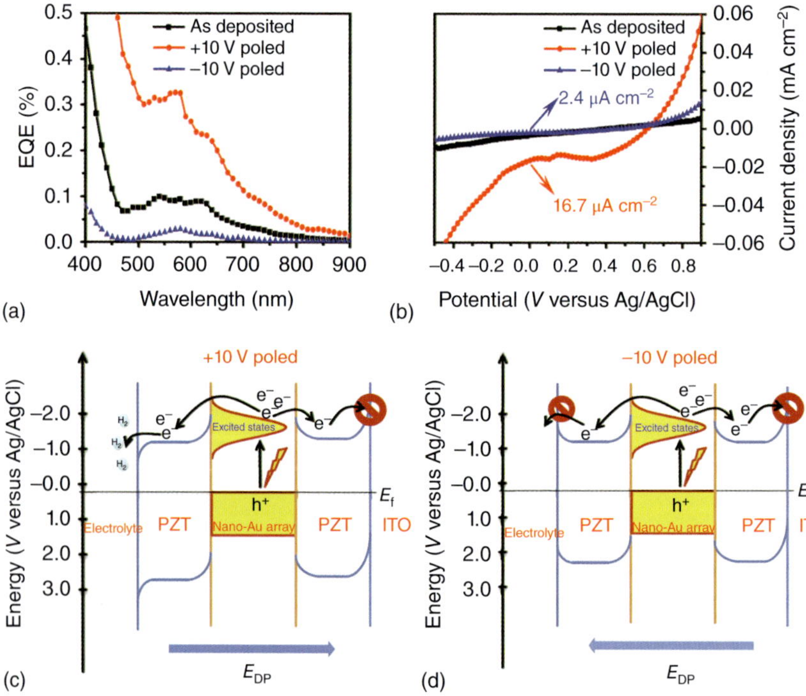

Figure 9.28 (a, b) EQE spectra and photocurrent–potential measurements of the as-grown (black), +10 V (red), and −10 V (blue) poled samples. (c, d) Schematic electronic band structure and mechanisms for the injected hot-electron transfer from Pb($Zr_{0.20}Ti_{0.80}$)O_3 films to the electrolyte for the two poling configurations [106].

and flexibly optimize the photoexcited charges for photoelectrochemical energy conversion.

Commonly, a combination of two or more simple materials for multifunctional composites has been one of the major strategies for achieving superior properties. Among the heterostructured materials, phase-ordered heteroepitaxial oxide nanocomposites have been widely applied due to their unique properties [111–113]. With respect to solar energy conversion applications, these self-organized systems have the following remarkable features, in addition to the generic characters of heterostructures: (i) Phase-separated components are usually vertically aligned pillars (1D) embedded in 3D matrix. Thus, these heterostructures have short diffusion pathways for photoinduced charge carriers in the radial direction before the charge recombination. (ii) Each component is of pseudo-single crystalline structure with significantly reduced grain boundaries, and hence the systems possess potentially reduced charge carrier trap sites. (iii) Epitaxial interfaces between two separated phases have less lattice defects than other heterojunction interfaces (e.g. polycrystalline–single crystalline or polycrystalline–polycrystalline heterojunctions). The three salient features give great prospects for improving the overall photo-generated charge transfer efficiency.

In fact, Quynh et al. addressed this topic and reported on spontaneously phase-ordered heteroepitaxial $BiFeO_3/\varepsilon\text{-}Fe_2O_3$ nanocomposite films that give rise to strongly enhanced photoelectrochemical performance [114], consistent with the scenario of enhanced photoinduced charge separation. The $BiFeO_3/\varepsilon\text{-}Fe_2O_3$ nanocomposite yields an enhanced photocurrent density of 0.19 mA cm^{-2} at 0.6 V versus Ag/AgCl (Figure 9.29), which is 3.4 times larger than that of plain $BiFeO_3$. Ultrafast dynamics spectroscopy and electrochemical impedance spectroscopy measurements revealed that the strongly enhanced photo-generated charge separation results from vertically aligned pseudo-single crystalline components, epitaxial heterojunctions, and a staggered band alignment of the components in the nanocomposite films. By enhancing the light absorption and charge separation, other "heterostructured" photoelectrodes have been widely investigated for their enhanced photoelectrochemical efficiencies, such as $BiFeO_3/SrTiO_3$ [115], $NiWO_4/WO_3$ [116], and $SrTiO_3/ZnFe_2O_4$ [117]. All these studies provide alternative pathways to the design of ferroelectric oxides-based photoelectrode for photoelectrochemical water-splitting.

9.7 Conclusion and Perspectives

In this chapter, a broad, yet brief, review on the integration of ferroelectrics into various photocatalytic systems and their promising applications has been presented, most likely with the authors' own bias. We are not in a position to cover every aspect of this topic, which is hard and not our intention here. We also apologize for any glaring omissions with the related work cited here, and any technical deficiencies in this chapter are, of course, our own.

We mainly discuss the general photochemical reactions, photocatalytic degradation, and photoelectrochemical water-splitting. The target applications

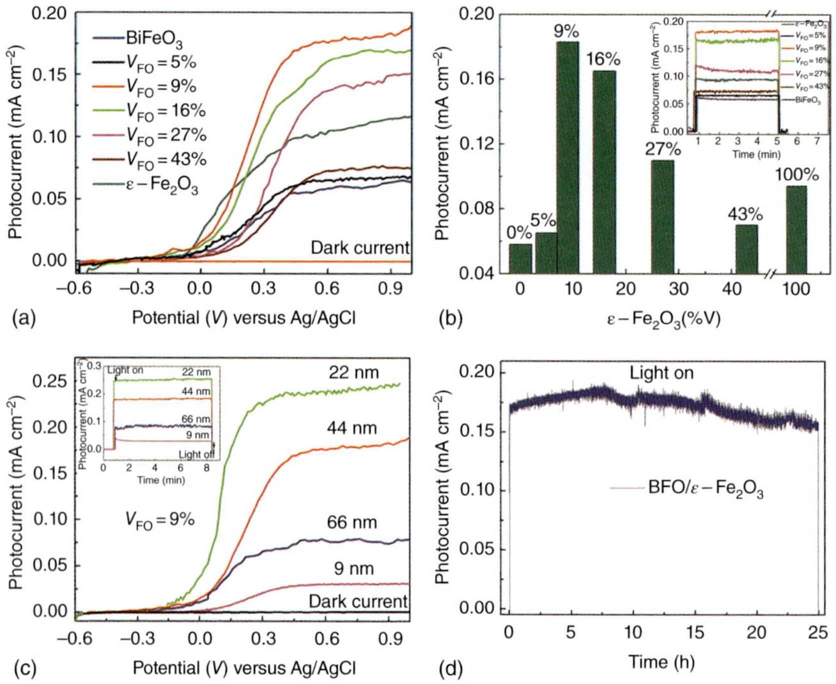

Figure 9.29 (a) Linear sweep voltammograms collected from $BiFeO_3$/ε-Fe_2O_3 electrodes with different percentages of ε-Fe_2O_3 in the dark and under light illumination. (b) Photocurrent density histograms for various $BiFeO_3$/ε-Fe_2O_3 photoelectrodes. The inset presents the curves of current density versus time at 0.6 V versus Ag/AgCl for all the films. The light is turned on and off at a time interval of 5 min. (c) Linear sweep voltammograms curves of $BiFeO_3$/ε-Fe_2O_3 electrodes with different thickness and the curves of current density versus time are shown in the inset (top left), and (d) long-term stability measurement of the photocurrent density versus time at 0.6 V versus Ag/AgCl for $BiFeO_3$/ε-Fe_2O_3 films. *Source:* Quynh et al. 2016 [114]. Reproduced with permission of John Wiley & Sons.

are closely in line with the ever-increasing demands for renewable/clean energy, solar energy harvesting, and pollutant decomposition, in a substantial effort to relieve the global energy crisis and environmental degradation. There are many parallels between photocatalysis and photovoltaics, in terms of the fundamental photo-physics, related photoactive materials, and device architectures. Photovoltaic effect in ferroelectrics has been reported for decades. However, it has not attracted sufficient attention in the scientific community until recent discoveries of highly promising performances in high-quality epitaxial ferroelectric thin films. The rejuvenated interest of photo-ferroelectrics also triggers intense research activities into the photocatalysis field. In light of this, research experiences can be readily borrowed from the photovoltaic counterpart.

The major ingredient for photo-ferroelectrics is their spontaneous polarization, which originates from the structural instability and symmetry-breaking on one hand, and their adjustable band gap on the other hand. Thanks to the asymmetry in electrical boundary conditions, ferroelectricity possesses spontaneous driving force for photo-generated carrier separation. This physical origin seems to underpin all photoelectric-related applications using ferroelectrics, such as photovoltaic, photocatalytic, and photoelectrochemical effects. In spite of the prominent progress, the details of the physical mechanisms for these effects yet leaves space for exploration, which is, however, critical for improving these effects. For example, regarding the ferroelectric photovoltaic effect, both bulk (anomalous) photovoltaic effect (bulk effect) and polarization-induced interface band bending (interfacial effect) have been proposed to explain the switchable photovoltaic outputs in ferroelectric solar cells. The bulk effect describes the phenomena as intrinsic physical behavior inherent to all non-centrosymmetric materials, due to the violation of the balancing principle of electron kinetics, resulting in a net photocurrent [118]. More recently, a shift-current model based on first-principles calculations was proposed to quantify the anomalous effects in ferroelectric photovoltaics, such as the above-band-gap photovoltage and light-polarization-dependent photocurrent [119]. Alternatively, conventional junction-based band bending model also prevails in explaining the polarization-mediated photovoltaic effects. In this latter model, ferroelectric polarization acts as a switchable gate that controls the internal field near the ferroelectric–metal interface and consequently the photocurrent flow direction. The bulk effect has been usually reported for single crystals and bulk ceramics with sample dimension from tens of micrometers to millimeters, whereas the interface model was widely employed in illustrating thin-film systems with submicron dimension. Therefore, it is highly possible that both bulk model and interfacial model are valid, and the sample dimension may be one of the conditions for which model is applicable. The same argument applies to other photoelectric-related phenomena in ferroelectrics. However, for real device applications, bulk samples will inevitably undermine the power conversion efficiency due to the energy loss during the charge transport across the whole sample. Furthermore, catalytic reactions usually have much better yield if the sample's surface-to-volume ratio is high to provide more active reaction sites. In this case, mesoscopic structures are naturally more favorable compared to planar structures. Consequently, interfaces between ferroelectrics and reactants play a more critical role in controlling the charge transfer process in these systems.

To conclude the chapter, it is also our intention to highlight relevant mechanisms and barriers currently existing in each of the steps throughout the whole photocatalysis process in ferroelectric-based materials. Possible solutions and future research directions are proposed with the aim of better understanding the underpinning physics and boosting the device performance. These perspectives are mainly raised from the viewpoint of fundamental photophysics due to the authors' physics background.

9.7.1 Light Absorption

Ferroelectrics usually possess wide band gaps (>3 eV), which makes them unfavorable for solar energy harvesting. This drawback can be mitigated by exploring narrow band gap ferroelectrics, an intensively addressed field in recent years. On one hand, multiferroics (magnetic ferroelectrics) come to our attention mainly because they have spontaneous polarization and a relatively small band gap due to the magnetic order. On the other hand, band gap engineering can be realized mainly by chemical substitution to create exotic electronic states close to the band edges, in which a couple of factors are closely involved, such as valence mismatch, defects, cation ordering, and so on.

It should however be noted that band gap reduction usually has an adverse effect on the ferroelectricity because of the internal screening of polarization by excessive free carriers. Therefore, it is important to maintain a compromise between the ferroelectric and the electronic properties. In the search for narrow band gap ferroelectrics, high-throughput combinatorial synthesis associated with theoretical prediction is highly desirable. Two notable examples are theory-directed band gap engineering in $KNbO_3$ system and cation ordering in $Bi(FeCr)O_3$ system [104, 120]. In both cases, the band gap can be reduced to 2.0 eV and below, a great improvement compared to their parent compounds. More intriguingly, a solar cell built on the latter materials strikes a record-high 8% power conversion efficiency, rivaling the mainstream photovoltaic materials. In addition to chemical alloying, strain is also an effective parameter for tailoring band structure in epitaxial thin films via its coupling to bond length and angle [121], which in turn affects the electronic orbital overlapping and thus band structures.

9.7.2 Carrier Separation and Transport

Ferroelectricity is most relevant in this step. Photoexcited electron–hole pairs are separated either by the coherent evolution of the electron and hole wave-functions or by the built-in field induced by the polarization bound charges. It is worth noting that owing to the low dimension of the ferroelectrics involved, space charge region could penetrate through the whole sample, depending on the carrier concentrations. Another factor that is commonly overlooked in photocatalytic process is the excitonic effect, which becomes prominent in low-dimension materials due to quantum confinement. Once dissociated, transport of the charge carriers becomes another big issue in ferroelectric materials, most of which are oxides with relatively low carrier mobility. Band-like or polaron-hopping transport mechanism needs to be carefully distinguished. Ferroelectric oxide semiconductors usually possess large lattice polarizability and significant amount of point defects, such as oxygen vacancies, both of which strongly favor the formation and trapping of the polarons, thus decreasing carrier mobility and promoting recombination. Regarding this issue, much more efforts are required to look into the defect chemistry of the ferroelectrics, in connection with the band structure engineering described above to alleviate the deep trap centers. Some insights into the polaronic transport can be borrowed from the organic–inorganic hybrid perovskite [122], whose liquid-like molecular dipoles are believed to facilitate the large polaron formation and protect the energetic carriers from recombination. Another commonly adopted method to circumvent this problem is to incorporate ferroelectrics with other superior electron/hole conductors. By selecting materials with appropriate band edges, electron/hole can be efficiently extracted out of the ferroelectrics, thus reducing the possibility for carrier recombination.

9.7.3 Carrier Collection/Reaction

When the photo-generated carriers survive the bulk recombination and reach the boundary of ferroelectrics, they will face solid–electrolyte interfaces or other heterogeneous interfaces, where charge transfer takes place. This step is probably the most complex and decisive part in the photocatalysis process. Owing to the electrostatic coupling, charged defects tend to accumulate near the surface of the ferroelectrics during the fabrication process. Besides, dangling bonds and surface reconstructions also prevail in the oxide surfaces. All of these high-density defects on the surface of ferroelectrics can act as recombination sites, posing great obstacles for effective charge transfer in the heterojunctions. Possible solution for it is to passivate the surface by chemical modification or by noble metal decoration. It can improve not only the carrier collection but also the chemical reaction through a better match between the metal work function and the redox potentials.

The "holy grail" of photocatalysis is to search for the "ideal" materials that possess strong visible light absorption, long carrier lifetime and diffusion length, large and balanced carrier mobilities, benign defect chemistry and, most importantly, bend edges that straddle the chemical reaction potentials. A single material that fulfills all the qualities is yet to be discovered and probably does not exist. Hence, heterogeneous structures with individual components playing separate roles in the photocatalytic process will continue to flourish in the near future. Although ferroelectric semiconductors may not show the best performance at the current stage, the introduction of polarization degree of freedom in the heterostructures did bring in a new twist to photocatalytic research. Further studies should keep focusing on understanding the fundamental physics and exploring novel materials, which would require intense interdisciplinary co-operations.

Acknowledgments

The authors would like to acknowledge the financial support from the National Natural Science Foundation of China (Grant Nos. 51431006 and 11774249), the National Key Research Program of China (Grant No. 2016YFA0300101), and the Natural Science Foundation of Jiangsu Province (BK20171209).

References

1 Fujishima, A. and Honda, K. (1972). Electrochemical photolysis of water at a semiconductor electrode. *Nature* 238: 37–38.
2 Hoffmann, M.R., Martin, S.T., Choi, W.Y., and Bahnemann, D.W. (1995). Environmental applications of semiconductor photocatalysis. *Chemical Reviews* 95: 69–96.
3 Gratzel, M. (2001). Photoelectrochemical cells. *Nature* 414: 338–344.
4 Kudo, A. and Miseki, Y. (2009). Heterogeneous photocatalyst materials for water splitting. *Chemical Society Reviews* 38: 253–278.

5 Chen, X.B. and Mao, S.S. (2007). Titanium dioxide nanomaterials: synthesis, properties, modifications, and applications. *Chemical Reviews* 107: 2891–2959.

6 Kudo, A., Omori, K., and Kato, H. (1999). A novel aqueous process for preparation of crystal form-controlled and highly crystalline $BiVO_4$ powder from layered vanadates at room temperature and its photocatalytic and photophysical properties. *Journal of the American Chemical Society* 121: 11459–11467.

7 Chen, D. and Ye, J.H. (2008). Hierarchical WO_3 hollow shells: dendrite, sphere, dumbbell, and their photocatalytic properties. *Advanced Functional Materials* 18: 1922–1928.

8 Kay, A., Cesar, I., and Graetzel, M. (2006). New benchmark for water photooxidation by nanostructured α-Fe_2O_3 films. *Journal of the American Chemical Society* 128: 15714–15721.

9 Li, L., Salvador, P.A., and Rohrer, G.S. (2014). Photocatalysts with internal electric fields. *Nanoscale* 6: 24–42.

10 Linsebigler, A.L., Lu, G.Q., and Yates, J.T. (1995). Photocatalysis on TiO_2 surfaces: principles, mechanisms, and selected results. *Chemical Reviews* 95: 735–758.

11 Subramanian, V., Wolf, E., and Kamat, P.V. (2001). Semiconductor-metal composite nanostructures. To what extent do metal nanoparticles improve the photocatalytic activity of TiO_2 films? *Journal of Physical Chemistry B* 105: 11439–11446.

12 Wang, H.L., Zhang, L.S., Chen, Z.G. et al. (2014). Semiconductor heterojunction photocatalysts: design, construction, and photocatalytic performances. *Chemical Society Reviews* 43: 5234–5244.

13 Lines, M.E. and Glass, A.M. (ed.) (1977). *Principles and Applications of Ferroelectrics and Related Materials*. Oxford: Clarendon Press, Ltd.

14 Morris, M.R., Pendlebury, S.R., Hong, J. et al. (2016). Effect of internal electric fields on charge carrier dynamics in a ferroelectric material for solar energy conversion. *Advanced Materials* 28: 7123–7128.

15 Tiwari, D. and Dunn, S. (2009). Photochemistry on a polarisable semi-conductor: what do we understand today? *Journal of Materials Science* 44: 5063–5079.

16 Kakekhani, A. and Ismail-Beigi, S. (2016). Ferroelectric oxide surface chemistry: water splitting via pyroelectricity. *Journal of Materials Chemistry A* 4: 5235–5246.

17 Khan, M.A., Nadeem, M.A., and Idriss, H. (2016). Ferroelectric polarization effect on surface chemistry and photo-catalytic activity: a review. *Surface Science Reports* 71: 1–31.

18 Kakekhani, A., Ismail-Beigi, S., and Altman, E.I. (2016). Ferroelectrics: a pathway to switchable surface chemistry and catalysis. *Surface Science* 650: 302–316.

19 Konstantinou, I.K. and Albanis, T.A. (2004). TiO_2-assisted photocatalytic degradation of azo dyes in aqueous solution: kinetic and mechanistic investigations – a review. *Applied Catalysis B: Environmental* 49: 1–14.

20 Chatterjee, D. and Dasgupta, S. (2005). Visible light induced photocatalytic degradation of organic pollutants. *Journal of Photochemistry and Photobiology C: Photochemistry Reviews* 6: 186–205.
21 Gaya, U.I. and Abdullah, A.H. (2008). Heterogeneous photocatalytic degradation of organic contaminants over titanium dioxide: a review of fundamentals, progress and problems. *JJournal of Photochemistry and Photobiology C: Photochemistry Reviews* 9: 1–12.
22 Tong, H., Ouyang, S.X., Bi, Y.P. et al. (2012). Nano-photocatalytic materials: possibilities and challenges. *Advanced Materials* 24: 229–251.
23 Maeda, K. (2011). Photocatalytic water splitting using semiconductor particles: history and recent developments. *Journal of Photochemistry and Photobiology C: Photochemistry Reviews* 12: 237–268.
24 Khaselev, O. and Turner, J.A. (1998). A monolithic photovoltaic-photoelectrochemical device for hydrogen production via water splitting. *Science* 280: 425–427.
25 Luo, J.S., Im, J.H., Mayer, M.T. et al. (2014). Water photolysis at 12.3% efficiency via perovskite photovoltaics and earth-abundant catalysts. *Science* 345: 1593–1596.
26 Hisatomi, T., Kubota, J., and Domen, K. (2014). Recent advances in semiconductors for photocatalytic and photoelectrochemical water splitting. *Chemical Society Reviews* 43: 7520–7535.
27 Levanyuk, A.P. and Sannikov, D.G. (1974). Improper ferroelectrics. *Soviet Physics Uspekhi* 17: 199–214.
28 Cheong, S.W. and Mostovoy, M. (2007). Multiferroics: a magnetic twist for ferroelectricity. *Nature Materials* 6: 13–20.
29 Subramanian, M.A., He, T., Chen, J.Z. et al. (2006). Giant room-temperature magnetodielectric response in the electronic ferroelectric $LuFe_2O_4$. *Advanced Materials* 18: 1737–1739.
30 Kobayashi, K., Horiuchi, S., Kumai, R. et al. (2012). Electronic ferroelectricity in a molecular crystal with large polarization directing antiparallel to ionic displacement. *Physical Review Letters* 108: 237601.
31 Yang, W.C., Rodriguez, B.J., Gruverman, A., and Nemanich, R.J. (2005). Photo electron emission microscopy of polarity-patterned materials. *Journal of Physics: Condensed Matter* 17: S1415–S1426.
32 Jones, P.M. and Dunn, S. (2007). Photo-reduction of silver salts on highly heterogeneous lead zirconate titanate. *Nanotechnology* 18: 185702.
33 Bhardwaj, A., Burbure, N.V., Gamalski, A., and Rohrer, G.S. (2010). Composition dependence of the photochemical reduction of Ag by $Ba_{1-x}Sr_xTiO_3$. *Chemistry of Materials* 22: 3527–3534.
34 Jones, P.M. and Dunn, S. (2009). Interaction of Stern layer and domain structure on photochemistry of lead-zirconate-titanate. *Journal of Physics D: Applied Physics* 42: 065408.
35 Nasby, R.D. and Quinn, R.K. (1976). Photoassisted electrolysis of water using a $BaTiO_3$ electrode. *Materials Research Bulletin* 11: 985–992.
36 Giocondi, J.L. and Rohrer, G.S. (2001). Spatial separation of photochemical oxidation and reduction reactions on the surface of ferroelectric $BaTiO_3$. *Journal of Physical Chemistry B* 105: 8275–8277.

37 Giocondi, J.L. and Rohrer, G.S. (2001). Spatially selective photochemical reduction of silver on the surface of ferroelectric barium titanate. *Chemistry of Materials* 13: 241–242.
38 Giocondi, J.L. and Rohrer, G.S. (2008). The influence of the dipolar field effect on the photochemical reactivity of $Sr_2Nb_2O_7$ and $BaTiO_3$ microcrystals. *Topics in Catalysis* 49: 18–23.
39 Cui, Y.F., Briscoe, J., and Dunn, S. (2013). Effect of ferroelectricity on solar-light-driven photocatalytic activity of $BaTiO_3$-influence on the carrier separation and Stern layer formation. *Chemistry of Materials* 25: 4215–4223.
40 Giocondi, J.L. and Rohrer, G.S. (2003). Structure sensitivity of photochemical oxidation and reduction reactions on $SrTiO_3$ surfaces. *Journal of the American Ceramic Society* 86: 1182–1189.
41 Ohara, K., Ohsawa, T., Koinuma, H., and Matsumoto, Y. (2006). Photochemical approach to analysis of ferroelectric transition in $Ba_xSr_{1-x}TiO_3$ epitaxial films. *Japanese Journal of Applied Physics* 45: L339–L342.
42 Kalinin, S.V., Bonnell, D.A., Alvarez, T. et al. (2002). Atomic polarization and local reactivity on ferroelectric surfaces: a new route toward complex nanostructures. *Nano Letters* 2: 589–593.
43 Jones, P.M., Dunn, S., and Gallardo, D.E. (2007). Photochemical growth of silver nanoparticles on c(−) and c(+) domains on lead zirconate titanate thin films. *Journal of the American Chemical Society* 129: 8724–8728.
44 Jones, P.M., Gallardo, D.E., and Dunn, S. (2008). Photochemical investigation of a polarizable semiconductor, lead-zirconate-titanate. *Chemistry of Materials* 20: 5901–5906.
45 Dunn, S., Sharp, S., and Burgess, S. (2009). The photochemical growth of silver nanoparticles on semiconductor surfaces-initial nucleation stage. *Nanotechnology* 20: 115604.
46 Shen, Z.K., Chen, G.P., Chen, Z.H. et al. (2011). Spatially selective photochemical reduction of silver on nanoembossed ferroelectric PZT nanowires. *Langmuir* 27: 5167–5170.
47 Chen, J., Lu, H.D., Liu, H.J. et al. (2013). Interface control of surface photochemical reactivity in ultrathin epitaxial ferroelectric films. *Applied Physics Letters* 102: 182904.
48 Zhang, M., Jiang, C.X., Dong, W. et al. (2013). Composition dependence of the photochemical reduction of Ag^+ by as-grown $Pb(Zr_xTi_{1-x})O_3$ films on indium tin oxide electrode. *Applied Physics Letters* 103: 102902.
49 Liu, X.Y., Kitamura, K., Terabe, K. et al. (2007). Photocatalytic nanoparticle deposition on $LiNbO_3$ nanodomain patterns via photovoltaic effect. *Applied Physics Letters* 91: 044101.
50 Sun, Y. and Nemanich, R.J. (2011). Photoinduced Ag deposition on periodically poled lithium niobate: wavelength and polarization screening dependence. *Journal of Applied Physics* 109: 104302.
51 Stock, M. and Dunn, S. (2012). Influence of the ferroelectric nature of lithium niobate to drive photocatalytic dye decolorization under artificial solar light. *Journal of Physical Chemistry C* 116: 20854–20859.
52 Clark, S.J. and Robertson, J. (2007). Band gap and Schottky barrier heights of multiferroic $BiFeO_3$. *Applied Physics Letters* 90: 132903.

53 Ihlefeld, J.F., Podraza, N.J., Liu, Z.K. et al. (2008). Optical band gap of $BiFeO_3$ grown by molecular-beam epitaxy. *Applied Physics Letters* 92: 142908.
54 Schultz, A.M., Zhang, Y.L., Salvador, P.A., and Rohrer, G.S. (2011). Effect of crystal and domain orientation on the visible-light photochemical reduction of Ag on $BiFeO_3$. *ACS Applied Materials & Interfaces* 3: 1562–1567.
55 Burbure, N.V., Salvador, P.A., and Rohrer, G.S. (2010). Photochemical reactivity of titania films on $BaTiO_3$ substrates: origin of spatial selectivity. *Chemistry of Materials* 22: 5823–5830.
56 Burbure, N.V., Salvador, P.A., and Rohrer, G.S. (2010). Photochemical reactivity of titania films on $BaTiO_3$ substrates: influence of titania phase and orientation. *Chemistry of Materials* 22: 5831–5837.
57 Zhang, Y.L., Schultz, A.M., Salvador, P.A., and Rohrer, G.S. (2011). Spatially selective visible light photocatalytic activity of $TiO_2/BiFeO_3$ heterostructures. *Journal of Materials Chemistry* 21: 4168–4174.
58 Kalinin, S.V., Bonnell, D.A., Alvarez, T. et al. (2004). Ferroelectric lithography of multicomponent nanostructures. *Advanced Materials* 16: 795–799.
59 Haussmann, A., Milde, P., Erler, C., and Eng, L.M. (2009). Ferroelectric lithography: bottom-up assembly and electrical performance of a single metallic nanowire. *Nano Letters* 9: 763–768.
60 Shen, Z.K., Qu, X.P., Chen, Y.F., and Liu, R. (2011). Nanoembossing induced ferroelectric lithography on PZT Films for silver particle patterning. *ACS Nano* 5: 6855–6860.
61 Carville, N.C., Manzo, M., Damm, S. et al. (2012). Photoreduction of SERS-active metallic nanostructures on chemically patterned ferroelectric crystals. *ACS Nano* 6: 7373–7380.
62 Seal, K., Rodriguez, B.J., Ivanov, I.N., and Kalinin, S.V. (2014). Anomalous photodeposition of Ag on ferroelectric surfaces with below-bandgap excitation. *Advanced Optical Materials* 2: 292–299.
63 Su, R., Shen, Y.J., Li, L.L. et al. (2015). Silver-modified nanosized ferroelectrics as a novel photocatalyst. *Small* 11: 202–207.
64 Lin, X.P., Xing, J.C., Wang, W.D. et al. (2007). Photocatalytic activities of heterojunction semiconductors $Bi_2O_3/BaTiO_3$: a strategy for the design of efficient combined photocatalysts. *Journal of Physical Chemistry C* 111: 18288–18293.
65 Huang, H.J., Li, D.Z., Lin, Q. et al. (2009). Efficient photocatalytic activity of PZT/TiO_2 heterojunction under visible light irradiation. *Journal of Physical Chemistry C* 113: 14264–14269.
66 Li, L., Zhang, Y.L., Schultz, A.M. et al. (2012). Visible light photochemical activity of heterostructured $PbTiO_3$–TiO_2 core–shell particles. *Catalysis Science & Technology* 2: 1945–1952.
67 Wang, R.X., Zhu, Q., Wang, W.S. et al. (2015). $BaTiO_3$–graphene nanocomposites: synthesis and visible light photocatalytic activity. *New Journal of Chemistry* 39: 4407–4413.
68 Gao, F., Chen, X.Y., Yin, K.B. et al. (2006). Visible-light photocatalytic properties of weak magnetic $BiFeO_3$ nanoparticles. *Advanced Materials* 19: 2889–2892.

69 He, J., Guo, R.Q., Fang, L. et al. (2013). Characterization and visible light photocatalytic mechanism of size-controlled $BiFeO_3$ nanoparticles. *Materials Research Bulletin* 48: 3017–3024.

70 Lu, X.M., Xie, J.M., Song, Y.Z., and Lin, J.M. (2007). Surfactant-assisted hydrothermal preparation of submicrometer-sized two-dimensional $BiFeO_3$ plates and their photocatalytic activity. *Journal of Materials Science* 42: 6824–6827.

71 Li, S., Lin, Y.H., Zhang, B.P. et al. (2010). Controlled fabrication of $BiFeO_3$ uniform microcrystals and their magnetic and photocatalytic behaviors. *Journal of Physical Chemistry C* 114: 2903–2908.

72 Huo, Y.N., Miao, M., Zhang, Y. et al. (2011). Aerosol-spraying preparation of a mesoporous hollow spherical $BiFeO_3$ visible photocatalyst with enhanced activity and durability. *Chemical Communications* 47: 2089–2091.

73 Fei, L.F., Yuan, J.K., Hu, Y.M. et al. (2011). Visible light responsive perovskite $BiFeO_3$ pills and rods with dominant $\{111\}_c$ facets. *Crystal Growth and Design* 11: 1049–1053.

74 Bharathkumar, S., Sakar, M., and Balakumar, S. (2016). Experimental evidence for the carrier transportation enhanced visible light driven photocatalytic process in bismuth ferrite ($BiFeO_3$) one-dimensional fiber nanostructures. *Journal of Physical Chemistry C* 120: 18811–18821.

75 Guo, R.Q., Fang, L., Dong, W. et al. (2010). Enhanced photocatalytic activity and ferromagnetism in Gd doped $BiFeO_3$ nanoparticles. *Journal of Physical Chemistry C* 114: 21390–21396.

76 Sakar, M., Balakumar, S., Saravanan, P., and Bharathkumar, S. (2015). Compliments of confinements: substitution and dimension induced magnetic origin and band-bending mediated photocatalytic enhancements in $Bi_{1-x}Dy_xFeO_3$ particulate and fiber nanostructures. *Nanoscale* 7: 10667–10679.

77 Feng, Y.D., Wang, H.C., Luo, Y.D. et al. (2013). Ferromagnetic and photocatalytic behaviors observed in Ca-doped $BiFeO_3$ nanofibres. *Journal of Applied Physics* 113: 146101.

78 Li, S., Lin, Y.H., Zhang, B.P. et al. (2009). $BiFeO_3/TiO_2$ core–shell structured nanocomposites as visible-active photocatalysts and their optical response mechanism. *Journal of Applied Physics* 105: 054310.

79 Guo, R.Q., Fang, L., Dong, W. et al. (2011). Magnetically separable $BiFeO_3$ nanoparticles with a γ-Fe_2O_3 parasitic phase: controlled fabrication and enhanced visible-light photocatalytic activity. *Journal of Materials Chemistry* 21: 18645–18652.

80 Niu, F., Chen, D., Qin, L.S. et al. (2015). Facile synthesis of highly efficient p–n heterojunction $CuO/BiFeO_3$ composite photocatalysts with enhanced visible-light photocatalytic activity. *ChemCatChem* 7: 3279–3289.

81 Yang, Y.C., Liu, Y., Wei, J.H. et al. (2014). Electrospun nanofibers of p-type $BiFeO_3$/n-type TiO_2 hetero-junctions with enhanced visible-light photocatalytic activity. *RSC Advances* 4: 31941–31947.

82 Fan, T., Chen, C.C., and Tang, Z.H. (2016). Hydrothermal synthesis of novel $BiFeO_3/BiVO_4$ heterojunctions with enhanced photocatalytic activities under visible light irradiation. *RSC Advances* 6: 9994–10000.

83 Moitra, D., Ghosh, B.K., Chandel, M., and Ghosh, N.N. (2016). Synthesis of a $BiFeO_3$ nanowire-reduced graphene oxide based magnetically separable nanocatalyst and its versatile catalytic activity towards multiple organic reactions. *RSC Advances* 6: 97941–97952.

84 Wang, X.F., Fan, J.C., Qian, F., and Min, Y.L. (2016). Magnetic $BiFeO_3$ grafted with MWCNT hybrids as advanced photocatalysts for removing organic contamination with a high concentration. *RSC Advances* 6: 49966–49972.

85 Ramezanalizadeh, H. and Manteghi, F. (2016). A novel MOF/$BiFeO_3$ composite as a catalyst for efficient and selective oxidation of alcohols. *RSC Advances* 6: 99096–99104.

86 Inoue, Y., Okamura, M., and Sato, K. (1985). A thin-film semiconducting TiO_2 combined with ferroelectrics for photoassisted water decomposition. *Journal of Physical Chemistry* 89: 5184–5187.

87 Inoue, Y., Sato, K., Sato, K., and Miyamat, H. (1986). Photoassisted water decomposition by ferroelectric lead zirconate titanate ceramics with anomalous photovoltatic effects. *Journal of Physical Chemistry C* 90: 2809–2810.

88 Inoue, Y., Hayashi, K., and Sato, K. (1990). Photocatalytic activities of potassium-doped lead niobates and the effect of poling. *Journal of the American Ceramic Society* 86 (12): 2277–2282.

89 Arney, D., Watkins, T., and Maggard, P.A. (2011). Effects of particle surface areas and microstructures on photocatalytic H_2 and O_2 production over $PbTiO_3$. *Journal of the American Ceramic Society* 94: 1483–1489.

90 Luo, J.H. and Maggard, P.A. (2006). Hydrothermal synthesis and photocatalytic activities of $SrTiO_3$-coated Fe_2O_3 and $BiFeO_3$. *Advanced Materials* 18: 514–517.

91 Gao, F., Yuan, Y., Wang, K.F. et al. (2006). Preparation and photoabsorption characterization of $BiFeO_3$ nanowires. *Applied Physics Letters* 89: 102506.

92 Li, S., Zhang, J.M., Kibria, M.G. et al. (2013). Remarkably enhanced photocatalytic activity of laser ablated Au nanoparticle decorated $BiFeO_3$ nanowires under visible-light. *Chemical Communications* 49: 5856–5858.

93 Li, L., Rohrer, G.S., and Salvador, P.A. (2012). Heterostructured ceramic powders for photocatalytic hydrogen production: nanostructured TiO_2 shells surrounding microcrystalline (Ba,Sr)TiO_3 cores. *Journal of the American Ceramic Society* 95: 1414–1420.

94 Park, S., Lee, C.W., Kang, M.G. et al. (2014). A ferroelectric photocatalyst for enhancing hydrogen evolution: polarized particulate suspension. *Physical Chemistry Chemical Physics* 16: 10408–10413.

95 Yang, W.G., Yu, Y.H., Starr, M.B. et al. (2015). Ferroelectric polarization-enhanced photoelectrochemical water splitting in TiO_2–$BaTiO_3$ core–shell nanowire photoanodes. *Nano Letters* 15: 7574–7580.

96 Chen, X.Y., Yu, T., Gao, F. et al. (2007). Application of weak ferromagnetic $BiFeO_3$ films as the photoelectrode material under visible-light irradiation. *Applied Physics Letters* 91: 022114.

97 Cao, D.W., Wang, Z.J., Nasori et al. (2014). Switchable charge-transfer in the photoelectrochemical energy-conversion process of ferroelectric

$BiFeO_3$ photoelectrodes. *Angewandte Chemie International Edition* 53: 11027–11031.

98 Liu, Q., Zhou, Y., You, L. et al. (2016). Enhanced ferroelectric photoelectrochemical properties of polycrystalline $BiFeO_3$ film by decorating with Ag nanoparticles. *Applied Physics Letters* 108: 022902.

99 Moniz, S.J.A., Quesada-Cabrera, R., Blackman, C.S. et al. (2014). A simple, low-cost CVD route to thin films of $BiFeO_3$ for efficient water photo-oxidation. *Journal of Materials Chemistry A* 2: 2922–2927.

100 Yilmaz, P., Yeo, D., Chang, H. et al. (2016). Perovskite $BiFeO_3$ thin film photocathode performance with visible light activity. *Nanotechnology* 27: 345402.

101 Maruyama, R., Sakamoto, W., Yuitoo, I. et al. (2016). Photocurrent enhancement of chemically synthesized Ag nanoparticle-embedded $BiFeO_3$ thin films. *Japanese Journal of Applied Physics* 55: 10TA14.

102 Ji, W., Yao, K., Lim, Y.F. et al. (2013). Epitaxial ferroelectric $BiFeO_3$ thin films for unassisted photocatalytic water splitting. *Applied Physics Letters* 103: 062901.

103 Rioult, M., Datta, S., Stanescu, D. et al. (2015). Tailoring the photocurrent in $BaTiO_3$/Nb:$SrTiO_3$ photoanodes by controlled ferroelectric polarization. *Applied Physics Letters* 107: 103901.

104 Nechache, R., Hamagea, C., Li, S. et al. (2015). Bandgap tuning of multiferroic oxide solar cells. *Nature Photonics* 9: 61–67.

105 Li, S., Alotaibi, B., Huang, W. et al. (2015). Epitaxial Bi_2FeCrO_6 multiferroic thin film as a new visible light absorbing photocathode material. *Small* 11 (32): 4018–4026.

106 Wang, Z.J., Cao, D.W., Wen, L.Y. et al. (2016). Manipulation of charge transfer and transport in plasmonic-ferroelectric hybrids for photoelectrochemical applications. *Nature Communications* 7: 10348.

107 Huang, Y.L., Chang, W.S., Van, C.N. et al. (2016). Tunable photoelectrochemical performance of Au/$BiFeO_3$ heterostructure. *Nanoscale* 8: 15795–15801.

108 Atwater, H.A. and Polman, A. (2010). Plasmonics for improved photovoltaic devices. *Nature Materials* 9: 205–213.

109 Takai, A. and Kamat, P.V. (2011). Capture, store, and discharge. Shuttling photogenerated electrons across TiO_2–silver interface. *ACS Nano* 5: 7369–7376.

110 Linic, S., Christopher, P., and Ingram, D.B. (2011). Plasmonic-metal nanostructures for efficient conversion of solar to chemical energy. *Nature Materials* 10: 911–921.

111 Moshnyaga, V., Damaschke, B., Shapoval, O. et al. (2003). Structural phase transition at the percolation threshold in epitaxial $(La_{0.7}Ca_{0.3}MnO_3)_{1-x}$:$(MgO)_x$ nanocomposite films. *Nature Materials* 2: 247–252.

112 Zheng, H., Wang, J., Lofland, S.E. et al. (2004). Multiferroic $BaTiO_3$–$CoFe_2O_4$ nanostructures. *Science* 303: 661–663.

113 MacManus-Driscoll, J.L., Foltyn, S.R., Jia, Q.X. et al. (2004). Strongly enhanced current densities in superconducting coated conductors of $YBa_2Cu_3O_{7-x}$ + $BaZrO_3$. *Nature Materials* 3: 439–443.
114 Quynh, L.T., Van, C.N., Bitla, Y. et al. (2016). Self-assembled $BiFeO_3$-ε-Fe_2O_3 vertical heteroepitaxy for visible light photoelectrochemistry. *Advance Energy Materials* 6: 1600686.
115 Cho, S., Jang, J.W., Zhang, W. et al. (2015). Single-crystalline thin films for studying intrinsic properties of $BiFeO_3$–$SrTiO_3$ solid solution photoelectrodes in solar energy conversion. *Chemistry of Materials* 27: 6635–6641.
116 Do, T.H., Van, C.N., Tsai, K.A. et al. (2016). Superior photoelectrochemical activity of self-assembled $NiWO_4$–WO_3 heteroepitaxy. *Nano Energy* 23: 153–160.
117 Cho, S., Jang, J.W., Li, L.G. et al. (2016). Self-assembled heteroepitaxial oxide nanocomposite for photoelectrochemical solar water oxidation. *Chemistry of Materials* 28: 3017–3023.
118 Fridkin, V.M. (2001). Bulk photovoltaic effect in noncentrosymmetric crystals. *Crystallography Reports* 46: 654–658.
119 Tan, L.Z., Zheng, F., Young, S.M. et al. (2016). Shift current bulk photovoltaic effect in polar materials – hybrid and oxide perovskites and beyond. *Npj Computer Materials* 2: 16026.
120 Grinberg, I., West, D.V., Torres, M. et al. (2013). Perovskite oxides for visible-light-absorbing ferroelectric and photovoltaic materials. *Nature* 503: 509–512.
121 Chu, K., Jang, B.K., Sung, J.H. et al. (2015). Enhancement of the anisotropic photocurrent in ferroelectric oxides by strain gradients. *Nature Nanotechnology* 10: 972–979.
122 Kojima, A., Teshima, K., Shirai, Y., and Miyasaka, T. (2009). *Journal of the American Chemical Society* 131: 6050–6051.

10

First-Principles Calculations on Ferroelectrics for Energy Applications

Gelei Jiang[1], *Weijin Chen*[1,2], *and Yue Zheng*[1]

[1] *Sun Yat-sen University, State Key Laboratory of Optoelectronic Materials and Technologies, School of Physics, No. 135, Xingang Xi Road, Guangzhou 510275, PR China*
[2] *Sun Yat-sen University, Sino-French Institute of Nuclear Engineering and Technology, Tang Jia Wan, Zhuhai 519082, PR China*

10.1 Introduction

Many energy conversion and storage techniques utilize ferroelectric materials for their fascinating coupling behaviors of electric polarization with external fields as well as their high dielectric performance, which is promising in the development of supercapacitors. In practical applications, one of the issues of most concern is energy conversion efficiency. Another is the energy storage capacity. To have a deep insight into these issues, the atomic mechanisms for related effects are worth exploring. While there remain many challenges, first-principles calculations based on density functional theory (DFT) and, more recently, first-principles-derived simulation methods (e.g. effective Hamiltonian method) have been proved powerful tools in accomplishing the goal.

In this chapter, we review the recent progress in first-principles calculations and first-principles-derived simulations on ferroelectrics for energy applications – energy conversion and energy storage. First, the basic idea of first-principles calculations and effective Hamiltonian method will be illustrated. Afterwards, we talk about the energy conversion based on electromechanical coupling. This issue has been revealed in many ferroelectric systems based on first-principles calculations, with fruitful information on the piezoelectric properties, flexoelectric properties, and their modifications by external mechanical and electrical fields. Abnormal electromechanical behaviors that are caused by novel phase transitions (e.g. polar–toroidal phase transition in nanostructures) will be emphasized.

Then, we turn to the photovoltaic effect in ferroelectrics. There has been extensive interest in the photoinduced effect in ferroelectric materials, particularly in perovskites with relatively low band gaps. The engineering of band gap and doping effect can be readily investigated by first-principles calculations. Also noteworthy is the potential in energy applications based on the pyroelectric and electrocaloric effects of ferroelectrics. These issues are now active and

Ferroelectric Materials for Energy Applications, First Edition.
Edited by Haitao Huang and James F. Scott.

have also been explored by first-principles calculations. Related aspects, e.g. the electric-field-induced change in temperature and polarization response in temperature-graded ferroelectrics, can be calculated at nonzero temperature by using first-principles-derived methods. Moreover, based on the dielectric properties of ferroelectrics and the process of energy storage and release during phase transition, ferroelectrics have great potential in high-density energy storage. The results of this issue predicted by first-principles calculations will be discussed. The content of this chapter should provide a general insight into the current state of first-principles calculations on ferroelectrics for energy applications.

10.2 Methods

10.2.1 First-Principles Calculations

First-principles methods, also known as *ab initio* methods, aim to predict material properties from the basic principle of quantum mechanics, without the use of empirical parameters or much simplifications, yet with high accuracy. With the development of DFT [1], the Schrodinger equation of multi-electron interaction can be readily solved under some reasonable approximation, and with the inputs of only some basic physical parameters, such as lattice constants, atomic species, and atom numbers. Consequently, the electronic structure and the stable atomic configuration of a system in its ground state can be obtained.

Since 1990s, first-principles calculations have been widely applied to investigate the properties of ferroelectrics. This is driven by several theoretical advances. First, the development of modern theory of polarization and magnetization based on Berry phase method [2, 3] provides access to quantities such as piezoelectric tensors [2, 4–6], Born effective charges [7], spontaneous polarization [8–11], magnetoelectric tensors [12], and magnetization [12]. Secondly, maximally localized Wannier functions [13], used in the construction of effective Hamiltonians [14, 15], broaden the application of DFT to the study of transport properties [16–18] of nanostructures. Wannier functions also help analyze the chemical bonds [19] in samples. Furthermore, a combination of DFT and perturbation theory called density functional perturbation theory (DFPT) [20] facilitates the calculation of a wide spectrum of important properties [21] such as, optical and static dielectric constants, Born effective charges, proper piezoelectric tensor, elastic constants, photovoltaic coefficients, and pyroelectric coefficients.

10.2.2 First-Principles-Derived Effective Hamiltonian Method

The effective Hamiltonian method can be used to study the characteristics of a system by building an effective Hamiltonian with a reduced number of degrees of freedom. In this chapter, the effective Hamiltonian method refers to that developed by Zhong, Vanderbilt, Rabe, and Waghmare et al. [14, 22, 23] in the 1990s, with a particular treatment on ferroelectric phase transition in the framework of group theory and lattice dynamics. In this method, the effective

Hamiltonian is written as a polynomial of reduced degrees of freedom. Taking perovskite ferroelectric $BaTiO_3$ (BTO) as an example, for a system with cells number of N (each cell is denoted by an index i), its effective Hamiltonian can be written as a function of the local mode vector $\mathbf{u}_i$, the local acoustic displacement vector $\mathbf{v}_i$, and the homogenous strain tensor $\eta_{H,l}$ ($l = 1$–6):

$$\begin{aligned} E^{\text{eff}} &= E^{\text{self}}(\{\mathbf{u}_i\}) + E^{\text{dpl}}(\{\mathbf{u}_i\}) + E^{\text{short}}(\{\mathbf{u}_i\}) \\ &\quad + E^{\text{elas}}(\{\mathbf{v}_i\}, \{\eta_{\text{H},l}\}) + E^{\text{int}}(\{\mathbf{u}_i\}, \{\mathbf{v}_i\}, \{\eta_{\text{H},l}\}) \end{aligned} \tag{10.1}$$

where E^{self} is the local mode self-energy, E^{dpl} is the long-range dipole–dipole interaction, E^{short} is the short-range interaction, E^{elas} is the elastic energy, and E^{int} is the local mode elastic coupling energy. Note that the effective Hamiltonian usually has the form of a Taylor expansion about the cubic perovskite phase, and the parameters are determined by fitting to the first-principles potential-energy surfaces. For a successful effective Hamiltonian, two conditions must be met: (i) The Taylor expansion must accurately reproduce the low-energy landscape, and (ii) anharmonic cross-terms involving neglected degrees of freedom must not have a significant effect on the thermodynamics of the system.

After construction of the effective Hamiltonian, the dipole state of the ferroelectric system is often solved by the molecular dynamics (MD) method [24]. Unlike the atomistic MD method, the degrees of freedom here are the local mode $\mathbf{u}_i$, the local displacement vector $\mathbf{v}_i$, and the homogenous strain tensor $\eta_{H,l}$. By adding the kinetic energy terms to Eq. (10.1), the total Hamiltonian of the system can be expressed as [25, 26]

$$\begin{aligned} H^{\text{eff}} &= \frac{M^*_{\text{dipole}}}{2} \sum_i \dot{\mathbf{u}}_i^2 + \frac{M^*_{\text{acoustic}}}{2} \sum_{R,\alpha} \dot{\mathbf{v}}_i^2 \\ &\quad + E^{\text{self}}(\{\mathbf{u}_i\}) + E^{\text{dpl}}(\{\mathbf{u}_i\}) + E^{\text{short}}(\{\mathbf{u}_i\}) \\ &\quad + E^{\text{elas}}(\{\mathbf{v}_i\}, \{\eta_{\text{H},l}\}) + E^{\text{int}}(\{\mathbf{u}_i\}, \{\mathbf{v}_i\}, \{\eta_{\text{H},l}\}) \end{aligned} \tag{10.2}$$

Besides the MD method, the effective Hamiltonian method is often combined with Monte Carlo (MC) method to simulate the equilibrium state of ferroelectric systems.

10.3 Energy Conversion

10.3.1 Piezoelectric and Flexoelectric Effect

Piezoelectricity refers to the phenomena that applied mechanical strain/stress can induce electric charges linearly (direct effect) in certain materials while the application of an electric field generates a mechanical deformation of the material proportional to the applied field (converse effect). Piezoelectricity thus provides a common mechanism for achieving the conversion between electrical and mechanical energy. To characterize piezoelectric effect, mathematical objects related to physical quantities that are called tensors are introduced. The existence of such tensors, as well as the magnitude and sign of their coefficients, has a critical impact on the functionality of piezoelectric devices. Since the mid-1990s, first-principles calculations have been applied to calculate the piezoelectric constants of many ferroelectric materials, with the aim of designing high-performance piezoelectric materials.

The direct piezoelectric effect can be described by a formula in terms of the applied stress σ_j ($j = 1–6$) and the resulting polarization P_i ($i = 1–3$) as

$$P_i = d_{ij}\sigma_j \tag{10.3}$$

where coefficients d_{ij} are the components of a third-order piezoelectric strain tensor, using the Voigt (or matrix) notation and the Einstein summation convention. Note that the number of nonzero piezoelectric coefficients can be much smaller than 27 due to the certain lattice symmetry of the material. Similarly, the converse piezoelectric effect can be described by the relationship between the applied electric filed E_i and the field-induced strain η_j as

$$\eta_j = d_{ij}E_i \tag{10.4}$$

Or one can connect stress tensor and electric field (at fixed strain) as

$$\sigma_j = e_{ij}E_i \tag{10.5}$$

where the piezoelectric stress coefficients e_{ij} also belong to a third-order tensor. In first-principles calculations, piezoelectric stress coefficients e_{ij} are often decomposed into homogeneous strain contributions $e_{ij,\,c}$ and internal strain contributions $e_{ij,\,\mathrm{int}}$:

$$e_{ij} = e_{ij,c} + e_{ij,int} \tag{10.6}$$

$$e_{ij,int} = \sum_k \frac{ea_i}{\Omega} Z^*{}_{k,ii}(k) \frac{du_i(k)}{dS_j} \tag{10.7}$$

where k is the amount of the atoms, e is the magnitude of the electron charge, a_i is the lattice constant along the i-direction, and Ω is the volume.

Generally speaking, there are two ways to compute piezoelectric properties of perovskite solid solutions via first-principles calculations. The most straightforward approach is to calculate using the supercell which consists of multiple primitive cells to simulate the real configuration of a solid solution. Another way is to use the virtual crystal approximation (VCA). By constructing a virtual cell to replace the true solid solution, the calculation can be completed in one unit cell. Although incapable of capturing local atomic features of real material, the use of the VCA method can represent average properties quite well and greatly save computing resources compared with the direct supercell method. Ramer and Rappe first [27, 28] applied the VCA method to predict the piezoelectric properties of the $Pb(Zr_{0.5}Ti_{0.5})O_3$ (PZT) solid solution. They developed new methods for more accurately portraying the electronic behavior of the VC atoms, with their results in good agreement with the superlattice calculations and experimental results.

An approach for the determination of piezoelectric coefficient was reported by Fu and Cohen [29] using first-principles calculation. For the first time, it was demonstrated that a large piezoelectric response can be driven by polarization rotation induced by an external electric field. The free energy, polarization, and strain response of a BTO single domain crystal were calculated as a function of an external electric field $\boldsymbol{E}$. The d_{ij} parameter was calculated as the slope of the strain-versus-electric field curve, in accord with Eq. (10.4). Along a specific polarization rotation path (the path of a → f → g → e shown in Figure 10.1A), the free energy change is small while the increase of *c*-axis polarization is large; a large piezoresponse can thus be obtained.

Actually, the giant electromechanical responses often exist close to some kind of criticality [32]. Using first-principles linear response calculations, Asthagiri et al. [30] concluded the relation between the piezoelectricity and the pressure-induced phase transitions in $PbTiO_3$ (PTO). A series of transitions from tetragonal (T) to monoclinic (M) to rhombohedral (R) and then to cubic were observed with increasing pressures, as depicted in the top of Figure 10.1B. The piezoelectric coefficients peaks in the phase transition region near the monoclinic phase (the middle and bottom of Figure 10.1B) indicate a large electromechanical response in this region. The giant piezoelectric responses of $Pb(Mg_{1/3}Nb_{2/3})O_3$-$PbTiO_3$ (PMN-PT)-type relaxor ferroelectrics near the morphotropic phase boundary have been attributed to reorientation of polar domains by the applied electric field and/or to field-induced polarization rotations [33–35].

It is of primary importance to predict finite-temperature piezoelectric properties of ferroelectrics, and the first-principles-derived effective Hamiltonian method developed for ferroelectrics is capable of doing this. In this method, an analytical expression of the total internal energy E^{tot} is derived in terms of a few degrees of freedom. The parameters of E^{tot} as shown in Eq. (10.1) are determined by the results of first-principles calculations on small supercells. Once the effective Hamiltonian is fully specified, it can be used in MC or MD simulations on large supercells to compute finite-temperature properties of ferroelectrics. The piezoelectric strain coefficients d_{ij} can be computed as the slope of the polarization-versus-stress curve or as the slope of the strain-versus-electric field curve according to Eqs. (10.4) and (10.5), respectively. In an early work, Bellaiche et al. [36] calculated the material characteristics of the PZT solid solution at a fixed temperature using MC simulations. A curve for d_{ij} changes with temperature is given in Figure 10.1C where solid lines and dashed lines correspond to the cases of including and neglecting the local energy E_{loc}, which includes the *on-site* effect of alloying on self-energy up to the fourth order in $\mathbf{u}_i$, and the intersite contributions. These results indicate that incorporating E_{loc} in the total energy leads to an enhancement of the d_{15} coefficient, and further results in an enhancement of the apparent d_{33} value in ceramic samples, which involves an average of the single crystal coefficients as

$$d_{33,c} = \int_0^{\pi/2} [(d_{31} + d_{15})\sin^2\theta + d_{33}\cos^2\theta] \sin\theta \cos\theta \, d\theta \tag{10.8}$$

The calculated $d_{33,c}$ of 163 $pC\,N^{-1}$ at room temperature is in excellent agreement with the experimental value of 170 $pC\,N^{-1}$ [37, 38], while neglecting E_{loc} leads to a smaller $d_{33,c}$ of 90 $pC\,N^{-1}$.

In addition to calculating the piezoelectric properties, first-principles calculations and the effective Hamiltonian method can also be applied to the calculation

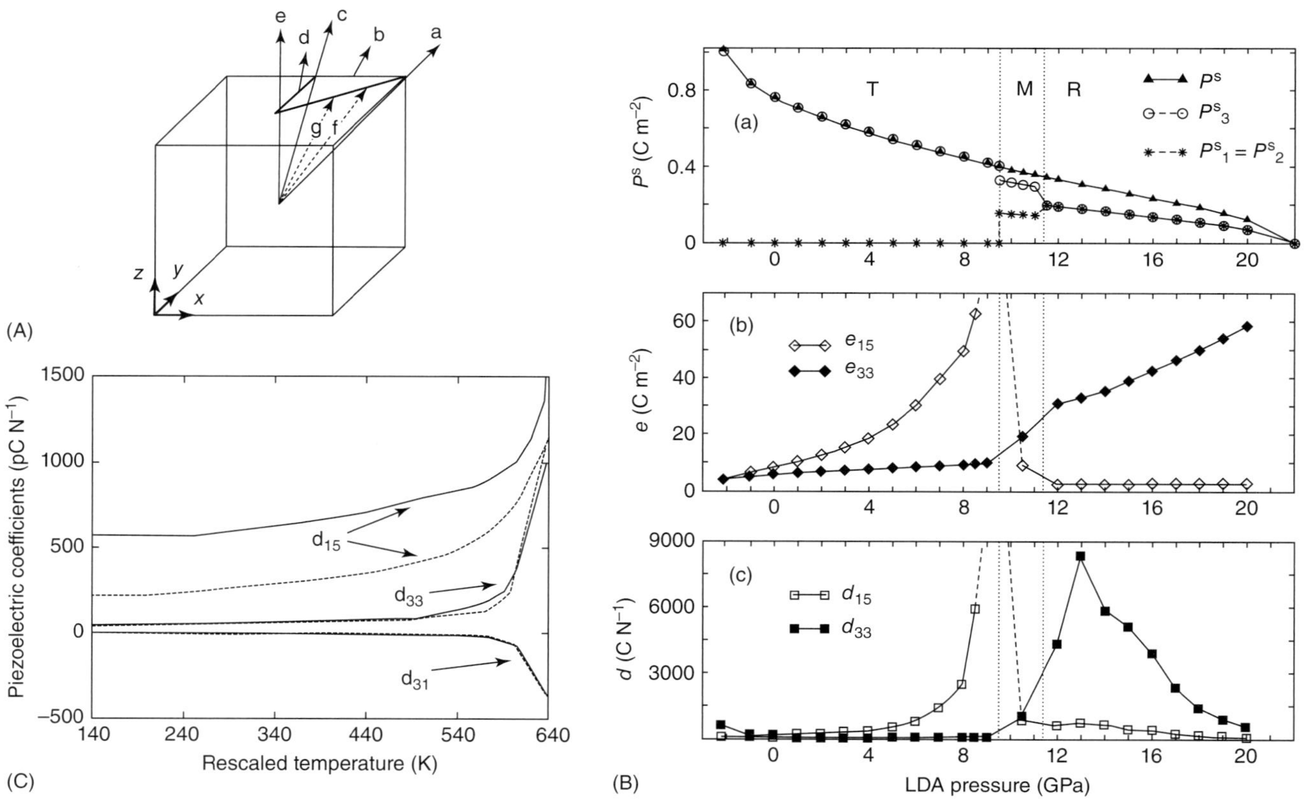

Figure 10.1 (A) Schematic illustration of the polarization directions. The lowest free energy path is found along the path a–f–g–e. *Source:* Fu and Cohen 2000 [29]. Reprinted with permission of Springer Nature. (B) Computed pressure dependence on piezoelectricity in $PbTiO_3$. From the top to the bottom are spontaneous polarization P^s, piezoelectric stress coefficients e_{15} and e_{33}, and piezoelectric strain coefficients d_{15} and d_{33}, respectively. *Source:* Asthagiri et al. 2006 [30]. Reprinted with permission of Taylor & Francis. (C) Piezoelectric coefficients as a function of temperature in disordered (4*mm*) $Pb(Zr_{0.5}Ti_{0.5})O_3$. Solid lines and dashed lines correspond to the cases of including and neglecting of E_{loc}. *Source:* Walizer et al. 2006 [31]. Reprinted with permission of American Physical Society.

of novel electromechanical behaviors such as giant piezoelectric resistance (GPR) effect and piezotoroidal effect. The former denotes the significant influence of mechanical stress on the electronic transport behavior especially found in ferroelectric and piezoelectric thin films [39–41]. The latter refers to a new kind of electromechanical effect related to the coupling between dipole toroidal moment and mechanical stress. A new form for mechanical–electrical energy conversion can be put forward. In order to go into this, we need to first remember the fact that ferroelectric nanoparticles tend to form a vortex structure for their electric dipoles under open-circuit boundary conditions [42, 43]. The vortex structure eliminates spontaneous polarization and is characterized by a toroidization, denoted by $\mathbf{g}$, and it is proportional to a cross product of the dipole moment $\mathbf{p}_i$ at site i with $\mathbf{r}_i$ the position vector:

$$\mathbf{g} = \frac{1}{2V}\sum_{i}\mathbf{r}_i \times \mathbf{p}_i \tag{10.9}$$

where V is the volume of the unit cell. Here, $\mathbf{g}$ is introduced in a manner similar to the definition of polarization, that is, the toroidal moment per unit volume. According to Dubovik et al. [44], the combination of all energetic terms linear in $\mathbf{g}$ provides a coupling of $-\mathbf{g}\cdot\mathrm{curl}\,\mathbf{E}$ with applied electric field $\mathbf{E}$, with curl$\mathbf{E}$ denoting the cross product $\nabla \times \mathbf{E}$. This describes how $\mathbf{g}$ appears in a thermodynamic potential Φ and provides a relevant energetic term, as shown in Eq. (10.10):

$$\Phi = -(\mathbf{g} - \mathbf{g}^{\mathrm{eq}}) \cdot (curl\mathbf{E}) - (\boldsymbol{\eta} - \boldsymbol{\eta}^{\mathrm{eq}}) \cdot \sigma + (curl\mathbf{E}) \cdot \mathbf{d}^{g} \cdot \sigma + \Phi_0 \tag{10.10}$$

where the "eq" superscript is used to indicate the value of the properties at equilibrium. Based on this formula, Prosandeev et al. [45] suggested that a tensor that connects curl$\mathbf{E}$ and σ should exist in toroidal systems, by analogy with the piezoelectric tensor $\mathbf{d}$ connecting $\mathbf{E}$ and strain in polar systems. This tensor is named the axial piezotoroidal tensor and is denoted by $\mathbf{d}_g$. Based on dissipation theorem [46, 47], it can be expressed as

$$d^{g}_{\alpha\beta} = \frac{V}{k_{\mathrm{B}}T}[\langle g_\alpha \eta_\beta\rangle - \langle g_\alpha\rangle\langle\eta_\beta\rangle] \tag{10.11}$$

where "⟨⟩" denotes the thermodynamic averaging. Moreover, an electric toroidal susceptibility tensor is proposed in analogy with the "normal" dielectric susceptibility tensor as

$$\chi^{g}_{\alpha\beta} = \frac{V}{\varepsilon_0 k_{\mathrm{B}}T}[\langle g_\alpha g_\beta\rangle - \langle g_\alpha\rangle\langle g_\beta\rangle] \tag{10.12}$$

where k_{B} is the Boltzmann constant and ε_0 is the dielectric permittivity of vacuum.

Prosandeev et al. [45] performed first-principles-based simulations on stress-free $PbZr_{0.6}Ti_{0.4}O_3$ nanoparticles under open-circuit conditions. Meanwhile, the fluctuation of the toroidization and strain at different MC sweeps are used to compute these tensors according to Eqs. (10.11) and (10.12), as shown in Figure 10.2. Beyond that, mechanisms of vortex switching, vortex rotation, and vortex-to-polarization transformation induced by various kinds of electric fields and mechanical loads in ferroelectric nanostructures have also been investigated by effective Hamiltonian simulations [48–53].

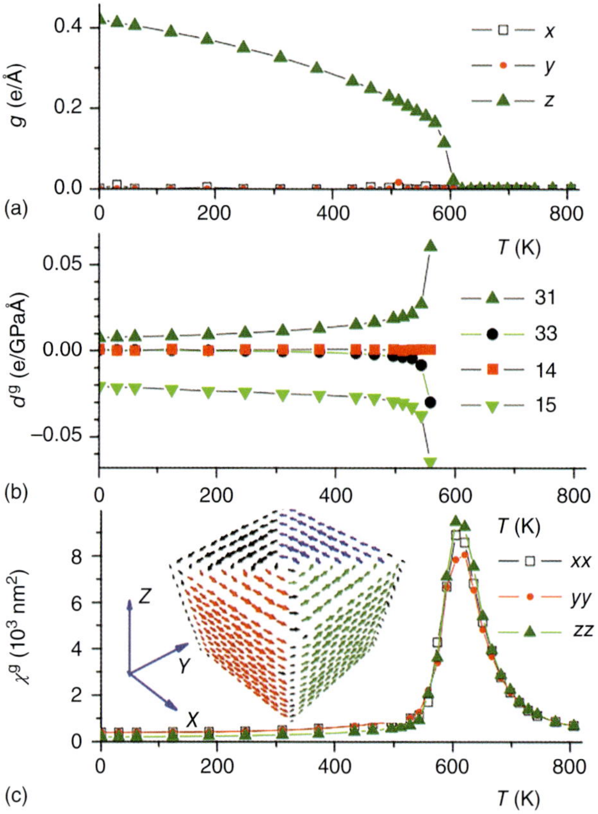

Figure 10.2 The temperature dependence of the Cartesian components of (a) the toroidal moment, (b) axial piezotoroidic coefficients, and (c) electric toroidal susceptibility elements, for a cubic nanodot of $PbZr_{0.6}Ti_{0.4}O_3$ (PZT) having 48 Å of lateral size, and being under stress-free and open-circuit boundary conditions, as predicted from the first-principles-based technique employed here and from the use of Eqs. (10.12) and (10.13). The inset of (c) shows the dipole vortex structure in the nanodot. The theoretical temperature has been rescaled to match the experimental value of the Curie temperature in bulk PZT. *Source:* Prosandeev et al. 2007 [45]. Reprinted with permission of American Physical Society.

In some specific nanosystems such as BTO nanowires and BTO nano-islands under uniaxial electrostatic screening, interesting dipolar states such as polar-toroidal multi-order (PTMO) states can exist [54–57]. Based on an effective Hamiltonian of $(Ba,Sr)TiO_3$ solid solution [31], interesting phenomena have been reported in a recent work by Chen et al. [58], where various dipole states in $(BaTiO_3)_m/(SrTiO_3)_n$ composite nanowires were found and can be tuned by the layer thickness, temperature, external strain, and electric fields. These states include the paraelectric state, a purely polar state, a PTMO state, a purely toroidal state, and a distorted toroidal state with the dipoles forming a "polydomain" pattern in the x–y plane. More importantly, in such a system, due to the coupling between the polar and toroidal orders, piezoelectric effects as well as piezotoroidal effects [59] can be observed. Shown in Figure 10.3a is their result for the

$(BaTiO_3)_{10}/(SrTiO_3)_n$ nanowires with $n = 2$, 3, and 4 as a function of temperature at zero external fields. A change of strain state accompanies the transition of dipole states. Particularly, large changes of axial strain happen near the phase boundaries, indicating large electromechanical responses in these regions. Figure 10.3b depicts the calculated piezoelectric and piezotoroidal coefficients, i.e. d_{33}^{P} and d_{33}^{g}, of the $(BTO)_{10}/(SrTiO_3)_2$ and $(BaTiO_3)_{10}/(SrTiO_3)_3$ nanowires based on the statistical correlation $d_{33}^{\mathrm{P}} = \frac{V}{k_{\mathrm{B}}T}\langle \Delta P_z \Delta \eta_{33} \rangle$ and $d_{33}^{\mathrm{g}} = \frac{V}{k_{\mathrm{B}}T}\langle \Delta g_z \Delta \eta_{33} \rangle$, where V is the supercell volume. Large peaks of the piezoelectric and piezotoroidal coefficients indeed occur, especially at the PTMO phase boundary. Figure 10.3c,d depict the strain state of the nanowire as a function of a curled field $\mathbf{E}_{\mathrm{C}} = S_{\mathrm{a}}\mathbf{e}_{\mathbf{z}} \times \mathbf{r}$ and the toroidization and polarization as functions of constrained strain η_{33}, respectively. The strain of the nanowire is sensitive to the curled field, showing significant inverse piezotoroidal effects. Meanwhile, a constraint on

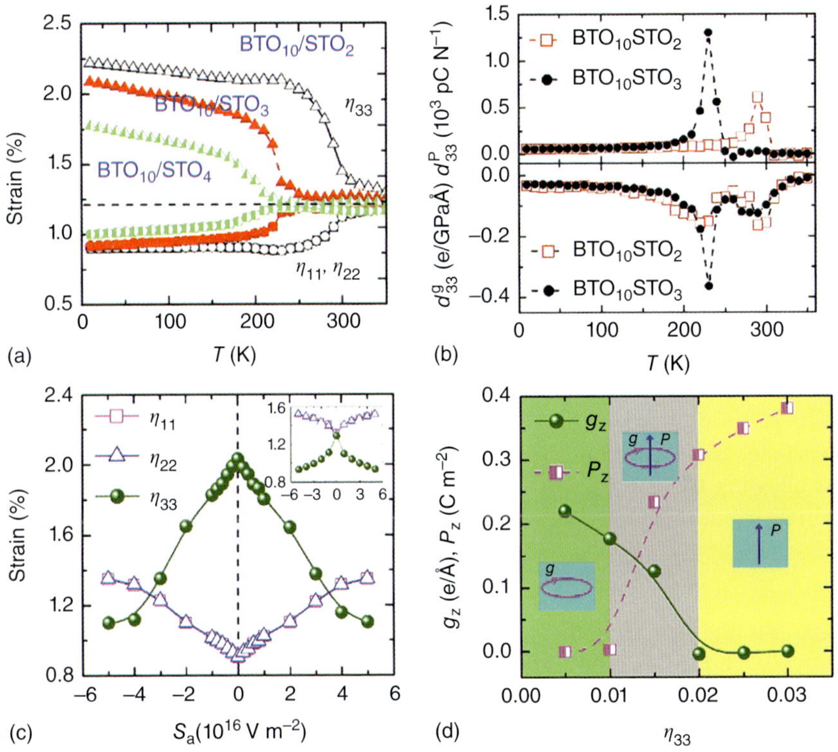

Figure 10.3 (a) The strain state of the $(BaTiO_3)_{10}/(SrTiO_3)_n$ nanowires with $n = 2$, 3, and 4 as a function of temperature at zero external fields. (b) The calculated piezoelectric d_{33}^{P} and piezotoroidal d_{33}^{g} coefficients of $(BaTiO_3)_{10}/(SrTiO_3)_2$ and $(BaTiO_3)_{10}/(SrTiO_3)_3$ nanowires. (c) Inverse and (d) direct piezotoroidal effect in $(BaTiO_3)_{10}/(SrTiO_3)_2$ nanowire at $T = 250$ K. The nanowire has an initial purely toroidal state. (c) The strain state of the nanowire as a function of a curled field $\mathbf{E}_{\mathrm{C}} = S_{\mathrm{a}}\mathbf{e}_{\mathbf{z}} \times \mathbf{r}$. The insert depicts the strain state of BTO nanodot as a function of S_{a} at $T = 250$ K. (d) The toroidization and polarization as functions of constrained strain η_{33}. *Source:* Chen et al. 2015 [58]. Reprinted with permission of Springer Nature.

strain η_{33} can lead to transformations between the PTMO state, purely toroidal state, and purely polar state in the nanowire, indicating a prominent direct piezotoroidal effect.

Also noteworthy is the effect of strain gradient on properties of ferroelectric films. Known as the so-called flexoelectric effect, electric polarization is coupled to the strain gradient, and the inhomogeneous strain can cause dielectric polarization in all insulators. The flexoelectric effect describes a linear response of electric polarization to a strain gradient, and can be written as

$$P_i = f_{ijkl} \frac{\partial \varepsilon_{ij}}{\partial x_k} \tag{10.13}$$

where P_i, f_{ijkl}, ε_{ij}, and x_k represent the flexoelectric polarization, flexoelectric coefficients, strain, and position coordination, respectively. Note that the number of nonzero flexoelectric coefficients can be reduced due to the certain lattice symmetry of the material. For example, for cubic crystals, due to the symmetry of their crystal structure, there are only three nonzero flexoelectric coefficients as shown in Figure 10.4a–c: the longitudinal f_{3333}, transverse f_{3311}, and shear f_{1313} flexoelectric coefficients [62, 63].

Methods including phenomenological approach [64–68], first-principles methods [60, 61, 69–71], and first-principles-derived methods [72] have been employed to theoretically predict the flexoelectric coefficients and study the flexoelectric effect of ferroelectrics at the atomistic level. The main difficulty in using direct first-principles methods for calculating flexoelectricity is the fact that such calculations require the use of supercells with periodic boundary conditions, whereas a strain gradient, by definition, results in an inhomogeneous distribution of lattice parameters and thus a breaking of the lattice periodicity. Hong et al. [61] determined the longitudinal flexoelectric coefficients f_{3333} of STO and BTO using direct first-principles calculations (as shown in Figure 10.4d–g). In order to overcome the problem of periodic boundary, they constructed an "accordion" supercell so that the strain gradient itself is periodic, thus allowing the recovery of periodic boundary conditions. In particular, the strain distribution along the z-direction is in cosine form:

$$\varepsilon(z) = \varepsilon_{\max} \cos\left(2\pi \frac{z}{h}\right) \tag{10.14}$$

where h is the height of the supercell and $\varepsilon_{\max}$ is the maximum strain in the supercell. Under this strain, the displacement of each atom in the supercell is given by

$$\delta(z) = \int_0^z \varepsilon(\xi)\, \mathrm{d}\xi = \frac{\varepsilon_{\max} h}{2\pi} \sin 2\pi \frac{z}{h} \tag{10.15}$$

After applying the strain gradient, the system is relaxed and the Born effective charges are used to calculate the local polarization in each unit cell. Subtly, the unit cell at $h/4$ has been chosen to calculate the flexoelectric coefficient, in order to have the maximum flexoelectricity and also to eliminate the piezoelectric contributions. However, there still remains orders of magnitude discrepancy between the theoretical results and the value obtained from experimental

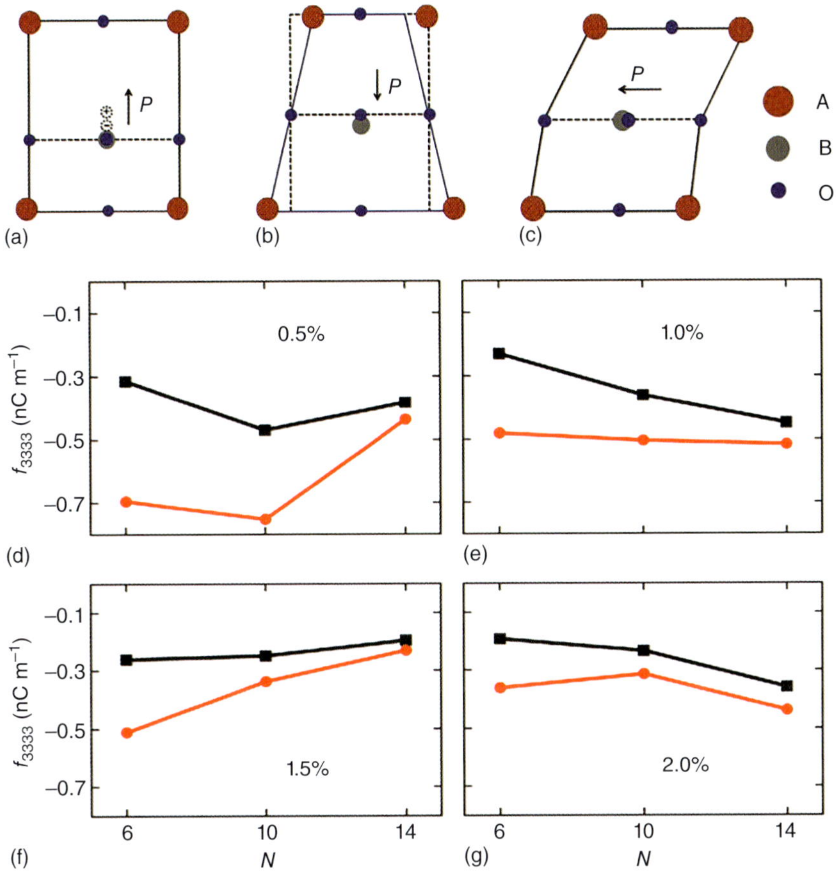

Figure 10.4 Schematic drawing of the coupling between the strain gradient and polarization for perovskite ABO_3 at the unit cell level: (a) longitudinal, (b) transverse, and (c) shear flexoelectricity. *Source:* Reprinted with permission from [60]. The longitudinal flexoelectric coefficient f_{3333} of rhombohedral $BaTiO_3$ for different supercell sizes ($N = 6, 10, 14$) under various strain gradient, ε_{max} = (d) 0.5%, (e) 1.0%, (f) 1.5%, (g) 2.0%. In each panel, the squares are for the unrelaxed strain, the dots are for the relaxed strain. The two values converge as the supercell size is increased. *Source:* Reprinted with permission from [61].

measurements [73]. The authors argued that the main reason for this discrepancy is that the characteristics such as phase and permittivity of the BTO at 0K in first-principles calculation are not the same as in the experiment. In a similar manner, Xu et al. [60] calculated the transverse and shear flexoelectric coefficients of BTO and STO.

Subsequently, Ponomareva et al. [72] developed an effective Hamiltonian which allows the study of flexoelectricity in $(Ba_{0.5}Sr_{0.5})TiO_3$ ferroelectric thin films at finite temperature and for different film thicknesses, with a special term that characterizes the contribution of inhomogeneous strain in the total energy density. Also, the flexoelectric response of a finite ferroelectric film should be essentially influenced by surface effects. This result recently received a direct confirmation based on DFPT from Stengel [70]. Other specific nanosystems such as ferroelectric nanotubes [74] and nanobelts [75] have also been investigated via first-principles calculations. Special radial polarizations and improper ferroelectricity were predicted. Nevertheless, there is still a long way to go to understand the flexoelectric effect in other materials in the first-principles framework.

10.3.2 Photovoltaic Effect

Developing photovoltaic devices is one of the effective ways to solve energy crisis and environmental problems. In a semiconducting material, the absorption of photons with energies above the band gap results in the excitation of electrons from the valence band to the conduction band. This process generates hole and electron carriers that can be separated by the built-in field and diffuse to opposite directions, leading to the appearance of a photocurrent. A number of designs have been implemented to achieve this process, among which the use of ferroelectric materials has received considerable interest recently as an effective substitute for maximizing charge separation and carrier collection. The internal depolarization field in ferroelectric materials, which enables the separation of charge carriers, can be controlled by the polarization and domain structure. Consequently, the photocurrent direction can be switched by changing the spontaneous polarization direction via an electric field, or the photocurrent can be controlled by the domain-wall orientation, allowing great flexibility in device design. Photovoltaic effect in ferroelectrics has been studied in the oxide perovskites such as lithium niobate ($LiNbO_3$) family [76, 77], BTO [78–80], lead zirconate titanate (PZT) family [81–84], and bismuth ferrite $BiFeO_3$ (BFO) family [85–87].

There are several explanations on the ferroelectric photovoltaic mechanism. Photovoltages in un-doped, single crystal samples of materials are reported as a bulk photovoltaic (BPV) effect and several models have been proposed in early years. Shift current models have been demonstrated for some ferroelectric materials including BFO [86] and BTO [80, 87], to explain their photocurrent, which may be closely related to the magnitude and the sign of polarization. For example, Young et al. [80] calculated the BPV response of the ferroelectrics BTO and PTO from first-principles by applying the shift current theory to the electronic structure. In this method, shift current response tensor has the form in Voigt notation

$$\sigma = \begin{bmatrix} 0 & 0 & 0 & 0 & \sigma_{yzY} & -\sigma_{yyY} \\ -\sigma_{yyY} & \sigma_{yyY} & 0 & \sigma_{yzY} & 0 & 0 \\ \sigma_{xxZ} & \sigma_{xxZ} & \sigma_{zzZ} & 0 & 0 & 0 \end{bmatrix} \tag{10.16}$$

where the electric field degrees of freedom have been condensed to a single dimension. A coefficient $\sigma_{rsq}(\omega)$ is introduced as the product of the transition intensity and the shift vector, which is described in detail in [80], in order to compute the shift current by

$$J_q = \frac{\sigma_{rrq}}{\alpha_{rr}} E_r^2 w = G_{rrq} I_r w \tag{10.17}$$

Here, α is the absorption coefficient, E is the electric field strength, which can be determined from the light intensity I, w is the width of the crystal surface exposed

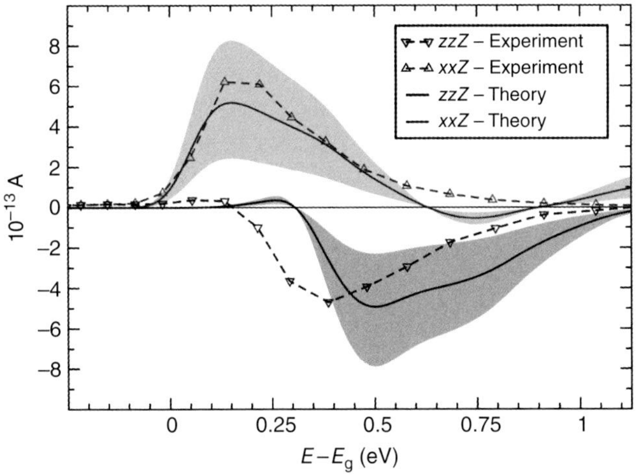

Figure 10.5 The experimental current and computed current of $BaTiO_3$ as a function of energy above their respective band gaps, for both transverse (*xxZ*) and longitudinal (*zzZ*) electric field orientation. *Source:* Young and Rappe 2012 [80]. Reprinted with permission of American Physical Society.

to illumination, and G_{rrq} is the Glass coefficient [76]. Figure 10.5 depicts the comparison between experimental current response from [88] and the response computed using shift current theory in this work. The agreement is quite good in both magnitude and spectrum profile, and for both transverse (xxZ) and longitudinal (zzZ) electric field orientations.

One of the major obstacles to ferroelectric oxide materials such as $LiNbO_3$, BTO, and PZT is their wide band gap (larger than 3 eV), which allows only a small fraction of the solar spectrum to be absorbed. Therefore, it is important to develop ferroelectric materials with a reduced band gap. BFO possesses a relatively small band gap (2.74 eV) that falls in the visible spectrum. This characteristic as well as the domain wall effect makes BFO extremely popular as a viable candidate for ferroelectric-based photovoltaic devices [87, 89–91]. The domain wall effect where the photovoltage in the BFO increases linearly with the total number of domain walls along the polarization direction was observed by Yang et al. [92] experimentally. It was demonstrated that the intrinsic potential drop at domain walls is the driving force for the dissociation of the electron–hole pairs and thus the illuminated domain walls can act as nanoscale photovoltage generators connected in series. However, researchers found it confusing to explain the photovoltaic behaviors in BFO solely by domain wall theory without considering BPV effect. To give a clear explanation, Young et al. [86] carried out a first-principles shift current theory in BFO considering the orientation of the domain polarization. The DFT+U method is used in the calculation, including an effective Hubbard $U_{eff} = U - J$ in the Hamiltonian, in order to better take account of the d-orbital electrons. The results indicate that the photocurrent, which comes from the domain wall effect, can be partially cancelled by the bulk effect. Thus, an enhanced energy conversion efficiency can be expected if the photovoltaic currents caused by the BPV effect and domain wall effect are designed to be in the same direction.

Anomalous photovoltaic (APV) effects, other than the domain walls effect, have been revealed to provide novel routes to convert light into electric energy [76, 92–95], such as Schottky-junction effect and depolarization field effect. When the ferroelectric semiconductor is in contact with the metal electrode to form a Schottky contact, there is a photocurrent under illumination driven by a local electric field, which is caused by band bending in the vicinity of the electrode. The resulting photocurrent is significantly determined by the Schottky barrier height and the depth of the depletion zone [96, 97]. Meanwhile, the depolarization field in ferroelectric films may be the dominating driving force for the separation of photogenerated charge-carrier pairs. Hence, it is suggested that the depolarization field may have a critical contribution to the APV effect [98–101]. Note that the Schottky barrier and the depolarization field are highly dependent on the properties of ferroelectric materials, the choice of electrode, and applied field such as strain. The utilization of first-principles calculation can accurately predict the height of Schottky barrier [102, 103] and the magnitude of depolarization field [16, 104], achieving an optimum design of the photovoltaic devices.

Moreover, first-principles calculations have been widely used in band-engineering to seek solutions such as doping to reduce the band gap [105–109]. The doping process consists in adding a controlled amount of impurities in the matrix material. By integrating impurities in the structure, the properties of the ferroelectric material can be drastically changed so that it can be better employed in photovoltaic devices. Results of the DFT+U calculations from Bennett et al. [105] demonstrated that substituting the B-site of PTO with metals Ni, Pd, and Pt indeed reduces the band gap while maintaining or even enhancing the ferroelectric polarization. Note that a proper value of Hubbard U is necessary for a more accurate band gap and to reproduce the experimental data. The calculations for M (where M = Ni, Pd, and Pt)-substituted PTO were performed on a supercell where one Ti in three is replaced and stabilized by an O vacancy to yield $PbTi_{2/3}M_{1/3}O_{8/3}$. As the d states of M fall in the PTO band gap, the band gap of the alloy is reduced. In another work from Bennett et al. [106], they performed Pd-substitution in $Ba(Ti_{1-x}Ce_x)O_3$ solid solutions. From the projected density of states (PDOS) shown in Figure 10.6a,b, it can be seen that the band gap decreases markedly for Pd-doped $Ba(Ti_{0.875}Ce_{0.125})O_3$ with the introduction of an O-vacancy stabilized Pd.

Other influences such as oxygen vacancy [90] and strain [112] can enhance the photovoltaic effect as well. This has been confirmed by first-principles calculations. For example, Matsuo et al. [110] investigated the influence of the oxygen vacancy ($V_O^{\bullet\bullet}$) distribution on the polarization properties of BFO capacitor with $SrRuO_3$ (SRO) being both electrodes. They found that a $V_O^{\bullet\bullet}$-rich layer tends to form near the interface. Figure 10.6c shows the total energy (E_{total}) of the (SrO-RuO_2)/(FeO-BiO_2) cell with respect to n (the distance of the site from the negatively charged interface RuO_2-FeO), calculated by the local spin density approximation (LSDA). The lowest E_{total} was obtained for the $n = 1$ cell, indicating that $V_O^{\bullet\bullet}$ is more likely to be located on the FeO layer adjacent to the

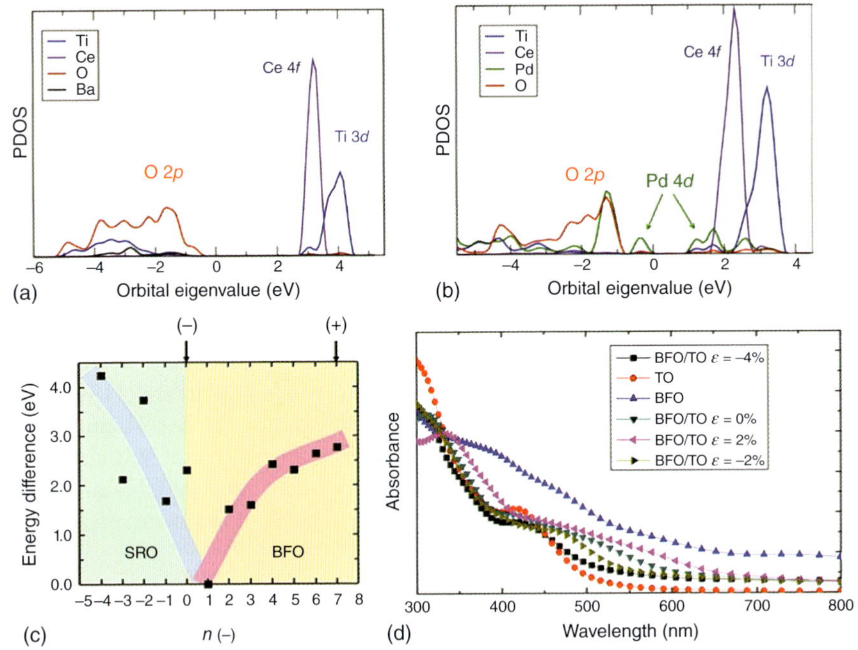

Figure 10.6 Projected density of states analysis shows that the electronic structure of (a) $Ba(Ti_{0.875}Ce_{0.125}O_3)$ differs greatly from (b) $Ba(Ti_{0.75}Ce_{0.125}Pd_{0.125})O_{3-\delta}$. While Ba (black) does not contribute to the states surrounding E_{gap}, combinations of Ti 3*d* (blue), Pd 4*d* (green), Ce 4*f* (magenta), and O 2*p* (red) states compose the various HOMOs and LUMOs. Values decrease markedly for (c) with the introduction of an O-vacancy stabilized Pd. *Source:* Bennett et al. [106]. Reprinted with permission of American Physical Society. (c) Dependence of the total energy on $V_O^{\bullet\bullet}$ site of $SrRuO_3$-$BiFeO_3$ (BFO) supercell. Number of *n* denotes relative position of the $V_O^{\bullet\bullet}$ site with respect to the negatively charged interface RuO_2-FeO. The (+) and (−) denote the positively and negatively charged interface, respectively. *Source:* Reprinted with permission from [110]. (d) The calculated optical absorption curves for BFO/TiO_2 (TO) heterostructures under different compressive and tensile strains, as well as the BFO and TO films. *Source:* Feng 2015 [111]. Reprinted with permission of Elsevier.

negatively charged interface. A large internal bias can be induced by using this $V_O^{\bullet\bullet}$-rich layer; thus an enhancement of photovoltaic conversion in the BFO capacitors is expected [110]. Also using first-principles calculation with LSDA, Feng showed that the photovoltaic effect of the BFO/TiO_2 heterostructures in the visible light region can be improved by applying epitaxial strain [111]. It can be seen in Figure 10.6d that the optical absorption increases monotonically with the strain, leading to the highest performance under tensile strain of 2% and the lowest one under compressive strain of −4% in the visible light region.

Besides focusing on the improvement of the photovoltaic conversion efficiencies of conventional perovskite ferroelectric oxides, efforts have been made to explore the photovoltaic effect in organic ferroelectric polymer PVDF–TrFE [113] and exploit novel materials such as $ZnSnO_3$ [114], $ZnGeO_3$ [115], $BaBi_2Ta_2O_9$ [116], $Na_{0.5}Bi_{0.3}TiO_3$ [117], and organic–inorganic halide perovskites [118–124]. It is a remarkable fact that among these new materials, organic–inorganic halide perovskites have exhibited great potential as photovoltaic materials, achieving remarkable photovoltaic conversion efficiency, currently over 20% in experiments [125–129]. For the organic–inorganic halide perovskites of current interest, the A-sites in the generic ABX_3 formula are universally occupied by organic groups, such as methylammonium ($CH_3NH_3^+$ or MA^+) or formamidinium ($NH_2CH{=}NH_2^+$ or FA^+). The cation B is typically a divalent metal ion (Pb_2^+ or Sn_2^+), while the anion X is a halide ion (I^-, Br^-, or Cl^-) [130]. In 2013, Brivio et al. [118] first adopted first-principles methods to calculate the dielectric constants and band gaps of halide perovskites $(CH_3NH_3)PbI_3$, finding them orientation dependent as active absorber layers of solar cells. What is more, due to the electric dipole of the methylammonium cation, a photoferroic effect may be accessible, which could enhance carrier collection. Soon after, better structural descriptions were obtained by Yun and coworkers [119] using the semi-local optB86b+vdWDF functional [131] including van der Waals (vdW) interaction. The vdW force plays an important role in a range of systems with weak interactions and can be achieved by the ionic bond between CH_3 and I. As can be seen in Figure 10.7a,b, the PDOS indicates that I 5p electrons, especially

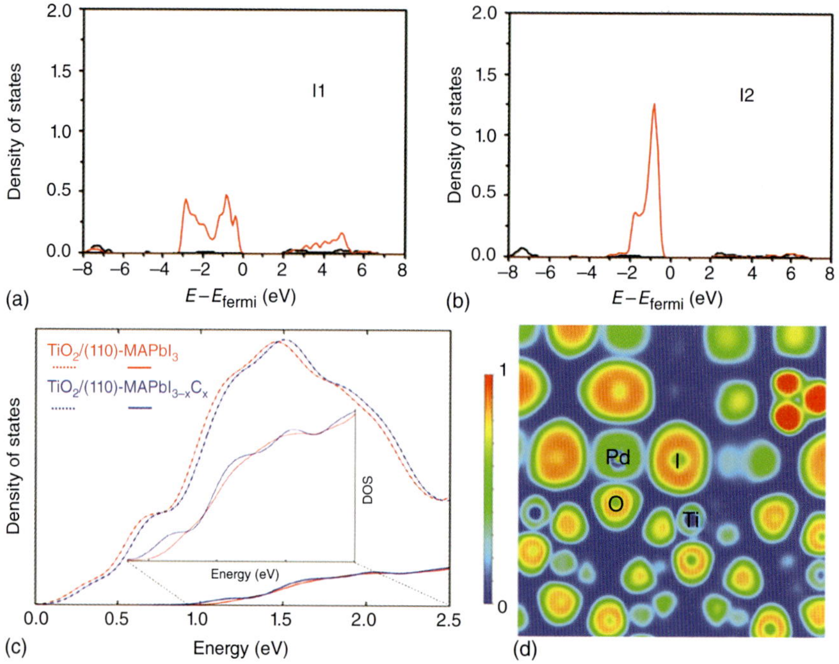

Figure 10.7 Partial density of states (PDOS) of (a) I1 and (b) I2 atoms in orthorhombic $CH_3NH_3PbI_3$ crystals using the optB86b+vdWDF functional. *Source:* Wang et al. 2014 [119]. Reprinted with permission of Royal Society of Chemistry. (c) PDOS with summed contributions from TiO_2 (dashed) and perovskite (solid) for the (110) $TiO_2/CH_3NH_3PbI_3$ (red) and $TiO_2/CH_3NH_3PbI_{3-x}Cl_x$ (blue) interfaces. The inset shows a zoom of the perovskites PDOS. The zero of the energy is set at the TiO_2 conduction band minimum. *Source:* Mosconi et al. 2014 [123]. Reprinted with permission of ACS. (d) Electron localization function contour plots with color scheme for the optimized $CH_3NH_3PbI_3/TiO_2$ interface. *Source:* Reprinted with permission from [124].

the I2 5p electrons, can be photoexcited to Pb 6p empty states. The I atoms in the Pb–I framework are not all equivalents. Thus, their contributions to the valence band are not exactly the same. As the organic–inorganic halide perovskites are often deposited on mesoporous TiO_2 to fabricate dye-sensitized solar cell devices in practical applications, Edoardo and coworkers [123] designed a $CH_3NH_3PbI_{3-x}Cl_x/TiO_2$ interface model using the first-principles method. They calculated the electronic structure and the total energy of the composite system. Their results demonstrated that $CH_3NH_3PbI_{3-x}Cl_x$ is more likely to grow along the (110) plane at the surface of TiO_2, due to the better lattice match of the halogen atoms I or Cl with Ti atoms in the materials. The interaction between $CH_3NH_3PbI_{3-x}Cl_x$ and TiO_2 controls the electronic structure at the interface, which leads to a stronger interface coupling and a slight upward shift of the TiO_2 conduction band, as shown in Figure 10.7c. Geng et al. [124] further investigated several different types of $CH_3NH_3PbI_3/TiO_2$ interfaces, suggesting that the interfacial Pb atoms play important roles in the structure stability and electronic properties. A charge transfer from Pb atoms to the O atoms of TiO_2 (Figure 10.7d) leads to a 0.4 eV higher band edge alignment of Pb-p than that of Ti-d, indicating a better carrier separation. It can be expected that further first-principles calculations will be performed to guide better material selection and design of ferroelectric–photovoltaic devices in the near future.

10.3.3 Pyroelectric and Electrocaloric Effect

Ferroelectrics always exhibit pyroelectric effect as the spontaneous polarization is temperature dependent. When sandwiched by electrodes and connected through an external circuit, the surface charge, which results from the change of polarization, can cause a net current to flow. In the absence of an applied electric field or applied stress, the pyroelectric coefficient Π thus can be defined as the rate of change of spontaneous polarization with temperature [132]:

$$\Pi = \frac{dP_s}{dT} \tag{10.18}$$

This coefficient can be quite high for ferroelectrics, and pyroelectric effect is typically stronger near the paraelectric/ferroelectric transitions. In the past decade, there has been a considerable upsurge of interest in the energy applications of the pyroelectric effect for the electrical energy harvesting from waste heat [133], to power autonomous sensors or improve the overall efficiency of combustion engines, for example [132].

It is obvious that investigations of the fundamental physics of pyroelectricity are difficult using only first-principles calculations, as computing properties as a function of temperature is still a large challenge. In this case, first-principles-derived methods are better choices. For example, Peng and Cohen [134] investigated the pyroelectricity in $LiNbO_3$ and built a shell model that contains fourth-order core–shell couplings, long-range Ewald interactions, and short-range interactions described by Rydberg potential. The parameters were fitted from previous first-principles results of total energies, forces, stresses, phonon frequencies, Born effective charges, and dielectric constants for ferroelectric structures. MD simulations were then carried out to compute the pyroelectric coefficient in Eq. (10.18). It was found that the pyroelectric effect can be understood from the changes in crystal structure with temperature as a simple anharmonic effect. Figure 10.8a gives the average value of the internal structural parameters z and w (fractional displacement of Li and O along polar axis, respectively) from the MD simulations, and the pyroelectric coefficient Π. The computed Π agrees with experiment and direct MD results. Another investigation on BTO was performed by Beckman et al. [136] to estimate its pyroelectric response with the effective Hamiltonian given in Eq. (10.2). The

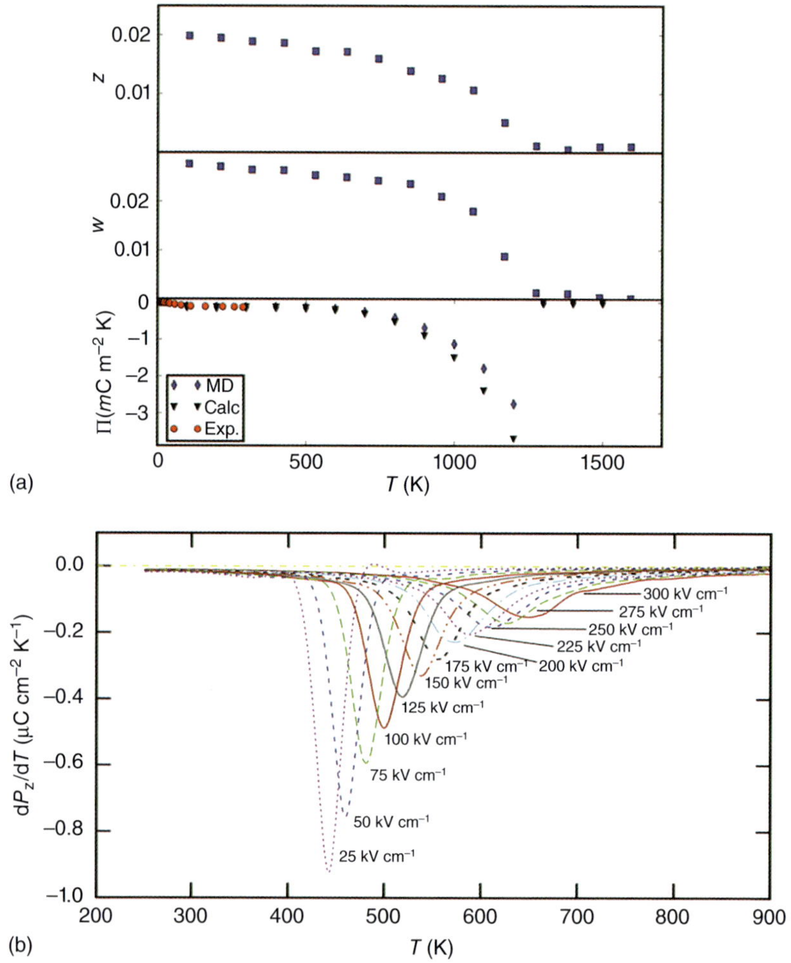

Figure 10.8 (a) The average value of the internal structural parameters z and w (fractional displacement of Li and O along polar axis, respectively) from the MD simulations. The pyroelectric coefficient Π (triangles) is compared with experiment (circles; Ref. [135]) and direct MD results (diamonds). *Source:* Reprinted with permission from [134]. (b) The determined temperature dependence of the pyroelectric coefficient. *Source:* Beckman et al. 2012 [136]. Reprinted with permission of Elsevier.

computed data shown in Figure 10.8b can be fitted with weighted splines for the determination of the pyroelectric coefficient $\partial P_z/\partial T$.

Another interesting property of ferroelectrics is the electrocaloric effect, which has attracted intensive interest inspired by the report in 2006 of a "giant electrocaloric effect" in a $PbZr_{0.95}Ti_{0.05}O_3$ thin film [137]. The electrocaloric effect, as the physical inverse of pyroelectric effect, refers to the adiabatic temperature change induced by the reversibly isentropic application of an electric field in ferroelectrics [138–140]. The strength of the electrocaloric effect can be determined by the temperature change ΔT for a material of density ρ with heat capacity C as

$$\Delta T = -\frac{1}{\rho}\int_{E_1}^{E_2} \frac{T}{C}\left(\frac{\partial P}{\partial T}\right)_E \mathrm{d}E \tag{10.19}$$

by using the Maxwell thermodynamic relation $\partial P/\partial T = \partial S/\partial E$ [137]. In applications, this temperature change in ferroelectrics can be put to use in refrigeration or energy scavenging [141–144]. However, the change of temperature is typically small in bulk ferroelectrics for a practical device to be efficient [139–141, 145]. A relatively large value of ΔT can be obtained in ferroelectric thin films because the application of a reasonable voltage can result in a relatively large electric field due to the small thickness [137, 141]. So it is natural to ask about the electrocaloric performance of low-dimensional ferroelectrics and the atomistic mechanism of electrocaloric effect. Answers to these questions require detailed investigations using accurate theoretical methods and many researchers choose to use first-principles-derived simulations.

In 2008, Prosandeev et al. [143] first combined first-principles-derived effective Hamiltonian with the nonequilibrium MD method to determine the electric-field-induced change in temperature in PZT bulk and nanostructure. To mimic the effect of the electric field on the physical properties of the system, a term that consists of the negative dot product between the electric dipole moments and the applied electric field **E**, i.e. $-\sum_i \mathbf{E}\cdot Z^*\mathbf{u}_i$, was added in the sum of the effective Hamiltonian energy (Eq. (10.2)). Moreover, the depolarization field denoted by $\langle \mathbf{E}_{\mathrm{dep}} \rangle$ was determined from the atomistic approach [146], and the screening of $\langle \mathbf{E}_{\mathrm{dep}} \rangle$ is mimicked by a parameter β (with $\beta = 1$ corresponding to the perfect screening and $\beta = 0$ corresponding to the perfect open circuit). Note also that the authors have used different expressions of the intrinsic alloy energy for PZT bulk [36, 147] and nanodots [42, 146], respectively. Then the total energy was used in MD with Newton's equations of motion being solved for all the degrees of freedom of the effective Hamiltonian approach. Based on the above method, the authors simulated the time dependence of the temperature in both ferroelectric bulk and nanodots under an AC electric field. The accuracy of this method was verified by comparing the numerical value of the electrocaloric coefficient α (which is defined as $\Delta T(t) = \alpha \Delta E(t)$) in PZT bulk with the experimental data [138]. In PZT nanodots, two types of equilibrium dipole patterns exist depending on the chosen value of the β parameter. A homogeneous and polarized state occurs when $\beta > 0.9$ and a mixed state with the coexistence of vortex and polar order occurs for $\beta < 0.9$, as shown in the inset of Figure 10.9a,b. Results of the time dependence of the temperature in these nanodots are shown in Figure 10.9a,b. Significant increase in α is obtained when $\beta = 0.95$ compared

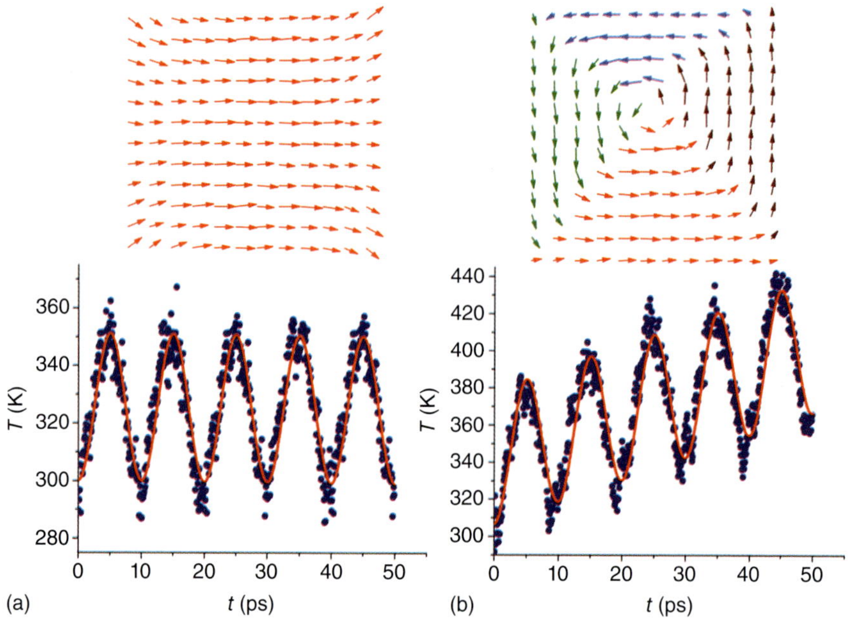

Figure 10.9 Predicted time dependence of the temperature in a cubic PZT nanodot under an AC electric field $E(t) = [0.6 - 0.5 \cos(2\pi \nu t)] \times 10^9$ V m^{-1} with the frequency $\nu = 100$ GHz for (panel (a)) $\beta = 0.95$ and (panel (b)) $\beta = 0.80$. The filled dots show the raw data while the solid line represents fitted $T(t)$. *Source:* Prosandeev et al. 2008 [143]. Reprinted with permission of APS.

with that of the bulk value. Such tuning arises from the fact that decreasing β leads to a decrease in the Curie temperature in the dot. When adjusting the value of β to 0.8, a more interesting phenomenon appears with a linear term adding to the temperature change ($\Delta T(t) = \alpha \Delta E(t) + \gamma t$). This is attributed to the field-induced dipole flips accompanying the expansion/compression of the vortex part. These results thus demonstrate that the electrocaloric performance of the nanodot can be tuned by adjusting the magnitude of the residual depolarization field.

Other influences such as phase transition [148], strain [149], stress [150], and crystal anisotropy [151] have been considered in simulations to reveal their effect on the temperature change ΔT. Either MD simulations [143, 149, 151] or MC simulations [148, 150, 152, 153] were performed under the atomistic framework of first-principles-derived effective Hamiltonian. The general steps can be summarized as follows. The first step involves conducting simulations at a constant temperature to equilibrate the sample. At the second step the simulations are usually switched to be at a constant total energy to catch the temperature change associated with the electric field applied [143, 148]. Note that an isoenthalpic process can be substituted for the process with constant total energy [150, 153]. Also, it is worth mentioning that in some works an extra degree of freedom, a "demon," is introduced that travels around the system, transferring energy as it changes the dynamical variables [153, 154].

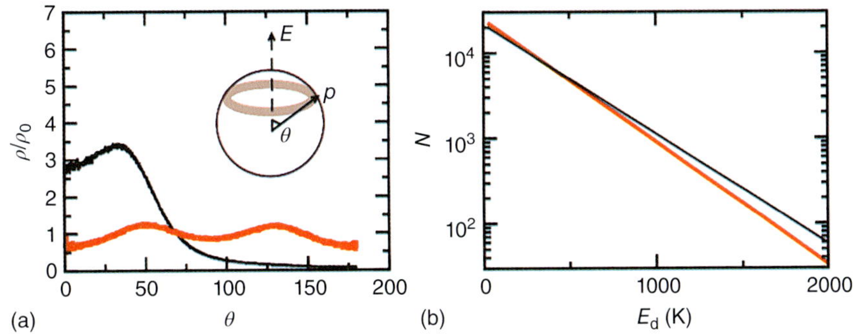

Figure 10.10 (a) The normalized dipole density ρ/ρ_0 as a function of the angle θ and (b) distributions of microcanonical demon energies under applied electric field of 500 kV cm^{-1} (thin line) and without electric field (thick line) at $T = 300$ K. *Source:* Lisenkov and Ponomareva 2009 [148]. Reprinted with permission of APS.

In fact, the driving force for electrocaloric effect is a change in entropy associated with the electric field variation that corresponds to the realignment of the electric dipole under an application or removal of electric field at atomistic level. For example, in the paraelectric phase, and in the absence of an electric field, the dipoles are completely disordered. Such disorder is associated with large entropy. When applying the electric field, the dipoles realign with the field, inducing polarization. Associated with the dipole alignment there is a decrease in the entropy part that should be compensated by an increase in the temperature to keep the total entropy unchanged as dictated by reversible adiabatic conditions. Lisenkov and Ponomareva [148] plotted the normalized dipole density ρ/ρ_0 as a function of the angles θ and distributions of microcanonical demon energies [154] for the same electric fields with fixed temperature $T = 300$ K in Figure 10.10. The demon energy here is related to the kinetic energy of the local mode or kinetic energy associated with the atomic vibrations in a unit cell. It can be seen that the application of the electric field results in a drastic increase in the dipole ordering, leading to an entropy decrease (Figure 10.10a), while the slope of the distribution decreases, indicating an increase in the disorder in the local mode kinetic energies (Figure 10.10b). These results indicate that the excess entropy due to the dipole ordering under the electric field is absorbed by the atomic vibrations. Thus, the whole process of electrocaloric effect can be taken as a "conversion" of spatially inhomogeneous dipole distribution into spatially inhomogeneous temperature distribution.

10.4 Energy Storage

Driven by the increasing demand for compact electronics, high energy storage capacitors become one of the key technologies in energy storage devices. Solid capacitors, as the most widely used electrochemical energy storage devices, can be divided into three categories according to the dielectric materials: ceramic capacitors, polymer capacitors, and composites-based solid capacitors. Many dielectric materials exhibit nonlinear polarization response and consequently the energy density is given by the integral [155]

$$J = \int_0^{D_{max}} E\,\mathrm{d}D \tag{10.20}$$

where E is the electric field and D_{max} is the electric displacement (D) at the highest field applied (E_{max}). For the dielectrics with high relative dielectric constant, the electric displacement D can be replaced by the polarization (P). Accordingly, the above formula can be written as

$$J = \int_0^{P_{max}} E\,\mathrm{d}P = \int_0^{E_{max}} P\,\mathrm{d}E \tag{10.21}$$

Evidently, based on Eq. (10.21), the value of J can be obtained by numerical integration of the area between the polarization axis and the curves of P–E loops. Figure 10.11 shows schematically the form of the dependence of D on E for (a) linear dielectrics with constant permittivity (e.g. Al_2O_3, glass), (b) antiferroelectric (AFE) with zero net remnant polarization (e.g. $PbZrO_3$ or PZO), (c) relaxor ferroelectrics with nanosized domains (e.g. (Pb,La)(Zr,Ti)O_3), and (d) ferroelectrics with spontaneous polarization (e.g. BTO, PTO) [158]. The shaded area in the first quadrant is the energy stored.

Note that there are also disadvantages such as a rather small maximum operating voltage (below 3.0 V), high leakage current in microamperes (thus a low energy efficiency), and limited cycling life ($\sim 10^5$) that prevent the application of dielectric capacitors in some advanced pulsed power systems [158]. Exploration of dielectric materials with improved electric energy storage performance has therefore attracted increasing attention. First-principles calculations, effective Hamiltonian method, and MD simulation are invaluable tools in the prediction of the energy storage properties and the design of new materials for energy storage application [24, 159–161]. The relevant works are introduced in this section.

Nishimatsu and Waghmare [159] have developed a fast MD method with the Hamiltonian presented in Eq. (10.2) and applied it to BTO thin film capacitors at short-circuited condition or under applied voltages. It was shown that properties of the ferroelectric capacitors are highly influenced by the interface between the ferroelectrics and the electrodes. An intermediate layer of low dielectric constant is formed near the interface, and is called "dead layer" [162–165]. To clarify the influence of dead layers, thickness of dead layers d are mimicked by making the amplitudes local soft-mode $\mathbf{u}$ to be zero ($\mathbf{u} = 0$) in the MD simulations. Note that the electrodes cannot fully screen the polarization of ferroelectrics, leading to a depolarization field that is automatically and implicitly included in the long-range dipole–dipole interaction $V^{dpl}(\{\mathbf{u}\})$. Strain effect is also considered in the simulation of hysteresis loops for ferroelectric thin-film capacitors with 1% in-plane biaxial compressive strain and without constraints of strain (namely, the "free" film). The temperature is maintained at 100 K in the simulations. Figure 10.12a,b demonstrate that the coercive field decreases as the film thicknesses decrease, which is consistent with the phenomenological results [166]. In the results, there is a large difference in coercive field between the epitaxially constrained film and the "free" film. Meanwhile, the hysteresis loops are "slimmer" and

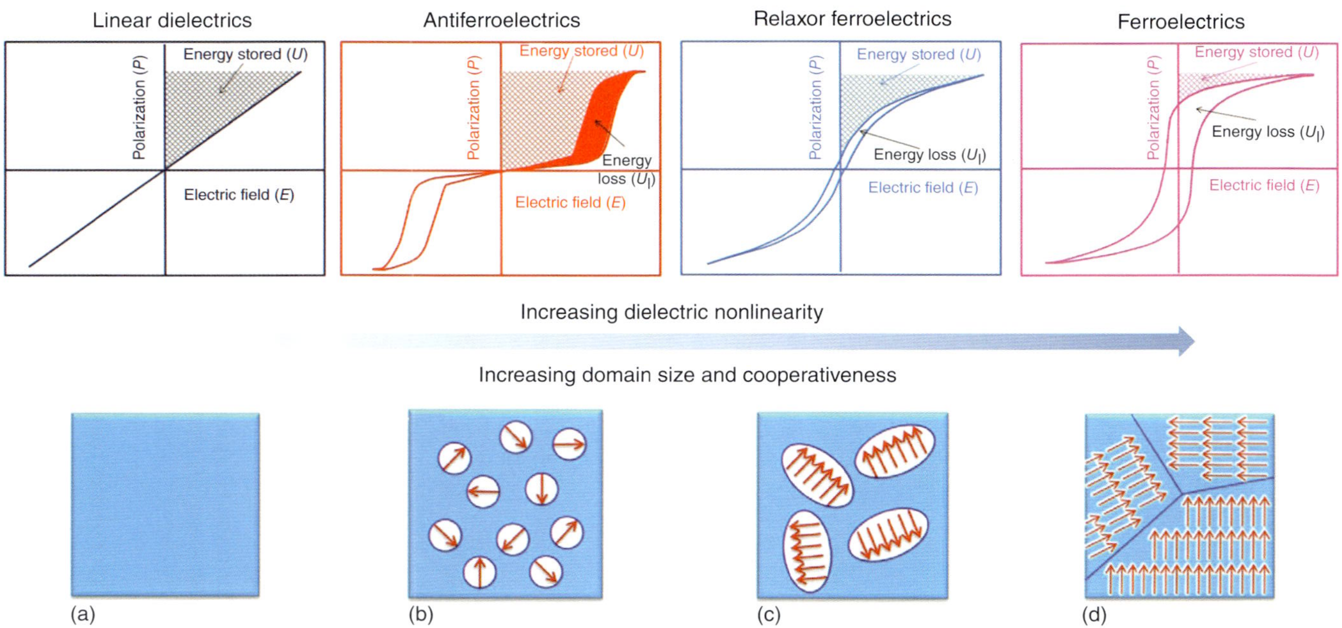

Figure 10.11 Schematic description of the energy storage characteristics (top panel) and ferroelectric domain structures (bottom panel) for (a) dielectric, (b) antiferroelectric, (c) relaxor ferroelectric and, (d) ferroelectric ceramics. *Source:* Reprinted with permission from [156, 157].

"shorter" with the existence of the dead layer, which is unavoidable in actual circumstances. Note that the hysteresis loops for "free" film capacitors with thickness $l = 63$ and $l = 127$ are very similar to the experimentally observed hysteresis loops that have "steps" and "plateaus" during polarization switching [167]. The authors attributed these phenomena to the occurrence of in-plane polarization as well as large in-plane components of the dipoles $Z^{*}\mathbf{u}(\boldsymbol{R})$ in the "free" film capacitors with dead layer ($d = 1$). Using this scheme, the energy density of bulk ferroelectric materials such as BTO, PTO, and $KNbO_3$ was also calculated [161]. It can be seen in Figure 10.12c that the integrated energy density of BTO bulk materials slowly increases with the rise of temperature and the increase becomes rapid when the temperature is higher than the Curie temperature. This can be explained by the structural phase transition and electric-field-induced phase transition.

In 1951, Kittel proposed the concept of AFE materials based on macroscopic phenomenological theory and predicted their basic characteristics [168]. Antiferroelectrics do not exhibit a macroscopic spontaneous polarization due to the antiparallel arrangement of the adjacent dipole moments. Nevertheless, phase transition from AFE phase to ferroelectric phase occurs under the external field and a double hysteresis loop appears. The ferroelectric phase can be switched back to the AFE phase by heating or pressing. This process is accompanied by large stress changes and the release of high-density electric charge. Antiferroelectrics thus are promising candidates for high-density energy storage capacitors [169–172]. As illustrated in Figure 10.11, compared with ferroelectrics, AFEs are more likely to be used for high energy storage because of their larger saturated polarization, smaller remnant polarization, and moderate breakdown field [158].

To date, there are more than 40 kinds of AFE materials that have been found, among which $Pb(Zr,Ti)O_3$-based compounds are the most valuable ones [173–178]. First-principles-based calculations on these AFEs were mainly focused on the atomic mechanisms of the competition and coexistence of their antiferroelectric orthorhombic (AFO) phase and rhombohedral ferroelectric (RF) phase [179–185], which are energetically close. In 2002, Leung et al. [182] first fitted an effective Hamiltonian that can capture the atomic structures of the AFO phase and the RF phase for $PbZr_{0.95}Ti_{0.05}O_3$ (PZT 95/5). In comparison with conventional effective Hamiltonian for ferroelectrics, the authors used an eighth-order Taylor expansion in powers of the instability vector components to reproduce the exact low-energy landscape. Another correction was to add a linear thermal expansion term to capture the effect of the coupling of the stable modes with strain into the Hamiltonian. In the finite-temperature simulations, the H_{eff} describes the phase transition in PZT 95/5. In Figure 10.13a there is an AFO phase below $T = 350$ K, a paraelectric phase above 898 K, and an RF phase in between.

Both the RF and AFO phases may be considered as distortions of the "perfect" cubic perovskite structure. However, these distortions are large in Zr-rich PZT and so these two phases are rather different from a structural point of view, even though they are energetically close. The AFO phase (Figure 10.13b) has large octahedral tilts [187, 188] while the RF phase (Figure 10.13c) of PZT has a different tilt pattern with large ferroelectric displacements of the atoms [189].

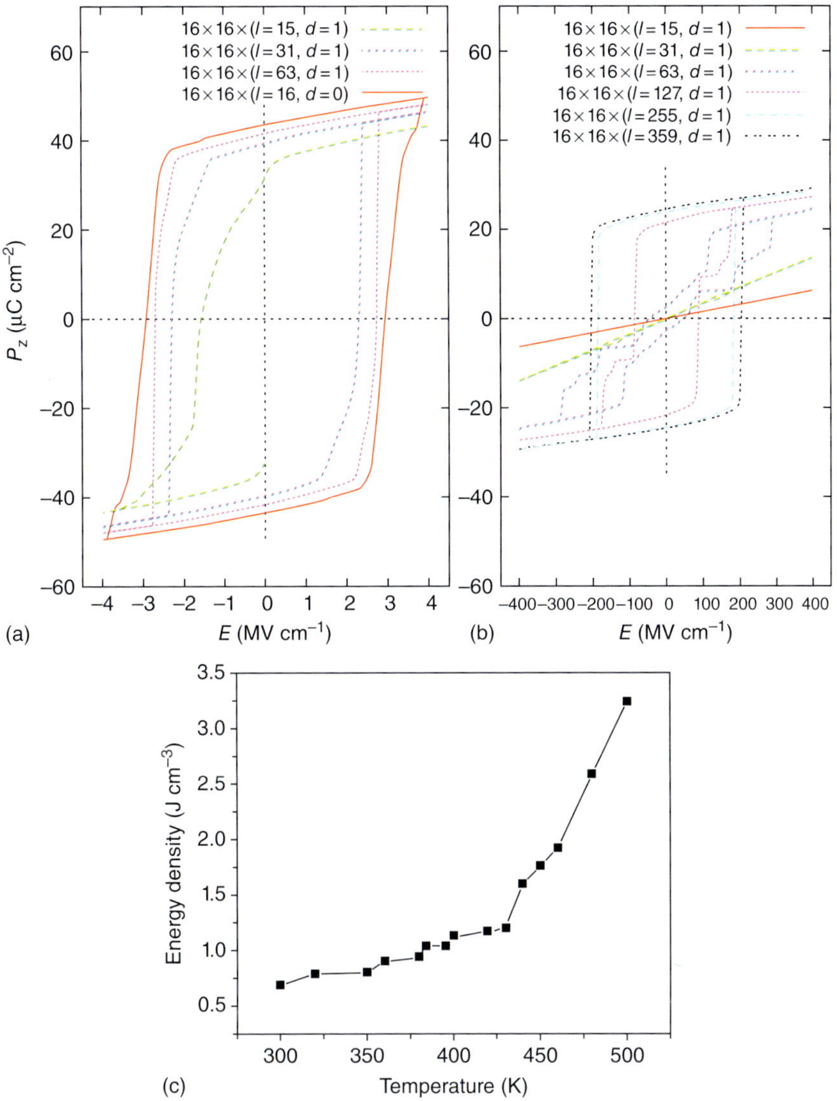

Figure 10.12 Calculated hysteresis loops for capacitors with (a) epitaxially constrained films and (b) "free" films of various thickness *l* and with dead layer *d*. *Source:* Nishimatsu and Waghmare 2008 [159]. Reprinted with permission of APS. (c) Temperature dependence of energy storage density of $BaTiO_3$. *Source:* Luo et al. 2016 [161]. Reprinted with permission of AIP Publishing.

Thus, the AFO and RF phases may exhibit quite different elastic behaviors, and as they are energetically close, the understanding of how strains affect the energetic balance between these two phases is important to the development of devices that utilize the AFE/ferroelectric switching. Kagimura [186] reported calculations of the elastic constants for both RF and AFO phases of PZO using a direct first-principles approach. Their results showed rather different elastic

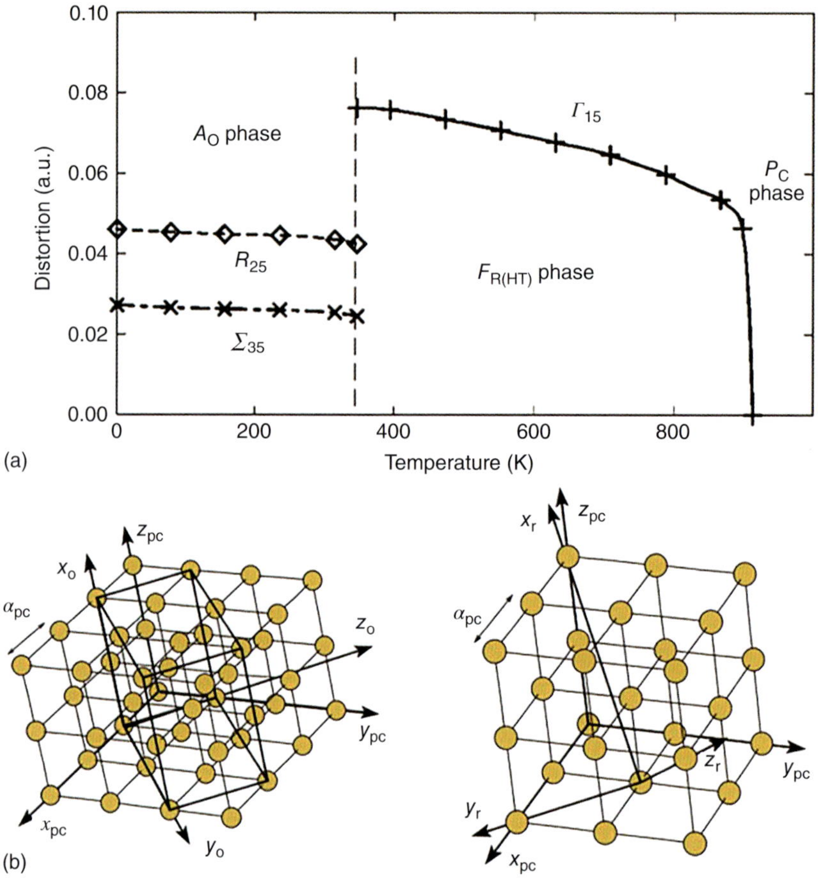

Figure 10.13 (a) Phase transition in $PbZr_{0.95}Ti_{0.05}O_3$ with mean amplitudes of various frozen-in modes, in units of a.u., as functions of temperature. *Source:* Leung et al. 2002 [182]. Reprinted with permission of APS. Schematic diagrams for the (b) orthorhombic and (c) rhombohedral structures and their relation to that of the pseudocubic cell in $PbZrO_3$. The indices pc, o, and f refer to the pseudocubic, orthorhombic, and rhombohedral cells, respectively. The solid circles represent the Pb atoms and a_{pc} is the pseudocubic lattice constant. *Source:* Kagimura 2008 [186]. Reprinted with permission of APS. (d) Energy (meV/f.u.) vs epitaxial strain (%) diagram. FE and AFE refer to the ferroelectric and antiferroelectric ground state, respectively. The "e" indicates epitaxial structures. Inset: Volume per f.u. (Å^3) vs epitaxial strain (%) in the AFE region. *Source:* Reyes-Lillo and Rabe 2014 [179]. Reprinted with permission of APS.

anisotropies of the two phases. Either of the phases can be the effectively stiffer one depending on the orientation, indicating a control via strain, which may favor one phase over the other. To illustrate the effect of epitaxial strain on the relative stability of various structures, Reyes-Lillo and Rabe [179] calculated the strain-dependent energy of the nonpolar ground state *Pbam* and the alternative polar phase *R*3*c* in $PbZrO_3$. For a given structure of relaxed unit cell with lattice vectors $\{\mathbf{t}_a, \mathbf{t}_b, \mathbf{t}_c\}$, and a given matching plane with out-of plane lattice vector $\mathbf{t}_j$, they estimated the equilibrium energy minimum under several epitaxial strains,

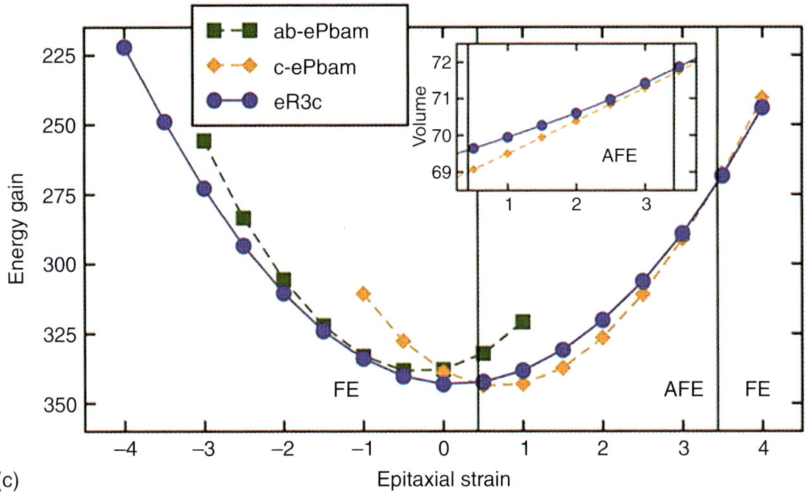

(c)

Figure 10.13 (*Continued*)

which can be expressed as

$$\sigma_j = 100 \times (1/2) \times \sum_i (\mathbf{t}_i - \mathbf{t}_{i0})/\mathbf{t}_{i0} \qquad (10.22)$$

where i denotes the two lattice vectors defining the matching plane, and the reference lattice vectors $\{\mathbf{t}_{a0}, \mathbf{t}_{b0}, \mathbf{t}_{c0}\}$ are the relevant linear combination of the cubic perovskite vectors with reference lattice constant $a_0 = 4.1$ Å. The resulting relationship between energy and epitaxial strain can be depicted as in Figure 10.13d. While ferroelectricity is favored for all compressive strains, the small energy difference between *Pbam* and *R3c* phase yields a robust AFE ground state for tensile strains. These works confirm the strain control to be a practical method to tune the competition between polar and nonpolar instabilities, and thus to engineer the antiferroelectric ground state.

References

1 Ghosez, P. and Junquera, J. (2006). First-principles modeling of ferroelectric oxide nanostructures. In: *Handbook of Theoretical & Computational Nanotechnology*, 623–728. American Scientific Publishers.

2 King-Smith, R. and Vanderbilt, D. (1993). Theory of polarization of crystalline solids. *Physical Review B* 47 (3): 1651.

3 Resta, R. (1994). Macroscopic polarization in crystalline dielectrics: the geometric phase approach. *Reviews of Modern Physics* 66 (3): 899.

4 Catti, M., Noel, Y., and Dovesi, R. (2003). Full piezoelectric tensors of wurtzite and zinc blende ZnO and ZnS by first-principles calculations. *Journal of Physics and Chemistry of Solids* 64 (11): 2183–2190.

5 Sághi-Szabó, G., Cohen, R.E., and Krakauer, H. (1998). First-principles study of piezoelectricity in $PbTiO_3$. *Physical Review Letters* 80 (19): 4321.

6 Bernardini, F. and Fiorentini, V. (2002). First-principles calculation of the piezoelectric tensor d of III–V nitrides. *Applied Physics Letters* 80 (22): 4145–4147.
7 Gonze, X. and Lee, C. (1997). Dynamical matrices, Born effective charges, dielectric permittivity tensors, and interatomic force constants from density-functional perturbation theory. *Physical Review B* 55 (16): 10355.
8 Neaton, J., Ederer, C., Waghmare, U. et al. (2005). First-principles study of spontaneous polarization in multiferroic $BiFeO_3$. *Physical Review B* 71 (1): 014113.
9 Zoroddu, A., Bernardini, F., Ruggerone, P., and Fiorentini, V. (2001). First-principles prediction of structure, energetics, formation enthalpy, elastic constants, polarization, and piezoelectric constants of AlN, GaN, and InN: comparison of local and gradient-corrected density-functional theory. *Physical Review B* 64 (4): 045208.
10 Wang, J., Neaton, J., Zheng, H. et al. (2003). Epitaxial $BiFeO_3$ multiferroic thin film heterostructures. *Science* 299 (5613): 1719–1722.
11 Vanderbilt, D. and King-Smith, R. (1993). Electric polarization as a bulk quantity and its relation to surface charge. *Physical Review B* 48 (7): 4442.
12 Resta, R. (2010). Electrical polarization and orbital magnetization: the modern theories. *Journal of Physics: Condensed Matter* 22 (12): 123201.
13 Marzari, N., Mostofi, A.A., Yates, J.R. et al. (2012). Maximally localized Wannier functions: theory and applications. *Reviews of Modern Physics* 84 (4): 1419.
14 Rabe, K. and Waghmare, U. (1995). Localized basis for effective lattice hamiltonians: lattice Wannier functions. *Physical Review B* 52 (18): 13236.
15 Zhang, F. and Rice, T. (1988). Effective Hamiltonian for the superconducting Cu oxides. *Physical Review B* 37 (7): 3759.
16 Luo, X., Wang, B., and Zheng, Y. (2011). Tunable tunneling electroresistance in ferroelectric tunnel junctions by mechanical loads. *ACS Nano* 5 (3): 1649–1656.
17 Bilc, D.I., Novaes, F.D., Íñiguez, J. et al. (2012). Electroresistance effect in ferroelectric tunnel junctions with symmetric electrodes. *ACS Nano* 6 (2): 1473–1478.
18 Calzolari, A., Marzari, N., Souza, I., and Nardelli, M.B. (2004). Ab initio transport properties of nanostructures from maximally localized Wannier functions. *Physical Review B* 69 (3): 035108.
19 Marzari, N. and Vanderbilt, D. (1997). Maximally localized generalized Wannier functions for composite energy bands. *Physical Review B* 56 (20): 12847.
20 Baroni, S., Giannozzi, P., and Testa, A. (1987). Green's-function approach to linear response in solids. *Physical Review Letters* 58 (18): 1861.
21 Wu, X., Vanderbilt, D., and Hamann, D. (2005). Systematic treatment of displacements, strains, and electric fields in density-functional perturbation theory. *Physical Review B* 72 (3): 035105.
22 Zhong, W., Vanderbilt, D., and Rabe, K.M. (1998). First-principles theory of ferroelectric phase transitions for perovskites: the case of $BaTiO_3$. *Ferroelectrics* 52 (1): 6301–6312.

23 Rabe, K.M. and Waghmare, U.V. (1995). Ferroelectric phase transitions: a first-principles approach. *Ferroelectrics* 164 (1): 15–32.
24 Nishimatsu, T., Waghmare, U.V., Kawazoe, Y., and Vanderbilt, D. (2008). Fast molecular-dynamics simulation for ferroelectric thin-film capacitors using a first-principles effective Hamiltonian. *Physical Review B* 78 (10): 104104.
25 Zhong, W., Vanderbilt, D., and Rabe, K. (1995). First-principles theory of ferroelectric phase transitions for perovskites: the case of $BaTiO_3$. *Physical Review B* 52 (9): 6301.
26 Waghmare, U. and Rabe, K. (1997). Ab initio statistical mechanics of the ferroelectric phase transition in $PbTiO_3$. *Physical Review B* 55 (10): 6161.
27 Ramer, N.J. and Rappe, A.M. (2000). Virtual-crystal approximation that works: locating a compositional phase boundary in Pb $(Zr_{1-x}Ti_x)O_3$. *Physical Review B* 62 (2): R743.
28 Ramer, N.J. and Rappe, A.M. (2000). Application of a new virtual crystal approach for the study of disordered perovskites. *Journal of Physics and Chemistry of Solids* 61 (2): 315–320.
29 Fu, H. and Cohen, R.E. (2000). Polarization rotation mechanism for ultrahigh electromechanical responsein single-crystal piezoelectrics. *Nature* 403 (6767): 281.
30 Asthagiri, A., Wu, Z., Choudhury, N., and Cohen, R.E. (2006). Advances in first-principles studies of transducer materials. *Ferroelectrics* 333 (1): 69–78.
31 Walizer, L., Lisenkov, S., and Bellaiche, L. (2006). Finite-temperature properties of (Ba, Sr) TiO_3 systems from atomistic simulations. *Physical Review B* 73 (14): 144105.
32 Kutnjak, Z., Petzelt, J., and Blinc, R. (2006). The giant electromechanical response in ferroelectric relaxors as a critical phenomenon. *Nature* 441 (7096): 956–959.
33 Park, S.E. and Shrout, T.R. (1997). Ultrahigh strain and piezoelectric behavior in relaxor based ferroelectric single crystals. *Journal of Applied Physics* 82 (4): 1804–1811.
34 Davis, M., Damjanovic, D., and Setter, N. (2006). Electric-field-, temperature-, and stress-induced phase transitions in relaxor ferroelectric single crystals. *Physical Review B* 73 (1).
35 Cohen, R.E., Heifets, E., and Fu, H. (2001). First-principles computation of elasticity of $Pb(Zr, Ti)O_3$: the importance of elasticity in piezoelectrics. 582 (1): 11–22.
36 Bellaiche, L., Garcia, A., and Vanderbilt, D. (2000). Finite-temperature properties of Pb $(Zr_{1-x}Ti_x)O_3$ alloys from first principles. *Physical Review Letters* 84 (23): 5427.
37 Du, X.H., Belegundu, U., and Uchino, K. (1997). Crystal orientation dependence of piezoelectric properties in lead zirconate titanate: theoretical expectation for thin films. *Japanese Journal of Applied Physics* 36 (9A): 5580–5587.
38 Berlincourt, D.A., Cmolik, C., and Jaffe, H. (1960). Piezoelectric properties of polycrystalline lead titanate zirconate compositions. *Proceedings of the IRE* 48 (2): 220–229.

39 Zheng, Y. and Woo, C.H. (2009). Giant piezoelectric resistance in ferroelectric tunnel junctions. *Nanotechnology* 20 (7): 075401.

40 Luo, X., Wang, B., and Zheng, Y. (2011). Tunable tunneling electroresistance in ferroelectric tunnel junctions by mechanical loads. *ACS Nano* 5 (5): 1649–1656.

41 Luo, X., Lin, S.P., Wang, B., and Zheng, Y. (2010). Impact of applied strain on the electron transport through ferroelectric tunnel junctions. *Applied Physics Letters* 97 (1): 012905.

42 Ponomareva, I., Naumov, I., and Bellaiche, L. (2005). Low-dimensional ferroelectrics under different electrical and mechanical boundary conditions: atomistic simulations. *Physical Review B* 72 (21): 214118.

43 Naumov, I.I., Bellaiche, L., and Fu, H. (2004). Unusual phase transitions in ferroelectric nanodisks and nanorods. *Nature* 432 (7018): 737–740.

44 Dubovik, V., Martsenyuk, M., and Saha, B. (2000). Material equations for electromagnetism with toroidal polarizations. *Physical Review E* 61 (6): 7087.

45 Prosandeev, S., Kornev, I., and Bellaiche, L. (2007). Tensors in ferroelectric nanoparticles: first-principles-based simulations. *Physical Review B* 76 (1): 012101.

46 GarcÍa, A. and Vanderbilt, D. (1998). Electromechanical behavior of $BaTiO_3$ from first principles. *Applied Physics Letters* 72 (23): 2981–2983.

47 Rabe, K., Cokayne, E. and Cohen, R. (1998). First-principles calculations for ferroelectrics. *Fifth Williamsburg Workshop*. AIP, Woodbury, NY, 156.

48 Prosandeev, S., Ponomareva, I., Kornev, I. et al. (2006). Controlling toroidal moment by means of an inhomogeneous static field: an ab initio study. *Physical Review Letters* 96 (23): 237601.

49 Prosandeev, S., Ponomareva, I., Kornev, I., and Bellaiche, L. (2008). Control of vortices by homogeneous fields in asymmetric ferroelectric and ferromagnetic rings. *Physical Review Letters* 100 (4): 047201.

50 Prosandeev, S. and Bellaiche, L. (2008). Controlling double vortex states in low-dimensional dipolar systems. *Physical Review Letters* 101 (9): 097203.

51 Naumov, I.I. and Fu, H. (2008). Cooperative response of $Pb(ZrTi)O_3$ nanoparticles to curled electric fields. *Physical Review Letters* 101 (19): 197601.

52 Naumov, I. and Fu, H. (2007). Vortex-to-polarization phase transformation path in ferroelectric $Pb(ZrTi)O_3$ nanoparticles. *Physical Review Letters* 98 (7): 077603.

53 Chen, W., Zheng, Y., and Wang, B. (2012). Vortex domain structure in ferroelectric nanoplatelets and control of its transformation by mechanical load. *Scientific Reports* 2 (796).

54 Wu, C.M., Chen, W.J., Zheng, Y. et al. (2014). Controllability of vortex domain structure in ferroelectric nanodot: fruitful domain patterns and transformation paths. *Scientific Reports* 4: 3946.

55 Pilania, G., Alpay, S.P., and Ramprasad, R. (2009). Ab initio study of ferroelectricity in $BaTiO_3$ nanowires. *Physical Review B: Condensed Matter* 80 (1).

56 Louis, L., Kornev, I., Geneste, G. et al. (2012). Novel complex phenomena in ferroelectric nanocomposites. *Journal of Physics. Condensed Matter* 24 (40): 402201.
57 Hong, J.W., Catalan, G., Fang, D.N. et al. (2009). Topology of the polarization field in ferroelectric nanowires from first principles. *Physical Review B: Condensed Matter* 81 (17): 1248–1248.
58 Chen, W.J., Zheng, Y., and Wang, B. (2015). Large and tunable polar-toroidal coupling in ferroelectric composite nanowires toward superior electromechanical responses. *Scientific Reports* 5: 11165.
59 Prosandeev, S., Ponomareva, I., Naumov, I. et al. (2008). Original properties of dipole vortices in zero-dimensional ferroelectrics. *Journal of Physics: Condensed Matter* 20 (19): 193201.
60 Xu, T., Wang, J., Shimada, T., and Kitamura, T. (2013). Direct approach for flexoelectricity from first-principles calculations: cases for $SrTiO_3$ and $BaTiO_3$. *Journal of Physics. Condensed Matter* 25 (41): 415901.
61 Hong, J., Catalan, G., Scott, J.F., and Artacho, E. (2010). The flexoelectricity of barium and strontium titanates from first principles. *Journal of Physics Condensed Matter An Institute of Physics Journal* 22 (11): 112201.
62 Zubko, P., Catalan, G., Buckley, A. et al. (2007). Strain-gradient-induced polarization in $SrTiO_3$ single crystals. *Physical Review Letters* 99 (16): 167601.
63 Ma, W. and Eric Cross, L. (2001). Observation of the flexoelectric effect in relaxor $Pb(Mg_1/3Nb_{2/3})O_3$ ceramics. *Applied Physics Letters* 78 (19): 2920–2921.
64 Tagantsev, A.K. (1986). Piezoelectricity and flexoelectricity in crystalline dielectrics. *Physical Review B: Condensed Matter* 34 (8): 5883–5889.
65 Sahin, E. and Dost, S. (1988). A strain-gradients theory of elastic dielectrics with spatial dispersion. *International Journal of Engineering Science* 26 (12): 1231–1245.
66 Maranganti, R. and Sharma, P. (2009). Atomistic determination of flexoelectric properties of crystalline dielectrics. *Physical Review B* 80 (5): 1956–1960.
67 Chen, W.J., Zheng, Y., Xiong, W.M. et al. (2014). Effect of mechanical loads on stability of nanodomains in ferroelectric ultrathin films: towards flexible erasing of the non-volatile memories. *Scientific Reports* 4: 5339.
68 Chen, W., Zheng, Y., Feng, X., and Wang, B. (2015). Utilizing mechanical loads and flexoelectricity to induce and control complicated evolution of domain patterns in ferroelectric nanofilms. *Journal of the Mechanics and Physics of Solids* 79: 108–133.
69 Stengel, M. (2013). Flexoelectricity from density-functional perturbation theory. *Physical Review B* 88 (17): 1352–1355.
70 Stengel, M. (2014). Surface control of flexoelectricity. *Physical Review B* 90 (20): 201112.
71 Yin, B. and Qu, S. (2014). An ab initio investigation of flexoelectric effect in ultrathin $BaTiO_3$ nanotubes. Journal of Applied Physics, 115 (7): 2069–2192.
72 Ponomareva, I., Tagantsev, A.K., and Bellaiche, L. (2012). Finite-temperature flexoelectricity in ferroelectric thin films from first principles. *Physical Review B: Condensed Matter* 85 (10): 1092–1097.

73 Ma, W. (2008). A study of flexoelectric coupling associated internal electric field and stress in thin film ferroelectrics. *Physica Status Solidi B* 245 (4): 761–768.

74 Yin, B. and Qu, S. (2014). An ab initio investigation of flexoelectric effect in ultrathin $BaTiO_3$ nanotubes. *Journal of Applied Physics* 115 (7): 074102.

75 Li, H., Zhang, G., Zheng, Y. et al. (2014). Ab initio study on mechanical-bending-induced ferroelectric phase transition in ultrathin perovskite nanobelts. *Acta Materialia* 76: 472–481.

76 Glass, A., Von der Linde, D., and Negran, T. (1974). High-voltage bulk photovoltaic effect and the photorefractive process in $LiNbO_3$. *Applied Physics Letters* 25 (4): 233–235.

77 Weis, R. and Gaylord, T. (1985). Lithium niobate: summary of physical properties and crystal structure. *Applied Physics A: Materials Science and Processing* 37 (4): 191–203.

78 Brody, P.S. (1975). High voltage photovoltaic effect in barium titanate and lead titanate-lead zirconate ceramics. *Journal of Solid State Chemistry* 12 (3-4): 193–200.

79 Koch, W., Munser, R., Ruppel, W., and Würfel, P. (1975). Bulk photovoltaic effect in $BaTiO_3$. *Solid State Communications* 17 (7): 847–850.

80 Young, S.M. and Rappe, A.M. (2012). First principles calculation of the shift current photovoltaic effect in ferroelectrics. *Physical Review Letters* 109 (11): 651–657.

81 Nonaka, K., Akiyama, M., Hagio, T., and Takase, A. (1995). Bulk photovoltaic effect in reduced/oxidized lead lanthanum titanate zirconate ceramics. *Japanese Journal of Applied Physics* 34 (5R): 2344.

82 Inoue, Y., Sato, K., Sato, K., and Miyama, H. (1986). Photoassisted water decomposition by ferroelectric lead zirconate titanate ceramics with anomalous photovoltaic effects. *Journal of Physical Chemistry;(United States)* 90 (13).

83 Nonaka, K., Akiyama, M., Xu, C.-N. et al. (2000). Enhanced photovoltaic response in lead lanthanum zirconate-titanate ceramics with A-site deficient composition for photostrictor application. *Japanese Journal of Applied Physics* 39 (9R): 5144.

84 Ichiki, M., Maeda, R., Morikawa, Y. et al. (2004). Photovoltaic effect of lead lanthanum zirconate titanate in a layered film structure design. *Applied Physics Letters* 84 (3): 395–397.

85 Alexe, M. and Hesse, D. (2011). Tip-enhanced photovoltaic effects in bismuth ferrite. *Nature Communications* 2: 256.

86 Young, S.M., Zheng, F., and Rappe, A.M. (2012). First-principles calculation of the bulk photovoltaic effect in bismuth ferrite. *Physical Review Letters* 109 (23): 236601.

87 Ji, W., Yao, K., and Liang, Y.C. (2010). Bulk photovoltaic effect at visible wavelength in epitaxial ferroelectric $BiFeO_3$ thin films. *Advanced Materials* 22 (15): 1763–1766.

88 Koch, W.T.H., Munser, R., Ruppel, W., and Würfel, P. (1976). Anomalous photovoltage in $BaTiO_3$. *Ferroelectrics* 13 (1): 305–307.

89 Ji, W., Yao, K., and Liang, Y.C. (2011). Evidence of bulk photovoltaic effect and large tensor coefficient in ferroelectric $BiFeO_3$ thin films. *Physical Review B: Condensed Matter* 84 (84): 3589–3594.

90 Guo, Y., Guo, B., Dong, W. et al. (2013). Evidence for oxygen vacancy or ferroelectric polarization induced switchable diode and photovoltaic effects in $BiFeO_3$ based thin films. *Nanotechnology* 24 (27): 275201.

91 Choi, T., Lee, S., Choi, Y.J. et al. (2009). Switchable ferroelectric diode and photovoltaic effect in $BiFeO_3$. *Science* 324 (5923): 63.

92 Yang, S.Y., Seidel, J., Byrnes, S.J. et al. (2010). Above-bandgap voltages from ferroelectric photovoltaic devices. *Nature Nanotechnology* 5 (2): 143.

93 Sturman, B.I. and Firdkin, V.M. (1992). The photovoltaic and photorefractive effects in non-centrosymmetric materials. *Ferroelectrics* 141 (1): 345–347.

94 Fridkin, V.M. and Popov, B.N. (1978). Anomalous photovoltaic effect in ferroelectrics. *Soviet Physics Uspekhi* 21 (12): 981–991.

95 Fridkin, V.M. (1979). *Photoferroelectrics*. Springer.

96 Zhang, J., Su, X., Shen, M. et al. (2013). Enlarging photovoltaic effect: combination of classic photoelectric and ferroelectric photovoltaic effects. *Scientific Reports* 3 (1): 2109.

97 Yi, H.T., Choi, T., Choi, S.G. et al. (2011). Mechanism of the switchable photovoltaic effect in ferroelectric $BiFeO_3$. *Advanced Materials* 23 (30): 3403–3407.

98 Qin, M., Yao, K., and Liang, Y.C. (2008). High efficient photovoltaics in nanoscaled ferroelectric thin films. *Applied Physics Letters* 93 (12): 122904.

99 Dong, W., Guo, Y., Guo, B. et al. (2013). Photovoltaic properties of $BiFeO_3$ thin film capacitors by using Al-doped zinc oxide as top electrode. *Materials Letters* 91(359–361.

100 Chen, B., Zuo, Z., Liu, Y. et al. (2012). Tunable photovoltaic effects in transparent $Pb(Zr_{0.53},Ti_{0.47})O_3$ capacitors. *Applied Physics Letters* 100 (17): 1671.

101 Chen, B., Li, M., Liu, Y. et al. (2011). Effect of top electrodes on photovoltaic properties of polycrystalline $BiFeO_3$ based thin film capacitors. *Nanotechnology* 22 (19): 195201.

102 Zhu, J., Chen, W.J., Zhang, G.H., and Zheng, Y. (2015). Exponential size-dependent tunability of strain on the transport behavior in ZnO tunnel junctions: an ab initio study. *Physical Chemistry Chemical Physics* 17 (38): 25583.

103 Zhang, Y., Liu, W., and Wang, Z.L. (2014). First principle simulations of piezotronic transistors. *Nano Energy* 14: 355–363.

104 Chen, W.J., Zheng, Y., Luo, X. et al. (2013). Ab initio study on the size effect of symmetric and asymmetric ferroelectric tunnel junctions: a comprehensive picture with regard to the details of electrode/ferroelectric interfaces. *Journal of Applied Physics* 114 (6): 1267-R.

105 Bennett, J.W., Grinberg, I., and Rappe, A.M. (2008). New highly polar semiconductor ferroelectrics through d8 cation-O vacancy substitution into $PbTiO_3$: a theoretical study. *Journal of the American Chemical Society* 130 (51): 17409–17412.

106 Bennett, J.W., Grinberg, I., Davies, P.K., and Rappe, A.M. (2010). Pb-free semiconductor ferroelectrics: a theoretical study of Pd-substituted Ba $(Ti_{1-x}Ce_x)O_3$ solid solutions. *Physical Review B* 82 (18): 184106.
107 Wang, W., Wang, R., Zhang, W. et al. (2013). A computer study and photoelectric property analysis of potassium-doped lithium niobate single crystals. *Physical Chemistry Chemical Physics Pccp* 15 (34): 14347.
108 Erhart, P., Träskelin, P., and Albe, K. (2013). Formation and switching of defect dipoles in acceptor-doped lead titanate: a kinetic model based on first-principles calculations. *Physical Review B* 88 (2): 024107.
109 Wang, F., Grinberg, I., Jiang, L. et al. (2015). Materials design of visible-light ferroelectric photovoltaics from first principles. *Ferroelectrics* 483 (1): 1–12.
110 Matsuo, H., Kitanaka, Y., Noguchi, Y., and Miyayama, M. (2016). Strong interaction between ferroelectric polarization and oxygen vacancy in $BiFeO_3$ thin film capacitors. *Journal of the Ceramic Society of Japan* 124 (6): 634–638.
111 Feng, H.-J. (2015). Photovoltaic and magnetic properties of $BiFeO_3/TiO_2$ heterostructures under epitaxial strain and an electric field. *Materials Chemistry and Physics* 153(405–409.
112 Zheng, Y. and Woo, C. (2010). Hyper-sensitive piezophotovoltaic effects in ferroelectric nanocylinders. *Journal of Applied Physics* 107 (10): 104120.
113 Duan, C.-G., Mei, W., Yin, W.-G. et al. (2003). Theoretical study on the optical properties of polyvinylidene fluoride crystal. *Journal of Physics: Condensed Matter* 15 (22): 3805.
114 Zhang, J., Yao, K., Liu, Z. et al. (2010). First-principles study of the ferroelectric and nonlinear optical properties of the $LiNbO_3$-type $ZnSnO_3$. *Physical Chemistry Chemical Physics* 12 (32): 9197–9204.
115 Zhang, J., Xu, B., Qin, Z. et al. (2012). Ferroelectric and nonlinear optical properties of the $LiNbO_3$-type $ZnGeO_3$ from first-principles study. *Journal of Alloys and Compounds* 514: 113–119.
116 Xu, B. and Yi, L. (2007). First-principle study of the ferroelectricity and optical properties of the $BaBi_2Ta_2O_9$. *Journal of Alloys and Compounds* 438 (1): 25–29.
117 Zeng, M., Or, S.W., and Chan, H.L.W. (2010). First-principles study on the electronic and optical properties of $Na_{0.5}Bi_{0.5}TiO_3$ lead-free piezoelectric crystal. *Journal of Applied Physics* 107 (4): 043513.
118 Brivio, F., Walker, A.B., and Walsh, A. (2013). Structural and electronic properties of hybrid perovskites for high-efficiency thin-film photovoltaics from first-principles. *APL Materials* 1 (4): 6050.
119 Wang, Y., Gould, T., Dobson, J.F. et al. (2014). Density functional theory analysis of structural and electronic properties of orthorhombic perovskite $CH_3NH_3PbI_3$. *Physical Chemistry Chemical Physics Pccp* 16 (4): 1424.
120 Haruyama, J., Sodeyama, K., Han, L., and Tateyama, Y. (2015). First-principles study of ion diffusion in perovskite solar cell sensitizers. *Journal of the American Chemical Society* 137 (32): 10048–10051.
121 Geng, W., Zhang, L., Zhang, Y.-N. et al. (2014). First-principles study of lead iodide perovskite tetragonal and orthorhombic phases for photovoltaics. *The Journal of Physical Chemistry C* 118 (34): 19565–19571.

122 Wang, L.-Z., Zhao, Y.-Q., Liu, B. et al. (2016). First-principles study of photovoltaics and carrier mobility for non-toxic halide perovskite $CH_3NH_3SnCl_3$: theoretical prediction. *Physical Chemistry Chemical Physics* 18 (32): 22188–22195.

123 Mosconi, E., Ronca, E., and De, A.F. (2014). First-principles investigation of the TiO_2/organohalide perovskites interface: the role of interfacial chlorine. *Journal of Physical Chemistry Letters* 5 (15): 2619. http://pubs.acs.org/doi/abs/10.1021/jz501127k.

124 Geng, W., Tong, C.-J., Liu, J. et al. (2016). Structures and electronic properties of different $CH_3NH_3PbI_3/TiO_2$ interface: a first-principles study. *Scientific Reports* 6.

125 Lee, M.M., Teuscher, J., Miyasaka, T. et al. (2012). Efficient hybrid solar cells based on meso-superstructured organometal halide perovskites. *Science* 1228604.

126 Im, J.H., Lee, C.R., Lee, J.W. et al. (2011). Efficient perovskite quantum-dot-sensitized solar cell. *Nanoscale* 3 (10): 4088–4093.

127 Burschka, J., Pellet, N., Moon, S.-J. et al. (2013). Sequential deposition as a route to high-performance perovskite-sensitized solar cells. *Nature* 499 (7458): 316–319.

128 Jeon, N.J., Noh, J.H., Yang, W.S. et al. (2015). Compositional engineering of perovskite materials for high-performance solar cells. *Nature* 517 (7535): 476–480.

129 Zhou, H., Chen, Q., Li, G. et al. (2014). Interface engineering of highly efficient perovskite solar cells. *Science* 345 (6196): 542–546.

130 Fan, Z., Sun, K., and Wang, J. (2015). Perovskites for photovoltaics: a combined review of organic–inorganic halide perovskites and ferroelectric oxide perovskites. *Journal of Materials Chemistry A* 3 (37): 18809–18828.

131 Klimeš, J., Bowler, D.R., and Michaelides, A. (2011). Van der Waals density functionals applied to solids. *Physical Review B* 83 (19): 195131.

132 Alpay, S.P., Mantese, J.V., Troliermckinstry, S. et al. (2014). Next generation electrocaloric and pyroelectric materials for solid state electrothermal interconversion. *MRS Bulletin* 39 (12): 1099–1111.

133 Guyomar, D., Pruvost, S., and Sebald, G. (2008). Energy harvesting based on FE–FE transition in ferroelectric single crystals. *IEEE Transactions on Ultrasonics Ferroelectrics & Frequency Control* 55 (2): 279–285.

134 Peng, Q. and Cohen, R.E. (2011). Origin of pyroelectricity in $LiNbO_3$. *Physical Review B* 83 (22): 220103.

135 Glass, A.M. and Lines, M.E. (1976). Low-temperature behavior of spontaneous polarization in $LiNbO_3$ and $LiTaO_3$. *Physical Review B* 13 (1): 180–191.

136 Beckman, S.P., Wan, L.F., Barr, J.A., and Nishimatsu, T. (2012). Effective Hamiltonian methods for predicting the electrocaloric behavior of $BaTiO_3$. 89 (25): 254–257.

137 Mischenko, A.S., Zhang, Q., Scott, J.F. et al. (2006). Giant electrocaloric effect in thin-film $PbZr_{(0.95)}Ti_{(0.05)}O_3$. *Science* 311 (5765): 1270–1271.

138 Thacher, P.D. (1968). Electrocaloric effects in some ferroelectric and antiferroelectric Pb (Zr, Ti)O_3 compounds. *Journal of Applied Physics* 39 (4): 1996–2002.
139 Strukov, B.A. (1967). Electrocaloric effect in single-crystal triglycine sulfate. *Soviet Physics Crystallography, USSR* 11 (6): 757.
140 Wiseman, G.G. and Kuebler, J.K. (1963). Electrocaloric effect in ferroelectric Rochelle salt. *Physical Review* 131 (5): 2023.
141 Akcay, G., Alpay, S.P., Mantese, J.V., and Rossetti, G.A. (2007). Magnitude of the intrinsic electrocaloric effect in ferroelectric perovskite thin films at high electric fields. *Applied Physics Letters* 90 (25): 252909.
142 Scott, J.F. (2007). Application of modern ferroelectrics. *Science* 315 (5814): 954–959.
143 Prosandeev, S., Ponomareva, I., and Bellaiche, L. (2008). Electrocaloric effect in bulk and low-dimensional ferroelectrics from first principles. *Physical Review B: Condensed Matter* 78 (78).
144 Moya, X., Karnarayan, S., and Mathur, N.D. (2014). Caloric materials near ferroic phase transitions. *Nature Materials* 13 (5): 439–450.
145 Nye, J.F. (1985). *Physical Properties of Crystals: Their Representation by Tensors and Matrices.* Oxford University Press.
146 Ponomareva, I., Naumov, I.I., Kornev, I. et al. (2005). Atomistic treatment of depolarizing energy and field in ferroelectric nanostructures. *Physical Review B* 72 (14): 140102.
147 Bellaiche, L., García, A., and Vanderbilt, D. (2002). Low-temperature properties of Pb $(Zr_{1-x}Ti_x)O_3$ solid solutions near the morphotropic phase boundary. *Ferroelectrics* 266 (1): 377–392.
148 Lisenkov, S. and Ponomareva, I. (2009). Intrinsic electrocaloric effect in ferroelectric alloys from atomistic simulations. *Physical Review B* 80 (14): 140102.
149 Marathe, M. and Ederer, C. (2014). Electrocaloric effect in $BaTiO_3$: a first-principles-based study on the effect of misfit strain. *Applied Physics Letters* 104 (21): 212902.
150 Lisenkov, S. and Ponomareva, I. (2012). Giant elastocaloric effect in ferroelectric $Ba_{0.5}Sr_{0.5}TiO_3$ alloys from first-principles. *Physical Review B* 86 (10): 104103.
151 Nishimatsu, T., Barr, J.A., and Beckman, S.P. (2013). Direct molecular dynamics simulation of electrocaloric effect in BaTiO3. *Journal of the Physical Society of Japan* 82 (11): 114605.
152 Lisenkov, S., Mani, B., Chang, C.-M. et al. (2013). Multicaloric effect in ferroelectric $PbTiO_3$ from first principles. *Physical Review B* 87 (22): 224101.
153 Ponomareva, I. and Lisenkov, S. (2012). Bridging the macroscopic and atomistic descriptions of the electrocaloric effect. *Physical Review Letters* 108 (16): 167604.
154 Creutz, M. (1983). Microcanonical Monte Carlo simulation. *Physical Review Letters* 50 (19): 1411.
155 Burn, I. and Smyth, D. (1972). Energy storage in ceramic dielectrics. *Journal of Materials Science* 7 (3): 339–343.

156 Yang, L., Li, X., Allahyarov, E. et al. (2013). Novel polymer ferroelectric behavior via crystal isomorphism and the nanoconfinement effect. *Polymer* 54 (7): 1709–1728.
157 Patel, S., Chauhan, A., and Vaish, R. (2014). Enhancing electrical energy storage density in anti-ferroelectric ceramics using ferroelastic domain switching. *Materials Research Express* 1 (4): 045502.
158 Hao, X. (2013). A review on the dielectric materials for high energy-storage application. *Journal of Advanced Dielectrics* 3 (01): 1330001.
159 Nishimatsu, T. and Waghmare, U.V. (2008). Fast molecular-dynamics simulation for ferroelectric thin-film capacitors using a first-principles effective Hamiltonian. *Physical Review B: Condensed Matter* 78 (10).
160 Nishimatsu, T. (2015). Molecular dynamics simulations of hysteresis loops for ferroelectric thin-film capacitors. *Ceramics Japan* 46: 456–461.
161 Luo, B., Wang, X., Tian, E. et al. (2016). First-principles effective Hamiltonian simulation of ABO_3-type perovskite ferroelectrics for energy storage application. *Journal of Applied Physics* 120 (7): 074106.
162 Miller, S.L., Nasby, R.D., Schwank, J.R. et al. (1990). Device modeling of ferroelectric capacitors. *Journal of Applied Physics* 68 (12): 6463–6471.
163 Lemanov, V.V. and Yarmarkin, V.K. (1996). Depolarization field and fatigue in ferroelectric thin films. *Physics of the Solid State* 38 (8): 1363–1367.
164 Jin, H.Z. and Zhu, J. (2002). Size effect and fatigue mechanism in ferroelectric thin films. *Journal of Applied Physics* 92 (8): 4594–4598.
165 Drougard, M.E. and Landauer, R. (1959). On the dependence of the switching time of barium titanate crystals on their thickness. *Journal of Applied Physics* 30 (11): 1663–1668.
166 Dawber, M., Chandra, P., Littlewood, P., and Scott, J. (2003). Depolarization corrections to the coercive field in thin-film ferroelectrics. *Journal of Physics: Condensed Matter* 15 (24): L393.
167 Scott, J.F. (2013). *Ferroelectric Memories*, vol. 3. Springer Science & Business Media.
168 Kittel, C. (1951). Theory of antiferroelectric crystals. *Physical Review* 82 (5): 729.
169 Love, G.R. (1990). ChemInform abstract: energy storage in ceramic dielectrics. *Journal of the American Ceramic Society* 21 (21): 323–328.
170 Lanagan, M.T., Kim, J.H., Jang, S.J., and Newnham, R.E. (2010). Microwave dielectric properties of antiferroelectric lead zirconate. *Journal of the American Ceramic Society* 71 (71): 311–316.
171 Hao, X., Zhai, J., and Yao, X. (2009). Improved energy storage performance and fatigue endurance of Sr-doped $PbZrO_3$ antiferroelectric thin films. *Journal of the American Ceramic Society* 92 (5): 1133–1135.
172 Hao, X., Zhai, J., Ling, B.K., and Xu, Z. (2014). A comprehensive review on the progress of lead zirconate-based antiferroelectric materials. *Progress in Materials Science* 63 (8): 1–57.
173 Zhiming, C., Jingyu, L., and Yongling, W. (1990). Phase transition studies of PSZT ceramics with vertical ferro-antiferroelectric phase boundary. *Ferroelectrics* 101 (1): 225–234.

174 Sharma, S., Singh, R., Goel, T.C., and Chandra, S. (2006). Synthesis, structural and electrical properties of La modified PZT system. *Computational Materials Science* 37 (1–2): 86–89.

175 Li, G., Furman, E., and Haertling, G.H. (1996). Fabrication and properties of PSZT antiferroelectric rainbow actuators. *Ferroelectrics* 188 (1): 223–236.

176 Dai, Z., Xi, Y., Zhuo, X. et al. (2008). Phase transition and dielectric properties of PbLa(Zr,Sn,Ti)O_3 antiferroelectric ceramics under hydrostatic pressure. *Journal of Electroceramics* 92 (1): 2663–2667.

177 Berlincourt, D., Krueger, H.H.A., and Jaffe, B. (1964). Stability of phases in modified lead zirconate with variation in pressure, electric field, temperature, and composition. *Journal of Physics and Chemistry of Solids* 2 (2): 659–674.

178 Berlincourt, D. and Krueger, H.H.A. (1959). Domain processes in lead titanate zirconate and barium titanate ceramics. *Journal of Applied Physics* 30 (11): 1804–1810.

179 Reyes-Lillo, S.E. and Rabe, K.M. (2014). Antiferroelectricity and ferroelectricity in epitaxially strained $PbZrO_3$ from first principles. *Physical Review B* 88 (18): 3895–3901.

180 Reyes-Lillo, S.E., Garrity, K.F., and Rabe, K.M. (2014). Antiferroelectricity in thin-film ZrO_2 from first principles. 90 (14).

181 Na, S. and Vanderbilt, D. (2000). First-principles study of ferroelectric and antiferrodistortive instabilities in tetragonal $SrTiO_3$. *Physical Review B* 62 (21): 13942–13950.

182 Leung, K., Cockayne, E., and Wright, A.F. (2002). Effective Hamiltonian study of $PbZr_{0.95}Ti_{0.05}O_3$. *Physical Review B* 65 (65): 392–397.

183 Lasave, J., Koval, S., Dalal, N.S., and Migoni, R.L. (2007). Origin of antiferroelectricity in $NH_4H_2PO_4$ from first principles. *Physical Review Letters* 98 (26): 267601.

184 Bousquet, E., Junquera, J., and Ghosez, P. (2010). First-principles study of competing ferroelectric and antiferroelectric instabilities in $BaTiO_3$/BaO superlattices. *Physical Review B* 82 (4): 045426.

185 Íñiguez, J., Stengel, M., Prosandeev, S., and Bellaiche, L. (2014). First-principles study of the multimode antiferroelectric transition in $PbZrO_3$. *Physical Review B* 90 (22).

186 Kagimura, R. (2008). First-principles investigations of elastic properties and energetics of antiferroelectric and ferroelectric phases of $PbZrO_3$. *Physical Review B* 77 (10).

187 Johannes, M.D. (2005). Crystal structure and electric field gradients of $PbZrO_3$ from density functional calculations. *Physical Review B* 71 (21): 2101.

188 Fujishita, H. and Katano, S. (1997). Re-examination of the antiferroelectric structure of $PbZrO_3$. *Journal of the Physical Society of Japan* 66 (11): 3484–3488.

189 Corker, D.L., Glazer, A.M., Whatmore, R.W. et al. (1998). A neutron diffraction investigation into the rhombohedral phases of the perovskite series? *Journal of Physics Condensed Matter* 10 (28): 6251–6269. (6219).

11

Future Perspectives

Haitao Huang

The Hong Kong Polytechnic University, Department of Applied Physics, Hung Hom, Kowloon, Hong Kong

The term ferroelectrics refers to a type of materials that exhibit switchable spontaneous polarizations under external electric field. Traditionally, they are studied for their application in high-permittivity capacitors [1, 2], positive temperature coefficient of resistivity (PTCR) thermistors [3], piezoelectric/electrostrictive transducers [3, 4], electro-optic devices [5], pyroelectric sensors [5, 6], and nonvolatile ferroelectric random access memories (NVFRAMs) [7–9].

Nowadays, ferroelectrics are being more and more studied and applied in energy-related applications. The previous chapters of this book show good examples for ferroelectrics being used in mechanical energy harvesting, photovoltaics, pyroelectric energy harvesting, electrocaloric cooling, electric energy storage, and photocatalysis. Those applications make use of the intrinsic properties of ferroelectrics, such as the high dielectric constant, temperature-dependent polarization, and the strong coupling between strain and polarization. In line with those applications, the future research trends in ferroelectrics can always be categorized into (i) developing new materials with better performance, lower cost, and more environmental friendliness, (ii) developing new composite materials or new solid solutions with additional functionalities and better performance, and (iii) designing new device structures to provide better performance and to satisfy the needs of new applications.

Apart from the above topics that have already been covered and summarized in the previous chapters, ferroelectrics can still find their niche applications in the energy-related areas where conventional ferroelectrics are not considered as the materials of choice, such as rechargeable batteries and organic solar cells. In those niche area applications, the internal electric field due to the spontaneous polarization of ferroelectric material is used to adjust charge-carrier density, separate electron–hole pairs, and modify the charge-transfer process at the surface/interface. A few typical examples will be presented and discussed in the following sections.

Ferroelectric Materials for Energy Applications, First Edition.
Edited by Haitao Huang and James F. Scott.

11.1 Enhanced Lithium Ion Transport in Polymer Electrolyte

The discovery of ionic conductivity in poly(ethylene oxide) (PEO) complexed with lithium salt has triggered intense research interest in polymer electrolyte for applications in solid-state lithium ion batteries (LIBs) [10, 11]. The advantages of polymer electrolyte over the liquid one are improved safety, flexibility, and processability. However, the poor ionic conductivity of the polymer electrolyte greatly restricts its applications in electrochemical devices, such as rechargeable batteries, supercapacitors, and fuel cells [12–14]. In order to enhance the ionic conductivity and lithium ion transfer number, several strategies have been developed, such as the use of plasticizers to facilitate the segmental motion of polymer chains and enhance the ion mobility [15] and the use of polar materials to dissociate ions and increase the charge-carrier concentration [16, 17].

As shown in Figure 11.1a, addition of a small amount (1.5 wt%) of ferroelectric $BaTiO_3$ ceramic fillers (0.5 μm in size) can greatly enhance the conductivity of PEO-based polymer electrolyte, while the addition of paraelectric $BaTiO_3$ ceramic fillers (0.1 μm) has little effect on the conductivity. The addition of paraelectric $SrTiO_3$ ceramic fillers may even have an adverse effect on the conductivity. Here, the suppression of ferroelectricity in nano-sized (0.1 μm) $BaTiO_3$ ceramic powders could be due to the existence of a large surface stress [18].

Figure 11.1b shows that the conductivity of PEO-based polymer electrolyte varies as a function of ferroelectric tetragonality. $BaTiO_3$ ceramic fillers in a size of 0.6–1.2 μm have sufficient dipole moment to dissociate the ions in the polymer matrix and hence increase the charge-carrier concentration. Too small

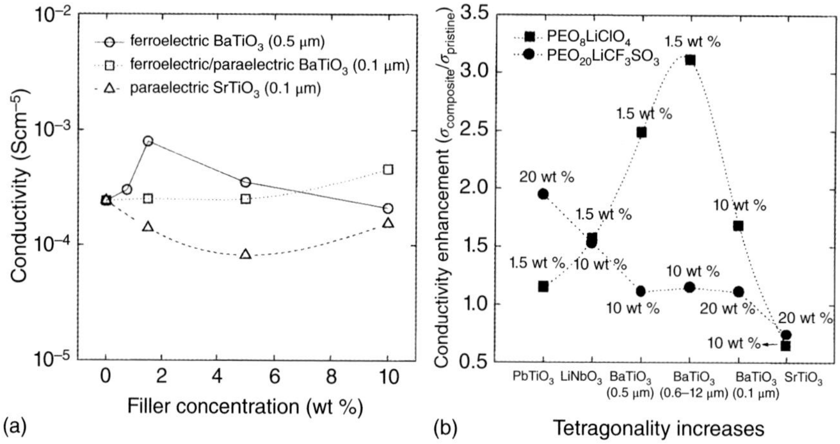

Figure 11.1 Conductivities of PEO-based polymer electrolytes with ceramic fillers. (a) Conductivity of PEO_8LiClO_4 with different ceramic fillers (ferroelectric $BaTiO_3$, paraelectric $BaTiO_3$, and paraelectric $SrTiO_3$) as a function of filler concentration. (b) The ratio of the conductivity of composite (PEO_8LiClO_4 and $PEO_{20}LiCF_3SO_3$ with different ceramic fillers) over the pristine one at 100 °C as a function of ferroelectric tetragonality. *Source:* Sun et al. 2000 [17]. Adapted with permission of Electrochemical Society.

a dipole moment is inadequate to modify the charge-carrier concentration too much. However, too large a dipole moment will make the PEO polymer chain stiff, decreasing the local conductivity at the polymer matrix/ceramic filler interface.

Compared with the conventional ceramic fillers, ferroelectric ceramic fillers also stabilize more effectively the interface between polymer matrix and ceramic particles, where the permanent dipoles play an important role [17]. The enhanced interface stability also results in better cyclic performance of the PEO-based polymer electrolyte [19]. The electrochemical stability window of the polymer electrolyte can be maintained at a potential higher than 4 V versus Li/Li^+ [17].

Ferroelectric $BaTiO_3$ was also used to tune the ionic conductivity of sodium superionic conductor (NASICON). It was found that the dipole moment of $BaTiO_3$ weakens the Na—O bond in $Na_4Zr_2Si_3O_{12}$ and enhances its ionic conductivity [20].

11.2 Enhanced Polysulfide Trapping in Li–S Batteries

Great success has been achieved in the past two decades for the commercialization of LIBs in consumer electronic market. However, the energy densities of the state-of-the-art LIBs are normally below 350 Wh kg^{-1}, which cannot catch up with the ever-increasing demand on energy density for electric vehicle applications [21–23]. Extensive research has gone beyond LIBs, searching for alternative technologies. Among them, lithium–sulfur battery (LSB) is one of the most promising alternatives due to its high theoretical energy density (~2600 Wh kg^{-1}), as well as low cost and environmental friendliness. Despite the great advantages of LSBs, there are also several challenges for them, such as low conductivity of sulfur, large volume changes during charge/discharge, and dissolution of polysulfides (Li_2S_n, $n = 4$–8) in electrolyte, with the last one being the major cause of capacity fading and poor cycling stability of LSBs.

Several strategies have been proposed to alleviate the abovementioned polysulfide shuttle effect, such as the geometrical confinement of polysulfides in microporous carbon [24–26] and chemical trapping in metal–organic frameworks (MOFs) [27, 28] and covalent organic frameworks (COFs) [29]. Considering the heteropolar nature of polysulfides, Xie et al. [30] recently proposed utilizing the built-in electric field of ferroelectric material to trap the polysulfides. In their approach, the hydrothermally synthesized 100-nm-sized $BaTiO_3$ (BTO) nanoparticles were simply mixed with carbon and S to form the composite cathode. X-ray diffraction (XRD) analysis shows that the BTO nanoparticles are in a tetragonal phase with a *c/a* ratio of 1.005, indicating a ferroelectric state. Raman spectrum and piezoresponse force microscopy (PFM) further confirm the ferroelectricity of synthesized BTO nanoparticles.

To illustrate the effect of polysulfide entrapment by the ferroelectric BTO, a composite cathode with carbon, S, and BTO mixed together (S content ~ 60%) was compared with the one without the addition of BTO. As shown in Figure 11.2, after two hours of discharge, the electrolyte color changes to yellow for the

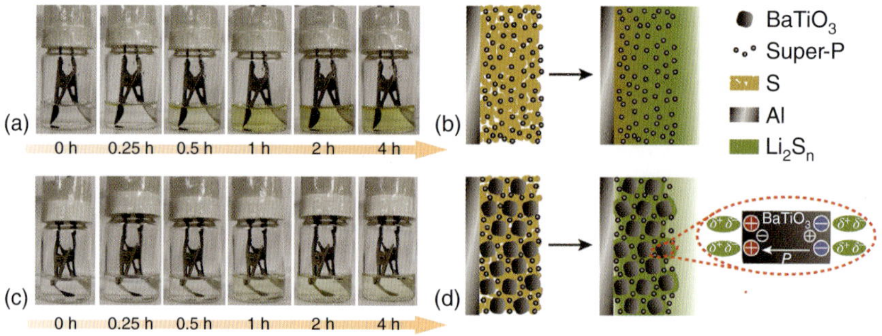

Figure 11.2 Photos showing the electrolyte color change during discharge/charge of the composite cathode at C/2 rate within a voltage window between open circuit voltage and 1.5 V: (a) C/S cathode and (c) C/S+BTO composite cathode. Schematics of the polysulfide entrapment mechanism: (b) C/S cathode and (d) C/S+BTO composite cathode. *Source:* Xie et al. 2017 [30]. Adapted with permission of John Wiley & Sons.

C/S cathode due to the dissolution of the polysulfide (Li_2S_n, $n = 4–8$) in the electrolyte (Figure 11.2a), which is the intermediate reaction product during discharge. In contrast, for the C/S+BTO composite cathode, the electrolyte color only changes faintly after two hours of discharge (Figure 11.2c), indicating suppressed polysulfide dissolution. Xie et al. also proposed a plausible mechanism for the trapping of polysulfides [30]. During discharge S will be reduced to polysulfides, which are heteropolar and can be absorbed on the polar surface of BTO through electrostatic interaction (Figure 11.2d).

As a result, the electrochemical performance of the composite cathode was greatly improved after the addition of ferroelectric BTO nanoparticles. As shown in Figure 11.3, with the addition of BTO, the cathode shows much higher specific capacity at all the different current densities measured. The long-term cycling stability was also improved with the suppressed dissolution of polysulfides. The initial discharge capacity of the cathode with BTO is 1143 mAh g^{-1} at 0.2 C. After 100 cycles, a reversible capacity of 835 mAh g^{-1} can still be maintained with a high and stable Coulombic efficiency (CE).

It is very important to note that the success of this facile method relies on two key issues: (i) BTO nanoparticles are electrochemically stable within the voltage window of 1.5–3.0 V, the operating voltage range of S cathode, and (ii) the ferroelectricity of BTO nanoparticles can be stably maintained during cycling.

11.3 Enhanced Dissociation of Excitons

Organic photovoltaic (OPV) cells are intensively studied in recent years due to their low cost, high performance, and mechanical flexibility that enables a roll-to-roll production [31–33]. High efficiencies over 10% have been realized by optimizing the morphology and aligning the electronic band structure of the polymer blends. Several factors limit the efficiency of OPV cells, such as insufficient photon absorption, misalignment of band structure, non-radiative recombination, low exciton diffusion length, and low carrier mobilities [34, 35].

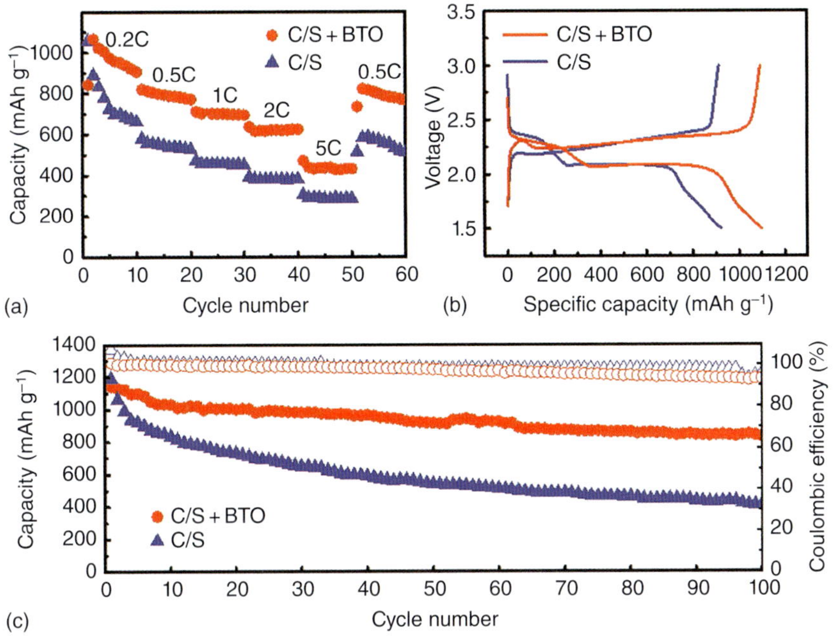

Figure 11.3 Electrochemical performance of the C/S composite cathode with and without the addition of ferroelectric BTO nanoparticles: (a) rate performance at different current densities, (b) charge/discharge curves, and (c) cycling performance at 0.2 C rate. *Source:* Xie et al. 2017 [30]. Adapted with permission of John Wiley & Sons.

The recombination of charge-transfer exciton (CTE, formed by Coulomb interaction of an electron at the acceptor and a hole at the neighboring donor) may contribute to more than 50% of the energy loss in polymer-fullerene OPV cells [36].

As shown in Figure 11.4, in an OPV cell, CTEs are formed due to photon-excited electron transfer. To reduce the efficiency loss of the device, a reverse bias field is normally required to separate the CTEs so that they are able to contribute to the photocurrent. According to the Onsager–Braun model [38], in most polymer-fullerene based OPVs, a reverse bias voltage greater than 5–10 V (electric field of 26–50 V μm^{-1}) is needed to separate most of the CTEs at room temperature. Yuan et al. proposed to use the large internal electric field produced by the permanent dipole moment of an ultrathin ferroelectric (FE) poly(vinylidene fluoride-trifluoroethylene) (P(VDF-TrFE)) film to dissociate the CTEs [37]. The P(VDF-TrFE) FE film was sandwiched between the P3HT:PC_{70}BM organic semiconductor film and the electrode (Figure 11.4b). A large polarization in the order of 10 $\mu C\ cm^{-2}$ can be generated in the FE thin film if it is properly polarized. The large polarization corresponds to a large surface charge density that generates a gradient in the electrostatic potential in the semiconductor film (Figure 11.4c,d). The induced electric field in a 150-nm-thick polymer blend semiconductor film is estimated to be around 50 V μm^{-1}, enough for the dissociation of CTEs.

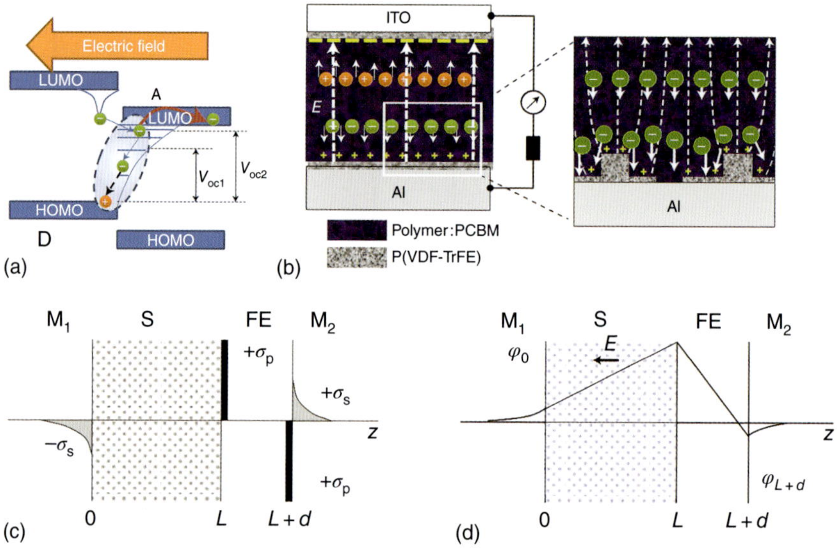

Figure 11.4 Schematics of the mechanism of FE-OPV cell. (a) The formation of a singlet charge-transfer exciton (CTE) at the donor–acceptor interface under the illumination of a photon and the dissociation of CTE can be achieved under a reverse bias electric field. (b) Structure of an OPV cell with thin ferroelectric P(VDF-TrFE) layers, showing the electric field distribution and charge extraction. (c, d) Charge density and electrostatic potential distributions, under zero bias electric field, in the structure with semiconductor (S) and ferroelectric (FE) layers sandwiched between two metal electrodes (M_1 and M_2). *Source:* Yuan et al. 2011 [37]. Adapted with permission of Springer Nature.

The dissociation of CTEs will increase the number of charges collected by the electrode, and, at the same time, the populated electrons will raise the quasi-Fermi level. As expected, the insertion of a thin FE film results in an increase not only in the short-circuit current (J_{sc}) but also in the open-circuit voltage (V_{oc}) (Figure 11.5a). Under reverse bias, the dark current of the cell with FE thin film was decreased by 2 orders of magnitude (Figure 11.5b). This is due to the higher barrier for charge injection induced by the FE thin film at reverse bias.

The effect of the ferroelectric polarization on the efficiency enhancement in OPVs is however, questioned by Asadi et al., who showed that the efficiency enhancement is independent of the direction of polarization [39]. The observed enhancement in efficiency is actually due to the non-Ohmic contact in non-optimized OPVs. In OPVs with standard LiF/Al electrode, the use of a ferroelectric thin layer may not be necessary.

11.4 New Materials

Since the discovery of ferroelectricity in Rochelle salt [40, 41], inorganic ferroelectrics with four main structures, i.e. perovskite, tungsten bronze, pyrochlore, and bismuth titanate layered structures, have played the central role in various

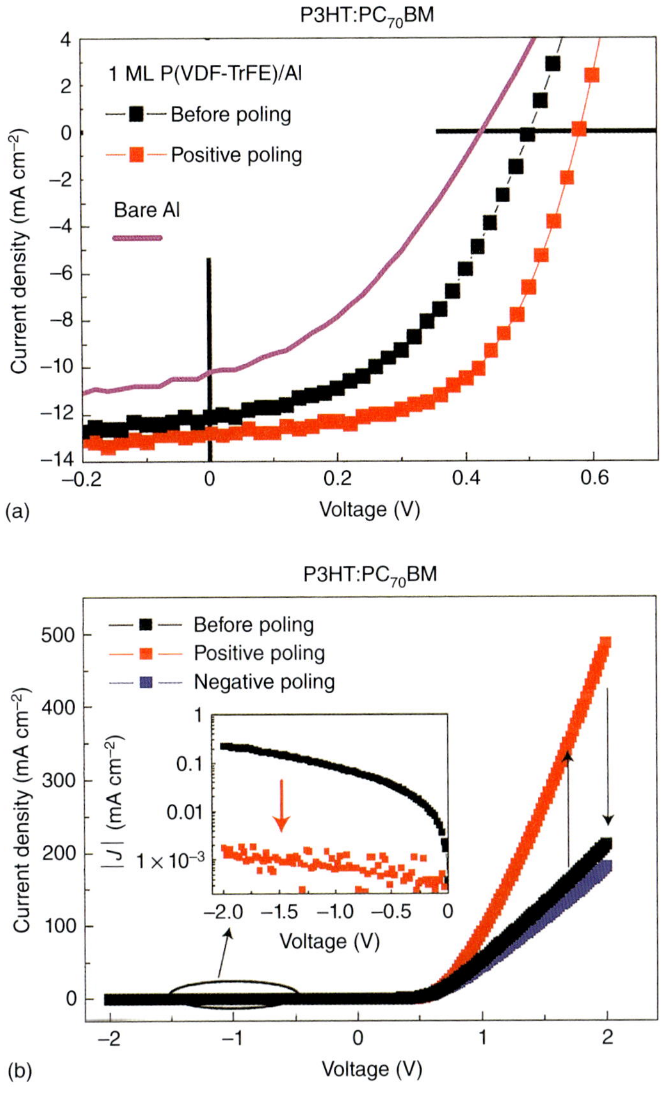

Figure 11.5 Device performance of the cells with and without the insertion of one monolayer of P(VDF-TrFE) FE film: (a) *I–V* curves and (b) dark current. The inset in (b) is the enlarged dark current. *Source:* Yuan et al. 2011 [37]. Adapted with permission of Springer Nature.

kinds of applications as sensors and actuators in smart and very smart structures [42]. All of the four groups have high dielectric constants, high refractive indices, and large electromechanical coupling coefficients, and all contain corner-linked rigid octahedral networks of highly polarizable active ions such as Ti^{4+}, Nb^{5+}, or other transition-metal ions. Those inorganic ferroelectrics normally require high processing temperature and are mechanically rigid. These features make those inorganic ferroelectrics unsuitable for a variety of modern applications, which require light weight, flexibility, and bio-friendliness.

Polymeric materials such as poly-vinylidene fluoride (PVDF) nylon and diisopropylammonium bromide (DIPAB) are known organic ferroelectrics that can be applied to bendable/wearable and implantable devices. However, their ferroelectricity is weak as compared with the conventional inorganic ferroelectrics. You et al. [43] have discovered a new organic–inorganic perovskite ferroelectric material, which possesses a large piezoelectric coefficient d_{33} that is comparable with $BaTiO_3$ piezoelectric ceramics. The newly discovered molecular ferroelectric is trimethylchloromethyl ammonium trichloromanganese (II) [$Me_3NCH_2ClMnCl_3$, (TMCM-$MnCl_3$)], which has a ferroelectric low-temperature phase (LTP) with a polar space group of *Cc* and a paraelectric high-temperature phase (HTP) with a space group of $P6_3/mmc$ (Figure 11.6). An order–disorder type of phase transition was observed between the LTP and HTP phases with a phase transition temperature at round 406 K, higher than that of the $BaTiO_3$ ceramics. At high temperatures, the organic cations are disordered, resulting in a centrosymmetric structure and a paraelectric state. At low temperatures, the cations are frozen, leading to macroscopic ordering of dipole moments.

Ferroelectricity in TMCM-$MnCl_3$ is unambiguously demonstrated by the polarization–electric field (*P–E*) hysteresis loop (Figure 11.7a). A maximum piezoelectric d_{33} coefficient of around 185 pC/N was obtained in the as-grown single crystalline TMCM-$MnCl_3$ along a direction close to [102] of the crystal, which is the highest reported value among the currently known organic

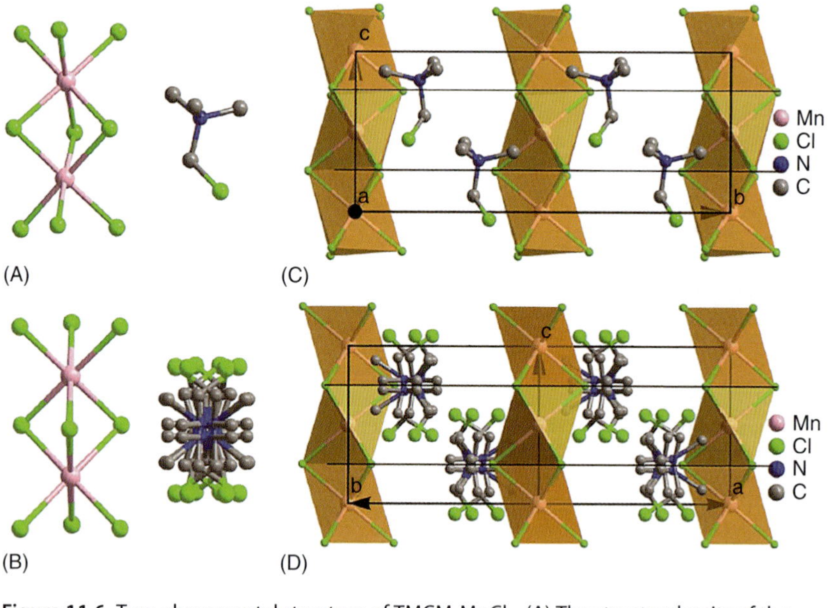

Figure 11.6 Two-phase crystal structure of TMCM-$MnCl_3$. (A) The structural units of the low-temperature phase: $MnCl_6$ octahedron and cationic structure. (B) The structural units of the high-temperature phase. (C) Projection of the low-temperature structure. (D) Projection of the high-temperature structure. H atoms are omitted for clarity. *Source:* You et al. 2017 [43]. Adapted with permission of AAAS.

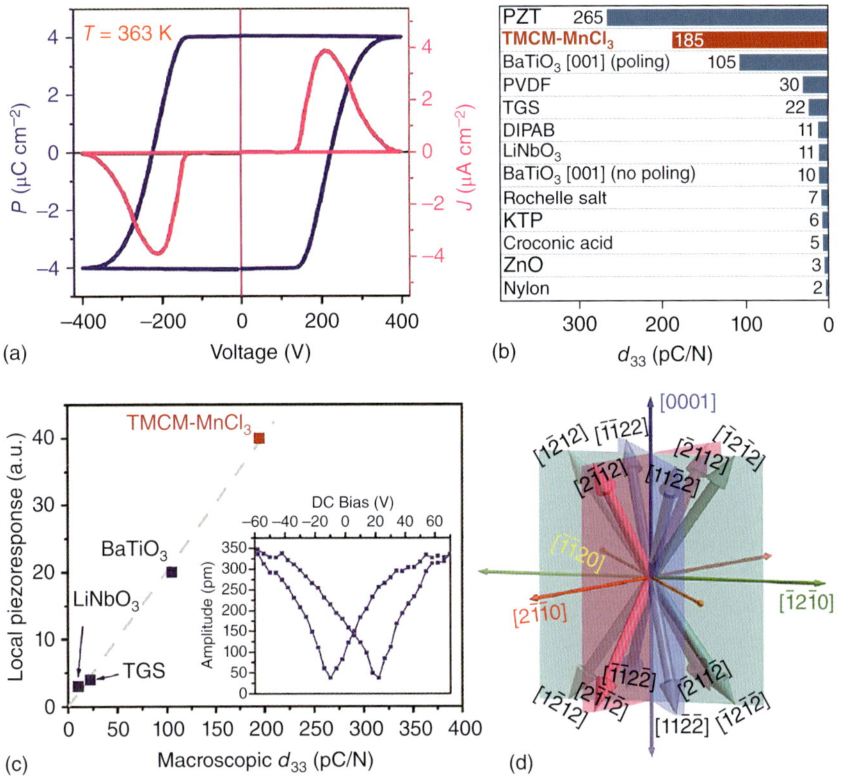

Figure 11.7 Ferroelectric properties of polarization directions of TMCM-$MnCl_3$. (a) The *P–E* hysteresis loop and current density versus bias voltage curves. (b) Piezoelectric d_{33} coefficients of some representative organic and inorganic ferroelectric materials. (c) Comparison of the local piezoelectric responses of some typical piezoelectric materials, measured by PFM. Inset: local piezoelectric response of TMCM-$MnCl_3$ as a function of applied voltage. (d) Twelve possible polarization directions in TMCM-$MnCl_3$. *Source:* You et al. 2017 [43]. Adapted with permission of AAAS.

flexible ferroelectrics (Figure 11.7b). PFM measurements show that the local piezoelectric response of the TMCM-$MnCl_3$ single crystal is much higher than that of conventional ferroelectrics, such as $BaTiO_3$, $LiNbO_3$, and triglycine sulfate (TGS), consistent with the macroscopic d_{33} measurements (Figure 11.7c). The TMCM-$MnCl_3$ single crystal in its LTP contains 12 possible polarization directions and the rotation of polarizations under external stress may contribute to the extremely large d_{33} coefficient observed (Figure 11.7d).

11.5 New Applications

Ferroelectrics are characterized by switchable polarization under external applied electric field. The spontaneous polarization is also responsive to external applied stress and heat. The variation of the spontaneous polarization under

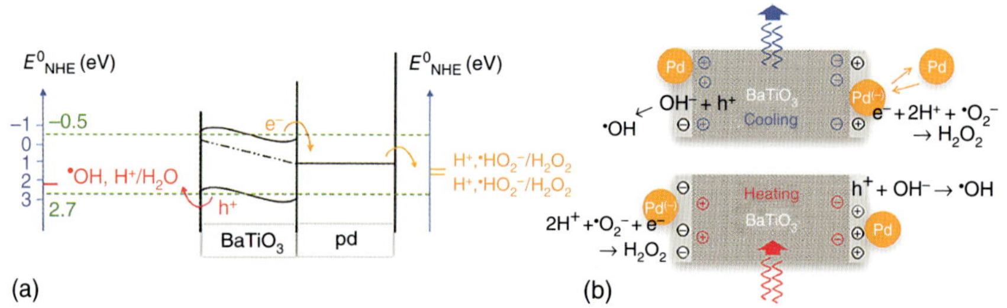

Figure 11.8 (a) Redox potentials between Pd nanoparticle catalyst, $BaTiO_3$, and neutral aqueous surrounding relative to standard hydrogen electrode are compared for selected species. (b) Schematic drawing showing the possible redox reactions at the polarized surface of $BaTiO_3$ particles under the pyroelectric effect. *Source:* Benke et al. 2015 [44]. Adapted with permission of ACS.

these external stimuli induces surface charges on the material surface. This property of a ferroelectric material is used in a traditional manner to sense a change in the environment and to elicit a useful response by using an external feedback control system, as in a combined sensor/actuator system. This is the reason that ferroelectric materials are called "smart" materials.

Reactive oxygen species (ROS), such as superoxide ($^{\cdot}O_2^-$), hydroxyl ($^{\cdot}OH$), singlet oxygen ($^1O_2$), and hydrogen peroxide (H_2O_2), are highly oxidative species used for disinfection due to their high oxidative potentials. Recently, the pyroelectric charge of $BaTiO_3$ was used to induce the generation of ROS. The polarization of $BaTiO_3$ particles is, in general, screened by solvated ions or dissociated water molecules in an aqueous solution. Owing to the pyroelectric effect, the change in temperature results in an imbalance between polarization and screened charges, which induces a change in surface potential. Although this potential change is smaller than 200 mV (which depends on the temperature change, dielectric permittivity, pyroelectric coefficient, and domain size of $BaTiO_3$), it is enough to affect the charge-transfer process in the redox reactions (Figure 11.8) [44].

A piezoelectric–catalytic effect was recently demonstrated in $BiOIO_3$ [45], which is a non-centrosymmetric piezoelectric material with a piezoelectric constant $d_{33} \sim 26\,\text{pm}\,\text{V}^{-1}$. The macroscopic polarization results from the well alignment of IO_3 polyhedra. The replacement of I^{5+} by V^{5+} distorts the polyhedra and decreases its size since the ionic radius of V^{5+} (68 pm) is less than that of I^{5+} (109 pm). The bandgap is also narrowed due to the downward shift of the conduction band, which is contributed by the V 3d orbital. The doping of V enhances the macroscopic polarization and promotes the charge separation in the piezoelectric–catalytic process (Figure 11.9), resulting in enhanced generation of $^{\cdot}O_2^-$ and $^{\cdot}OH$ radicals under the actuation of ultrasonic wave [45].

In summary, the renaissance of ferroelectrics will bring about more applications in energy conversion and storage devices. Particularly, the coupling of ferroelectric polarization with mechanical, thermal, magnetic, and optical signals may enable new concepts to emerge in the design of energy devices toward integrated or smart energy device systems. Moreover, new ferroelectric materials will always provide impetus to drive the development and offer new opportunities of ferroelectrics in energy-related applications.

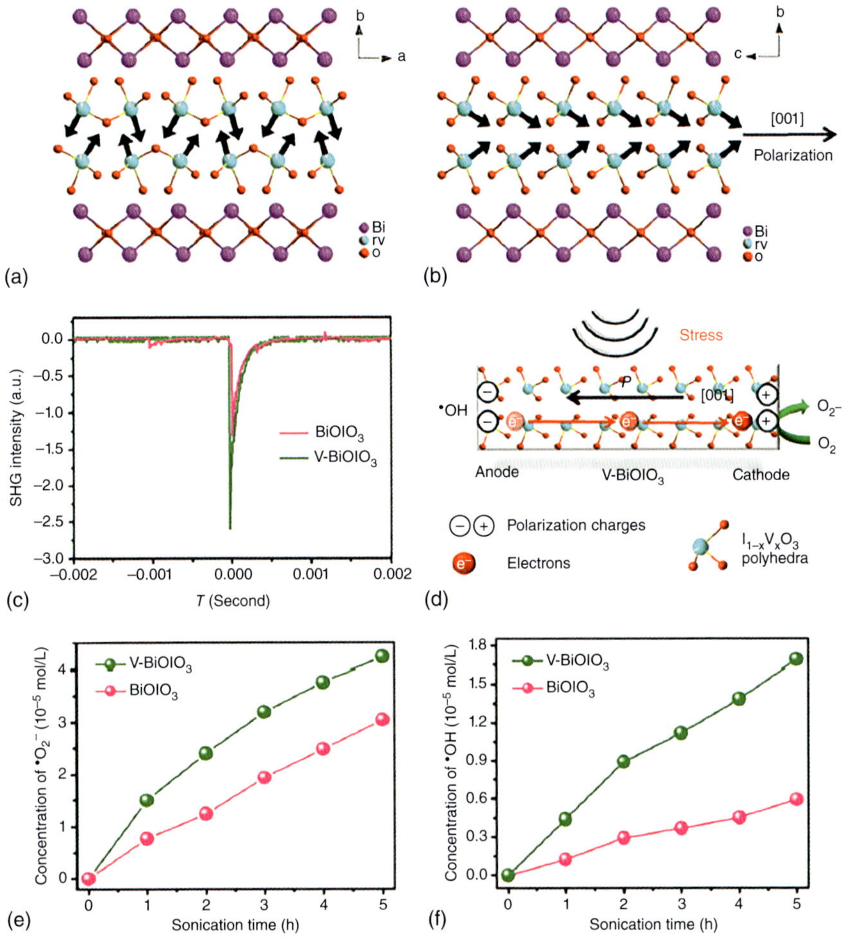

Figure 11.9 Crystal structure of V-doped $BiOIO_3$ ($BiOI_{0.926}V_{0.074}O_3$) in the (a) *ab* and (b) *bc* planes. The polarization direction is indicated by black arrows. (c) Second harmonic generation (SHG) for $BiOIO_3$ and $BiOI_{0.926}V_{0.074}O_3$. (d) Schematic showing the mechanism of ROS generation under mechanical actuation. (e) $^{\bullet}O_2^-$ and (f) $^{\bullet}OH$ evolution curves of $BiOIO_3$ and $BiOI_{0.926}V_{0.074}O_3$ under ultrasonication (40 kHz, 300 W). *Source:* Huang et al. 2017 [45]. Adapted with permission of John Wiley & Sons.

References

1 Mulvihill, M.L., Cross, L.E., and Uchino, K. (1995). *Journal of the American Ceramic Society* 78: 3345.

2 Baek, J.-G., Isobe, T., and Senna, M. (1997). *Journal of the American Ceramic Society* 80: 973.

3 Cahn, R.W., Haasen, P., and Kramer, E.J. (1994). *Materials Science and Technology*, vol. 11. Weinheim; New York; Basel; Cambridge; Tokyo: Wiley-VCH.

4 Moulson, A.J. and Herbert, J.M. (1990). *Electroceramics*. Chapman and Hall.

5 Haertling, G.H. (1999). *Journal of the American Ceramic Society* 82: 797.

6 Bell, A., Huang, Y., Paul, O. et al. (1995). *Integrated Ferroelectrics* 6: 231.
7 de Araujo, C.A.P., Cuchiaro, J.D., Mcmillan, L.D. et al. (1995). *Nature* 374: 627.
8 Scott, J.F. and de Araujo, C.A.P. (1989). *Science* 246: 1400.
9 Lee, S.J., Kang, K.Y., Han, S.K. et al. (1998). *Applied Physics Letters* 72: 299.
10 Fenton, D.E., Parker, J.M., and Wright, P.V. (1973). *Polymer* 14: 589.
11 Abraham, K.M. (1993). *Application of Electroactive Polymers* (ed. B. Scrosati), 75. London: Chapman & Hall.
12 Long, L., Wang, S., Xiao, M., and Meng, Y. (2016). *Journal of Materials Chemistry A* 4: 10038.
13 Senthilkumar, S.T., Wang, Y., and Huang, H. (2015). *Journal of Materials Chemistry A* 3: 20863.
14 Li, X.H., Yu, Y.F., Liu, Q.F., and Meng, Y.Z. (2012). *ACS Applied Materials & Interfaces* 4: 3627.
15 Huq, R., Farrington, G.C., Koksbang, R., and Tonder, P.E. (1992). *Solid State Ionics* 57: 277.
16 Sun, H.Y., Sohn, H.-J., Yamamoto, O. et al. (1999). *Journal of the Electrochemical Society* 146: 1672.
17 Sun, H.Y., Takeda, Y., Imanishi, N. et al. (2000). *Journal of the Electrochemical Society* 147: 2462.
18 Huang, H., Sun, C.Q., Tianshu, Z., and Hing, P. (2001). *Physical Review B: Condensed Matter* 63: 184112.
19 Li, Q., Takeda, Y., Imanish, N. et al. (2001). *Journal of Power Sources* 97–98: 795.
20 Kandkar, A. and Wagner, J.B. (1986). *Solid State Ionics* 20: 267.
21 Xu, M., Fei, L., Lu, W. et al. (2017). *Nano Energy* 35: 271.
22 Xu, M., Fei, L., Zhang, W. et al. (2017). *Nano Letters* 17: 1670.
23 Chen, Y., Li, X., Park, K. et al. (2016). *Angewandte Chemie International Edition* 55: 15831.
24 Ji, X.L., Lee, K.T., and Nazar, L.F. (2009). *Nature Materials* 8: 500.
25 Manthiram, A., Fu, Y.Z., Chung, S.H. et al. (2014). *Chemical Reviews* 114: 11751.
26 Zhang, C., Lv, W., Tao, Y., and Yang, Q.H. (2015). *Energy & Environmental Science* 8: 1390.
27 Zhou, J., Li, R., Fan, X. et al. (2014). *Energy & Environmental Science* 7: 2715.
28 Zheng, J., Tian, J., Wu, D. et al. (2014). *Nano Letters* 14: 2345.
29 Yoo, J., Cho, S.J., Jung, G.Y. et al. (2016). *Nano Letters* 16: 3292.
30 Xie, K., You, Y., Yuan, K. et al. (2017). *Advanced Materials* 29: 1604724.
31 Yu, G., Gao, J., Hummelen, J.C. et al. (1995). *Science* 270: 1789.
32 Sariciftci, N.S., Smilowitz, L., Heeger, A.J., and Wudl, F. (1992). *Science* 258: 1474.
33 Blom, P.W.M., Mihailetchi, V.D., Koster, L.J.A., and Markov, D.E. (2007). *Advanced Materials* 19: 1551.
34 Veldman, D., Meskers, S.C.J., and Janssen, R.A.J. (2009). *Advanced Functional Materials* 19: 1939.
35 Vandewal, K., Tvingstedt, K., Gadisa, A. et al. (2009). *Nature Materials* 8: 904.

36 Kirchartz, T., Taretto, K., and Rau, U. (2009). *Journal of Physical Chemistry C* 113: 17958.
37 Yuan, Y., Reece, T.J., Sharma, P. et al. (2011). *Nature Materials* 10: 296.
38 Onsager, L. (1934). *Journal of Chemical Physics* 2: 599.
39 Asadi, K., de Bruyn, P., Blom, P.W.M., and de Leeuw, D.M. (2011). *Applied Physics Letters* 98: 183301.
40 Valasek, J. (1920). *Physical Review* 15: 537.
41 Valasek, J. (1921). *Physical Review* 17: 475.
42 Newnham, R.E. (1997). *MRS Bulletin* 22 (5): 20.
43 You, Y.-M., Liao, W.-Q., Zhao, D. et al. (2017). *Science* 357: 306.
44 Benke, A., Mehner, E., Rosenkranz, M. et al. (2015). *Journal of Physical Chemistry C* 119: 18278.
45 Huang, H., Tu, S., Zeng, C. et al. (2017). *Angewandte Chemie International Edition* 56: 11860.

Index

Ferroelectric Materials for Energy Applications, First Edition.
Edited by Haitao Huang and James F. Scott.